Food Bioactive Ingredients

Series Editor

Seid Mahdi Jafari
Food Materials Process Design
Gorgan University
Gorgan, Iran

The Food Bioactive Ingredients Series covers recent advances and research on the science, properties, functions, technology, engineering and applications of food bioactive ingredients and their relevant products. The series also covers health-related aspects of these bioactive components, which have been shown to play a critical role in preventing or delaying different diseases and to have many health-improving properties. The books in this series target professional scientists, academics, researchers, students, industry professionals, governmental organizations, producing industries and all experts performing research on functional food development, pharmaceuticals, cosmetics and agricultural crops.

More information about this series at http://www.springer.com/series/16307

M. Ali Aboudzadeh
Editor

Emulsion-based Encapsulation of Antioxidants

Design and Performance

Springer

Editor
M. Ali Aboudzadeh
Donostia International Physics Center
Donostia-San Sebastian, Spain

ISSN 2661-8958 ISSN 2661-8966 (electronic)
Food Bioactive Ingredients
ISBN 978-3-030-62051-6 ISBN 978-3-030-62052-3 (eBook)
https://doi.org/10.1007/978-3-030-62052-3

This Springer imprint is published by the registered company Springer Nature Switzerland AG
The registered company address is: Gewerbestrasse 11, 6330 Cham, Switzerland

To my father
HASHEM

Preface

Oxidative stress develops inside all organisms in an oxygenated medium. Formation of free radicals by oxidative stress causes serious health problems in humans. Antioxidants have a key role in preventing lipid oxidation, since they can delay, control, or inhibit oxidation reactions. Therefore, various antioxidant systems have been developed by biological systems to protect against oxidation reactions induced by environmental conditions. The sources of food from biological tissues generally contain considerable endogenous antioxidant systems but at very low concentration that could be removed undesirably during food processing. Thus, it is common to add exogenous antioxidants to processed foods.

Due to several health-promoting impacts of bioactive compounds with antioxidant activity, consumer demand for them has been increasing rapidly over time. Therefore, the food industry has made serious attempts to incorporate antioxidants into various food products. However, low water solubility of these compounds and their high sensitivity to environmental factors such as light, oxygen, and heat restrict their incorporation into the foods. These limitations present a tremendous challenge in the developing process and formulation of new drug products. Therefore, in recent years, there has been growing interest in establishing methods to overcome these challenges and to protect these sensitive compounds with the final aim of increasing the public sanitation grades.

Encapsulation processes of antioxidants have widely been used for that aim, which allow their protection and delivery in a controlled manner and within a controlled environment. Miscellaneous encapsulation techniques serve as useful platform to integrate antioxidant compounds into the food matrix. The most common methods used to deliver or encapsulate antioxidants are emulsions, colloidal dispersions, suspensions, coacervation, and liposomes.

Among encapsulation technologies, emulsion-based systems are particularly interesting because they can easily be created from food-grade ingredients using relatively simple processing protocols. It is one of the favorable delivery systems applied to increase the solubility of phytochemicals, nutraceuticals, and food additives. Furthermore, some emulsion-encapsulated antioxidants presented even higher biological activities compared with pure free molecules. Therefore, novel formulations have been developed so far using a combination of hydrophilic and lipophilic emulsifiers to encapsulate them in oil or water droplets and disperse them in aqueous or oil systems, respectively.

This book contributes to advance and diversify current knowledge in the field of colloid science, through a specific investigation of the effects of formulation and process parameters on the emulsion production, as well as to deepen comprehension of the technological and biological aspects of the incorporation of antioxidants in food matrices and explication of their activity. First chapter of the book mainly addresses the reasons why antioxidants are encapsulated in emulsions (regardless of the preparation method) and their final fate prior to performing their biological action. The next two chapters discuss preparation methods of antioxidant-loaded emulsions that can broadly be classified as high-energy or low-energy approaches. In the following and in individual chapters, encapsulation process of antioxidants through different aqueous dispersion systems such as nanoemulsion, microemulsion, double emulsion, and pickering emulsion are explained by presenting several case studies. Another chapter devotes to understanding the relevance of oxidative stability of emulsions. The final chapter of the book focuses on characterization of the emulsion-based systems, from the terms of stability, rheology, cytotoxicity, or sterilization of the whole emulsion to the specific study of properties such as morphology, structure, or size of the dispersed phase; special emphasis has been placed on the antioxidant activity of the carriers being the key advantage of these systems.

The main aim of the book is to inspire and guide fellow scientists and students in this field. In this context, the book can be used also as a practical handbook or graduate textbook, and it includes plenty of illustrations, figures, practical examples, and historical perspectives. For the industrial establishments, the book also presents easy-to-achieve approaches that have been developed so far and could create a platform for industrial pharmaceutical production.

The editor expresses his appreciation to all contributors from different parts of the world who have cooperated in the preparation of this book. In this context, this international book gives the active reader different perspectives on the subject and encourages him/her to read the entire book.

Donostia-San Sebastian, Spain M. Ali Aboudzadeh

Contents

Contributors

M. Ali Aboudzadeh Donostia International Physics Center (DIPC), Donostia–San Sebastián, Spain

María Rosa Aguilar Group of Biomaterials, Department of Polymeric Nanomaterials and Biomaterials, Institute of Polymer Science and Technology, ICTP-CSIC, Madrid, Spain

Networking Biomedical Research Centre in Bioengineering, Biomaterials and Nanomedicine, CIBER-BBN, Madrid, Spain

Aqsa Akhtar School of Food and Agricultural Sciences, University of Management and Technology, Lahore, Pakistan

Océane Alliod Université Claude Bernard Lyon 1, CNRS, Villeurbanne, France

Rakesh Kumar Ameta School of Chemical Sciences, Central University of Gujarat, Gandhinagar, India

SMMPISR, Kadi SarvaVishwavidhyalaya, Gandhinagar, India

Sadia Aslam School of Food and Agricultural Sciences, University of Management and Technology, Lahore, Pakistan

Carlos Bravo-Díaz Facultad de Química, Departamento de Química-Física, Universidad de Vigo, Vigo, Spain

Catherine Charcosset Université Claude Bernard Lyon 1, CNRS, Villeurbanne, France

Rocío Díaz-Ruiz Department of Chemical and Environmental Engineering, University of Oviedo, Oviedo, Spain

Pelin Günç Ergönül Department of Food Engineering, Faculty of Engineering, Manisa Celal Bayar University, Manisa, Turkey

F. Javier Espejo-Carpio Department of Chemical Engineering, University of Granada, Granada, Spain

Eva Espinosa-Cano Group of Biomaterials, Department of Polymeric Nanomaterials and Biomaterials, Institute of Polymer Science and Technology, ICTP-CSIC, Madrid, Spain

Networking Biomedical Research Centre in Bioengineering, Biomaterials and Nanomedicine, CIBER-BBN, Madrid, Spain

Daniel Fernández-Villa Group of Biomaterials, Department of Polymeric Nanomaterials and Biomaterials, Institute of Polymer Science and Technology, ICTP-CSIC, Madrid, Spain

Anna Froelich Poznan University of Medical Sciences, Chair and Department of Pharmaceutical Technology, Poznań, Poland

Pedro J. García-Moreno Department of Chemical Engineering, University of Granada, Granada, Spain

Ana Gledović Department of Pharmaceutical Technology and Cosmetology, University of Belgrade – Faculty of Pharmacy, Belgrade, Serbia

Emilia M. Guadix-Escobar Department of Chemical Engineering, University of Granada, Granada, Spain

Gemma Gutiérrez Department of Chemical and Environmental Engineering, University of Oviedo, Oviedo, Spain

Shaghayegh Hamzehlou POLYMAT and Kimika Aplikatua Saila, Kimika Fakultatea, University of the Basque Country UPV-EHU, Joxe Mari Korta Zentroa, Tolosa Hiribidea, Donostia-San Sebastian, Spain

Miguel Huerta-Madroñal Group of Biomaterials, Department of Polymeric Nanomaterials and Biomaterials, Institute of Polymer Science and Technology, ICTP-CSIC, Madrid, Spain

Nauman Khalid School of Food and Agricultural Sciences, University of Management and Technology, Lahore, Pakistan

Sonia Losada-Barreiro Facultad de Química, Departamento de Química-Física, Universidad de Vigo, Vigo, Spain

REQUIMTE-LAQV, Departamento de Química e Bioquímica. Faculdade de Ciências, Universidade do Porto, Porto, Portugal

Tamara Major Center for Stem Cell Research and Drug Development, University of Belgrade - Faculty of Pharmacy, Belgrade, Serbia

Parth Malik School of Nano Sciences, Central University of Gujarat, Gandhinagar, India

Ali Marefati Department of Food Technology, Engineering, and Nutrition, Lund University, Lund, Sweden

Maria Matos Department of Chemical and Environmental Engineering, University of Oviedo, Oviedo, Spain

Ines Nikolić Department of Pharmaceutical Technology and Cosmetology, University of Belgrade – Faculty of Pharmacy, Belgrade, Serbia

Tomasz Osmałek Poznan University of Medical Sciences, Chair and Department of Pharmaceutical Technology, Poznań, Poland

Zeynep Aksoylu ÖzbeK Department of Food Engineering, Faculty of Engineering, Manisa Celal Bayar University, Manisa, Turkey

Fátima Paiva-Martins REQUIMTE-LAQV, Departamento de Química e Bioquímica. Faculdade de Ciências, Universidade do Porto, Porto, Portugal

A. Raúl Pérez-Gálvez Department of Chemical Engineering, University of Granada, Granada, Spain

Gloria María Pontes-Quero Group of Biomaterials, Department of Polymeric Nanomaterials and Biomaterials, Institute of Polymer Science and Technology, ICTP-CSIC, Madrid, Spain

Alodia Farmacéutica SL, Madrid, Spain

Networking Biomedical Research Centre in Bioengineering, Biomaterials and Nanomedicine, CIBER-BBN, Madrid, Spain

Nor E. Rahmani-Manglano Department of Chemical Engineering, University of Granada, Granada, Spain

Marilyn Rayner Department of Food Technology, Engineering, and Nutrition, Lund University, Lund, Sweden

Rao Sanaullah Khan School of Food and Agricultural Sciences, University of Management and Technology, Lahore, Pakistan

Snežana Savić Department of Pharmaceutical Technology and Cosmetology, University of Belgrade – Faculty of Pharmacy, Belgrade, Serbia

Bakht Ramin Shah University of South Bohemia in Ceske Budejovice, Faculty of Fisheries and Protection of Waters, South Bohemian Research Centre of Aquaculture and Biodiversity of Hydrocenoses, Institute of Aquaculture and Protection of Waters, České Budějovice, Czech Republic

Man Singh School of Chemical Sciences, Central University of Gujarat, Gandhinagar, India

Slobodanka Tamburić Cosmetic Science Research Group, London College of Fashion, University of the Arts London, London, United Kingdom

Blanca Vázquez-Lasa Group of Biomaterials, Department of Polymeric Nanomaterials and Biomaterials, Institute of Polymer Science and Technology, ICTP-CSIC, Madrid, Spain

Networking Biomedical Research Centre in Bioengineering, Biomaterials and Nanomedicine, CIBER-BBN, Madrid, Spain

Chapter 1
Why Encapsulate Antioxidants in Emulsion-Based Systems, Where They Are Located, and How Location Affects Their Efficiency

Sonia Losada-Barreiro, Carlos Bravo-Díaz, and Fátima Paiva-Martins

1.1 Introduction

Administration of bioactives to humans is challenging because their release into the surrounding medium needs to be controlled to improve their bioavailability and because the bioactive compound may need to be protected from environment to maintain its therapeutic properties (Garti 2008; Nowak et al. 2019; de Melo Barbosa et al. 2019). Research strategies and directions in the administration and release of bioactives are mostly inspired by the modeling of biological systems. For example, cell walls protect the interior components from the extracellular environment, controlling the exchange of matter with the surrounding medium. Based on these observations, researchers aim to define formulations and strategies to protect the bioactives of interest from external attacks (light, oxygen, other chemicals), to deliver them to target areas without losing their bioactivity or, simply, aim to design formulations to solubilize them in a medium to finally improve their uptake and release (McClements 2018).

Research on the encapsulation of polyunsaturated fatty acids (PUFAs) and polyphenolic antioxidants has attracted great interest in the functional food, nutraceutical and pharmaceutical industries, due to their health benefits to humans as energy

S. Losada-Barreiro
Universidad de Vigo, Facultad de Química, Departamento de Química – Física, Vigo, Spain

REQUIMTE-LAQV, Departamento de Química e Bioquímica. Faculdade de Ciências. Universidade do Porto, Porto, Portugal

C. Bravo-Díaz (✉)
Universidad de Vigo, Facultad de Química, Departamento de Química – Física, Vigo, Spain
e-mail: cbravo@uvigo.es

F. Paiva-Martins
REQUIMTE-LAQV, Departamento de Química e Bioquímica. Faculdade de Ciências. Universidade do Porto, Porto, Portugal

M. A. Aboudzadeh (ed.), *Emulsion-based Encapsulation of Antioxidants*, Food Bioactive Ingredients, https://doi.org/10.1007/978-3-030-62052-3_1

sources and in the prevention of some diseases (Lin et al. 2017; de Roos et al. 2019). Emulsions have a significant potential in delivering polyphenols or other bioactives such as PUFAs (McClements 2018) and emulsion encapsulation is considered as one of the most promising techniques for protection and delivery of polyphenolic antioxidants. Among others, economical and practical reasons, including its low production costs, high-efficiency, and maintenance of the chemical stability of the capsule, and the controlled release of the encapsulated bioactive make them very attractive. In addition, a variety of emulsion-based delivery systems for polyphenols can be prepared, and single-, multiple- and nano-emulsions have been described and are widely used. In spite of this, they may have some disadvantages, mainly related to their chemical stability, because important emulsion components such as PUFAs and stabilizing proteins are prone to oxidation when they are in contact with air or pro-oxidant molecules (Frankel 2005; Schaich et al. 2013).

Several reviews describing methods commonly employed to encapsulate polyphenols in oil in water (O/W) emulsions are available in the literature (Belščak-Cvitanović et al. 2011; Bartosz and Irene 2016; Ballesteros et al. 2017; Fanga and Chandaria 2010). These methods include, among others, homogenization, homogenization-solvent removal, emulsion-cooling and interfacial polycondensation, with varying efficiency. However, much less attention has been paid to describe the fate of the polyphenols once they are loaded in the emulsions, and fundamental questions are raised such as how they distribute between the oil, water and interfacial regions of the emulsions, what are the effective concentrations in a particular region, and how environmental and formulation conditions (e.g., acidity, temperature, oil nature, oil to water ratio, etc.) affects their partitioning and their efficiency.

The interfacial oil-water layer plays a crucial role in emulsified systems because it is a highly anisotropic region composed of oil, water and surfactant, with solvent properties between those of the bulk water and oil (Berton-Carabin et al. 2018; Bravo-Díaz et al. 2015). Antioxidants partition thermodynamically according to their relative solubility in the oil, water and interfacial regions, and this differential partitioning is an important phenomenon that requires a particular attention in attempting to control the oxidative stability of the emulsions because the effective concentration of antioxidants in the interfacial region correlates directly with the antioxidant efficiency (Freiría-Gándara et al. 2018b; Costa et al. 2020a, b; Raimúndez-Rodríguez et al. 2019). The bioefficiency of antioxidants depends not only on their effective concentration at the interfacial region of the emulsion but also on a number of other parameters that, at the end, control their effective concentration. Those parameters include the chemical properties of the antioxidants, the nature of the food emulsion (emulsifier and oil types) and the environmental conditions (acidity, temperature, electrolyte concentration, etc.). All of them have an enormous impact on antioxidant performance, since the rate of any reaction in which they are involved depends, among others, on their effective concentrations at the reaction site (Raimúndez-Rodríguez et al. 2019).

In this book chapter, we will focus on the performance of polyphenols in O/W emulsions. We will mainly focus on the reasons why antioxidants are encapsulated in emulsions and on their final fate prior to performing their biological action: where they are located and why. Once we have analyzed their differential locations and distributions, some discussion on how the location affects their efficiency in inhibiting the harmful and undesirable lipid oxidation reaction will be included. Description of methods for the preparation of emulsion-based delivery systems or capsules will not be covered, and the interested reader is referred to specialized reviews or to other chapters in the present book.

Bearing in mind the above considerations, we have divided the chapter in three sections for practical purposes. Section 1.2 summarizes some of the current knowledge on lipid oxidation reactions and emphasizes on the role of polyphenolic antioxidants in preventing the oxidation of lipids and on their effects on the human health, highlighting some of the reasons why antioxidants are encapsulated and loaded in emulsion-based systems. Section 1.3 centers on the fate of the antioxidants once they have been intentionally added to emulsions, focusing on how they distribute between the different domains of the emulsion, what is the driven force for their partitioning, highlighting the effects of partitioning on their efficiency in inhibiting the oxidation of lipid-based emulsions. Finally, Sect. 1.4 summarizes what has been learned after reading the chapter and provides some future perspectives and directions.

Indeed, lipid-based delivery systems have a great potential in parenteral nutrition as energy sources and, at the same time, for the delivery of multiple nutrients (Raman et al. 2017; Mundi et al. 2017; Spray 2016). Thus, knowledge on how one can control the antioxidant distributions and how partitioning influences their efficiency is challenging to the food and pharmaceutical industries. This knowledge is also crucial for developing specific strategies to prepare healthier and nutritional foods with longer shelf-lives and tailored properties, for example to meet predesigned specifications. The use of encapsulated polyphenols, instead of free compounds, can effectively overcome solubility problems and, at the same time, preserve the stability, bioactivity and bioavailability of the bioactive ingredient.

1.2 Role of Polyphenolic Antioxidants in Human Health and in Controlling Lipid Oxidation

1.2.1 Oxidative Stress, Neurodegenerative Diseases and Antioxidants

Neurodegenerative disorders (NDs) are a heterogeneous group of nervous system diseases, including the brain, spinal cord, and peripheral nerves that have many different etiologies (Chang 2011; Hardiman and Doherty 2011). They represent significant medical, social, and financial burden on the society due to their prevalence, morbidity, and mortality. Probably the main effect of NDs is the degradation of regions of the brain as consequence of hereditary, toxic, metabolic or infectious

Table 1.1 Main characteristic features of some common neurodegenerative diseases

Neurodegenerative disease	Clinical features
Alzheimer	Dementia, deterioration of language skills, perception
Amyotrophic Lateral Sclerosis	Atrophy of skeletal muscles, progressive weaknes
Parkinson	Tremor, rigidity, slowness of movement,
Huntington	Cognitive impairments, personality changes, depression

Adapted from literature (Losada-Barreiro and Bravo-Díaz 2017)

Table 1.2 Main radical and non-radical ROS and RNS

ROS		RNS	
Free radicals	*Non-radicals*	*Free radicals*	*Non-radicals*
Hydroxyl $HO^{\bullet}$	Hydrogen peroxide H_2O_2	Nitric Oxide $HO^{\bullet}$ $^{\bullet}NO$	Nitrogen dioxide
Superoxide anion $O_2^{\bullet -}$	Singlet oxygen $^1O_2$	Nitrous acid HNO_2	Dinitrogen tetraoxide
Lipid peroxyl $ROO^{\bullet}$	Ozone O_3		Peroxynitrite $ONOO^-$
Thiyl $^{\bullet}RS$	Lipid peroxide ROOH		Nitrosothiols SNOs

processes. Table 1.1 summarizes some clinical features of the major neurodegenerative diseases.

Despite of increasing the current knowledge of the neurobiology of NDs, the exact mechanism of NDs development is still not clear. It is currently believed that the oxidative stress, an imbalance resulting from metabolic reactions that use oxygen, shifts the equilibrium between oxidative - antioxidative reactions in living organisms, playing a central role in their development. The delicate balance between beneficial and harmful effects of reactive oxygen and nitrogen species, ROS and RNS, respectively, is a very important aspect for the protection of living organisms. This balance controls various oxidative stresses and the maintenance of the "redox homeostasis" by controlling the redox regulation in vivo (Wang et al. 2013). Table 1.2 presents some ROS and RNS, including charged and neutral species.

Methods currently employed in the management of NDs include the use of preventive strategies, disease-modifying schemes, hormone replacement- and immunetherapy (Feng and Wang 2012). The preventive strategy is, probably, the main therapy currently employed because it is inspired by the endogenous defense mechanisms developed by biological systems to protect themselves against free radical induced cell damage, to maintain the redox homeostasis and to reduce the secondary pathologies of the disease (Di Domenico et al. 2015; Feng and Wang 2012). These protective mechanisms include scavenging or detoxification of reactive oxygen species (ROS), blocking ROS production, sequestration of transition metals, as well as enzymatic and nonenzymatic antioxidant defenses produced in the body, which are endogenous and other antioxidants supplied with the diet, namely, exogenous ones (Kim et al. 2010; Benfeito et al. 2013; Malar and Devi 2014).

ROS may also act as signaling molecules in the maintenance of physiological functions — a process termed redox biology – by increasing the ROS levels to activate signaling pathways that initiate biological processes (Ray et al. 2012; Schieber et al. 2014). The effects of ROS in cell signaling have often been attributed to changes in the redox potential of the cells, and have functional consequences, which mediate various pathological processes. Recent studies point to endogenous ROS participating in several signal transduction pathways in cells, although the sensitive pathways has just begun to be identified.

In all these protective mechanisms, polyphenolic antioxidants play a crucial role either by chelating metals, minimizing the possibility of formation of radicals, or as radical scavengers. Scheme 1.1 shows some of the diseases where the action of polyphenols has been shown to be beneficial.

1.2.2 Significance of the Lipid Oxidation Reactions and Their Control

Important cell components, particularly proteins, nucleic acids and lipids are prone to oxidation due to the action of molecular oxygen, leading to oxidative damage of the structures where they are located (Bhattacharya 2015).

Oxygen is an essential compound for living organisms, yet it is very dangerous because of its radical nature. The reaction of molecular oxygen with organic molecules is a process of considerable interest in chemistry and biology (Litwinienko and Ingold 2007). The auto-oxidation of organic materials such as edible oils, fats, and lipid-containing foods and living organisms is particularly interesting reaction as well, which historically has been known as lipid peroxidation (or auto-oxidation) reaction.

The most stable and abundant form of oxygen has, in its ground state, a triplet electronic configuration with two unpaired electrons in different antibonding (π*)

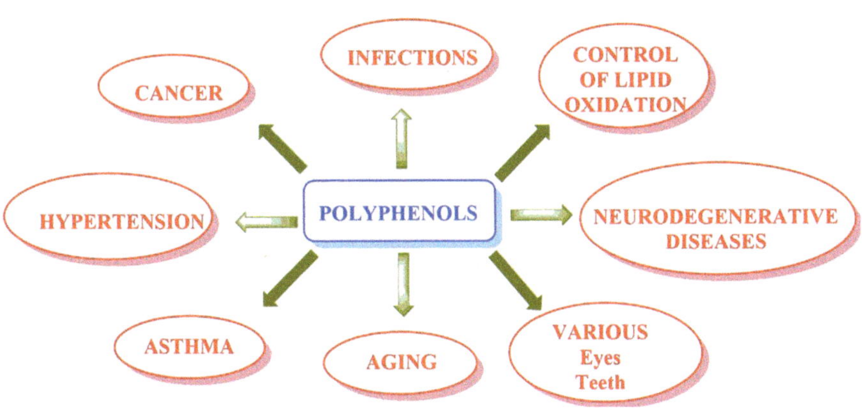

Scheme 1.1 Polyphenolic antioxidants have been proved to be beneficial for the treatment of a number of diseases and to control the oxidation of lipids

orbitals at parallel spins (Scheme 1.2) (Atkins and de Paula 2010). Molecular oxygen cannot react with double bonds of unsaturated lipids because the double bond needs to be excited into a triplet state, which requires large amounts of energy (*E*a ≈ 300–500 kJ mol^{-1}). Although this reaction can take place to some extent, it is usually negligible in practice, and initiators or catalysts are required to start the lipid oxidation process by removing an electron from either the lipid or oxygen or by changing the electron spin of the oxygen.

The mechanism of the lipid oxidation reaction can be understood in terms of a chain process consisting of initiation, propagation, and termination steps as shown in Scheme 1.3 (Schaich et al. 2013; Kamal-Eldin and Min 2008; Frankel 2005, Ingold and Pratt 2014). In the initiation step, the key step is the formation of the lipid radical, R•, whose formation in biological systems (e.g., cellular membranes) can be induced by exogenous or endogenous reagents such as oxygen, metals, UV-light or ionizing radiations, food pigments such as chlorophylls or riboflavins or enzymes such as lipoxygenases and cyclooxygenases (Schaich et al. 2013; Frankel 2005). In vitro, oxidation of lipids can be induced by a variety of methods including the use of metals, enzymes or free radical initiators (Frankel 2005; Yin et al. 2011).

Whatever is the mechanism of generating the first free radical, the chain reaction is initiated and new carbon and oxygen-centered free radicals are formed, and the radical process is propagated. The subsequent course of the lipid oxidation reaction depends on several factors including hydrogen bond accepting and anion-solvation abilities of the solvent, the electron affinities and reactivities (dissociation enthalpies) of radicals. ROS react with polyunsaturated fatty acids, PUFAs, which surround the cells and other cellular structures, generating lipid carbon centered radicals which, in turn, readily trap molecular oxygen under physiological conditions to form lipid peroxyl radicals ROO• acting as effective chain carriers in the lipid chain auto-oxidation.

Peroxidation and destruction of the *cis*-double bonds (mostly converted to the *trans* configuration), may lead to a reduction in the membrane fluidity and the formation of hydroperoxides, alcohols and electrophilic α-β-unsaturated aldehydes among others. All these secondary products are cytotoxic and partially account for

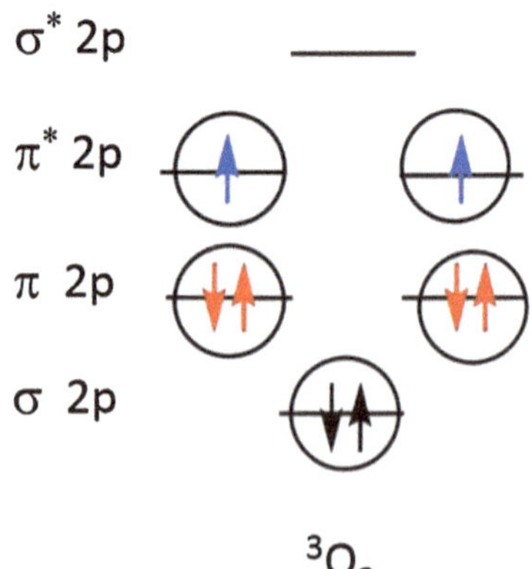

Scheme 1.2 Electronic distribution of ground state molecular oxygen as commonly found in nature

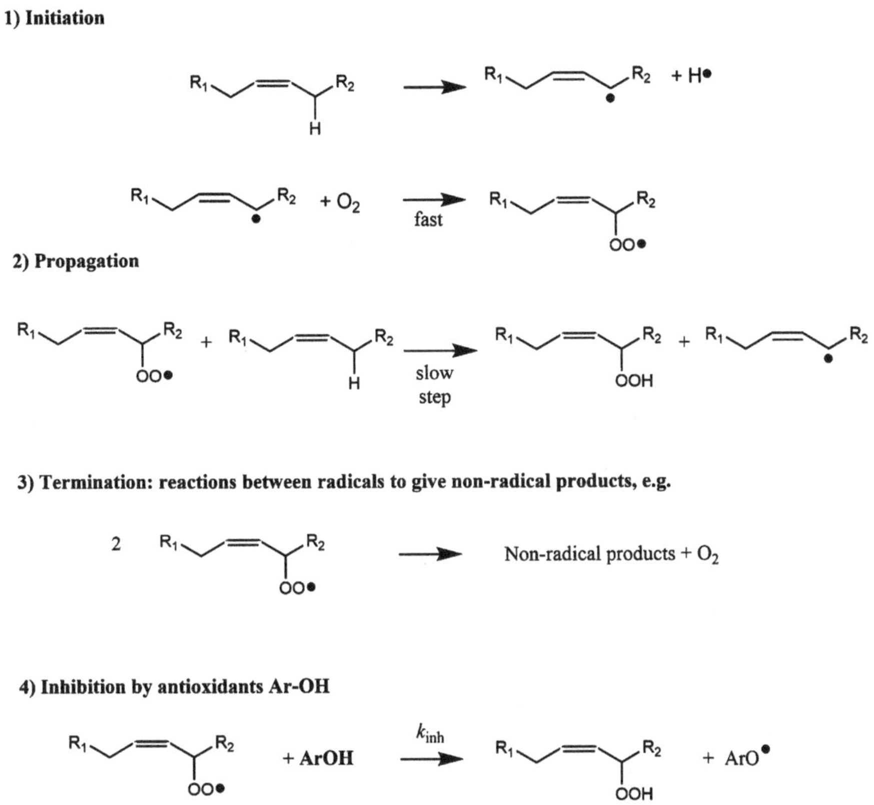

Scheme 1.3 The three main steps of the free radical oxidation of unsaturated lipids and their inhibition by radical scavenger antioxidants ArOH. The rate-determining step of the reaction is the reaction of the peroxyl radical ROO• with the lipid substrate (reaction 2) to yield the corresponding hydroperoxyde and a lipid radical that may undergo a variety of processes (propagation step). The reaction terminates through various pathways, all of them leading to the formation of non-radical products. In the presence of antioxidants, reaction 4 becomes competitive and may halt or minimize step 2

the severe pathological effects of ROS, e.g., neurodegenerative diseases, atherosclerosis, and cell apoptosis (Bhattacharya 2015).

To minimize, in as much as possible, the lipid oxidation reaction, antioxidants are added to the system. Most of the antioxidants are phenolic in nature, working as radical scavenging radicals. They react with the peroxyl radicals, reaction 4 in Scheme 1.3, but they may also limit the formation of radicals in the initiation step (secondary antioxidants, reaction 1 in Scheme 1.3) (Ross et al. 2003; Frankel 2005). Efficient antioxidants are those molecules whose rate of trapping radicals (reaction 4 in Scheme 1.3) is equal to, or higher than, the rate of production of radicals (reaction 2 in Scheme 1.3) (Raimúndez-Rodríguez et al. 2019; Costa et al. 2017; Freiría-Gándara et al. 2018b; Ferreira et al. 2018).

1.2.3 Role and Properties of Antioxidants in Controlling Radical Production

Polyphenolic antioxidants have been used for centuries in preventive medicine and human epidemiological studies that support the idea that there is an inverse relationship between antioxidant levels and intake and cognition function and development of neurodegenerative diseases (Choi et al. 2012; Ebrahimi and Schluesener 2012). The multifunctionality of polyphenols is indeed very attractive for food and pharmaceutical formulators because they constitute sets of molecules with the same reactive moieties, reducing in a great extent the risk of bio-incompatibility.

As a consequence, polyphenolic antioxidants are among the most widely employed natural compounds to protect cellular components from the oxidative damage caused by ROS and transition metals (Roleira et al. 2015; Nimse and Palb 2015; Chavarria et al. 2015; Vladimir-Knežević et al. 2012; Brewer 2011) because they contribute to the delay or prevention of the oxidation of biomolecules.

Briefly, polyphenolic antioxidants operate by preventing, or slowing, the progression of oxidative damage by inhibiting and minimizing the effects of oxidation reactions. The reactivity of polyphenolic antioxidants against ROS and their bioavailability appear to be a crucial parameter on evaluating their efficiency (Ross et al. 2003; Rice-Evans et al. 1996; Rein et al. 2012; D'Archivio et al. 2010). Both parameters are largely influenced by the chemical structure of the polyphenol, which varies widely from one compound to another, and the most abundant ones in our diet are not necessarily those that have the best bioavailability profile (Manach et al. 2004, 2005; Williamson and Manach 2005; Rein et al. 2012).

All polyphenolic compounds possess one common structural feature: an aromatic ring bearing at least one hydroxyl substituent. From this basic structure, several thousand of naturally occurring compounds have been described, from simple phenolic acids to highly polymerized compounds such as tannins (Hemingway and Laks 2012; Watson et al. 2013). Polyphenols are usually classified according to their chemical structure as a function of the number of phenol subunits and the number and position of substituents in the aromatic rings. Some of the most popular polyphenols are shown in Scheme 1.4.

On the basis of their mode of action, polyphenolic antioxidants can be classified in two broad groups, those who reduce or delay the attack of ROS usually by trapping chain-propagating, O-centered, free radicals (*chain-breaking or primary* antioxidants) and those that prevent the attack of ROS on a substrate (*preventive* or *secondary* antioxidants). Primary (chain-breaking) antioxidants inactivate free radicals by three main mechanisms, all playing important roles in determining radical scavenging activities depending on the particular environmental conditions: (1) transferring H-atoms to peroxyl radicals (Hydrogen atom transfer, HAT, mechanism); (2) single electron transfer-proton transfer (SETPT mechanism); and (3) sequential proton loss-electron transfer (SPLET) mechanism (Ingold and Pratt 2014; Jodko-Piórecka et al. 2018).

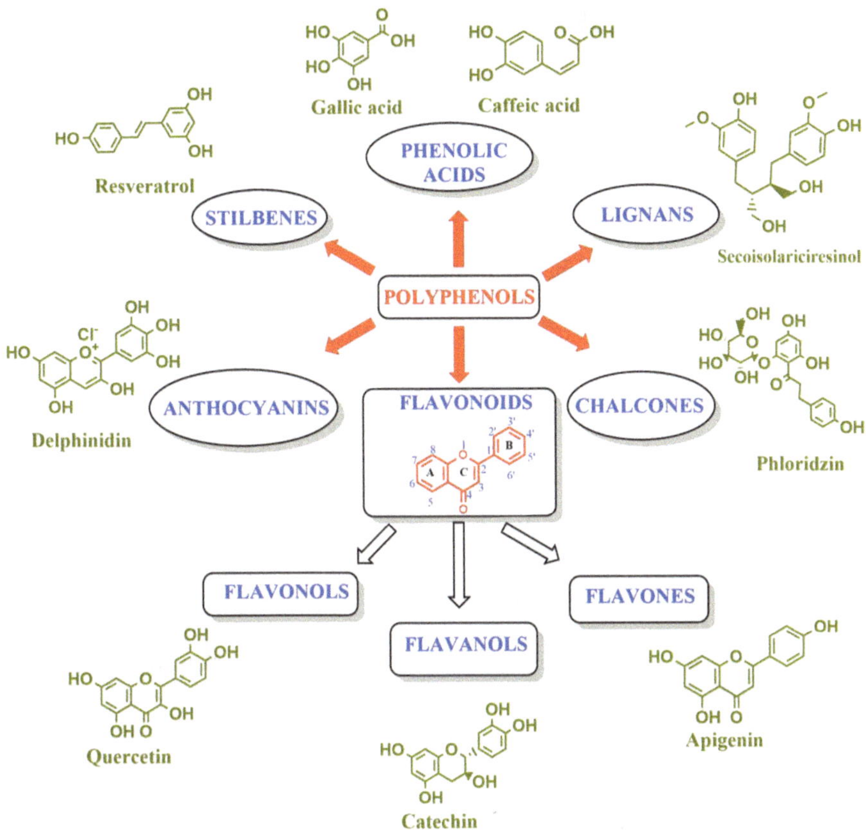

Scheme 1.4 Typical classification of polyphenolic antioxidants and some relevant chemical structures. (Reprinted from the reference (Losada-Barreiro and Bravo-Díaz 2017), with permission of Elsevier)

1.2.4 So... Why Encapsulate Antioxidants?

The human body produces endogenous antioxidant molecules, whose concentration is usually quite low. The concentrations of polyphenols measured in vivo are often at least one order of magnitude lower than the levels that appear to be effective when in vitro (Fanga and Chandaria 2010; Parisi et al. 2014; Ballesteros et al. 2017). This means that, frequently, their minimum effective concentration (i.e., the plasma concentration below which a patient's response is too small for a clinical benefit) is low and to increase their concentration to get adequate levels, it is often necessary to get them from external sources such as foods and/or supplements (Embuscado 2015).

The main source of exogenous phenolic antioxidants is dietary, particularly from fruits and vegetables, and it is gaining worldwide attention as nutraceuticals in the prevention of several diseases. Recently, research on antioxidant strategies with the aim of minimizing neurodegenerative and some other diseases has increased notably. Moreover, consumers have the strong perception of acquiring polyphenolic antioxidants from fruits and vegetables as natural and healthy products, as they constitute one of the most important groups of secondary metabolites widely distributed in the plant kingdom (Ballesteros et al. 2017; Munin and Edwards-Lévy 2011).

Encapsulated polyphenols are frequently employed to improve their bioavailability and their chemical stability both in vivo and in vitro instead of free compounds. Encapsulation of polyphenols has also other advantages as it can provide extra chemical stability of the own lipid-based formulations, for example those widely employed in parenteral nutrition. Loading lipid-based emulsions with polyphenols provide extra protection against lipid oxidation and, at the same time, can maintain the properties of the bioactives to deliver them in the required moment.

1.2.5 Encapsulation and Its Application in the Food and Pharmaceutical Industries

Encapsulation is a concept commonly employed to describe the process of entrapping bioactive molecules within a carrier material to improve their delivery to specific sites where the molecules are beneficial (Jan Zuidam and Nedovic 2010; Jafari et al. 2015; Ray et al. 2016). The materials used for the design of the delivery systems must be food-grade, biodegradable and capable of forming a barrier between the internal region and the surrounding medium. Proteins, lipids and polysaccharides are appropriate materials for encapsulation (Nedovic et al. 2011).

The food and pharmaceutical industries employ encapsulation processes for several reasons (Jafari et al. 2015; Sukhorukov 2014; Andersson Trojer et al. 2013). First, encapsulated bioactive molecules such as antioxidants, vitamins, fatty acids, etc. can be delivered into the human organism at specific areas where the bioactive component can be released over prolonged periods under specific conditions. Second, encapsulation is used to allow an easier handling of the bioactive compounds to provide adequate physiological concentrations. Finally, the entrapped bioactive compounds can also be used to protect the capsule itself from the action of external agents such as molecular oxygen.

Antioxidants are added to O/W parenteral emulsions to protect the fatty acids of the emulsion from the action of oxygen, which reacts with the fatty acids leading to the formation of lipid radicals (lipid peroxidation reaction) (Raman et al. 2017; Mundi et al. 2017). Functional bioactive compounds are also used to control the texture, color and flavor properties of the carrier.

Capsules and microcapsules – hereafter we will refer, for the sake of simplicity, to all them as capsules - can be characterized by their nature as reservoirs, matrixes and coated matrixes (Fig. 1.1). The reservoir type has a layer around the core material (also called capsule). The matrix type has the active agent dispersed over the carrier or located in the surface of the material. A combination of reservoir type and

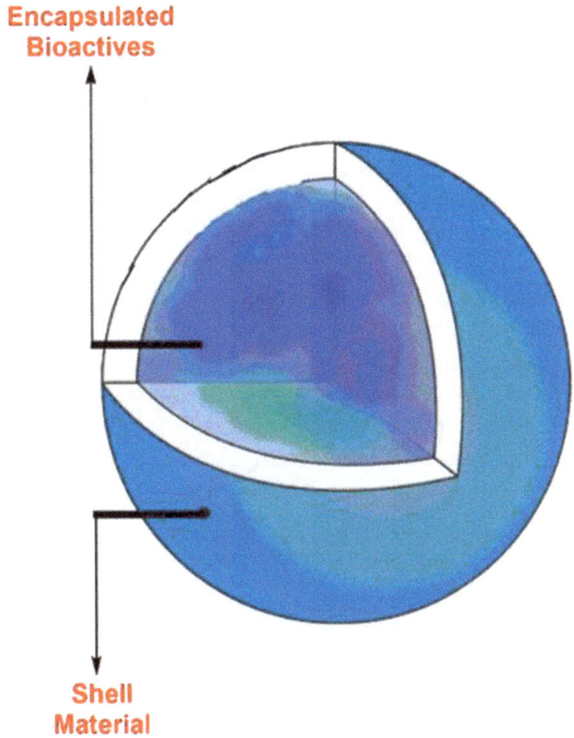

Fig. 1.1 Basic illustration of a spherical capsule. The capsule diameter may range from hundreds of nanometers to micrometers. The thickness and hardness of the shell may be tailored to control the entrance and exit of the encapsulated material

matrix type gives a third encapsulate system commonly called coated matrix, in which the active agent is a capsule covered by an additional layer (Ray et al. 2016).

The wall material plays a central role in the stability and applications of the capsule and must be chemically inert. In addition, it is desirable that the wall material possess, among others, good rheological properties, good ability to stabilize kinetically the capsule (for example, when emulsions are produced) and capability of completely release the solvent or other materials used during the process of encapsulation under desolvation conditions.

1.2.6 Some Encapsulation Techniques

The choice of the encapsulation method is crucial for the practical applications of the capsule and the encapsulated biomaterial (Ray et al. 2016; Jan Zuidam and Nedovic 2010). The production of capsules should be efficient, cheap, and easily incorporated to the target system (for example, food). Since bioactives have different molecular structures, no single universal procedure for encapsulation is available, but a number of them have been proposed including spray-drying, extrusion, fluid-bed-coating, and molecular inclusion with liposomes and cyclodextrins. For the sake of clarity, we will describe briefly only some of them, and the interested reader is referred to specialized reviews (Jan Zuidam and Nedovic 2010; Lin et al. 2017; Bartosz and Irene 2016; Andersson Trojer et al. 2013).

The produced capsules can then be characterized by a number of parameters such as their size and polydispersity, surface charge, the chemical nature of its components and the distribution of the added bioactives. All these parameters play a role on the properties, on the physical stability and on the release mechanisms. When the bioactive is liquid, a variety of drying technologies such as spray-bed-drying, fluid-bed-drying and freeze-drying can be employed to encapsulate the bioactive compounds. These methods are usually flexible, continuous, and produce particles of good quality, which lead to capsule sizes less than 40 μm (Nedovic et al. 2011).

Extrusion methods are also widely employed. Encapsulation by extrusion involves dispersion of the core material in a molten carbohydrate mass. This mixture is forced through a die into a dehydrating liquid, which hardens the coating to trap the core material. Typically, it consists of dropping droplets of aqueous polymer solutions (generally 0.6–3% by weight of sodium alginate) into a gelling bath that produces particles down to 50 μm (Jan Zuidam and Nedovic 2010; Nedovic et al. 2011). Extrusion provides true encapsulation in that the core material is completely surrounded by the wall material. When the material contacts the dehydrating liquid and the wall is hardened, all residual oil or core material is removed from the surface. The absence of residual surface oil and the complete encapsulation gives the manufactured products an excellent shelf life. The dripping tool can be simply a pipette, a syringe, vibrating nozzle, spraying nozzle, etc. Electrostatic extrusion is especially effective for production of very small particles, down to 50 nm. An alternative extrusion technology is co-extrusion. It might be utilized to prepare spherical microbeads with a hydrophobic core and a hydrophilic or hydrophobic shell (Jan Zuidam and Nedovic 2010; Nedovic et al. 2011).

Most probably, the carbohydrate-based molecular inclusion is the most expensive one and therefore less exploited, however they are suitable for industrial applications since they are biocompatible and degradable. Cyclodextrins (shown in Fig. 1.2) have a hydrophobic inner cavities of different sizes, ranging 0.57–0.95 nm, in which hydrophobic bioactive molecules can be reversibly entrapped. However, the outermost part of the cyclodextrin is hydrophilic and, thus, the hydrophobic bioactive can be located in a bulk aqueous environment. It is worth noting that the relatively small inner cavities limit critically their bioactive loading capacity because only bioactive compounds with appropriate size can be hosted inside their cavities.

Lipid or surfactant-based systems are composed of food-grade components such as liposomes and emulsions. They might be formed in different structures, and have advantageous characteristics including easy to scale-up and low preparation/modification costs. Liposomes, shown in Fig. 1.2, are droplets with sizes ranging 30–30,000 nm formed with phospholipids with a polar inner core where polar bioactives can be encapsulated while amphiphilic and hydrophobic bioactive molecules are solubilized within the phospholipid bilayer according to their polarity. Liposomes have the distinct advantage of being both nontoxic and biodegradable as they are formed of naturally occurring substances. The encapsulated material in liposomes is

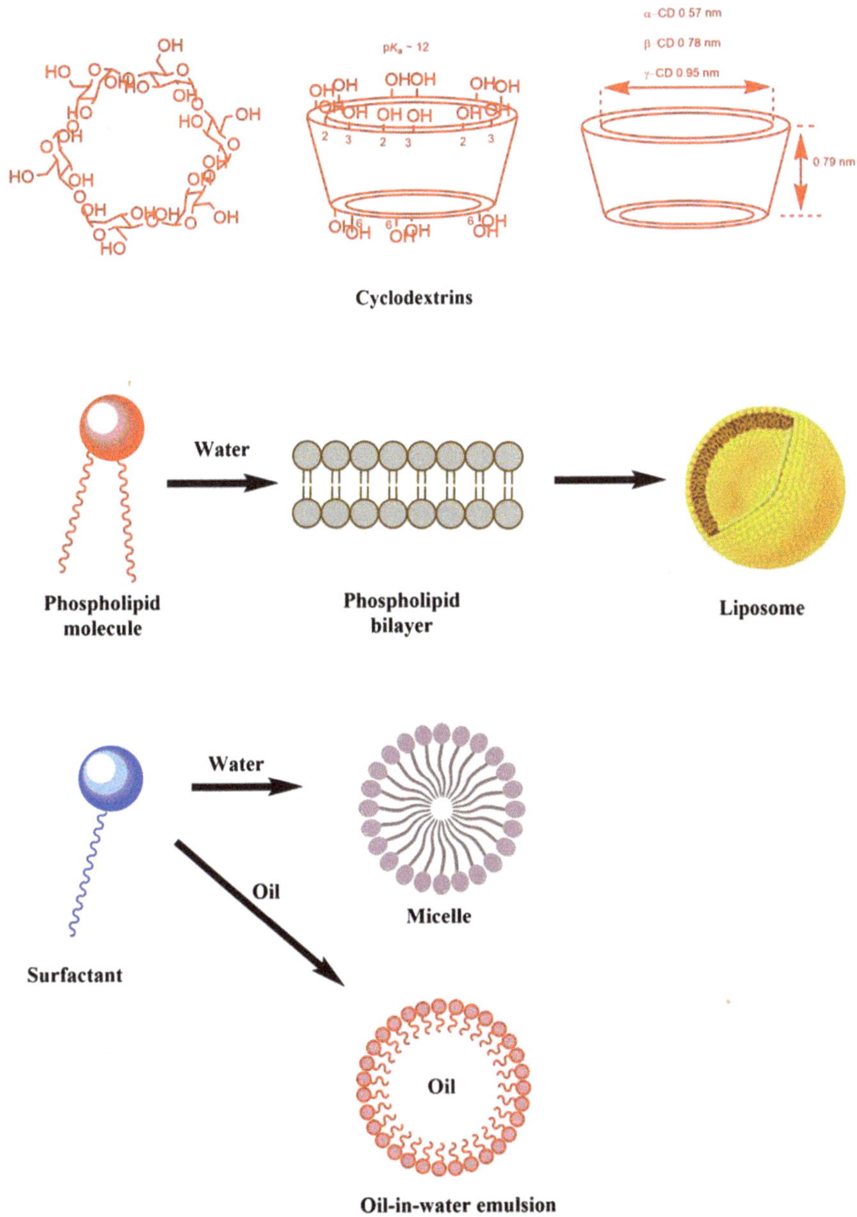

Fig. 1.2 Some carbohydrate- and lipid-based capsules capable of hosting bioactive compounds

protected to varying extents from immediate dilution and degradation, making them attractive encapsulation systems for delivery of hydrophilic and hydrophobic drugs.

Emulsification is one of the most popular methods for encapsulation consisting in the preparation of oil-in-water (O/W), water-in oil (W/O) or water-oil-water

(W/O/W) double emulsions. This is one of the most promising techniques for protection and delivery of polyphenols because they are cost-effective and because of their high-efficiency encapsulation capability, maintenance of chemical stability and controlled release. O/W emulsions (shown in Fig. 1.2) can be dried by employing any of the drying methods mentioned before to produce a powder that can encapsulate bioactive compounds or employed as instant formulations for numerous food products.

1.3 Performance of Antioxidants in Emulsions

1.3.1 Single O/W Emulsions: Preparation, Domain Division and Ion Exchange

The reason why surfactant-based aggregates such as emulsions are widely employed as mimetic-systems is that amphiphilic molecules are found in living systems and constitute the basic structural elements of biological membrane bilayers of cells compartments, for example, mithochondria, chloroplasts and vacuoles (Romsted 2012). Lipid membranes separate living cells from the external environment. These biological membranes act as efficient barriers selectively preventing - or enabling - the passage of a variety of molecules into and out of cells as well as intracellular organelles found within eukaryotic cells. This allows for the compartmentalization of biological processes - a fundamental feature of living organisms.

Moreover, emulsions are important components of a wide variety of different materials including foods, pharmaceuticals, agro-chemicals and cosmetics (McClements 2015). They can be classified according to the relative organization of the oil and aqueous phases as O/W emulsions, where the oil droplets are dispersed in an aqueous phase, and W/O emulsions, where water droplets are dispersed in the oil. However, more complex systems such as multiple emulsions including oil-in-water-in-oil (O/W/O) or water-in-oil-in-water (W/O/W) can also be prepared.

The oil type influences the nutritional profile of food-grade nanoemulsions and can be formulated using different nonpolar compounds, such as triglycerides, mineral oils, flavor oils, or essential oils. The composition of the aqueous phase plays a major role in determining the physicochemical properties of nanoemulsions and a variety of water-soluble constituents, including minerals, acids, bases, flavors, preservatives, vitamins, sugars, surfactants, proteins, and polysaccharides, can be added to the aqueous phase to change its properties, for instance the pH and the ionic strength.

Surfactants can be cationic (positive), anionic (negative), non-ionic (neutral), or zwitterionic (both positive and negative), Scheme 1.5, and play a major role on the formation, stability, and functional properties of the emulsion.

Cationic surfactants are scarcely used in the food industry, with the exception of lauric arginate, which has strong antimicrobial properties (Gu et al. 2013; Gao et al.

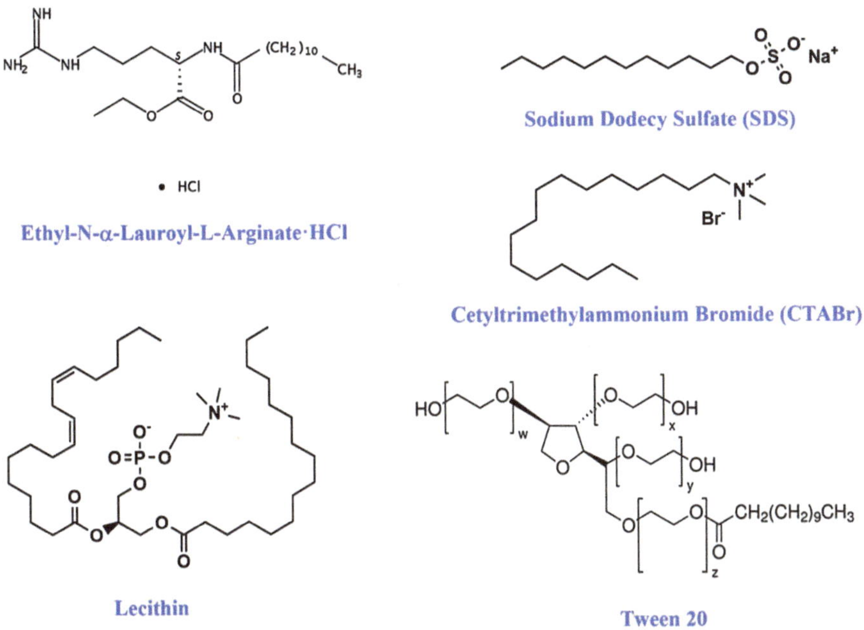

Scheme 1.5 Some common surfactants employed in the preparation of emulsions

2013). Zwitterionic surfactants are molecules with two or more ionizable groups with opposite charges on the same molecule. The two most commonly used types of zwitterionic emulsifiers for food applications are lecithins and proteins. Anionic surfactants include sodium dodecyl sulfate (SDS), sodium lauryl sulfate, citric acid esters of monoglycerides, and diacetyl tartaric acid esters of monoglycerides (Qian and McClements 2011). Nonionic surfactants include sucrose esters, such as sorbitan monooleate or sucrose monopalmitate, polyoxyethylene sorbitan esters of monoglycerides (Tweens), and polyoxyethylene ether surfactants (Brij). Nonionic surfactants are commonly used for nanoemulsion production (Salvia-Trujillo et al. 2017) and are not expected to develop an electrical charge on the droplets.

Coarse emulsions loaded with antioxidants are prepared by shaking together the appropriate amounts of oil, water, antioxidant and surfactant, with some form of agitation leading to the formation of micrometer-sized (0.1 – and 100 μm) spherical droplets (Fig. 1.3). The surface of each droplet is an interface between hydrophobic and hydrophilic molecules, and the free energy needed to increase the surface area between the oil and water phases is positive. Thus, emulsions are inherently thermodynamically unstable systems (as opposed to microemulsions). To form kinetically stable emulsions for relatively long periods of time (minutes to years), surfactants (Scheme 1.5) are added prior to homogenization. High-pressure homogenizers are used to reduce the size of the droplets of coarse emulsions, and spherical droplets of sizes less than 80 nm can be obtained depending on the particular conditions employed (Fig. 1.3).

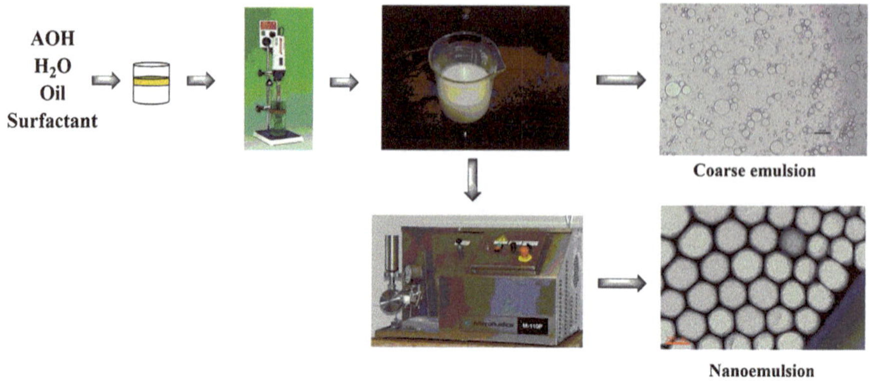

Fig. 1.3 Basic illustration of the preparation of coarse emulsions and nanoemulsions composed of antioxidants (AOH), water, oil and surfactant. Nanoemulsions are prepared by passing the coarse emulsion through a high-pressure homogenizer. (Reproduced from Barreiro-Losada (2014), with permission)

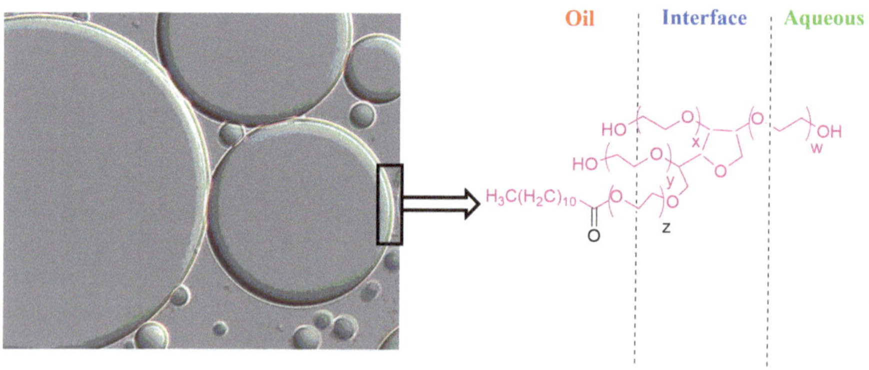

Fig. 1.4 Optical microscope photograph of an O/W emulsion and conceptual division of the domains with different solvent properties. (Reproduced from Barreiro-Losada (2014), with permission)

Conceptually, it is convenient to divide an emulsion into three distinct regions: the interior of the droplets, the continuous phase, and the interfacial region (Fig. 1.4). The interface consists of a narrow region surrounding each emulsion droplet that is comprised mainly of surface-active molecules but that may also contain some oil and water molecules as well as any other type of molecules that are attracted towards it.

The droplet interface is a narrow region (typically 2–20 nm thick) that surrounds each emulsion droplet and contains a mixture of oil, water, and emulsifier molecules. The surfactant monolayer that is formed around the droplets facilitates the emulsification process and aids in the minimization of coalescence (McClements 2015). The properties of this surfactant monolayer (e.g., fluidity, cohesiveness, charge density, etc.) often affect the overall characteristics of the emulsion (e.g., stability, droplet size, total interfacial area, etc.). When surfactants adsorb at the interface, the interfacial tension is decreased and the surfactant molecules act as a barrier delaying the coalescence of droplets by electrostatic and/or steric repulsion (McClements 2015).

If the surface of the droplets is charged, Scheme 1.6, the distribution of molecules is quite complex because the existing electrostatic forces play an important role that needs to be taken into consideration. For example, in cetrimonium bromide (CTAB) stabilized emulsions, addition of electrolytes such as NaBr, increase the interfacial H^+ concentration via the Donnan equilibrium. However, anionic emulsions displace H^+ from the interfacial region, reducing its concentration via the ion exchange equilibrium (Gu et al. 2013; Gao et al. 2013).

It is worthwhile to note that, frequently, the interfacial region of an emulsion does not contribute significantly to the total volume of an emulsion unless the droplet size is smaller than approximately 1 mm (Bravo-Díaz et al. 2015; Romsted and Bravo-Díaz 2013). However, the interfacial layer has a central role in the properties and utility of emulsions because many important chemical, physical, biological and technological processes are influenced by the presence of the interfacial region (Berton-Carabin et al. 2014; Berton et al. 2011).

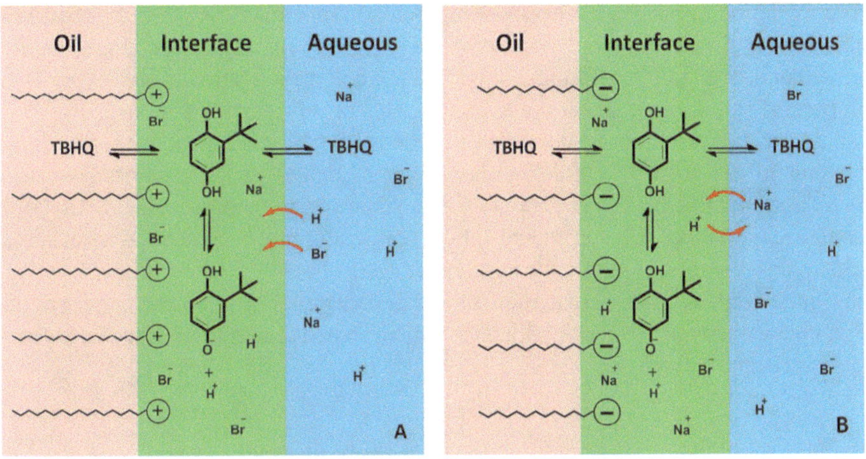

Scheme 1.6 Illustration of the different equilibria existing in cationic and anionic emulsions in the presence of an added salt NaBr. (**a**) Donnan equilibrium in emulsions stabilized with cationic surfactants. (**b**) Ion exchange in emulsions stabilized with anionic surfactants. The ionization equilibria of a representative antioxidant, tert-butylhydroquinone (TBHQ), is also displayed. (Reproduced from reference (Gu et al. 2013), with permission of Elsevier)

1.3.2 Methods to Measure the Relative Efficiency of Antioxidants

To estimate the antioxidant capacity in a given system containing lipids, samples with and without antioxidants are subjected to an accelerated oxidation test under standardized conditions. An appropriate end point is chosen to determine the appropriate levels of oxidative deterioration. The determination of the increase in the time necessary for samples containing antioxidant molecules to reach the limit of acceptability (for example, before a significant decomposition of conjugated dienes takes place) in comparison to samples without these molecules, allows evaluating their antioxidant capacity.

The acceleration of the oxidation process can be carried out by increasing the temperature (40–140 °C), by adding metallic catalysts (1–100 ppm), by increasing the oxygen pressure (3–165 psi) or by radical initiators. The choice of the most appropriate method and oxidation conditions is critical in interpreting the results obtained.

Temperature is an important factor to be considered when assessing the antioxidant activity of edible oils, since the oxidation mechanism may change with temperature. In addition, at high temperatures, the rate of lipid oxidation may even depend on oxygen pressure, because its solubility decreases and may become limiting during reaction progress. Hydroperoxides decompose at different speeds upon changing temperature, altering the type and concentration of the oxidation products. The oxidation rate increases exponentially with temperature, and thus the shelf life of a lipidic system decreases logarithmically with increasing temperature. On the other hand, at high temperatures, polymerization and cyclization reactions are also important, although they are not significant for normal storage temperatures. Consequently, results from stability tests obtained at high temperature can be misleading in assessing the effectiveness of antioxidants. Usually, its effectiveness increases as the temperature decreases and, sometimes, the pro-oxidant activity of some antioxidants can be promoted at high temperatures. The Schaal Oven test uses the mildest conditions within the accelerated tests ($T < 60$ °C) (Table 1.3), correlating quite well with the actual shelf times. Several studies show that there are no significant changes in the oxidation mechanisms when this test is performed at temperatures up to 60 °C.

1.3.2.1 Schaal Oven Test

This method is one of the most used methods for evaluating antioxidant capacity since the test conditions of this method is very similar to the storage conditions of the food. In this test, food samples (oil, emulsion, etc.) containing different antioxidants at the same concentration are subjected to temperatures up to 60 °C (Rajalakshmi and Narasimhan 1995), and their oxidation status is examined at regular intervals, through sensory measurements or by the determination of the sample

Table 1.3 Overview of some accelerated oxidation laboratory tests (Madhavi et al. 1996; Rajalakshmi and Narasimhan 1995)

Accelerated test	Storage conditions	Comments
Non-accelerated test	Room temperature, Atmospheric pressure.	Long periods of testing
Light	Room temperature, Atmospheric pressure.	Different mechanism
Metal catalyst, usually Fe(III)	Room temperature, Atmospheric pressure.	More peroxide decomposition
Weight increase	T = 30–80 °C, Atmospheric pressure	Difficulty in determine the endpoint
Schaal Oven Test	T = 20–60 °C, Atmospheric pressure	Similar oxidation mechanism, May have long periods of testing
Oxygen consumption	T = 80–100 °C, Atmospheric pressure	Different mechanism
Oxytest	T = 99 °C, 65–110 psi O_2	Different mechanism
Active oxygen method (AOM)	T = 98 °C	Different mechanism
Rancimat	110–140 °C	Questionable endpoint, Different mechanism

contents in oxidation products (Williams 1996). Primary changes can be assessed by the amount of hydroperoxides (e.g. test of conjugated dienes) and secondary changes by the formation of carbonyl compounds, monaldehyde and other aldehydes, among others (e.g. p–anisidine test) (Williams 1996). The results provided by this test usually have a good correlation with the evaluation carried out under the normal storage conditions for oils and fats (Gray 1978; Schaich 2005). This method has also been applied to emulsified samples (Paiva-Martins et al. 2006).

A number of methods to monitor the relative efficiency of antioxidants in inhibiting oxidations, e.g., the lipid oxidation in foods, have been described in the literature. They can be classified into five groups based on what they measure: (1) absorption of oxygen, (2) loss of initial substrates, (3) formation of free radicals, (4–5) formation of primary (hydroperoxides) and secondary (aldehydes, ketones, hydrocarbons, etc.) oxidation products. As expected, there is a variety of experimental techniques and chemical tests available to measure antioxidant effectiveness (Wang and Joseph 1999; Frankel 2005; Laguerre et al. 2010; Brewer 2011; Pinchuk et al. 2012; de Macêdo et al. 2017; Mozuraityte et al. 2016).

One of the most simple methods, yet reliable, exploits the formation of primary oxidation products (conjugated dienes, CDs) (Frankel 2005; Shahidi and Zhong 2005; Laguerre et al. 2010; Pinchuk et al. 2012). CDs are formed during the formation of hydroperoxides from unsaturated fatty acids due to the rearrangement of the double bonds. The resulting conjugated dienes exhibit an intense absorption at 234 nm and thus can be detected by employing UV/VIS spectroscopy (Fig. 1.5). An increase in UV absorption with time theoretically reflects the formation of primary oxidation products in fats and oils. The method is, therefore, simple, fast, and

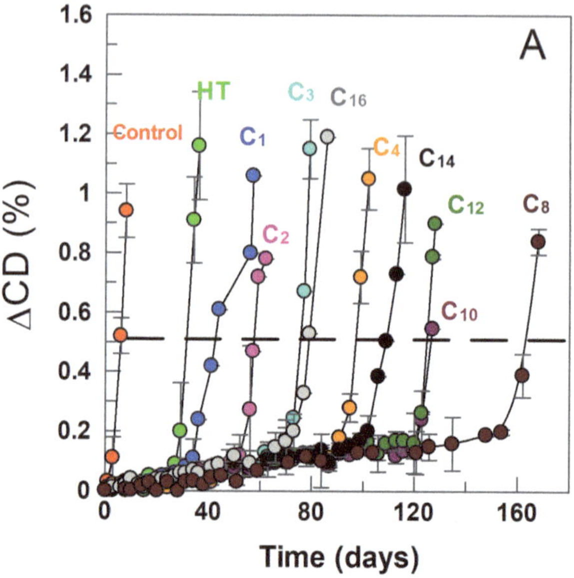

Fig. 1.5 Typical oxidation kinetic plots of olive O/W lipids in the presence of hydroxytyrosol (HT) and some alkyl ester derivatives (C1–C16) of different hydrophobicity. Oxidation develops slowly initially (initiation) and the induction period is a measure of the time required for a sudden change in the rate of oxidation, i.e, the propagation step. The extent of oxidation is assessed by monitoring the formation of conjugated dienes (CDs) at early stages of oxidation. The relative efficiency of HT derivatives is evaluated by measuring the time required to increase the amount of conjugated dienes, CDs, in a given amount (here 0.5%CD) after the propagation step of the radical reaction is reached (dotted line). (Figure extracted from the reference (Almeida et al. 2016), Copyright American Chemical Society)

requires no chemical reagents and only small amounts of samples are needed. Measurement of conjugated dienes is a sensitive method to follow the early stages of the oxidation process because at later stages, the formed secondary oxidation products overlap in the same UV detection range (Frankel 2005). One limitation of this method is its strong dependence on fatty acid composition in the studied sample. Oils containing high amount of PUFAs will have a faster increase in conjugated diene content compared to oils with less PUFAs. Consequently, the method cannot be used to compare oxidation in oils with different composition of fatty acids. Moreover, the method is only useful for measurement of changes in oils that contain substantial amounts of linoleate or more highly unsaturated fatty acids because conjugated diene are produced from abstraction of hydrogen in unsaturated fatty acids (Frankel 2005).

1.3.3 *Effects of Compartimentalization: Location of Antioxidants in Emulsions*

The efficiency of antioxidants depends on a number of factors, including their structural characteristics, concentration at the reaction site, acidity of the medium, nature of the oxidizable substrate, the physical state of the system and the presence of pro-oxidants and synergists. In general, antioxidants may protect against oxidation by one or a combination of various mechanisms, the predominant mechanism in a particular situation determines, largely, the efficiency of the antioxidant in inhibiting oxidation.

The chemical structure of an antioxidant determines its reactivity towards free radicals and other ROS and hence its antioxidant activity. In addition, the efficiency of antioxidants depends strongly on their concentrations at the reaction site, which does not need to be the same as the stoichiometric one, especially in multiphasic systems such as cells, emulsions or microemulsions, because antioxidants partition between the different regions of the system (Fig. 1.6). All this parameters need to be taken into consideration to optimize the efficiency of antioxidant and thus their effects need to be analyzed in detail.

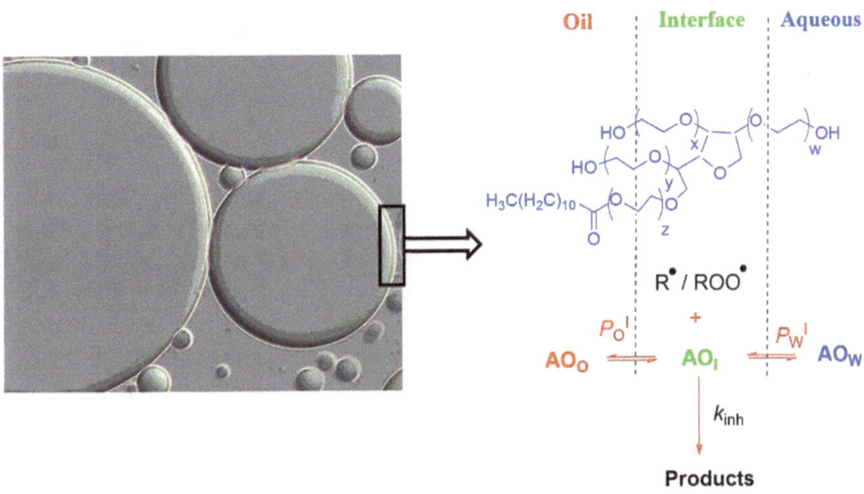

Fig. 1.6 Left: Optical microscope image of the droplets of an olive oil O/W emulsion taken as a model of emulsified lipid-based foods (see text). Olive oil composition includes oleic, linoleic and linolenic fatty acids in different proportions. Right: Magnified two-dimensional representation of a small portion of the oil, interfacial, and aqueous regions of the emulsion showing the surfactant employed to stabilize kinetically the emulsion and the distribution of antioxidants between these regions. In lipid-based emulsions, the interfacial region is believed to be the reaction site where lipid oxidation takes place (Frankel 2005; Shahidi 2015). (Reproduced from Barreiro-Losada (2014), with permission)

1.3.4 Distribution of Antioxidants in Emulsions: Where They Are Located?

Determining the distribution of antioxidants in emulsions is not a trivial experiment because, physically, it is impossible to separate the interfacial region from the oil and aqueous ones without disturbing the existing equilibria. Thus, it needs to be evaluated in the intact emulsions (Bravo-Díaz et al. 2015), making common analytical methods employed in binary oil-water systems (in the absence of any surfactant) useless for determining partition constants in emulsions (Berthod and Carda-Bosch 2004; Bravo-Díaz et al. 2015). The distribution of antioxidants (generally any reactant) can be only attained with the use of chemical methods through the use of chemical probes, as discussed elsewhere (Bravo-Díaz et al. 2015). A hydrophobic chemical probe, 4-hexadecylbenzenediazonium, 16-ArN_2^+, was chosen to assess the distribution of antioxidants because arenediazonium ions react with virtually all antioxidants. In addition, 16-ArN_2^+ probe bears a long alkyl chain which makes it to be water insoluble, but it is oil insoluble due to its cationic head group (Scheme 1.7).

Thus, the reactive $-N_2^+$ group is localized at the interface, where it reacts with the antioxidants located in that region (Scheme 1.7). The partition constants are obtained from the changes of the measured rate constant, k_{obs}, with the concentration of surfactant (expressed as the surfactant volume fraction $\Phi_I = V_{surf}/V_{emulsion}$), which are interpreted on the grounds of the pseudophase kinetic model. The relevant equations, their derivatization, and the procedure to determine the partition constants of the antioxidants in the intact emulsions, as well as the distribution of the antioxidants can be found elsewhere (Bravo-Díaz et al. 2015).

Table 1.4 displays some of the partition constants reported in the literature together with the chemical structures of the investigated antioxidants. In all cases, the partition constant values $P > 1$ are obtained, indicating the preference of antioxidants to be located in the interfacial region. Note that the Gibbs free energy of transfer ($\Delta G = -RT\ln P$) of antioxidants from the aqueous or oil phase to the interfacial region is negative, indicating the natural tendency of antioxidants to incorporate to the interfacial region. However, as expected, such tendency is different from one antioxidant to another and depends, for emulsions of the same composition, on the chemical nature of the antioxidant. For example, P_O^I values increase upon increasing the hydrophobicity of the antioxidant up to a maximum after which decreases, indicating that the tendency of the most hydrophobic antioxidants to be incorporated in the interfacial region decreases.

Determining the distribution of a given antioxidant once their partition constants are known is straightforward and can be assesed by employing Eqs. 1.1, 1.2, and 1.3.

$$\%AO_I = \frac{100\Phi_I P_W^I P_O^I}{\Phi_O P_W^I + \Phi_I P_W^I P_O^I + \Phi_W P_O^I} \tag{1.1}$$

$$\%AO_W = \frac{100\Phi_W P_O^I}{\Phi_O P_W^I + \Phi_I P_W^I P_O^I + \Phi_W P_O^I} \tag{1.2}$$

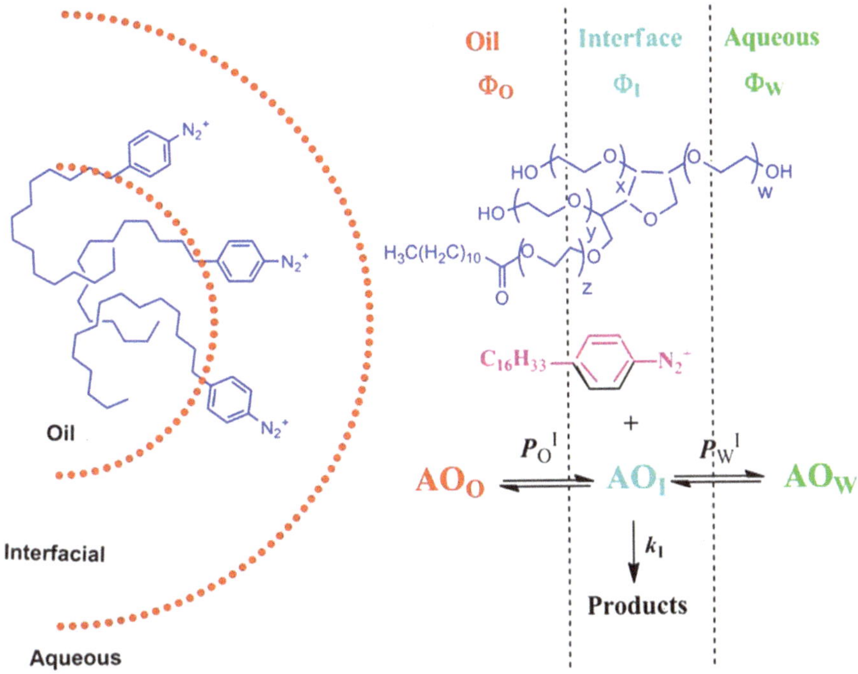

Scheme 1.7 Left: Symbolic representation of an O/W emulsion showing the presumed location of the reactive group ($-N_2^+$) of the chemical probe employed to assess the distribution of antioxidants in intact emulsions (Bravo-Díaz et al. 2015). Right: Pictorial representation of the three domains of an emulsion showing the partitioning of an antioxidant in these domains. The reactive $-N_2^+$ group, anchored in the interfacial region, reacts with antioxidants present in this region, and the observed rate constant k_{obs} of the reaction depends on the interfacial concentration of the antioxidant, which in turn depends on the partition constants between the oil-interfacial, P_O^I, and aqueous-interfacial, P_W^I, regions of the emulsion. (Right picture reproduced from the reference (Almeida et al. 2016) with permission of American Chemical Society)

$$\%AO_O = \frac{100\Phi_O P_W^I}{\Phi_O P_W^I + \Phi_I P_W^I P_O^I + \Phi_W P_O^I} \tag{1.3}$$

Figues 1.7 and 1.8 show some examples of the dependency of the percentage of antioxidant in the interfacial region on both the nature of the antioxidant and the surfactant volume fraction (Fig. 1.7) and on the acidity and the surfactant volume fraction (Fig. 1.8). The general picture from a number of published partitioning experiments for antioxidants in different emulsions is that the percentage of the antioxidant in the interfacial region increases upon increasing the surfactant volume fraction. Therefore, at the highest concentration typically used, $\Phi_I \approx 0.04$, a large fraction (>80–90%) of the antioxidant is located in the interfacial region.

Analyses of the effects of some factors controlling the distribution of antioxidants in emulsions (e.g., temperature, acidity, oil to water ratio, and type of oil) suggest that the surfactant volume fraction is the main parameter controlling their

Table 1.4 Values for the partition constants between the oil-interfacial, P_O^I, and aqueous-interfacial, P_W^I, regions of emulsions for different antioxidants in olive oil O/W emulsions. (Data extracted from the references (Losada-Barreiro et al. 2013a; Costa et al. 2015, 2017; Almeida et al. 2016; Meireles et al. 2020)). The chemical structures of the antioxidants are indicated in the left row, where R stands for the akly chain grafted to the parent antioxidant

Structure	-R	P_O^I	P_W^I
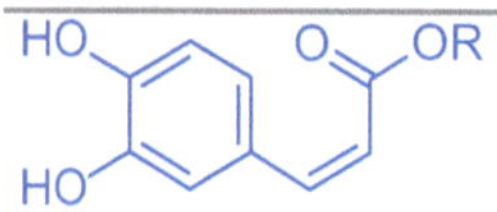 Caffeic acid	-H (CA)	---	349
	$-CH_3$ (C1)	312	720
	$-CH_2 CH_3$ (C2)	405	3156
	$-(CH_2)_2 CH_3$ (C3)	454	---
	$-(CH_2)_7 CH_3$ (C8)	502	---
	$-(CH_2)_{15} CH_3$ (C16)	376	---
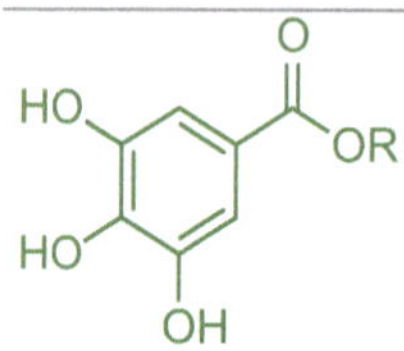 Gallic acid	-H (GA)	---	101
	$-(CH_2)_2 CH_3$ (C3)	449	328
	$-(CH_2)_7 CH_3$ (C8)	27	---
	$-(CH_2)_{11} CH_3$ (C12)	17	---
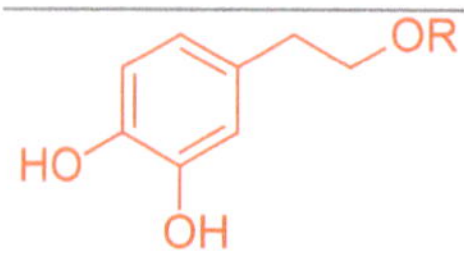 Hydroxytyrosol	-H (HT)	---	53
	$-CO-CH_3$ (C2)	320	93
	$-CO-(CH_2) CH_3$ (C3)	197	123
	$-CO-(CH_2)_2 CH_3$ (C4)	171	373
	$-CO-(CH_2)_5 CH_3$ (C7)	184	---
	$-CO-(CH_2)_7 CH_3$ (C9)	296	---
	$-CO-(CH_2)_9 CH_3$ (C11)	125	---
	$-CO-(CH_2)_{11} CH_3$ (C13)	82	---
	$-CO-(CH_2)_{15} CH_3$ (C17)	52	---
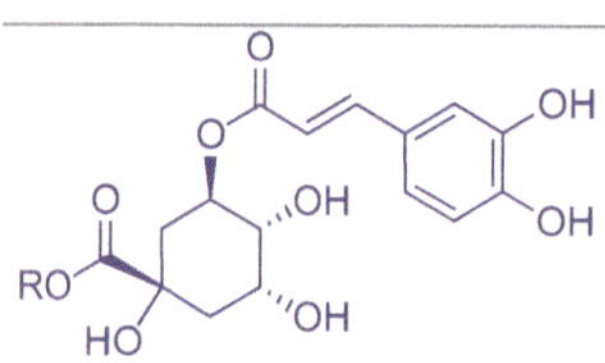 Chlorogenic	-H (CGA)	---	40
	$-CH_2 CH_3$ (C2)	---	78
	$-(CH_2)_3 CH_3$ (C4)	---	141
	$-(CH_2)_7 CH_3$ (C8)	111	---
	$-(CH_2)_9 CH_3$ (C10)	124	---
	$-(CH_2)_{11} CH_3$ (C12)	159	---
	$-(CH_2)_{15} CH_3$ (C16)	89	---

distribution and that other parameters have only a minor effect, except the acidity of the solution for phenolic acids (Galan et al. (2016), Losada-Barreiro 2009) or Vitamin C (Pastoriza-Gallego et al. 2012).

Figure 1.7b shows that antioxidants are mostly distributed in the interface of the emulsion, despite the fact that the volume of this region is only a minor fraction (as much as 4%) of the total emulsion volume. $\%AO_I$ increases upon increasing Φ_I so that at $\Phi_I = 0.04$, more than 85% of all antioxidants are located in that region. Moreover, Fig. 1.7b shows an outstanding result: at any surfactant concentration,

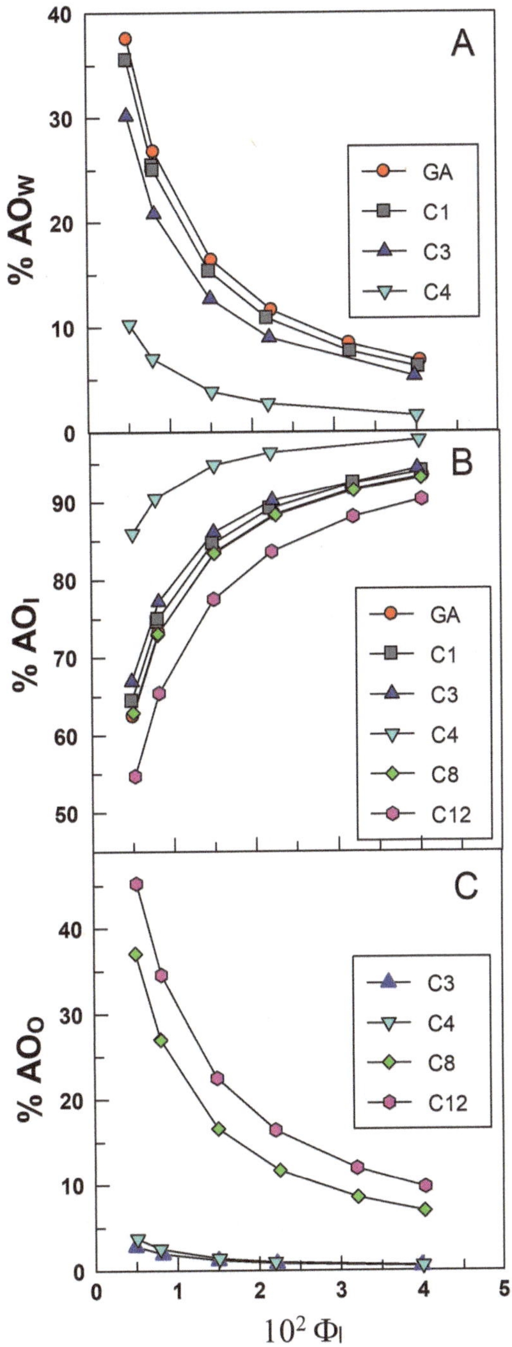

Fig. 1.7 Variation of the percentage of gallic acid (as representative antioxidant) and its hydrophobic derivatives (see Table 1.4 for their chemical structures) in the aqueous (**a**) interfacial (**b**) and oil (**c**) regions of 1:9 soybean O/W emulsions at T = 25 °C (pH = 3). (Reproduced from the reference (Mitrus et al. 2019) with permission of American Chemical Society)

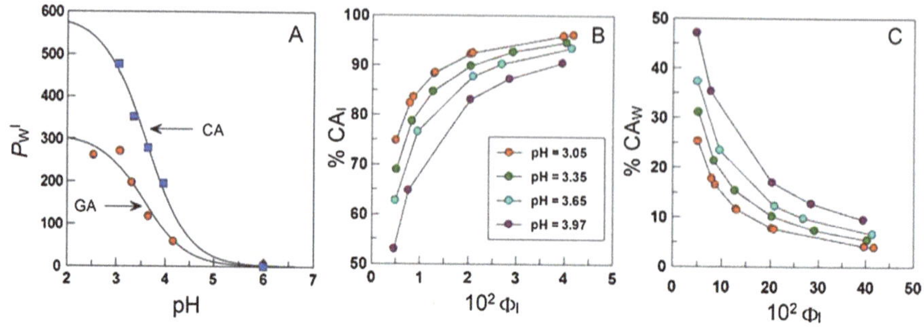

Fig. 1.8 (**a**) Variations of P_W^I with pH for gallic acid (GA) and caffeic acid (CA). (**b**) and (**c**) Effects of acidity on the distribution of CA between the interfacial (**b**) and aqueous (**c**) regions of 1:9 corn oil emulsions. Solid lines drawn to aid the eye. (Reproduced from the reference (Losada-Barreiro et al. 2015a), with permission of Wiley & Sons)

$\%AO_I$ does not relate with the hydrophobicity of the antioxidants, following, for the gallic acid (GA) derivatives, the order: C12 < GA ≈ C8 < C2 < C3 < C4 (see Table 1.4 for chemical structures). In other words, the antioxidant with the highest percentage in the interfacial region (butyl gallate, C4, in the present case) is not the most hydrophobic one. This observation was already reported in literature (Losada-Barreiro et al. 2013b; Freiría-Gándara et al. 2018b), and is related with the balance of the different hydrophobic-hydrophilic forces involved in the partitioning process, including the solvation and hydrogen bonding ability of antioxidants in the various regions of the emulsion (Bravo-Díaz et al. 2015). Figure 1.7c shows that only C8 (octyl gallate) and C12 (laury gallate) are present in a significant percentage in the oil region ($\%AO_O$). In both cases $\%AO_O$ decreases upon increasing Φ_I, for example $\%C12_O$ decreases from ~45% $\Phi_I = 0.005$ to ~10% $\Phi_I = 0.04$ (C12 is the most hydrophobic antioxidant of the homologous series of antioxidants from gallic acid).

Figure 1.8a shows the sigmoidal variation of P_W^I with pH. At low acidities (pH << pK_a), phenolic acids are mostly protonated, i.e., they are neutral, and the P_W^I values at such acidities should be independent of pH as shown in Fig. 1.8a. Upon lowering the acidity, phenolic acids deprotonate and their carboxylic groups become anionic. At pH ~ pK_a, phenolic acids are about 50% ionized and the P_W^I values decrease because of the higher solubility of the anionic forms in water with respect to that of the neutral antioxidant. At pH > pK_a, phenolic acids are fully deprotonated, and their carboxylates are fully dissolved in the aqueous region (Losada-Barreiro et al. 2015a).

Figure 1.8b shows that, at any acidity, $\%AO_I$ increases with increasing Φ_I, $\%CA_I \approx 50$ ($\Phi_I = 0.005$, pH = 3.97) to $\%CA_I \approx 90$ at $\Phi_I = 0.04$. Figure 1.8b also shows that at any given Φ_I, an increase in acidity leads to a substantial increase in the percentage of phenolic acids in the interfacial region of the emulsions.

1.3.5 Why Antioxidants Are Spontaneously Incorporated Into the Interfacial Regions of Emulsions? Driving Forces for the Hydrophobic Effect That Partition Antioxidants

Relevant thermodynamic parameters (ΔG, ΔH and ΔS) for the transfer of components between the regions of the emulsion can be obtained from the variations of the partition constants with temperature (Losada-Barreiro et al. 2015b; Raimúndez-Rodríguez et al. 2019). Knowledge about thermodynamic parameters for the transfer of components might be useful to get insights into the transfer mechanism, i.e., on the driving forces (enthalpic, entropic, spontaneity of the transfer process, the hydrogen bond interactions between the component and the solvent). Knowledge of thermodynamic transfer parameters may be also useful to predict the partitioning behavior of structurally similar compounds.

The Gibbs free energy for the transfer of antioxidants from the aqueous and oil regions to the interfacial one can be determined from the P_W^I and P_O^I values reported in Table 1.4, which are in the range 50–320, indicating that the transfer of antioxidants from the oil to the interfacial and from the aqueous to the interfacial regions is spontaneous (i.e., $\Delta G_{transfer} < 0$). That is, all antioxidants have a latent tendency to be incorporated into the interfacial region, but in a different extent, which is controlled by the P_W^I and P_O^I values. At equilibrium, the chemical potentials of the antioxidants in the oil, aqueous and interfacial regions are equal to each other, and the Gibbs free energies for the transfer of 1 mol of the antioxidant from the oil to the interfacial region, $\Delta G_T^{0,O\rightarrow I}$, and from the aqueous to the interfacial region, $\Delta G_T^{0,W\rightarrow I}$, are given by Eqs. 1.4 and 1.5.

$$\Delta G_{\mathrm{T}}^{0,\mathrm{O}\rightarrow\mathrm{I}} = \mu_C^{0,\mathrm{I}} - \mu_C^{0,\mathrm{O}} = RT\ln\frac{V_{\mathrm{m}}^{\mathrm{O}}}{P_{\mathrm{O}}^{\mathrm{I}}V_{\mathrm{m}}^{\mathrm{I}}} \tag{1.4}$$

$$\Delta G_{\mathrm{T}}^{0,\mathrm{W}\rightarrow\mathrm{I}} = \mu_C^{0,\mathrm{I}} - \mu_C^{0,\mathrm{W}} = RT\ln\frac{V_{\mathrm{m}}^{W}}{P_{W}^{\mathrm{I}}V_{\mathrm{m}}^{\mathrm{I}}} \tag{1.5}$$

$$\Delta H_T^{0,W\rightarrow I} = R\left[\frac{\partial\left(\ln P_W^I\right)}{\partial\left(\frac{1}{T}\right)}\right]_P \tag{1.6}$$

$$\Delta G_T^{0,W\rightarrow I} = \Delta H_T^{0,W\rightarrow I} - T\Delta S_T^{0,W\rightarrow I} \tag{1.7}$$

where V_m^O, V_m^W and V_m^I are the molar volumes of oil, water and emulsifier, whose values can be obtained from literature density values. Calculations are done under

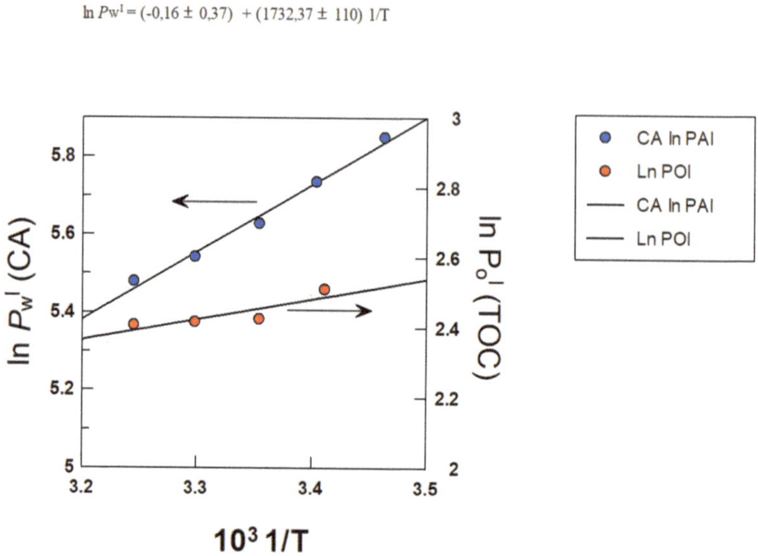

Correlation coefficient: 0.9940

Parameter	Value	Std. Error
a (intercept)	-0.1622	0.3703
b (gradient)	1732.3694	110.4172

Fig. 1.9 Plots of the variation of ln P_W^I and ln P_O^I versus 1/T for the transfer of the hydrophilic caffeic acid (CA) and the hydrophobic α-tocopherol (α-TOC) antioxiadnts according to the van't Hoff Eq., from which $\Delta H_T^{0,W\rightarrow I}$and $\Delta H_T^{0,O\rightarrow I}$are obtained. (Reproduced from the reference (Losada-Barreiro et al. 2015b), with permission of Royal Society of Chemistry)

the assumption that the variation of the molar volumes of oil, water and the surfactant are essentially constant over the temperature range commonly employed (290–310 K).

The thermodynamic parameters $\Delta H_T^{0,W\rightarrow I}$and $\Delta H_T^{0,O\rightarrow I}$ for transfer of a given antioxidant can be obtained from the variation of lnP_W^I and lnP_O^I values with T (Fig. 1.9) (van't Hoff Eq. 1.6), and the entropic values are obtained from the Gibbs Eq. 1.7.

The results obtained for the transfer of hydrophobic antioxidants such as α-tocopherol (water insoluble) and hydrophilic antioxidants such as caffeic acid (oil insoluble) are consistent with the transfer of the antioxidant to the interfacial region being driven by an increase in disorder, that is, an increase in entropy.

1.3.6 Phenomenological Observations on the Efficiency of Polyphenolic Antioxidants: Are All Antioxidants with the Same Reactive Moiety Equally Efficient in Emulsions?

More than twenty-five years ago, Porter and coworkers (Porter 1993) published the first phenomenological observation, coined as the polar paradox, which was an attemps to rationalize what makes antioxidants capable of protecting lipid-based emulsified systems from oxidation and how to predict their efficiency. They particularly investigated the effects of the hydrophilic-lipophilic balance (HLB) of the antioxidant on the oxidative stability of a number of lipid-based systems. They noticed that hydrophilic antioxidants tend to be more active than their hydrophobic counterpart in bulk oils; however the opposite was observed in lipid dispersions such as O/W emulsions, membranes, and even whole tissues. This "rule of thumb" has been used since then in attemps to design the most efficient antioxidants to prevent or minimize the oxidation of lipid dispersions.

Later in 1994, Frankel and coworkers (Frankel et al. 1994; Frankel 2005) assumed that the main reaction site of (lipid) oxidation reactions in compartimentalized systems was the interfacial region and postulated that the polar paradox was a consequence of the involved interfacial phenomena due to the differential partitioning of antioxidants in the system. Accordingly, in O/W emulsions, nonpolar antioxidants would concentrate in the oil-water interface and inhibit oxidation more efficiently than polar antioxidants that partition into the water phase, where they would be less effective. In contrast, in bulk oils, they assumed that the increased effectiveness of hydrophilic antioxidants was due to their ability to concentrate at the air-oil interface where oxidation is prevalent, whereas lipophilic antioxidants were solubilized in the oil region.

The relative hydrophilic/hydrophobic characteristics of an antioxidant are important in affecting its solubility, absorption, distribution, metabolism, and excretion (ADME). Highly polar or highly hydrophilic antioxidants do not pass over the cell membranes of the gut wall quickly. One way round this is to inject them, but they cannot be used against intracellular targets, as they will not cross cell membranes. The hydrophobic character of a drug can be measured experimentally by testing the drug's relative distribution in an appropriate oil-water mixture. Hydrophobic molecules will prefer to dissolve in the oil phase whereas hydrophilic molecules will prefer the aqueous layer. Unfortunately, the most frequently used partition constants to assess the hydrophobicity of a molecule, usually determined in octanol-water binary systems, cannot be extrapolated to oil systems and thus need to be determined for the particular oil employed (Freiría-Gándara et al. 2018a).

Based on the polar paradox and Frankel's ideas, a large number of experimental reports have been published in the search for a better understanding of the factors than control the antioxidant activity (Frankel 2005; Mozuraityte et al. 2016; Berton-Carabin et al. 2014; Bravo-Díaz et al. 2015; Almeida et al. 2016; Losada-Barreiro et al. 2015b; Costa et al. 2015). In general, an increase in the hydrophobicity of the antioxidant was considered advantageous to improve its efficiency, but a number of subsequent reports indicated that this might not be necessarily true, as structurally similar antioxidants do not always have similar efficiencies. For example, the hydrophobicity of antioxidants can be modified by grafting alkyl chains of different length while maintaining the reactive moiety, but an increase the number of C atoms in the alkyl chains of antioxidants not only changes their HLB but also affects their relative solubility in the oil, water, and interfacial regions (Bravo-Díaz et al. 2015; Losada-Barreiro et al. 2015b) of the emulsion and consequently affects the oxidation/reduction potential of antioxidant (Teixeira et al. 2013). On the other hand, an increase in the HLB of the antioxidants may lead to a decrease in their concentrations at the reaction site (interfacial region) (Costa et al. 2015, 2016; Almeida et al. 2016; Losada-Barreiro et al. 2015a; Bravo-Díaz et al. 2015).

Recent investigations on the effects of the HLB for several series of homologous antioxidants show that, in O/W emulsions, their efficiency does not increase linearly with increasing the number of C atoms in their alkyl chain but is parabolic-like, with a maxima at an intermediate (C4–C12, see Table 1.4 for chemical structures) chain length (Fig. 1.10) (Bravo-Díaz et al. 2015; Costa et al. 2015; Losada-Barreiro et al. 2013a; Laguerre et al. 2012; Medina et al. 2009). This parabolic dependence of antioxidant efficiency upon chain length is also observed in the biological activity of homologous series of antioxidants, which increase up to a critical point after which their activity decreases. The phenomena, known as the "cut-off" effect, was reported for the first time more than a century ago, it is now commonly observed for various antioxidant, biological, and cytotoxic activities of almost every homologous series of amphiphiles tested (Balgavý and Devínsky 1996; Laguerre et al. 2009; Muñoz-Marín et al. 2013; Calderón-Montaño et al. 2013; Mateos et al. 2008; Trujillo et al. 2006).

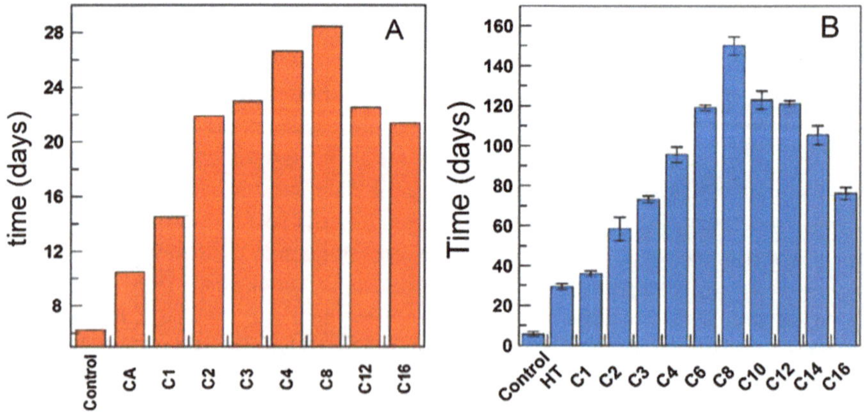

Fig. 1.10 Parabolic-like variations on the oxidative stability of emulsions as a function of the alkyl chain length of caffeic acid derivatives (**a**) and hydroxytyrosol derivatives (**b**). (Reproduced from the references (Costa et al. 2015) and (Almeida et al. 2016), with permission of Elsevier (**a**) and American Chemical Society (**b**))

1.3.7 The Fate of Bioantioxidants in O/W Emulsions: Relationships Between Partitioning and Efficiency

Several hypotheses have been put forward to rationalize the cut-off effect on the antioxidative and biological activities of series of antioxidants (Balgavý and Devínsky 1996), including: (a) changes in reactant diffusivity; (b) changes in the self-aggregation ability of the long-chain derivatives; (c) differential solubility of the antioxidants in the interfacial region and (d) differences in reactivity could be accounted for changes in physical properties of the reactants, e.g., chain length induced changes in reactivity orientation in the interfacial region (Aliaga et al. 2016). However, none of them has been proved experimentally.

Recently, our group proposed that the cut-off effect for emulsions can be naturally explained on the grounds of the relative solubility of antioxidants in the oil, interfacial and aqueous regions of an emulsion and demonstrated that antioxidant efficiency correlates with the fraction of the antioxidant in the interfacial region (Costa et al. 2015; Losada-Barreiro et al. 2013a; Lisete-Torres et al. 2012). Our approach was to determine the distribution of an antioxidant between the oil, aqueous and interfacial regions of intact emulsions by using a chemical kinetic method that exploits the overall bimolecular reaction between the hydrophobic 4-hexadecylbenzenediazonium, 16-ArN_2^+, ion - located completely in interfacial region of the emulsion - and antioxidants (Scheme 1.7). The interpretation of the experimental results in emulsions is based on the pseudophase kinetic models widely employed to interpret reactivity in colloidal systems (Bravo-Díaz et al. 2015; Romsted 2012; Bunton and Savelli 1986) and allows to determine, in the intact emulsion, the partition constants of the antioxidant between the oil-interfacial, P_O^I, and the water-interfacial, P_W^I, regions of the emulsion. Experimental P_O^I and P_W^I values obtained for some representative antioxidants are displayed in Table 1.4.

The seemingly complex variation in P_W^I and P_O^I for the antioxidants with the length of the alkyl chain is more apparent than real. The percentages of antioxidant in the aqueous, interfacial and oil regions can be calculated from the partition constant values and were plotted as a function of the alkyl chain length at different Φ_I. Figure 1.11 shows that the $\%AO_I$ follows a parabola-like form for CA and its esters with increasing alkyl chain length with a maximum at C8.

Thus, the percentage of antioxidant in the interfacial region increases up to a maximum at the C4–C8, depending on the nature of the antioxidant, and then gradually decreases with increasing its hydrophobicity. Results on the distribution of antioxidants in emulsions show, therefore, that: (i) the fraction of antioxidants in the interfacial regions of the emulsions do not correlate directly with the hydrophobicity of the antioxidant; and (ii) that the fraction of antioxidant in the interfacial region reaches a maximum for antioxidant with intermediate polarity or intermediate alkyl chain length. The C4–C8 derivatives represent antioxidants with an optimal balance of hydrophobicity and hydrophilicity (head group polarity) required for maximum solubility of the antioxidant in the interfacial region of the emulsion and therefore maximum antioxidant efficiency.

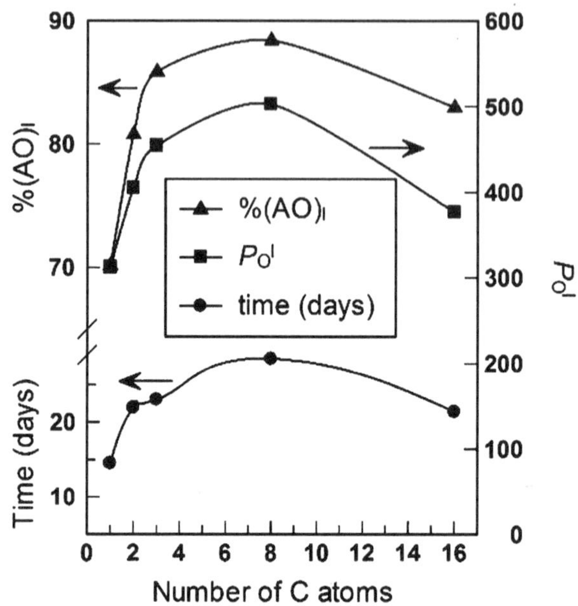

Fig. 1.11 Changes in the percentage of antioxidant in the interfacial region (%AO_I), in the partition constant of the antioxidants between the oil and interfacial region (P_O^I), and in the time necessary to increase 1% the percentage of conjugated dienes (which asses the relative antioxidant efficiency) all as a function of hydrophobicity of caffeic acid derivatives. (Reproduced from the reference (Costa et al. 2015) with permission of Elsevier)

1.4 Summary and Future Perspective

Delivery of bioactive compounds to an specific location and chemical preservation of the capsule from external agents is of great interest for the food, medical and pharmaceutical sectors. Lipid-based delivery systems such as emulsions, composed of biodegradable lipids, have a clear advantage above other encapsulation methods because they are cost-efficient, require minimal manipulation and they are, usually, easy to scale-up. However, one of the main drawbacks of these systems is their chemical unstability, specially when polyunsaturated fatty acids are employed in their formulations.

Polyphenols are a major group of highly effective antioxidants and epidemiological studies conclude that their intake has potential health benefits. They are widely used to protect the lipids from oxidation, since they exhibit potent free radical scavenging capability and are able to chelate pro-oxidants such as transition metals. However, the chemical instability and low bioavailability of polyphenols greatly limit their potential health benefits in preventing ageing, cancer, inflammation and neurodegenerative diseases. Using encapsulated polyphenols instead of free molecules improves both the stability and bioavailability of the molecules in vitro and in vivo.

Nowadays, either free polyphenols or encapsulated compounds are mainly used to prevent lipid oxidation and in the treatment of some human diseases, even though some reports suggest that there is no clear relationship between the intake of polyphenols and their health benefits. Research on a better understanding of the location and reaction mechanisms of polyphenols in emulsion-based delivery systems (for example, emulsions for parenteral nutrition), to control the release of incorporated bioactives in specific sites will be crucial in increasing the efficiency of encapsulated polyphenols in the biology, medical, pharmaceutic and cosmetic areas. Certainly, a deep knowledge on the location and distribution of the bioactives in the delivery systems will certainly help in the development of new methods not only as nutritional supplements but also with potential benefits to the human health. Worldwide interest in endogenous and exogenous antioxidants has triggered enormous attention since reactive oxygen species and the oxidative stress play an important role in the aetiology and progression of major human degenerative diseases.

Changes in the composition and/or preparation of the lipid-based system results in the preparation of a range of emulsions with new properties, all of them having (in principle) a great potential in delivery of polyphenols or other bioactive compounds. Consequently, many emulsion-based delivery systems for polyphenols have been well established. For example, the use of unsaturated fatty acids as the oil phase in delivery systems would provide a valuable energy source for patients and at the same time, an excellent vehicle to deliver highly hydrophobic drugs. However, the location and performance of the antioxidants or other bioactives in the delivery system is crucial for their practical applications. In contrast to solid particles, emulsion droplets have deformable surfaces, emulsion components and other added bioactive compounds diffuse, and adsorb and desorb from the continuous and disperse liquid regions to the interface region. Thus, the efficiency of antioxidants in inhibiting the lipid oxidation reaction can be compromised because the added polyphenols distribute between different regions according to their polarity, and therefore, research on methods to determine the location of antioxidants and their distribution are crucial and of great importance.

Acknowledgement Financial support of the following institutions is acknowledged: Xunta de Galicia (REDUSO, Grant number ED431D 2017/18), UID/QUI/50006/2019 and University of Vigo. S. L-B thanks Xunta de Galicia for a postdoctoral grant (POS-B/2016/012) and University of Vigo (Plan de Retención de Talento 2018). This book chapter was prepared during a sabbatical leave supported by the University of Vigo.

References

Aliaga C, Bravo-Moraga F, Gonzalez-Nilo D, Márquez S, Lürh S, Mena G, Rezende MC (2016) Location of TEMPO derivatives in micelles: subtle effect of the probe orientation. Food Chem 192:395–401

Almeida J, Losada-Barreiro S, Costa M, Paiva-Martins F, Bravo-Díaz C, Romsted LS (2016) Interfacial concentrations of hydroxytyrosol and its lipophilic esters in intact olive oil-in-water emulsions: effects of antioxidant hydrophobicity, surfactant concentration, and the oil-to-water ratio on the oxidative stability of the emulsions. J Agric Food Chem 64:5274–5283

Andersson Trojer M, Nordstierna L, Nordin M, Nydén M, Holmberg K (2013) Encapsulation of actives for sustained release. Phys Chem Chem Phys 15:17727–17741

Atkins P, De Paula J (2010) Physical chemistry. W. H. Freeman, New York

Balgavý P, Devínsky F (1996) Cut-off effects in biological activities of surfactants. Adv Colloid Interface Sci 66:23–63
Ballesteros LF, Ramirez MJ, Orrego CE, Teixeira JA, Mussatto SI (2017) Encapsulation of antioxidant phenolic compounds extracted from spent coffee grounds by freeze-drying and spray-drying using different coating materials. Food Chem 237:623–631
Barreiro-Losada S (2014), PhD Thesis, Universidad de Vigo, Spain
Bartosz T, Irene T (2016) Polyphenols encapsulation – application of innovation technologies to improve stability of natural products. Phys Sci Rev 20150005–20150010
Belščak-Cvitanović A, Stojanović R, Manojlović V, Komes D, Cindrić IJ, Nedović V, Bugarski B (2011) Encapsulation of polyphenolic antioxidants from medicinal plant extracts in alginate–chitosan system enhanced with ascorbic acid by electrostatic extrusion. Food Res Int 44:1094–1101
Benfeito S, Oliveira C, Soares P, Fernandes C, Silva T, Teixeira J, Borges F (2013) Antioxidant therapy: still in search of the 'magic bullet'. Mitochondrion 13:427–435
Berthod A, Carda-Bosch S (2004) Determination of liquid-liquid partition coefficients by separation methods. J Chromatog A 1037:3–14
Berton C, Ropers MH, Viau M, Genot C (2011) Contribution of the interfacial layer to the protection of emulsified lipids against oxidation. J Agric Food Chem 59:5052–5061
Berton-Carabin CC, Ropers M-H, Genot C (2014) Lipid oxidation in oil-in-water emulsions: involvement of the interfacial layer. Compr Rev Food Sci Food Safety 13:945–977
Berton-Carabin CC, Sagis L, Schroën K (2018) Formation, structure, and functionality of interfacial layers in food emulsions. Annu Rev Food Sci Technol 9:551–587
Bhattacharya S (2015) Reactive oxygen species and cellular defense system. In: Rani V, Yadav UCS (eds) Free radicals in human health and disease. Springer India, New Delhi. https://doi.org/10.1007/978-81-322-2035-0_2
Bravo-Díaz C, Romsted LS, Liu C, Losada-Barreiro S, Pastoriza-Gallego MJ, Gao X, Gu Q, Krishnan G, Sánchez-Paz V, Zhang Y, Ahmad-Dar A (2015) To model chemical reactivity in heterogeneous emulsions, think homogeneous microemulsions. Langmuir 31:8961–8979
Brewer MS (2011) Natural antioxidants: sources, compounds, mechanisms of action, and potential applications. Compr Rev Food Sci Food Saf 10:221–247
Bunton CA, Savelli G (1986) Organic reactivity in aqueous micelles and similar assemblies. Adv Phys Org Chem 22:213
Calder6N-Montaño JM, Madrona A, Burgos-Mor6n E, Luis Orta M, Mateos S, Espartero JL, L6Pez-LáZaro M (2013) Selective cytotoxic activity of new lipophilic hydroxytyrosol alkyl ether derivatives. J Agric Food Chem 61:5046–5053
Chang RC-C (2011) Advanced understanding of neurodegenerative diseases. InTech. https://doi.org/10.5772/2451
Chavarria D, Silva T, Martins D, Bravo J, Summavielle T, Garrido J, Borges F (2015) Exploring cinnamic acid scaffold: development of promising neuroprotective lipophilic antioxidants. Med Chem Commun 6:1043–1053
Choi D-Y, Lee Y-J, Hong JT, Lee H-J (2012) Antioxidant properties of natural polyphenols and their therapeutic potentials for Alzheimer's disease. Brain Res Bull 87:144–153
Costa M, Losada-Barreiro S, Paiva-Martins F, Bravo-Díaz C, Romsted LS (2015) A direct correlation between the antioxidant efficiencies of caffeic acid and its alkyl esters and their concentrations in the interfacial region of olive oil emulsions. The pseudophase model interpretation of the "cut-off" effect. Food Chem 175:233–242
Costa M, Losada-Barreiro S, Paiva-Martins F, Bravo-Diaz C (2016) Optimizing the efficiency of antioxidants in emulsions by lipophilization: tuning interfacial concentrations. RSC Adv 6:91483–91493
Costa M, Losada-Barreiro S, Paiva-Martins F, Bravo-Díaz C (2017) Physical evidence that the variations in the efficiency of homologous series of antioxidants in emulsions are a result of differences in their distribution. J Sci Food Agric 97:564–571

Costa M, Freiría-Gándara J, Losada-Barreiro S, Paiva-Martins F, Bravo-Díaz C (2020a) Effects of droplet size on the interfacial concentrations of antioxidants in fish and olive oil-in-water emulsions and nanoemulsions and on their oxidative stability. J Colloid Interface Sci 562:352–362

Costa M, Losada-Barreiro S, Bravo-Díaz C, Vicente AA, Monteiro LS, Paiva-Martins F (2020b) Influence of AO chain length, droplet size and oil to water ratio on the distribution and on the activity of gallates in fish oil-in-water emulsified systems: emulsion and nanoemulsion comparison. Food Chem 310:125716

D'archivio M, Filesi C, Varì R, Scazzocchio B, Masella R (2010) Bioavailability of the polyphenols: status and controversies. Int J Mol Sci 11:1321–1342

De Macêdo IYL, Garcia LF, Oliveira Neto JR, De Siqueira Leite KC, Ferreira VS, Ghedini PC, De Souza Gil E (2017) Electroanalytical tools for antioxidant evaluation of red fruits dry extracts. Food Chem 217:326–331

De Melo Barbosa R, Severino P, Finkler CLL, De Paula E (2019) Chapter 11 – Lipid-based colloidal carriers for transdermal administration of bioactives. In: Holban A-M, Grumezescu AM (eds) Materials for biomedical engineering. Elsevier, Cambridge, MA

De Roos B, Aura A-M, Bronze M, Cassidy A, Conesa M-TG, Gibney ER, Greyling A, Kaput J, Kerem Z, Knežević N, Kroon P, Landberg R, Manach C, Milenkovic D, Rodriguez-Mateos A, Tomás-Barberán FA, Van De Wiele T, Morand C (2019) Targeting the delivery of dietary plant bioactives to those who would benefit most: from science to practical applications. Eur J Nutr 58:65–73

Di Domenico F, Barone E, Perluigi M, Butterfield DA (2015) Strategy to reduce free radical species in Alzheimer's disease: an update of selected antioxidants. Expert Rev Neurother 15:19–40

Ebrahimi A, Schluesener H (2012) Natural polyphenols against neurodegenerative disorders: potentials and pitfalls. Ageing Res Rev 11:329–345

Embuscado ME (2015) Spices and herbs: natural sources of antioxidants – a mini review. J Funct Foods 18, Part B:811–819

Fanga Z, Chandaria B (2010) Encapsulation of polyphenols: a review. Trends Food Sci Technol 21:510–523

Feng Y, Wang X (2012) Antioxidant therapies for Alzheimer's disease. Oxid Med Cell Longev 2012:1–17. https://doi.org/10.1155/2012/472932

Ferreira I, Costa M, Losada-Barreiro S, Paiva-Martins F, Bravo-Díaz C (2018) Modulating the interfacial concentration of gallates to improve the oxidative stability of fish oil-in-water emulsions. Food Res Int 112:192–198

Frankel EN (2005) Lipid oxidation. The Oily Press, PJ Barnes & Associates, Bridgwater

Frankel EN, Huang SW, Kanner J, German JB (1994) Interfacial phenomena in the evaluation of antioxidants: bulk oils vs emulsions. J Agric Food Chem 42:1054–1059

Freiría-Gándara J, Losada-Barreiro S, Paiva-Martins F, Bravo-Díaz C (2018a) Differential partitioning of bioantioxidants in edible oil–water and octanol–water systems: linear free energy relationships. J Chem Eng Data 63:2999–3007

Freiría-Gándara J, Losada-Barreiro S, Paiva-Martins F, Bravo-Díaz C (2018b) Enhancement of the antioxidant efficiency of gallic acid derivatives in intact fish oil-in-water emulsions through optimization of their interfacial concentrations. Food Funct 9:4429–4442

Galan A, Losada-Barreiro S, Bravo-Díaz C (2016) A physicochemical study of the effects of acidity on the distribution and antioxidant efficiency of trolox in olive oil-in-water emulsions. Chem Phys Chem 17:296–304

Gao X, Bravo-Díaz C, Romsted LS (2013) Interpreting ion-specific effects on the reduction of an arenediazonium ion by *t*-butylhydroquinone (TBHQ) using the pseudophase kinetic model in emulsions prepared with a zwitterionic sulfobetaine surfactant. Langmuir 29:4928–4933

Garti N (2008) Delivery and controlled release of bioactives in foods and nutraceuticals. Woodhead Publishing, Cambridge

Gray JI (1978) Measurement of lipid oxidation: a review. J Am Oil Chem Soc 55:539–546

Gu Q, Bravo-Díaz C, Romsted LS (2013) Using the pseudophase kinetic model to interpret chemical reactivity in ionic emulsions: determining antioxidant partition constants and interfacial rate constants. J Colloid Interface Sci 400:41–48
Hardiman O, Doherty CP (2011) Neurodegenerative disorders: a clinical guide. Springer, London
Hemingway RW, Laks PE (2012) Plant polyphenols: synthesis, properties, significance. Springer/Plenum Press, New York
Ingold KU, Pratt DA (2014) Advances in radical-trapping antioxidant chemistry in the 21st century: a kinetics and mechanisms perspective. Chem Rev 114:9022–9046
Jafari SM, Fathi M, Mandala I (2015) Chapter 13 - Emerging product formation. In: Galanakis CM (ed) Food waste recovery. Academic Press, San Diego
Jan Zuidam N, Nedovic VA (2010) Encapsulation technologies for active food ingredients and food processing. Springer, New York
Jodko-Piórecka J, Cedrowski J, Litwinienko G (2018) Physico-chemical principles of antioxidant action, including solvent and matrix dependence and interfacial phenomena. In: Apak R, Capanoglu E, Shahidi F (eds) Measurement of antioxidant activity & capacity: recent trends and applications. J. Wiley & Sons, West Sussex
Kamal-Eldin A, Min DB (2008) Lipid oxidation pathways, vol 2. AOCS Press, Champaign
Kim J, Lee HJ, Lee KW (2010) Naturally occurring phytochemicals for the prevention of Alzheimer's disease. J Neurochem 112:1415–1430
Laguerre M, López-Giraldo LJ, Lecomte J, Figueroa-Espinoza MJ, Baréa B, Weiss J, Decker EA, Villeneuve P (2009) Chain length affects antioxidant properties of chlorogenate esters in emulsion: the cut-off theory behind the polar paradox. J Agric Food Chem 57:11335–11342
Laguerre M, Decker EA, Lecomte J, Villeneuve P (2010) Methods for evaluating the potency and efficacy of antioxidants. Curr Opin Clin Nutr Metab Care 13:518–525
Laguerre M, Bayrasy C, Lecomte J, Chabi B, Decker EA, Wrutniak-Cabello C, Cabello G, Villeneuve P (2012) How to boost antioxidants by lipophilization? Biochimie 95:1–7
Lin C-H, Chen C-H, Lin Z-C, Fang J-Y (2017) Recent advances in oral delivery of drugs and bioactive natural products using solid lipid nanoparticles as the carriers. J Food Drug Anal 25:219–234
Lisete-Torres P, Losada-Barreiro S, Albuquerque H, Sánchez-Paz V, Paiva-Martins F, Bravo-Díaz C (2012) Distribution of hydroxytyrosol and hydroxytyrosol acetate in olive oil emulsions and their antioxidant efficiency. J Agric Food Chem 60:7318–7325
Litwinienko G, Ingold KU (2007) Solvent effects on the rates and mechanisms of reaction of phenols with free radicals. Acc Chem Res 40:222–230
Losada-Barreiro S (2009) PhD Thesis, Universidad de Vigo
Losada-Barreiro S, Bravo-Díaz C (2017) Free radicals and polyphenols: the redox chemistry of neurodegenerative diseases. Eur J Med Chem 133:379–402
Losada-Barreiro S, Bravo Díaz C, Paiva Martins F, Romsted LS (2013a) Maxima in antioxidant distributions and efficiencies with increasing hydrophobicity of gallic acid and its alkyl esters. The pseudophase model interpretation of the Cut-off effect. J Agric Food Chem 61:6533–6543
Losada-Barreiro S, Sánchez Paz V, Bravo-Díaz C (2013b) Effects of emulsifier hydrophile–lipophile balance and emulsifier concentration the distributions of gallic acid, propyl gallate, and α-tocopherol in corn oil emulsions. J Colloid Interface Sci 389:1–9
Losada-Barreiro S, Bravo Díaz C, Romsted LS (2015a) Distributions of phenolic acid antioxidants between the interfacial and aqueous regions of corn oil emulsions: effects of pH and emulsifier concentration. Eur J Lipid Sci Technol 117:1801–1813
Losada-Barreiro S, Sánchez-Paz V, Bravo-Díaz C (2015b) Transfer of antioxidants at the interfaces of model food emulsions: distributions and thermodynamic parameters. Org Biomol Chem 13:876–885
Madhavi DL, Deshpande NR, Salunkhe DK (1996) Food antioxidants, technological, toxicological and health perspectives. Marcel Dekker, New York
Malar DS, Devi KP (2014) Dietary polyphenols for treatment of Alzheimer's disease– future research and development. Curr Pharm Biotechnol 15:330–342

Manach C, Scalbert A, Morand C, Rémésy C, Jiménez L (2004) Polyphenols: food sources and bioavailability. Am J Clin Nutr 79:727–747

Manach C, Williamson G, Morand C, Scalbert A, Rémésy C (2005) Bioavailability and bioefficacy of polyphenols in humans. I. Review of 97 bioavailability studies. Am J Clin Nutr 81(1 suppl):230S–242S

Mateos R, Trujillo M, Pereira-Caro G, Madrona A, Cert A, Espartero JL (2008) New lipophilic tyrosyl esters. Comparative antioxidant evaluation with hydroxytyrosyl esters. J Agric Food Chem 53:10960–10966

Mcclements DJ (2015) Food emulsions, principles, practices and techniques. CRC Press, Boca Raton

Mcclements DJ (2018) Enhanced delivery of lipophilic bioactives using emulsions: a review of major factors affecting vitamin, nutraceutical, and lipid bioaccessibility. Food Funct 9:22–41

Medina I, Lois S, Alcántara D, Lucas L, Morales JC (2009) Effect of lipophilization of hydroxytyrosol on its antioxidant activity in fish oils and fish oil-in-water emulsions. J Agric Food Chem 57:9773–9779

Meireles M, Losada-Barreiro S, Costa M, Paiva-Martins F, Bravo-Díaz C, Monteiro LS (2020) Control of antioxidant efficiency of chlorogenates in emulsions: modulation of antioxidant interfacial concentrations. J Sci Food Agric 99:3917–3925

Mitrus O, Żuraw M, Losada-Barreiro S, Bravo-Díaz C, Paiva-Martins F (2019) Targeting antioxidants to interfaces: control of the oxidative stability of lipid-based emulsions. J Agric Food Chem 67:3266–3274

Mozuraityte R, Kristinova V, Rustad T (2016) Oxidation of food components. In: Encyclopedia of food and health. Academic Press, Oxford

Mundi MS, Salonen BR, Bonnes SL, Hurt RT (2017) Parenteral nutrition – lipid emulsions and potential complications. Pract Gastroenterol 41:32–37

Munin A, Edwards-Lévy F (2011) Encapsulation of natural polyphenolic compounds; a review. Pharmaceutics 3:793–829

Muñoz-Marín J, De La Cruz JP, Guerrero A, López-Leiva I, López-Villodres JA, Reyes JJ, Espartero JL, Madrona A, Labajos MT, González-Correa JA (2013) Cytoprotective effect of hydroxytyrosyl alkyl ether derivatives after oral administration to rats in a model of glucose–oxygen deprivation in brain slices. J Agric Food Chem 60:7659–7667

Nedovic V, Kalusevic A, Manojlovic V, Levic S, Bugarski B (2011) An overview of encapsulation technologies for food applications. Procedia Food Sci 1:1806–1815

Nimse SB, Palb D (2015) Free radicals, natural antioxidants, and their reaction mechanisms. RSC Adv 5:27986–28006

Nowak E, Livney YD, Niu Z, Singh H (2019) Delivery of bioactives in food for optimal efficacy: what inspirations and insights can be gained from pharmaceutics? Trends Food Sci Technol 91:557–573

Paiva-Martins F, Santos V, Mangericão H, Gordon MH (2006) Effects of copper on the antioxidant activity of olive polyphenols in bulk oil and oil-in-water emulsions. J Agric Food Chem 54:3738–3743

Parisi OI, Puoci F, Restuccia D, Farina G, Iemma F, Picci N (2014) Chapter 4 - Polyphenols and their formulations: different strategies to overcome the drawbacks associated with their poor stability and bioavailability. In: Polyphenols in human health and disease. Academic Press, San Diego

Pastoriza-Gallego MJ, Losada-Barreiro S, Bravo Díaz C (2012) Effects of acidity and emulsifier concentration on the distribution of vitamin C in a model food emulsion. J Phys Org Chem 25:908–915

Pinchuk I, Shoval H, Dotan Y, Lichtenberg D (2012) Evaluation of antioxidants: scope, limitations and relevance of assays. Chem Phys Lipids 165:638–647

Porter WL (1993) Paradoxical behavior of antioxidants in food and in biological systems. Princeton, Priceton Scientific

Qian C, Mcclements DJ (2011) Formation of nanoemulsions stabilized by model food-grade emulsifiers using high-pressure homogenization: factors affecting particle size. Food Hydrocoll 25:1000–1008

Raimúndez-Rodríguez EA, Losada-Barreiro S, Bravo-Díaz C (2019) Enhancing the fraction of antioxidants at the interfaces of oil-in-water emulsions: a kinetic and thermodynamic analysis of their partitioning. J Colloid Interface Sci 555:224–233

Rajalakshmi D, Narasimhan S (1995) Food antioxidants: sources and methods of evaluation, in food antioxidants. CRC Press, Boca Raton

Raman M, Almutairdi A, Mulesa L, Alberda C, Beattie C, Gramlich L (2017) Parenteral nutrition and lipids. Nutrients 9:388–349

Ray PD, Huang B-W, Tsuji Y (2012) Reactive oxygen species (ROS) homeostasis and redox regulation in cellular signaling. Cell Signal 24:981–990

Ray S, Raychaudhuri U, Chakraborty R (2016) An overview of encapsulation of active compounds used in food products by drying technology. Food Biosci 13:76–83

Rein MJ, Renouf M, Cruz-Hernandez C, Actis-Goretta L, Thakkar SG, Silva Pinto M (2012) Bioavailability of bioactive food compounds: a challenging journey to bioefficacy. Br J Clin Pharmacol 75:588–602

Rice-Evans CA, Miller NJ, Paganga G (1996) Structure-antioxidant activity relationships of flavonoids and phenolic acids. Free Radic Biol Med 20:933–956

Roleira FMF, Tavares-Da-Silva EJ, Varela CL, Costa SC, Silva T, Garrido J, Borges F (2015) Plant derived and dietary phenolic antioxidants: anticancer properties. Food Chem 183:235–258

Romsted LS (2012) Introduction to surfactant self-assembly. In: Gale PA, Steed JW (eds) Supramolecular chemistry: from molecules to nanomaterials. J. Wiley & Sons Ltd, New York

Romsted LS, Bravo-Díaz C (2013) Modelling chemical reactivity in emulsions. Curr Opin Colloid Interface Sci 18:3–14

Ross L, Barclay C, Vinqvist MR (2003) Phenols as antioxidants. In: Rappoport Z (ed) The chemistry of phenols. J. Wiley & Sons, West Sussex

Salvia-Trujillo L, Soliva-Fortuny R, Rojas-Graü MA, Mcclements DJ, Martín-Belloso O (2017) Edible nanoemulsions as carriers of active ingredients: a review. Annu Rev Food Sci Technol 8:439–466

Schaich KM (2005) Lipid oxidation: theoretical aspects. In: Shahidi F (ed) Bailey's industrial oil and fat products. J. Wiley & Sons, New York

Schaich KM, Shahidi F, Zhong Y, Eskin NAM (2013) Chapter 11 - Lipid oxidation. In: Biochemistry of foods, 3rd edn. Academic Press, San Diego

Schieber M, Chandel, Navdeep s. (2014) ROS function in redox signaling and oxidative stress. Curr Biol 24:R453–R462

Shahidi F (2015) Handbook of antioxidants for food preservation, 1st edn. Woodhead Pub, Sawston, Cambridge

Shahidi F, Zhong Y (2005) Lipid oxidation: measurement methods. In: Shahidi F (ed) Bailey's industrial oil and fat products, 6th edn. John Wiley & Sons, Hoboken

Sonia Barreiro-Losada (2014), PhD Thesis, Universidad de Vigo, Spain

Spray JW (2016) Review of intravenous lipid emulsion therapy. J Infus Nurs 39:377–380

Sukhorukov GB (2014) Editorial overview: new technologies: how to put everything you need in a tiny pack and track its delivery? Curr Opin Pharmacol 18:vii–ix

Teixeira J, Gaspar A, Garrido EM, Garrido J, Borges F (2013) Hydroxycinnamic acid antioxidants: an electrochemical overview. Bio Med Theatr Res Int 2013:1–11. https://doi.org/10.1155/2013/251754

Trujillo M, Mateos R, Collantes De Teran L, Espartero JL, Cert R, Jover M, Alcudia F, Bautista J, Cert A, Parrado J (2006) Lipophilic hydroxytyrosyl esters. Antioxidant activity in lipid matrices and biological systems. J Agric Food Chem 54:3779–3785

Vladimir-Knežević S, Blažeković B, Štefan MB, Babac M (2012) Plant polyphenols as antioxidants influencing the human health. In: Rao V (ed) Phytochemicals as nutraceuticals – global approaches to their role in nutrition and health. InTech, London. https://doi.org/10.5772/2375

Wang H, Joseph JA (1999) Quantifying cellular oxidative stress by dichlorofluorescein assay using microplate reader. Free Radic Biol Med 27:612–616

Wang K, Zhang T, Dong Q, Collins Nice E, Huang C, Wei Y (2013) Redox homeostasis: the linchpin in stem cell self-renewal and differentiation. Cell Death Dis 4:e537. https://doi.org/10.1038/cddis.2013.5

Watson RR, Preedy VR, Zibadi S (2013) Polyphenols in human health and disease. Elsevier Science, Cambridge, MA

Williams KA (1996) Oils, fats and fatty foods. J & A. Churchill, London

Williamson G, Manach C (2005) Bioavailability and bioefficacy of polyphenols in humans. II. Review of 93 intervention studies. Am J Clin Nutr 81(1 suppl):243S–255S

Yin H, Xu L, Porter NA (2011) Free radical lipid peroxidation: mechanisms and analysis. Chem Rev 111:5944–5972

Chapter 2
High-Energy Emulsification Methods for Encapsulation of Lipid-Soluble Antioxidants

Zeynep Aksoylu Özbek, Pelin Günç Ergönül, and M. Ali Aboudzadeh

2.1 Introduction

Any disperse system which is composed of two immiscible liquids (mainly oil and water) is called "emulsion" (McClements 2016). Depending on droplet size, emulsions are divided into three groups: macroemulsion (0.5–100 μm), miniemulsion (100–1000 nm) and nanoemulsion (1–100 nm) (Santana et al. 2013). Emulsion formation requires an energy input that allows reaching a dispersed transient state in which the Gibbs free energy is greater than in the separated phase counterpart. The free energy required (ΔG) to form an emulsion is given by the following equation (Eq. 2.1):

$$\Delta G = \gamma \Delta A - T \Delta S \tag{2.1}$$

where, the $\gamma \Delta A$ term is the necessary free energy to extend the oil–water interface (γ is the interfacial tension and A is the interfacial area) and the $T\Delta S$ term is the free energy related to increasing the number of possible patterns of droplets in an emulsion (where T is the temperature and S is the entropy) compared to the separated phases. In most cases, $\Delta G > 0$, it means that the formation of emulsions is nonspontaneous and the system is thermodynamically unstable (Fig. 2.1) and a certain amount of energy is needed to be applied on the mixture, because change in entropy is not great enough to overcome the free energy required to increase the interface. This free energy can be supplied by mechanical forces (in high-energy methods) or

Z. Aksoylu Özbek (✉) · P. Günç Ergönül
Department of Food Engineering, Faculty of Engineering, Manisa Celal Bayar University, Manisa, Turkey
e-mail: zeynep.aksoylu@cbu.edu.tr

M. A. Aboudzadeh
Donostia International Physics Center (DIPC), Donostia–San Sebastián, Spain

M. A. Aboudzadeh (ed.), *Emulsion-based Encapsulation of Antioxidants*, Food Bioactive Ingredients, https://doi.org/10.1007/978-3-030-62052-3_2

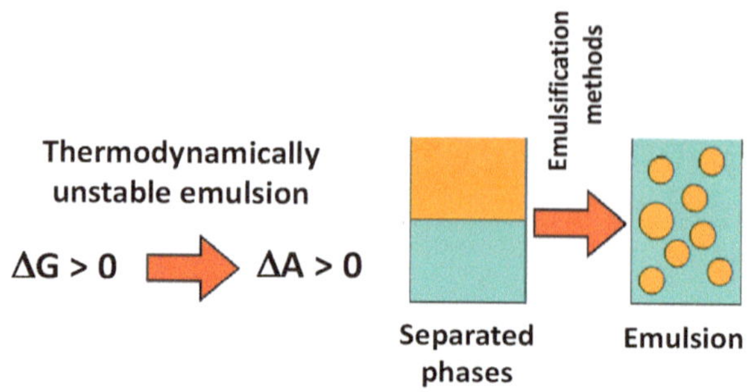

Fig. 2.1 Schematic illustration of an emulsion system formation

by the chemical potential of the system (in low-energy methods) (Komaiko and McClements 2016).

Emulsion-based systems for delivery and encapsulation of different bioactive components are widely used in food technology. These emulsions can be formulated from common food ingredients using low-energy emulsification techniques such as spontaneous emulsification (SE), phase inversion temperature (PIT), emulsion phase inversion (EPI) and membrane emulsification (ME) or high-energy emulsification methods including high-speed mixer (HSM), colloid mill (CM), high-pressure homogenizer (HPH), microfluidizer (MF), and ultrasonic homogenizer (USH) (Leal-Calderon et al. 2007; Santana et al. 2013; Rayner 2015; McClements 2015). Low-energy emulsification approaches require continuous simple stirring which makes them inexpensive, whereas high-energy emulsification approaches that form stable emulsions from various materials can be easily scaled up for industrial production (McClements and Rao 2011; Mayer et al. 2013). Type of homogenization is one of the key predictors of emulsion droplet size together with emulsification temperature, composition, and physicochemical characteristics of oil and water phases. Following in this chapter first the basic concepts of high-energy emulsification methods, including mechanisms and involved equipment are reviewed. Then, focusing particularly on the type of compound with antioxidant properties, its encapsulation through employment of high-energy emulsification methods in the most relevant studies is discussed.

2.2 High-Energy Emulsification Methods

The process of emulsion formation needs some free energy input to expand the oil–water interface (Tadros et al. 2004). In high-energy emulsification approaches, the size of droplets are reduced and two immiscible liquids are emulsified by intense mechanical energy arising from stirring, shearing, cavitation, or ultrasonic waves (Rayner 2015). The most common mechanical devices which can generate intensive disruptive forces for an effective emulsification process are HSM, CM, HPH, MF, and USH (McClements 2016). The efficiency of emulsification in these high-energy equipment is mainly influenced by process parameters (e.g. the quantity of mechanical energy, treatment time, pressure etc.) and/or material-based factors (e.g. composition of continuous and dispersed phases and their physicochemical characteristics) (Santana et al. 2013).

HSMs are fundamentally rotor-stator type mixers. They are inexpensive and versatile, available from lab-scale to industrial-scale and with high emulsification efficiency (Pacek et al. 2013). The droplet size of dispersed phase can be decreased until the limit of 200–300 nm depending on the increase of rotation speed (Koroleva and Yurtov 2012). Emulsification time, viscosity of dispersed and continuous phases, type and ratio of surfactant, design of rotor/stator, volume, and ratio of both phases are the other factors which influence the emulsification effectiveness of these equipments (Maa and Hsu 1996). As shown in Fig. 2.2 (Rosdi et al. 2018), a combination of longitudinal, rotational and radial velocity gradients is developed in the structured gap between stator and rotor as a result of fast rotation of the mixing head. The interface between oil and aqueous phases is ruptured and larger droplets are broken into smaller ones by this force (McClements 2016; Saravacos and Kostaropoulos 2016). CM is a kind of rotor-stator device in which bigger droplets in coarse emulsions consisting of intermediate or high viscosity materials are broken into smaller ones by shear stresses (up to 10^7 s^{-1}) formed in a narrow ($\leq$0.1 mm) conical gap between a rapid (up to 20.000 rpm) rotating disc (rotor) and a static disc (stator) (Fig. 2.3) (Dickinson 1994; Kohler and Schuchmann 2015; McClements 2016).

HPHs are designed for continuous operations. Droplets are disrupted as a result of inertial and shearing forces in turbulent flow and cavitation and shear stresses in laminar extension flow (Stang et al. 2001; Schultz et al. 2004). Coarse emulsion is pushed by high-pressure force (50–500 MPa) through a narrow valve in which large droplets are exposed to a combination of destructive forces leading to size reduction (Fig. 2.4) (McClements and Rao 2011; Loh et al. 2014; Yadav and Kale 2019). Similar to CM, HPH is an appropriate choice for producing fine emulsions from coarse emulsions rather than combining two immiscible liquids (McClements 2015). Emulsions with droplet sizes between 50 nm and 5 μm may be fabricated using HPH (Leal-Calderon et al. 2007). The efficiency of droplet disruption in HPH can be modified by the type of nozzle used in the equipment (McClements 2016).

MF which is similar to HPH does not contain any moving part in the droplet destruction zone (Rayner 2015). Coarse emulsion is introduced to the inlet of MF using a pump at relatively high-pressure (up to 276 MPa), which forces the emulsion to pass through an interaction chamber containing fixed-geometry (Y- or Z-type chamber, Fig. 2.5) narrow channels (Olson et al. 2004; Monroy-Villagrana et al. 2015; McClements 2015). Cavitation, shear, and impact forces generated in the narrow channels are responsible for the droplet size reduction in MF (Maa and Hsu 1999).

Sound waves ranging from 20 kHz to 1 MHz are known as "power ultrasound" (high-intensity, low-frequency), while "diagnostic ultrasound" (low-intensity, high-frequency) has a frequency above 1 MHz (Kentish and Ashokkumar 2011). The

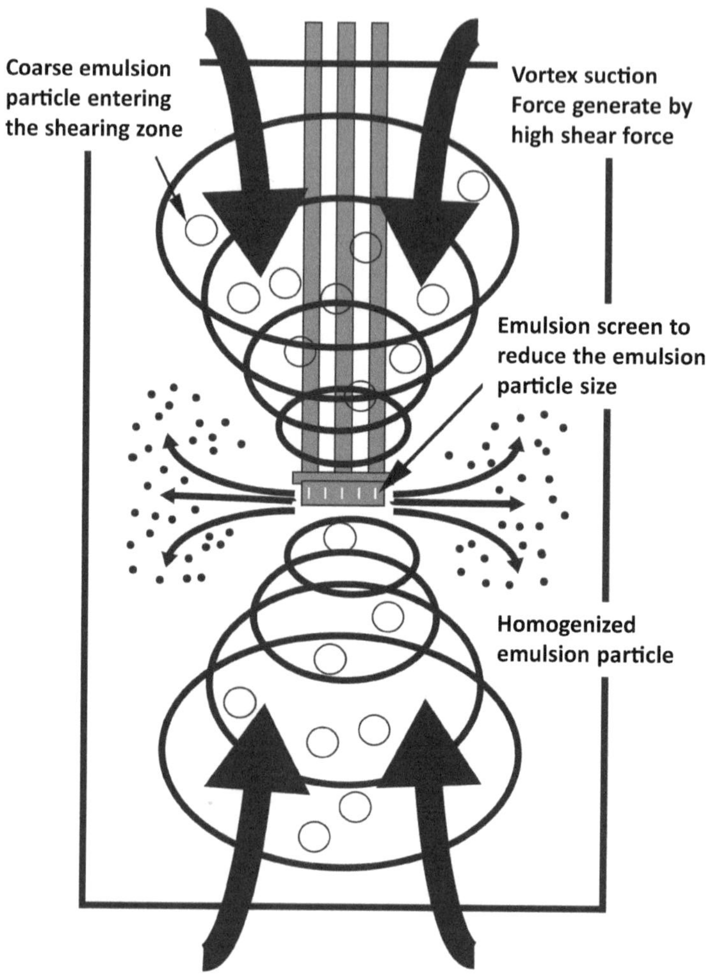

Fig. 2.2 Flow pattern in HSM. (Adopted from the reference Rosdi et al. (2018) with permission from Elsevier)

ultrasound power is inversely proportional to ultrasound frequency (Canselier et al. 2002). Application of high-intensity powerful ultrasounds leads to physical and chemical changes when they interact with the material (Abbas et al. 2013). Therefore, powerful ultrasound technology is generally used in destruction and break-down processes such as homogenization, depolymerization of macromolecules and deflocculating droplets (Assadpour and Jafari 2019). When the liquids are exposed to ultrasounds, droplets are disrupted by acoustic cavitation. However, as cavitation collapse generates intense local heating, continuous application of ultrasound can result in over-heating of sample (Hielscher 2005; Abbas et al. 2013; McClements 2016). Droplets with a mean size of 1–2 μm can be produced by USH technique (Saravacos and Kostaropoulos 2016).

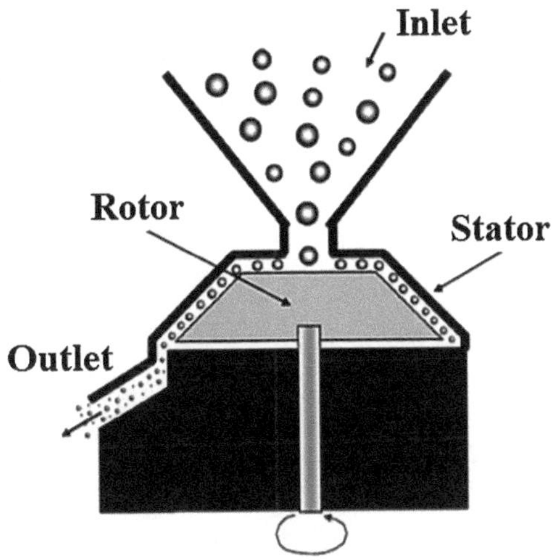

Fig. 2.3 CM used to homogenize and emulsify the liquids with intermediate or high viscosity. (Adopted from the reference McClements (2016) with permission from Taylor and Francis)

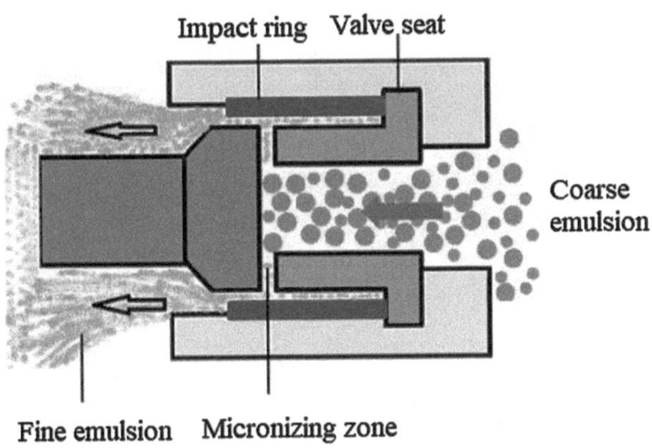

Fig. 2.4 Schematic diagram of HPH. (Adopted from the reference Loh et al. (2014) with permission from Elsevier)

2.3 Encapsulation of Lipid-Soluble Antioxidants

Vitamin E isomers, ubiquinones, carotenoids, retinoids, and flavonoids are the most well-known members of lipid-soluble antioxidants group (Ong and Packer 1992). However, the highly hydrophobic nature, sensitivity to oxidation reactions, and lower bioaccessibility of these compounds restrict their incorporation into the foods (Odriozola-Serrano et al. 2014). In order to overcome these challenges, these compounds should be encapsulated. Although many encapsulation systems (complex coacervation, extrusion, molecular inclusion, nanoprecipitation, liposomes, ionic

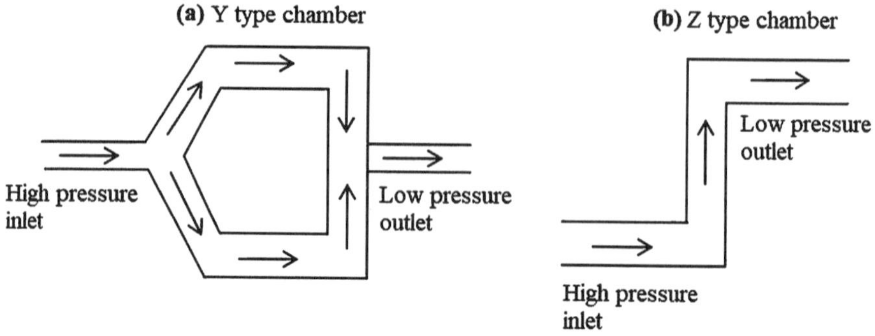

Fig. 2.5 (**a**) Y-type, and (**b**) Z-type interaction chambers used in MF. (Adopted from the reference Monroy-Villagrana et al. (2015) with permission from Springer Nature)

gelation, spray- or freeze-drying) have been developed for lipophilic bioactives, emulsion-based ones have become prominent due to their simple fabrication from easily accessible ingredients through widely used equipment in the food industry (Dima et al. 2015; McClements et al. 2016). Molecular characteristics including molecular weight (MW), functional groups, structure, charge, and polarity are the main determinants of physicochemical properties of lipophilic antioxidants, which require a specific emulsion-based system for each bioactive compound (McClements et al. 2007). Therefore, different emulsion-based delivery and encapsulation systems such as conventional emulsions, double emulsions, filled hydrogel particles, and solid lipid nanoparticles have been designed so far (McClements and Li 2010).

2.3.1 Vitamin E (VE)

In plants, naturally occurring lipid-soluble 6-hydroxychroman compounds are called "Vitamin E". It consists of α-, β-, γ-, δ-tocopherol and tocotrienol isomers (Packer and Obermüller-Jevic 2002; Eitenmiller and Lee 2004).

Edible oils of cereal grains, beans, nuts, and several seeds are the richest sources of VE (Ball 2004). The dietary reference intakes for VE (mg α-tocopherol/day) are summarized in Table 2.1 (Institute of Medicine 2006).

VE is a non-specific chain-breaking antioxidant and due to its high antioxidant activity, it has protective effects on many chronic and age-related diseases related to oxidative stress including cancer, cardiovascular disorders, cataract, macular degenerations, central neurodegenerative disorders (Alzheimer's and Parkinson's disease), diabetes and inflammatory diseases (Institute of Medicine 2000; Eitenmiller and Lee 2004). In addition to health benefits resulting from antioxidant properties of VE, it functions as cofactor for enzymes, inhibits platelet adhesion and aggregation and activates protein phosphatase 2A and diacylglycerol kinase (Birringer et al. 2019). However, VE isomers are sensitive to oxygen, high temperature (under aerobic conditions) and ultraviolet (UV) light (Zhou et al. 2012; Gregory III 2017).

Table 2.1 Dietary reference intakes (mg α-tocopherol/day) for VE

	Estimated Average Requirement (EAR)	Recommended Daily Allowance (RDA)	Adequate Intake (AI)	Tolerable Upper Intake Level (UL)
Infants				
0–6 months	–	–	4	–
7–12 months	–	–	5	–
Children				
1–3 years	5	6	–	200
4–8 years	6	7	–	300
9–13 years	9	11	–	600
14–18 years	12	15	–	800
Adults				
19–30 years	12	15	–	1000
31–50 years	12	15	–	1000
51–70 years	12	15	–	1000
>70 years	12	15	–	1000
Pregnancy				
≤18 years	12	15	–	800
19–50 years	12	15	–	1000
Lactation				
≤18 years	16	16	–	800
19–50 years	19	19	–	1000

Moreover, hydrophobic character of VE limits its incorporation into foods. Therefore, several studies dealing with protection (against environmental conditions) and delivery of VE isomers have been carried out so far. A major portion of these investigations involves encapsulation through emulsion-based systems. These studies will be discussed under the next subsection.

2.3.1.1 High-Energy Emulsification Approaches to Encapsulate VE

An overview of most relevant studies on the encapsulation of VE through high-energy emulsification approaches is presented in Table 2.2. In general, different homogenization devices are successively used to obtain emulsions with fine droplets. However, some researchers successfully use one homogenization technique for formation of stable emulsions. For instance, VE-enriched mini-emulsions stabilized with octenyl succinic anhydride modified starch (OSA-MS) or maltodextrin (MD) were formed using only HSM. The authors did not observe any phase separation in the emulsions for 24 h (Gangurde et al. 2017). Accordingly, microfluidization technique yielded VE-loaded nanoemulsions that were stable at 4 °C during 12 weeks (Lv et al. 2018). The authors also recorded that the storage of nanoemulsions did not cause a change in VE bioaccessibility. However, it was reported that bioaccessibility

Table 2.2 An overview of the most relevant studies on the encapsulation of VE in emulsion-based systems using high-energy emulsification methods

Emulsification technique(s)	Emulsifier(s)[a]	Carrier oil	Emulsification conditions	Main findings	References
HSM	MD or various OSA-MS	PEG-40 castor oil	10.000 rpm–5 min	All emulsions physically stable for 24 h	Gangurde et al. (2017)
HPH	Native, hydrolysed or acylated SFP	N/A	50 MPa–2 passes	(a) Degree of hydrolysis ↑, oil droplet size and retention efficiency ↓ (b) Acylation of sunflower protein: retention efficiency ↑	Nesterenko et al. (2013)
MF	QS	Corn oil	14.000 psi –1 pass	VE level in oil phase ↑, the rate and extent of lipid digestion ↓, and bioaccessibility of VE ↓	Lv et al. (2018)
MF	Gum arabic, QS or WPI	Corn oil	12.000 psi – 3 passes	(a) Saponin and gum arabic-stabilized nanoemulsions at pH 2-8: resistant to droplet aggregation (b) WPI-stabilized nanoemulsions: flocculation at around pH 5 (c) Highest bioaccessibility of VE → WPI-stabilized nanoemulsions	Lv et al. (2019)
1. HSM + 2. HPH	NaCas	Low and high temperature melting point triacylglycerols	1. 10.000 rpm–10 min 2. 600 bar – 6 passes	Storage stability of VE → HMT (47–77%) > LMT (27–46%)	Relkin et al. (2009)
1. HSM + 2. HPH	Lecithin or QS	Orange oil	1. Rotation speed N/A ⟶2 min 2. 12.000 psi–3 passes	(a) Surfactant concentration ↑, mean particle diameter ↓ (b) Instability in nanoemulsions at pH = 2 (c) Stable nanoemulsions between 30 °C and 90 °C (pH = 7) (d) >100 mM NaCl → Instability in lecithin stabilized systems (e) ≥400 mM NaCl → Instability in QS-stabilized system	Ozturk et al. (2014)

1. HSM + 2. HPH	Various OSA-MS	MCT oil	1. 14.000 rpm –5 min 2. 120 MPa–5 passes	High degree of substitution, low average MW, and low interfacial tension = emulsification ability of OSA-MS ↑	Hategekimana et al. (2015)
1. HSM + 2. HPH	Tween 80 and soybean lecithin	RBD canola oil	1. 15.000 rpm–10 min 2. 75, 95, 115, 135 or 155 MPa–7 passes	(a) Homogenization pressure ↑, droplet size ↓ (b) Surfactant concentration ↑, emulsion stability ↑	Mehmood (2015)
1. HSM + 2. HPH	WPI	Sunflower oil	1. 10.000 rpm–1 min 2. 50 MPa–1 min	(a) 0.01 wt% WPI addition → Maximum encapsulation efficiency (b) High levels of ascorbic acid addition → decomposition of α-tocopherol (c) Resveratrol addition → stability of α-tocopherol ↑	Wang et al. (2016)
1. HSM + 2. HPH	WPI	Orange oil	1. Rotation speed N/A–2 min 2. 25 MPa + 8 MPa	Mild denaturation of WPI by thermal process → emulsion stability ↑	Raikos (2017)
1. HSM + 2. HPH	OSA-MS	Soybean oil	1. 10.000 rpm–10 min 2. 80 MPa–3 passes	Stable emulsion with droplet size of 0.13–0.16 μm	Huang et al. (2019a)
1. HSM + 2. MF	DMO and soybean lecithin	MCT oil	1. 9000 rpm–15 min 2. 1000 kg/cm^2 – number of passes N/A	(a) VE loading >30 wt% → considerable droplet aggregation (b) Nanoemulsification → VE bioavailability ↑	Hatanaka et al. (2010)
1. HSM + 2. MF	QS or Tween 80	MCT oil	1. Rotation speed N/A –2 min 2. 9.000 psi – 4 passes	(a) Tween 80 produced smaller droplets at low VE loadings (≤40 wt%) (b) QS produced smaller droplets at high VE loading (60–80 wt%)	Yang and McClements (2013a)

(continued)

Table 2.2 (continued)

Emulsification technique(s)	Emulsifier(s)[a]	Carrier oil	Emulsification conditions	Main findings	References
1. HSM + 2. MF	QS	Corn oil (LCT) and MCT oil	1. Rotation speed N/A –2 min 2. 9.000 psi – 4 passes	(a) MCT emulsions: Greater lipid digestion (b) LCT emulsions: Better VE bioaccessibility	Yang and McClements (2013b)
1. HSM + 2. MF	Bile salt	MCT or LCT oil	1. Rotation speed N/A –2 min 2. 9.000 psi – 4 passes	Addition of calcium ions to simulated small intestinal fluid (SSIF): Lipid digestion and bioaccessibility of VE in LCT emulsions ↑ and bioaccessibility of VE in MCT emulsions ↓	Yang et al. (2015)
1. HSM + 2. MF	QS	Corn oil or MCT oil	1. Rotation speed N/A –2 min 2. 9.000 psi– 4 passes	Bioaccessibility of VE → LCT > MCT	Yang et al. (2017)
1. HSM + 2. MF	OSA-MS	N/A	1. 10.000 rpm –2 min 2. 20.000 ps i–3 passes	(a) Number of homogenization passes ↑, particle size and stabilizing ability of starch ↓ (b) Oil-to-starch ratio ↓, particle size ↓ (c) At pH > 8, oil load efficiency ↓, particle size ↑	Qiu et al. (2015)
1. HSM + 2. MF	QS	Sunflower oil	*Conventional emulsion:* 1. 15.500 rpm –5 min 2. 1000 psi –1 pass *Nanoemulsion:* 1. 15.500 rpm –5 min 2. 12.000 psi–4 passes	3-fold higher bioavailability of VE in nanoemulsion than conventional emulsion	Parthasarathi et al. (2016)

1. HSM + 2. MF	OSA–MS or Tween 80	Cold pressed flaxseed (LCT) oil or MCT oil	1. 18.000 rpm –3 min 2. 100 MPa –3 passes	(a) Tween 80 → Physical stability ↑ (b) MCT → Oxidative stability ↑	Sharif et al. (2017a)
1. HSM + 2. MF	OSA-MS	Flaxseed oil	1. 18.000 rpm –5 min 2. 110 MPa –6 passes	VE loading up to 4 wt% → successful encapsulation and good physical stability for 4 weeks at 25 °C	Sharif et al. (2017b)

[a]Abbreviations for emulsifiers: *DMO* decaglyceryl monooleate, *MD* maltodextrin, *NaCas* sodium caseinate, *OSA-MS* octenyl succinic anhydride modified starch, *QS* Quillaja saponin, *RBD* refined, bleached and deodorised, *SFP* sunflower protein, *WPI* whey protein isolate

of VE in nanoemulsions homogenized with microfluidizer was mainly influenced by the type of emulsifier (Lv et al. 2019).

Currently it is very common to use a combination of high-energy techniques for a complete emulsification. In a study, the effect of lipid phase composed of triacylglycerols (TAG) with different melting temperatures on physicochemical properties and thermal degradation of α-tocopherol nanoemulsions was investigated. The coarse emulsions were fabricated with a HSM, while further nanoemulsions were produced by HPH. The authors suggested that triacylglycerol composition, crystalline fat content and polymorphism properties of lipid phase are more efficient than oil droplet size in protecting the core materials encapsulated in the emulsion (Relkin et al. 2009). Likewise, the composition of aqueous phase also plays a major role in determining the emulsion characteristics. In this context, in another study the effect of physicochemical and molecular properties on emulsification ability of OSA-MS was investigated. Using HSM for coarse emulsions and then, HPH for fine emulsions, the authors stated that OSA-MS with high degree of substitution, low MW, and low interfacial tension exhibited better emulsification properties in VE-loaded nanoemulsions (Hategekimana et al. 2015). Similarly, the suitability of OSA-MS in VE-encapsulated nanoemulsions obtained through HSM, and HPH was confirmed in another study as well. The authors attributed the stabilisation ability of modified starch to the presence of hydrophobic OSA group in its highly branched molecular structure (Huang et al. 2019a). Analogous to carbohydrates, proteins are also preferred as surfactant in oil-in-water (O/W) emulsions. For example, whey protein isolate (WPI) successfully entrapped the tocopherols mixture in orange O/W beverage emulsions. In this study, the coarse emulsions were initially formed by using HSM, and then they were subjected to HPH. Beverage emulsions exposed to high temperature treatment for short time were the most stable ones during storage at 4 °C for 28 days and VE retention in all heat-treated emulsion samples were higher than 85%. Therefore, beverage emulsions displayed high potential as delivery systems for lipophilic bioactive compounds (Raikos 2017). Likewise in another report, nanoemulsion served as reliable system for delivery of D-α-tocopherol, in which the in vitro cytotoxicity of emulsions obtained by HSM and HPH was evaluated (Teixeira et al. 2017).

In addition to carbohydrates and proteins, natural and/or artificial surfactants can be used as components of aqueous phase. A surfactant mixture consisting of Tween 80/lecithin (3:1) was successfully used to encapsulate VE in nanoemulsions. The coarse emulsions were homogenized by HSM accompanied with HPH. The optimized conditions that led to nanoemulsions with minimum droplet size and maximum stability were determined as following: 135 MPa homogenization pressure, 6.18 wt% canola oil concentration (as the oil phase), 6.39 wt% surfactant concentration, and 1 wt% VE concentration (Mehmood 2015). The emulsion-forming ability of quillaja saponin (another natural surfactant) was reported to be comparable with that of lecithin in VE-loaded nanoemulsions (up to 50 wt% VE content in oil phase) obtained through HSM and HPH, respectively. The scientists recommended the use of lower amounts of quillaja saponin than those of lecithin. Quillaja saponin-stabilized VE nanoemulsions exhibited good stability at wide ranges of pH (pH 3–8), and salt concentrations (0–300 mM NaCl), but flocculation occurred at lower pH and higher salt levels (Ozturk et al. 2014).

Another high-energy technique frequently coupled with HSM is MF. For instance, VE was encapsulated into OSA-MS based nanoemulsions using combination of HSM and MF. It was reported that as the amount of VE increased, droplet sizes and polydispersity index (PDI) values increased from 179 to 208 nm and from 0.09 to 0.12, respectively. Furthermore, a slight increase in droplet size of the emulsion that contained 4 wt% VE (the only selected formulation by the authors) was observed during storage at 25 °C for 4 weeks; however, this change did not lead to any physical instability in the emulsion (Sharif et al. 2017b). Similar to emulsions formed by HPH, some parameters should be optimized to fabricate stable emulsions by homogenization techniques coupled with MF. The negative effects of increased number of MF passes and free OSA content on the particle size and stabilizing ability of VE-loaded emulsions stabilized by OSA were reported elsewhere. The mean droplet sizes of VE emulsions decreased from 0.221 to 0.122 μm when the number of MF passes increased from 3 to 12 (Fig. 2.6). However, the stabilizing ability of OSA-MS impaired because probably increasing the number of passes resulted in cleavage of hydrophobic bound OS groups and producing free OS groups during microfluidization. Therefore, the adsorption ability of OSA-MS to the oil-water interface reduced (Qiu et al. 2015). In addition to process conditions, modifying the physical properties of both oil and aqueous phase may influence the efficiency of emulsification process by MF. For instance, high viscosity of VE can be decreased by incorporation of VE into any oil, while viscosity of aqueous phase can be increased by addition of glycerol in order to obtain emulsions with smaller droplets (Yang and McClements 2013a).

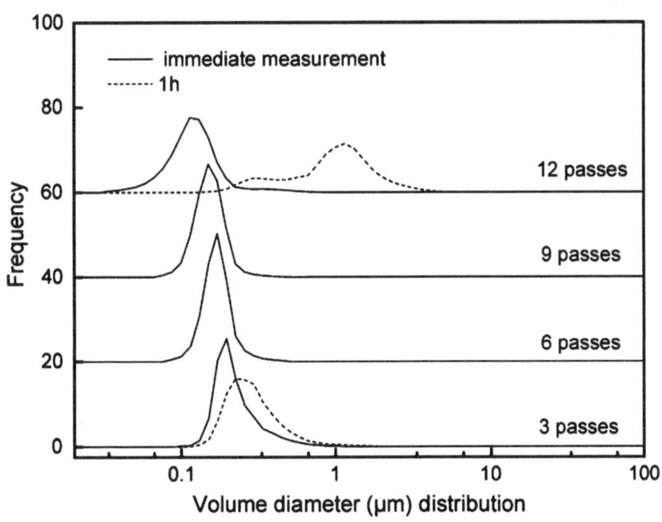

Fig. 2.6 Effects of number of passes on the particle size of VE emulsions. (Reproduced from the reference Qiu et al. (2015) with permission from John Wiley and Sons)

Protection of sensitive core materials against environmental conditions is one of the main aims of encapsulation. Therefore, storage stability of entrapped core materials should be tracked. In a study, α-tocopherol and β-carotene (βC) were co-encapsulated in nanoemulsions containing flax seed oil or medium-chain triglyceride (MCT) as the oil phase and OSA-MS or Tween 80 as emulsifiers. Coarse emulsions were formed using HSM and subsequently they passed through a MF. After passing 4 weeks, higher degradation rates of α-tocopherol and βC from the emulsions stored at 40 °C was observed in comparison to those stored at 25 °C. Surprisingly, the authors found out that α-tocopherol served as antioxidant for the protection of βC in MCT-based emulsions, however, it did not have any protection effect on the degradation of βC in flax seed oil-based emulsions. Furthermore, the addition of eugenol as natural antioxidant to nanoemulsions improved the stability of both α-tocopherol and βC throughout storage time (Sharif et al. 2017a). Similarly, use of ascorbic acid and resveratrol as natural antioxidants in α-tocopherol-loaded O/W nanoemulsions stabilized with various concentrations of WPI (0.001–1 wt%) were tested in another work. It was shown that the incorporation of ascorbic acid into the aqueous phase up to ascorbic acid/WPI molar ratios <5 or addition of resveratrol into oil phase, retard the decomposition of encapsulated α-tocopherol (Wang et al. 2016).

Nanoemulsions obtained through a combination of HSM and MF techniques are considered as effective delivery systems to increase bioavailability and bioaccessibility of lipophilic bioactives. For instance, a study demonstrated a significant reduction in lipoperoxidant levels in different organs of diabetic rats that were orally administered with α-tocopherol-loaded nanoemulsions. These nanoemulsions were fabricated using HSM and MF techniques. Moreover, the authors reported 2.6-fold increase in the bioavailability of encapsulated α-tocopherol in nanoemulsion systems compared to control sample (mixture of α-tocopherol and oil) (Hatanaka et al. 2010). However, the bioaccessibility of active materials entrapped in nanoemulsions are influenced by several factors including droplet surface area, interfacial composition, carrier oil type, calcium ions, and phospholipids. For instance, in the absence of any salt, higher lipid digestion rates were obtained in VE-loaded emulsions containing MCT as carrier oil compared to those that include long-chain triglyceride (LCT), which was related to differences in the water-dispersibility of the medium and long chain fatty acids generated during lipolysis (Yang et al. 2015). Addition of calcium ions to the simulated small intestinal fluids (SSIF) highly increased the extent of lipid digestion for LCT-emulsions, but had minor influence on MCT-emulsions, which was associated to the capability of calcium ions to remove long-chain fatty acids from droplet surfaces. Also, bioaccessibility of VE increased in the presence of calcium and phospholipids in LCT-emulsions, but decreased it in MCT-emulsions (Yang et al. 2015). The same scientists reported that LCT is a more efficient carrier lipid than MCT in increasing the total bioavailability and bioaccessibility of encapsulated VE in O/W emulsions formed by HSM and MF techniques (Yang and McClements 2013b; Yang et al. 2017). These scientists also compared the solubilization capacity of VE and VE acetate in the mixed micelle solutions and they found higher capacity for VE than for VE acetate, which was attributed to differences in the ability of the vitamin molecules to be incorporated into the micelle structures (Yang and McClements 2013c).

2.3.2 *Carotenoids*

Along with chlorophylls, carotenoids are one of the most dominant pigments of crude oils. However, refining process causes severe losses in these compounds (Winkler-Moser and Mehta 2015). Most carotenoids have 40-carbon skeleton (C40 carotenoid) and are usually tetraterpenes consisting of eight C_5 isoprene units (Rodriguez-Amaya 2019). The conjugated double bonds of carotenoids are responsible for the yellow, orange or red colour of many plant products (Rodriguez-Amaya 2016; Mozafar 2018). On the other side, the presence of these conjugated double bonds makes them sensitive against oxygen, light, heat, oxidizing reagents and strong acids (Liaaen-Jensen 2004; Belitz et al. 2009). Oxidation mechanisms of carotenoids and the formed products are illustrated in Fig. 2.7 (Boon et al. 2010).

Carotenoids are biosynthesized by plants, however, animals cannot synthesize these compounds (Schwartz et al. 2017). Therefore, plant-based foods are the best sources for carotenoids intake in human diet. Carotenoids are mainly divided into

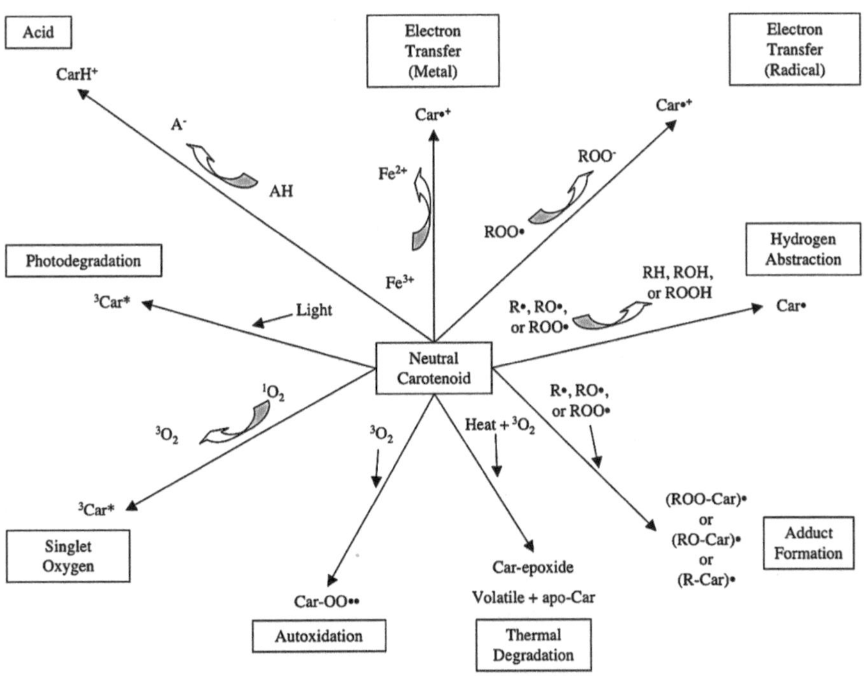

Fig. 2.7 Common oxidation mechanisms of carotenoids. (Reproduced from the reference Boon et al. (2010) with permission from Taylor & Francis)

two main groups: the hydrocarbon carotenoids containing only hydrogen and carbon atoms (α-carotene, β-carotene (β C), γ-carotene, lycopene) and the oxygenated xanthophylls containing at least one oxygen atom (astaxanthin, canthaxanthin, zeaxanthin, lutein, bixin). In general, most of the carotenoids are naturally found in all-*trans* form (Stahl and Sies 2003; Wong 2018). However, mainly heat treatment and, to a minor extent, light exposure promote the formation of *cis* isomers with lower antioxidant and vitamin A (VA) activities (Pénicaud et al. 2011). The xanthophylls have higher polarity than carotenes and this polar structure contributes to the improved bioavailability of xanthophylls (Bohn 2008). The relative VA activities, weighted average of antioxidant activities and sources of common carotenoids are given in Table 2.3.

The bioaccessibility and bioavailability of carotenoids are influenced by several factors including other dietary components (dietary lipids, dietary fibre, phytosterols, phospholipids), chemical structure and polarity of carotenoid species, functional groups linked to carotenoids, VA status of the person, host-related factors (age, gender, pathologies) and genetic factors (Yonekura and Nagao 2007; Desmarchelier and Borel 2017). Nevertheless, the absorption of carotenoids is dominated by quantity and type of dietary lipids as the presence of fat in the intestine determines the formation of mixed micelles (van het Hof et al. 2000). Mixed micelles are essential to provide lipid digestion in the body, because their main function is to carry the lipids to the surface of enterocytes by incorporating these nutrients into their structure (Read et al. 1981; Binder and Reuben 2009). On the other hand, dietary fibres reduce the carotenoid absorption in different ways (Riedl et al. 1999). In their natural state, carotenoids are generally complexed with proteins in plants. Additionally, they are packed in rigid cell walls that should be destructed by food processing techniques including cooking or homogenization. Thus, formulated carotenoids (e.g. natural or synthetic carotenoids dissolved in oil, water-dispersible beadlet) have high carotenoid bioavailability, while processed fruit and vegetables (e.g. tomato juice with oil, tomato paste with oil, mildly cooked spinach and carrot) have mild bioavailability and raw fruits and vegetables (e.g. carrot, spinach, tomato and pepper) have low bioavailability (van het Hof et al. 2000; Yeum and Russell 2002; Nagao 2014; Saini et al. 2015) Encapsulation is a process that improves bioavailability of carotenoids by enhancing stability against isomerization and degradation, increasing solubility and bioaccessibility in the gut, modifying release kinetics in the small intestine and minimizing the detrimental effects of other dietary constituents such as dietary fibres (Soukoulis and Bohn 2018).

2.3.2.1 Beta-Carotene (βC)

βC, the dominant hydrocarbon carotenoid of plants, has an extreme lipophilic character (Rodriguez-Amaya 2016). Besides being provitamin A, βC has diverse health functions such as antioxidant activity, protection against chronic diseases and photoprotection of skin against ultraviolet lights (Burton and Ingold 1984; Rodriguez-Amaya 2016). Humans and animals cannot synthesize carotenoids, so they need to

Table 2.3 VA activity, weighted mean of antioxidant activity and sources of dietary carotenoids

Carotenoid	VA activity (%)	Weighted average of antioxidant activity[a]	Sources	References
β-carotene	100	1.4	Carrot, paprika, red pepper, kale, parsley, pumpkin, spinach, apricot, broccoli, cherry, lettuce, melon, green pepper, sweet potato, virgin olive oil, gac oil, palm oil, sunflower oil, rapeseed oil, pumpkin seed oil, peanut oil, pequi oil, cold-pressed seed oils of marionberry, boysenberry, red raspberry, blueberry	Pattee and Purcell (1967), Bauernfeind (1972), Manorama and Rukmini (1992), Matus et al. (1993), Gandul-Rojas and Minguez-Mosquera (1996), Parry et al. (2005), Monde et al. (2009), Nhung et al. (2010), Ribeiro et al. (2012), Kreps et al. (2014), Toti et al. (2018)
α-carotene	50–54	1.4	Carrot, red pepper, pumpkin, winter squash, peach, celery, orange, green lettuce, palm oil, pumpkin seed oil	Bauernfeind (1972), Matus et al. (1993), Sommerburg et al. (1998), Ping and Lian (2005), Monde et al. (2009), Schwartz et al. (2017), Toti et al. (2018)
γ-carotene	42–50	–	Grapefruit, carrot, tomato, persimmon, saffron, olive, *Ginkgo biloba*, sea-buckthorn, Barbados cherry, mango	Bauernfeind (1972), Barbosa-Filho et al. (2008)
Lycopene	No activity	1.9	Tomato, grapefruit, papaya, red watermelon, guava, pineapple, orange, apple (red), pumpkin, rhubarb, mango, apricot, gac oil	Bauernfeind (1972), Sommerburg et al. (1998), Maiani et al. (2009), Toti et al. (2018)
Lutein	No activity	1.4	Broccoli, kale, lettuce, parsley, pea, green pepper, red pepper, pistachio, sage, pumpkin, spinach, egg yolk, cucumber, carrot, basil, avocado, banana, apricot, virgin olive oil, peanut oil, pumpkin seed oil, cold-pressed seed oils of marionberry, boysenberry, red raspberry, and blueberry	Pattee and Purcell (1967), Bauernfeind (1972), Matus et al. (1993), Gandul-Rojas and Minguez-Mosquera (1996), Parry et al. (2005), Maiani et al. (2009), Nhung et al. (2010), Toti et al. (2018)

(continued)

Table 2.3 (continued)

Carotenoid	VA activity (%)	Weighted average of antioxidant activity[a]	Sources	References
Astaxanthin	No activity	0.9	Salmon, trout, krill, shrimp, cray fish, crustaceans, algae, yeast	Bauernfeind (1972), Seabra and Pedrosa (2010), Ambati et al. (2014)
Canthaxanthin	No activity	0.7	Mushroom (*Cantherellus cinnabarinus*), cray fish, sea trout, birds, marine algae, bacteria	Bauernfeind (1972), Gupta et al. (1985)
Zeaxanthin	No activity	1.4	Corn, egg yolk, orange pepper, scallion, honeydew, mango, orange, nectarine, alfalfa, red grape (seedless), green grape, cucumber, peach, cold-pressed seed oils of marionberry, boysenberry, red raspberry, blueberry	Bauernfeind (1972), Sommerburg et al. (1998), Parry et al. (2005), Sajilata et al. (2008), Perry et al. (2009)
Bixin	No activity	1.3	Annatto seed	Bauernfeind (1972), Rodrigues et al. (2014)

[a]Weighted means of values were calculated by using results of the α-tocopherol/Trolox equivalent antioxidant capacity (TEAC), ferric reducing antioxidant power (FRAP), and luminol-chemiluminescence based peroxyl radical scavenging capacity (LPSC) assays (Müller et al. 2011)

obtain them through their diet (Shegokar and Mitri 2012). However, βC is very sensitive to oxidative degradation induced by light, oxygen, enzymes and hydroperoxides (Zhou et al. 2012). Due to the hydrophobic and sensitive chemical nature of βC, its utilization in the foods has been limited. In order to enhance the chemical stability and bioavailability of βC, it is widely entrapped into food-grade emulsion-based systems. Nevertheless, both these characteristics may change depending on various factors such as hydrophobicity of carotenoid, pH, amounts of bile lipids and interactions between different carotenoids which regulate the transfer rate of carotenoids from the emulsion droplets to mixed micelles in digestion system (Tyssandier et al. 2001). Therefore, considerable efforts have been made to determine optimum conditions to develop βC-loaded emulsions with desirable properties.

High-Energy Emulsification Approaches to Encapsulate βC

HSM and HPH are conventional high-energy emulsification techniques that are commonly used for production of βC-loaded emulsions. Table 2.4 summarizes the most relevant studies that aimed to encapsulate βC in emulsion-based systems through various high-energy emulsification techniques.

Table 2.4 An overview of the most relevant studies on the encapsulation of βC in emulsion-based systems using high-energy emulsification methods

Emulsification technique(s)	Emulsifier(s)[a]	Carrier oil	Emulsification conditions	Main findings	References
HSM	Lecithin	Colza oil	Homogenization conditions N/A	βC bioaccessibility from: emulsions > oil-based formulation > diet enriched with βC in colza oil	Grolier et al. (1995)
HSM	BSA	Neutral oil	13.000 rpm −30 s + 16.000 rpm −2 min	Stable emulsions for at least 5 days	Wackerbarth et al. (2009)
HSM	Tween 80	Stearic acid	1000, 2929, 5000, 10.000, 15.000 or 17.071 rpm −1.17, 2, 4, 6, or 6.38 min	(a) Homogenization speed ↑, droplet size ↓ (b) Homogenization time ↑, droplet size ↓	Flores-Miranda et al. (2015)
HSM	WPI and lecithin	Soybean oil	12.000 rpm −2 min	Eugenol addition → βC stability against heating up to 80 °C and UV radiation ↑ and physical emulsion stability ↑	Guan et al. (2016)
Dual-channel MF	QS or WPI	Corn oil	9-19 kpsi −1 pass	(a) Homogenization pressure ↑, droplet size ↓ (b) WPI-stabilized nanoemulsions → Better physical stability at 4, 25, or 55 °C (c) QS-stabilized emulsions at 55 °C → Aggregation (d) Storage temperature ↑, βC degradation ↑	Luo et al. (2017)
1. HSM + 2. HPH	Tween 20, Tween 40, Tween 60 or Tween 80	MCT oil	1. 5000 rpm −10 min 2. 60, 80, 100, 120, or 140 MPa −1, 2, 3, 4, or 5 passes	(a) Tween 20 → Smallest droplet size (b) Homogenization pressure and number of cycles ↑, droplet size of Tween 20-stabilized emulsions ↓ (c) βC degradation in Tween 20-stabilized emulsions → 25 °C > 4 °C	Yuan et al. (2008)
1. HSM + 2. HPH	DML, Tween 20, WPI, OSA-MS or Tween 20/WPI	MCT oil	1. 5000 rpm – homogenization time N/A 2. 20 MPa, 80 MPa, or 140 MPa −3 passes	(a) Tween 20 and DML → Droplet size and emulsion stability ↓ (b) WPI → Improved βC stability at 55 °C for 12 days (c) Tween 20/WPI (1:1) → Emulsion stability ↑	Mao et al. (2009)

(continued)

Table 2.4 (continued)

Emulsification technique(s)	Emulsifier(s)[a]	Carrier oil	Emulsification conditions	Main findings	References
1. HSM + 2. HPH	WPI	HPKO	1. Rotation speed N/A– 30 s 2. 850 bar –1 pass	βC degradation rate → solid lipid particles > liquid lipid particles	Cornacchia and Roos (2011a)
1. HSM + 2. HPH	SSPS and chitosan	MCT oil	1. 10.000 rpm – homogenization time N/A 2. 60 MPa –3 passes	(a) Chitosan concentration → ζ-potential, particle size, and rheological properties of emulsions (b) Chitosan concentration of 0.5 wt% → Minimum droplet size and the best βC chemical stability (c) Adsorption of chitosan → Improved physicochemical stability of βC emulsions	Hou et al. (2010)
1. HSM + 2. HPH	SPSS and chitosan products with different MW	MCT oil	1. 10.000 rpm– homogenization time N/A 2. 60 MPa –3 passes	(a) Chitosan MW → ζ -potential, particle size and rheological properties of emulsion, and physicochemical stability of βC (b) SPSS-stabilized emulsions with medium MW chitosan → βC stability against heat and light ↑	Hou et al. (2012)
1. HSM + 2. HPH	SPSS, WPI or DML	MCT oil	1. 10.000 rpm – homogenization time N/A 2. 60 MPa–3 passes	WPI-stabilized emulsions → maximum micellarization and release rate of βC	Hou et al. (2014)
1. HSM + 2. HPH	SPSS, WPI or DML	MCT oil	1. 10.000 rpm – homogenization time N/A 2. 60 MPa–3 passes	(a) Type of emulsifier influenced the droplet size, particle electric charge and microstructure change throughout digestion of βC (b) DML-stabilized emulsions → maximum βC release rate	Liu et al. (2012)
1. HSM + 2. HPH	Gum arabic	MCT oil	1. 10.000 rpm – homogenization time N/A 2. 60 MPa –3 passes	α-tocopherol addition → the best βC stability against light and temperature	Liu et al. (2015)

1. HSM + 2. HPH	α-La or Tween 20	MCT or corn oil	1. Rotation speed N/A – 2 min 2. 12.000 psi–5 passes	(a) Addition of EGCG → βC degradation in MCT and corn oil emulsions ↓, but ineffective for lipid oxidation (b) α-La-stabilized corn oil emulsions → Stability of βC ↑	Liu et al. (2016a)
1. HSM + 2. HPH	LF, conjugate of CA/LF, physical mixture or conjugate of CA/LF/PD	MCT oil	1. Rotation speed N/A–2 min 2. 900 psi–3 passes	CA/LF/PD conjugate-stabilized emulsion → Resistant to flocculation at pH 8-9, and bioaccessibility of βC ↑	Liu et al. (2017)
1. HSM + 2. HPH	Tween 20	HPO, cacao butter (CB) or HPO/CB (50:50)	1. Homogenization conditions N/A 2. 9000 psi–3 passes	Aggregation and βC degradation → liquid lipid nanoparticles < solid lipid nanoparticles	Qian et al. (2013)
1. HSM + 2. HPH	SMP and lysolecithin	Corn oil and lemon oil	1. Homogenization conditions N/A 2. 9000 psi–3 passes	Amount of corn oil ↑, βC bioaccessibility ↑	Rao et al. (2013)
1. HSM + 2. HPH	LC, LC/EGCG physical complex or conjugate of LC/EGCG	MCT oil	1. 10.000 rpm –10 min 2. 60 MPa –3 passes	LC/EGCG conjugate → Droplet size ↓, physical stability ↑, and βC degradation ↓	Lei et al. (2014)
1. HSM + 2. HPH	OSA-MS with different contents of amylopectin	Canola oil	1. 9500 rpm –20 min2. 200 bar/50 bar – 6 passes	(a) Amylopectin content ↑, droplet size ↓ (b) Amount of amylopectin with high degree of branching → Oxidative stability of βC ↑	Sweedman et al. (2014)
1. HSM + 2. HPH	HPMCs with different contents of methyl and hydroxypropyl or gum acacia	MCT oil	1. Homogenization conditions N/A 2. 69 MPa–3 passes	(a) Gum acacia → Droplet size ↓ (b) HPMC with high methyl:hydroxypropyl ratio → emulsion physical stability ↑	Akinosho and Wicker (2015)

(continued)

Table 2.4 (continued)

Emulsification technique(s)	Emulsifier(s)[a]	Carrier oil	Emulsification conditions	Main findings	References
1. HSM + 2. HPH	PHC	Olive oil	1. Rotation speed N/A – 10 min 2. 100 MPa –1 pass	Concentration of PHC ↑, βC bioaccessibility ↑	Verrijssen et al. (2015a)
1. HSM + 2. HPH	Citrus pectin with different DM degree and/or PHC	Olive oil	1. Rotation speed N/A – 10 min 2. 100 MPa–1 pass	βC bioaccessibility in citrus pectin-stabilized emulsions depended on the presence of PHC.	Verrijssen et al. (2015b)
1. HSM + 2. HPH	α-La, NaCas, α-La & LC/EGCG bilayer, or NaCas & LC/EGCG bilayer	MCT oil	1. 10.000 rpm –3 min 2. 60 MPa –3 passes	(a) Chemical stability of βC against heat and UV light → primary emulsions < bilayer emulsion (b) NaCas & LC/EGCG bilayer coated emulsion → The best physicochemical characteristics	Wei and Gao (2016a)
1. HSM + 2. HPH	Chitosan, physical or covalent complex of chitosan/CA	MCT oil	1. 10.000 rpm–10 min 2. 60 MPa – 3 passes	Chitosan/CA covalent complex → The smallest droplet size and the highest oxidative stability	Wei and Gao (2016b)
1. HSM + 2. HPH	WPI, NaCas or Tween 80	Sunflower oil	1. 10.000 rpm –1 min2. 20 or 70 MPa – number of passes N/A	(a) WPI → Maximum βC bioaccessibility (b) NaCas → Maximum cellular uptake of βC (c) Bioavailability of βC → NaCas > Tween 80 > WPI	Lu et al. (2017)
1. HSM + 2. HPH	UFP, gum arabic or beet pectin	Soybean oil	1. 26.000 rpm –3 min 2. 75 MPa–3 passes	(a) UFP → Maximum βC bioaccessibility (b) EDTA or α-T addition → βC stability ↑	Shao et al. (2017a)
1. HSM + 2. HPH	UFP, gum arabic or beet pectin	Soybean oil	1. 26.000 rpm –3 min 2. 75 MPa –3 passes	(a) Up to 3 wt% UFP → Droplet size ↓ (b) UFP → Better emulsifying properties and improved physical stability of emulsions	Shao et al. (2017b)
1. HSM + 2. HPH	NaCas or NaCas/MD	MCT oil	1. 10.000 rpm –2 min 2. 600 bar –10 passes	Retention of βC in emulsions with MD > retention of βC in emulsions without MD	Zhang et al. (2017)
1. HSM + 2. HPH	EWP, mixture or conjugate of CT/EWP	Sunflower oil	1. 10.000 rpm –2 min 2. 60 MPa –3 passes	CT/EWP conjugates → maximum physicochemical stability of emulsions	Gu et al. (2017)

1. HSM + 2. HPH	WPI	RBD palm olein	1. Rotation speed N/A – 2 min 2. 500/100 bar –2 passes	(a) Fresh emulsions → More βC in oil phase (b) 30 day-stored emulsions → More βC in aqueous phase (c) Oxidation rate of βC → oil phase > aqueous phase	Fahmi Wan Mohamad et al. (2017)
1. HSM + 2. HPH	Tea polyphenols	Corn oil	1. Rotation speed N/A – 2 min 2. 14 kpsi –5 passes	Tea polyphenols → Stability and bioaccessibility of βC ↑	Meng et al. (2019)
1. HSM + 2. HPH	Glucose-fructose syrup	Sunflower oil	1. 13.000 rpm –2 min2. 60 MPa + 15 MPa– two-step homogenization	Polarity of carotenoids ↓, physical stability of emulsions ↑	Domian and Szczepaniak (2020)
1. HSM + (2) HPH or MF	Tween 20, DML, WPI or OSA-MS	MCT oil	1. 5000 rpm – Homogenization time N/A 2. 40, 60, 80, 100 or 120 MPa –1, 2, 3, 4 or 5 passes	(a) MF → Smaller droplets (b) Emulsifiers with large molecule size (WPI and OSA-MS) → droplet size ↑ (c) WPI-stabilized emulsion → highest βC stability at storage	Mao et al. (2010)
1. HSM + 2. MF	SPI	Soybean oil	1. Homogenization conditions N/A 2. 40 MPa –4 passes	Physicochemical properties of emulsified lipophilic bioactives influence their incorporation into mixed micelles	Nik et al. (2011)
1. HSM + 2. MF	WPI, mixture or conjugate of WPI/beet pectin	MCT oil	1. 10.000 rpm –3 min 2. 50 MPa –3 passes	WPI/beet pectin conjugates → The smallest droplet size and improved freeze-thaw stability, the best storage stability of βC	Xu et al. (2012)
1. HSM + 2. MF	β-Lg or Tween 20	Corn oil	1. Rotation speed N/A–2 min 2. 9000 psi –3 passes	(a) β-Lg → colour fading ↓ (b) Antioxidant effect → EDTA > ascorbic acid (c) Antioxidant effect → coenzyme Q10 > VE	Qian et al. (2012a)

(continued)

Table 2.4 (continued)

Emulsification technique(s)	Emulsifier(s)[a]	Carrier oil	Emulsification conditions	Main findings	References
1. HSM + 2. MF	β-Lg or Tween 20	Orange oil	1. Rotation speed N/A–2 min 2. 9000 psi–3 passes	(a) Storage temperature ↑, βC degradation ↑ (b) β-Lg-stabilized emulsions → βC degradation ↓	Qian et al. (2012b)
1. HSM + 2. MF	Tween 20	MCT, corn or orange oils	1. Rotation speed N/A–2 min 2. 9000 psi –3 passes	βC bioaccessibility → corn oil > MCT > orange oil	Qian et al. (2012c)
1. HSM + 2. MF	Tween 20	Different ratios of MCT and LCT oils	1. 10.000 rpm –2 min 2. 9000 psi –3 passes	LCT content ↑, βC bioaccessibility↑	Salvia-Trujillo et al. (2013)
1. HSM + 2. MF	WPI	MCT oil	1. 10.000 rpm –3 min 2. 50 MPa –3 passes	(a) Lower chemical stability of βC at acidic conditions (pH 3–4) (b) Addition of EDTA or α-tocopherol → βC stability ↑ (c) Antioxidant efficiency → α-tocopherol > EDTA	Xu et al. (2013a)
1. HSM + 2. MF	Mixture or conjugates of WPI/ beet pectin	MCT oil	1. Rotation speed N/A – 3 min 2. 62 MPa–3 passes	(a) WPI/beet pectin conjugates → Stability of βC ↑ (b) 200 ppm of α-T addition → Improved stability of βC in WPI/beet pectin conjugate-stabilized emulsion	Xu et al. (2013b)
1. HSM + 2. MF	NaCas	Corn oil	1. Rotation speed N/A – 2 min 2. 10.3, 31, 62.1 or 103.4 MPa –7 passes	(a) Homogenization pressure ↑ (to 100 MPa), droplet size ↓ (b) Nanoemulsion formation at 100 MPa (c) Droplet size ↓, βC bioaccessibility↑ (d) Oxidation stability of βC → conventional emulsions > nanoemulsions	Yi et al. (2014)
1. HSM + 2. MF	α-La or catechin/ α-La conjugates	Corn oil	1. Rotation speed N/A – 2 min 2. 103.4 MPa–7 passes	Retention of emulsified βC → catechin/α-La conjugates > α-La	Yi et al. (2016)

1. HSM + 2. MF	WPI or WPI/dextran conjugates	Corn oil	1. Rotation speed N/A – 2 min 2. 103.4 MPa–7 passes	(a) Emulsion droplet size → WPI/dextran conjugates < WPI (b) WPI/dextran conjugates → pH stability of emulsion ↑ (c) Molecular weight of dextran ↑, βC release ↓	Fan et al. (2017)
1. HSM + 2. MF	Mandarin fibre	Corn oil	1. 9500 rpm–2 min 2. 30.000 psi–5 passes	(a) Amount of mandarin fibre ↑, droplet size ↑ (b) Mandarin fibre >1.5 wt% → Droplet aggregation (c) Up to 1 wt% of mandarin fibre → βC bioaccessibility ↑	Gasa-Falcon et al. (2017)
1. HSM + 2. MF	OSA-MS or Tween 80	Flax seed oil or MCT oil	1. 18.000 rpm –3 min 2. 100 MPa–5 passes	(a) Adding eugenol → Oxidative stability and βC retention ↑ (b) VE → Acts as antioxidant in MCT oil-emulsions (c) Chemical stability of βC → OSA-MS > Tween 80	Sharif et al. (2017a)
1. HSM + 2. MF	Casein	Vegetable oil	1. Rotation speed N/A – 1 min 2. 12.000 psi–3 passes	(a) βC solubility in oil phase → Cold-sonication application (1.73%) > conventional heating application (1.26%) (b) βC degradation in the emulsions → Cold-sonication-assisted dissolution (4%) < temperature-assisted dissolution (19%)	Chen et al. (2017)
1. HSM + 2. MF	Span 60 and Tween 60	Cocoa butter	1. 10.000 rpm–1 min 2. 100 MPa–5 passes	Physical state (solid or liquid) of triacylglycerols influences solubilization of lipophilic bioactives during digestion	Hart et al. (2018)
1. HSM + 2. MF	SGPIs	LCT or MCT oils	(a) 10.000 rpm–2 min (b) 12.000 psi–5 passes	(a) βC bioaccessibility → LCT > MCT (b) Cellular uptake of βC → LCT > MCT	Han et al. (2019)
1. HSM + 2. MF	SGPIs	MCT or LCT oils	1. 11.000 rpm–2 min 2. 12.000 psi–5 passes	SGPI-stabilized LCT oil emulsions → Improved stability against environmental conditions and better oxidative stability of βC	Han et al. (2020)

(continued)

Table 2.4 (continued)

Emulsification technique(s)	Emulsifier(s)[a]	Carrier oil	Emulsification conditions	Main findings	References
1. HSM + 2. ME	WPC, Tween 20, BSA, WPC/Tween 20 or BSA/ Tween 20	Sunflower oil	1. 15.500 rpm –2 min2. 500 or 900 kPa – 4 or 5 passes	(a) Tween 20 → Maximum βC degradation at 35 °C (b) WPC/Tween 20 → Chemical stability of βC at 22 °C and 35 °C ↑	Trentin et al. (2011)
1. HSM + 2. USH	Zein or WPC	Soybean oil	1. 6000 rpm –5 min 2. 20 kHz –2 min	(a) USH → Physical stability of emulsions and microencapsulation efficiency ↑ (b) USH influenced physical properties of WPC-stabilized emulsions	Gómez-Mascaraque et al. (2017)
1. HSM + 2. USH	Tween 80 and soybean lecithin	Olive oil	1. 8000 rpm –7 min 2. 20 kHz –2.98, 4, 5.5, 7 and 8.02 min	Optimized conditions: 5.82% surfactant concentration, ultrasonic homogenization for 4 min, and 6.50% oil concentration	Mehmood et al. (2018)

[a]Abbreviations for emulsifiers: *α-La* alpha-lactalbumin, *β-Lg* beta-lactoglobulin, *BSA* bovine serum albumin, *CA* chlorogenic acid, *CA/LF* chlorogenic acid/ lactoferrin, *CA/LF/PD* chlorogenic acid/lactoferrin/polydextrose, *CT/EWP* catechin/egg white protein, *DML* decaglycerol monolaurate, *EGCG* (-)-epigallocatechin-3-gallate, *EWP* egg white protein, *HPO* hydrogenated palm oil, *HPMC* hydroxypropyl methylcellulose, *LC* low molecular weight chitosan, *LC/EGCG* low molecular weight chitosan/EGCG conjugates, *LF* lactoferrin, *MD* maltodextrin, *NaCas* sodium caseinate, *OSA-MS* octenyl succinic anhydride modified starch, *QS* Quillaja saponins, *PHC* L-α-phosphatidylcholine, *RBD* refined, bleached and deodorised, *SMP* sucrose monopalmitate, *SPI* soy protein isolate, *SSPS* soybean soluble polysaccharides, *UFP Ulva fasciata* polysaccharide, *WPC* whey protein concentrate, *WPI* whey protein isolate

Physicochemically stable emulsions can be fabricated by employing only one homogenization technique. For instance, quillaja saponins- or WPI-stabilized βC monodisperse nanoemulsions with mean droplet diameter of 0.14–0.16 μm were fabricated after a single pass through dual-channel MF (Luo et al. 2017). Alternative to common protein-based wall materials (e.g. WPI or whey protein concentrate, WPC), bovine serum albumin (BSA)-stabilized emulsions produced with HSM successfully entrapped βC. In this work, the carotenoid was first bound to BSA and then the βC/BSA complex was used to prepare emulsions, which were stable for minimum 5 days (Wackerbarth et al. 2009). Similar to any food process, effects of homogenization parameters on the quality of final product should be examined and (if necessary) be optimized. In a study, using HSM, the effects of homogenization speed (5000–15.000 rpm) and time (2–6 min) and oil phase-to-aqueous phase ratio (0.1:99.9–1:99) on the physicochemical properties of Tween 80-stabilized βC-loaded nanoemulsions were investigated. The authors produced nanoemulsions with droplet sizes ranging from 0.418 to 1.7 μm. An increase in homogenization speed led to an increase in droplet size, whereas an increase in the homogenization time reduced the droplet size (Flores-Miranda et al. 2015). Use of natural antioxidants can retard the degradation of βC in nanoemulsions. For instance, the addition of eugenol (10 wt% of soybean oil) remarkably improved the stability of βC during ambient storage and heating (up to 80 °C) and UV radiation in nanoemulsions produced by HSM (Guan et al. 2016). Likewise, α-tocopherol, tertiary butyl hydroquinone (TBHQ) and ascorbyl palmitate (0.01, 0.05 or 0.10 wt%) were incorporated into gum arabic-stabilized βC emulsions which were homogenized with HSM and HPH. The impact of these antioxidants on the thermal stability of βC at 25 °C was depending on their polarity and increased in the following order: α-tocopherol > TBHQ > ascorbyl palmitate. α-tocopherol in a ratio of 0.10 wt% served as the most effective antioxidant to prevent the βC oxidation induced by light (Liu et al. 2015). In another study, the efficiency of different ratios (0.0025, 0.005, 0.01, 0.02, or 0.05 wt%) of epigallocatechin-3-gallate (EGCG) on βC degradation in α-lactalbumin (α-La)-stabilized MCT oil emulsions were investigated. After a 7-day storage at 55 °C, the percentage of βC remaining in the emulsions was 81.30% and 97.20% when 0.0025 wt% and 0.05 wt% EGCG was added, respectively (Liu et al. 2016a).

As stated before, combination of two homogenization techniques is a more common way to form fine emulsions. In one of pioneering studies, O/W nanoemulsions of βC were produced by HSM followed by HPH. The impact of emulsifying conditions including homogenization pressure, temperature and cycle on the properties and stability of the nanoemulsions were studied (Yuan et al. 2008; Zhang et al. 2017).

In a study, reduced droplet size and improved physical stability were observed when an additional USH step was applied on HSM-treated emulsions containing βC (Fig. 2.8) (Gómez-Mascaraque et al. 2017). Nevertheless, the same authors mentioned the drawback of USH technique for thermolabile components, which may damage them (Gómez-Mascaraque and López-Rubio 2016). Moreover, prior to homogenization process, use of sonication was more efficient (1.73 wt%) in dissolving βC in the oil phase than conventional heating method (1.21 wt%) (Chen et al. 2017).

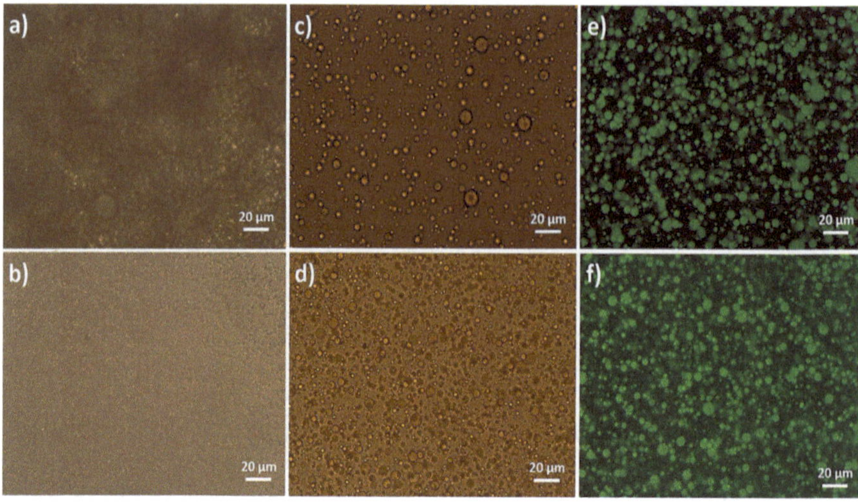

Fig. 2.8 Optical micrographs of the βC-loaded emulsions prepared using WPC (**a**, **b**) and zein (**c**, **d**), and fluorescence micrographs of zein emulsions containing βC (**e**, **f**). Emulsions **a**, **c** and **e** were prepared by HSM and emulsions **b**, **d** and **f** were prepared applying an additional USH step. (Reproduced from the reference Gómez-Mascaraque et al. (2017) with permission from Elsevier)

MF process is generally coupled with HSM for preparation of βC-encapsulated emulsions. For instance, a combination of HSM and MF yielded βC-loaded nanoemulsions which were physically stable at 55 °C for 15 days (Qian et al. 2012a). Despite the similarity in terms of the mechanism of action between these two processes, it has been shown that HPH produced βC emulsions with larger droplets than MF under the same experimental conditions (Fig. 2.9) (Mao et al. 2010). On the other hand, MF parameters may influence the initial droplet size and physical stability of emulsions. In this context, it was reported that in NaCas-stabilized βC emulsions, as the pressure (1500 psi–15.000 psi) or number of passes (0–7 passes) in MF increase, the droplet diameter decreases linearly or exponentially, respectively. As a consequence of smaller emulsion droplet diameter, bioaccessibility which was defined by the amount of βC recovered in the aqueous phase after ultracentrifugation, increased (Yi et al. 2014).

High-energy emulsification techniques can be combined with low-energy techniques in food applications. Utilization of HSM along with ME allowed the production of WPC-, BSA- and Tween 20-stabilized βC emulsions. For all emulsions, the most pronounced reduction in droplet size was observed during the 1st cycle of ME. However, no further decrease in the droplet size of emulsions stabilized with WPC or BSA occurred for the subsequent ME cycles, while the droplet size of emulsions containing Tween 20 reduced until the 3rd cycle. Moreover, WPC led to higher membrane fouling compared to BSA due to differences in their molecular structures and conformations. On the other side, when both proteins were coupled with Tween 20, a reduction in membrane fouling was observed. Additionally, WPC/Tween 20 combination yielded emulsions with smaller droplets that exhibited

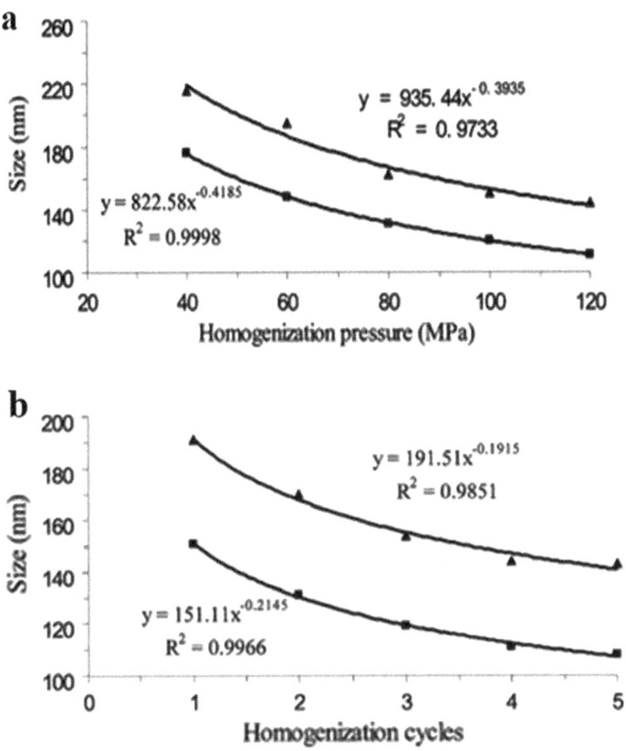

Fig. 2.9 Effects of (**a**) homogenization pressures on droplet size of βC nanoemulsions: valve homogenization for three cycles (▲); microfluidization for three cycles (■); (**b**) homogenization cycles on droplet size of βC nanoemulsions: valve homogenization at 100 MPa (▲); microfluidization at 100 MPa (■). (Reproduced from the reference Mao et al. (2010) with permission from Taylor & Francis)

higher physicochemical stability during storage at 22 °C and 35 °C than those of containing BSA/Tween 20. The authors concluded that using appropriate surfactant mixtures, βC can be successfully protected in O/W emulsions formed by ME process (Trentin et al. 2011).

In addition to the type of homogenization process, the nature of surfactants and carrier oils is important for determining the physicochemical properties and stability of emulsions containing carotenoids (Hejri et al. 2013; Igielska-Kalwat 2018). For instance, addition of MD to NaCas-stabilized emulsions (homogenized by HSM and HPH, respectively) increased the retention ratio of βC to more than 92.1% after 90 days storage at 4 °C under dark conditions (Zhang et al. 2017). In another study, O/W emulsions incorporating βC within droplets stabilized by multiple-layer of carbohydrate-based emulsifiers consisting of soybean soluble polysaccharides (SSPS) and chitosan were fabricated using the combination of HSM and HPH techniques. The mean particle diameter of the emulsions considerably reduced at chitosan concentrations higher than 0.33 wt% and it reached to the smallest value

(0.79 μm) at chitosan concentration of 0.5 wt%. At this concentration of chitosan, the least degradation occurred in βC-loaded emulsions, which were stored at different storage temperatures (4, 30, 50, and 70 °C) (Hou et al. 2010). Furthermore, the influence of MW of chitosan on the chemical stability of SSPS-stabilized emulsions containing βC was investigated by the same research group. As result, the particle size increased with the rise of chitosan MW at high chitosan concentrations (>0.2 wt%). The authors recommended the use of medium-MW chitosan to improve the heat and light stability of βC in emulsion systems (Hou et al. 2012).

In addition to whey protein and casein, other proteins are often utilized as emulsifying agents in foods and can inhibit the oxidation of dispersed phase by preventing the penetration of pro-oxidants within the emulsified droplet. For instance, it is reported that β-lactoglobulin (β-Lg) was more effective surfactant to improve chemical stability of emulsified βC than Tween 20. The authors attributed this effect to (a) antioxidant activity of proteins through chelating transition metals or scavenging free radicals, (b) ability of proteins to form molecular complexes with carotenoids by means of hydrophobic interactions, (c) performance of β-Lg as a physical barrier which blocks the contact of any pro-oxidants found in aqueous phase or (d) smaller oil-water interfacial area of larger droplets in β-Lg-stabilized emulsions (Qian et al. 2012b). Through similar mechanisms, scallop gonad protein isolates (SGPI) successfully increased the chemical stability of βC under different storage conditions (4, 25, and 37 °C up to 30 days) (Han et al. 2020).

A blend of surfactants was found to improve the stability of βC-loaded nanoemulsion. In one of early studies, Tween 20, WPI, OSA-MS, decaglycerol monolaurate (DML) and series of blends of Tween 20 and WPI (at different ratios) were examined as emulsifiers in βC nanoemulsions homogenized by HSM and HPH. Consequently, Tween 20 and DML were able to decrease interfacial tension and yield emulsions with smaller droplets, whereas WPI and OSA-MS produced more stable emulsions with larger droplets. Emulsions containing WPI preserved βC better than OSA-MS-stabilized nanoemulsions when they were stored at 55 °C for 12 days. Therefore, the authors suggested the use of Tween 20/WPI blend (at the ratio of 1:1) to enhance the stability of βC-loaded emulsions (Mao et al. 2009). However, a study demonstrated a close direct correlation between WPI oxidation and βC loss. The authors explained this finding by stabilization effect of proteins on lipids and triggering of protein oxidation by lipid (oil phase) oxidation products (peroxyl radicals). Hence, the authors suggested the incorporation of metal chelators and/or free radical scavengers to improve βC stability in protein-stabilized emulsions (Xu et al. 2013a).

Conjugates of surfactants (through covalent linkage between them) exhibited improved emulsifying activity and stability of βC emulsions. For example, covalent conjugates of WPI/beet pectin and WPI/dextran formed through Maillard reaction under dry-heating conditions, and covalent conjugates of low MW chitosan (LC)/EGCG and α-La/catechin synthesized by free radical grafting approach were successfully prepared and used in βC-loading emulsions (Xu et al. 2012, 2013b; Lei et al. 2014; Yi et al. 2016; Fan et al. 2017). Similarly, catechin/egg white protein

(EWP) conjugates were developed as food-grade antioxidant emulsifier for βC-loaded emulsions that were homogenized by combination of HSM and HPH. The authors attributed the higher physical and chemical stability of emulsions (during storage time) to an increase in interfacial thickness (thereby, increasing the steric repulsion) and higher antioxidant capacity of conjugates (Gu et al. 2017). Besides, LC/EGCG covalent conjugates formed through free radical grafting reaction were effectively used as secondary layer in bilayer βC emulsions stabilized with NaCas or α-La (Wei and Gao 2016a). The same authors examined and compared the ability of both physical and covalent complexes of chitosan and chlorogenic acid (CA), a polar antioxidant, in stabilizing βC emulsions formed by HSM and HPH. Chitosan/CA covalent complex was synthesized according to carbodiimide-mediated coupling method. The obtained results exhibited that emulsions including chitosan/CA covalent complex had higher viscosity. In addition, interfacial concentration fraction of CA in the emulsions coated by the covalent complex was higher than for those stabilized with the physical complex. These differences led to less degradation rate of βC in the emulsions stabilized by chitosan/CA covalent complex and brought about them the highest physical stability (Wei and Gao 2016b).

Physicochemical properties of the oil phase, where the lipophilic bioactive is dissolved in, are of vital importance as well. For example, βC emulsions containing solid hydrogenated palm kernel oil (HPKO) or liquid HPKO with identical compositions exhibited different chemical stabilities during a 28-day storage at 15 °C under dark conditions. A fully solid carrier increased βC degradation as a result of exclusion of βC toward the interface where the oxidation reactions may take place. On the other hand, use of a liquid carrier could be limited due to lower physical stability of these systems. Consequently, a partially solid matrix was recommended to isolate βC into small domains within efficient solid lipid barrier (Cornacchia and Roos 2011a). Similarly, the same researchers showed that a partially solid lipid carrier (HPKO) was more effective than a fully liquid lipid carrier (sunflower oil) protecting βC at 20 °C due to entrapment of βC in isolated domains within the solid lattice and thus, blocking its contact with reactive species in the surroundings (Cornacchia and Roos 2011b). Utilization of essential oil as carrier oil in βC emulsions was investigated too. For example it was shown that peppermint oil provided increased βC stability against thermal processing (60 or 80 °C for 0.5, 1, 2, 4, 8 and 16 h) and UV radiation (254 or 302 nm up to 16 h) due to its high antioxidant capacity arising from its unique components such as limonene, menthone, and menthol (Chen and Zhong 2015).

Multi-layer emulsions can be used as functional components to enhance the nutritional features of foods and beverages. In accordance with this purpose, in a very recent study, βC-enriched tertiary emulsion consisting of lactoferrin/alginate/ε-poly-L-lysine were developed. The tertiary emulsion exhibited a good physical stability up to 60 °C and provided higher encapsulation efficiency (96.06%) than primary emulsion composed of only lactoferrin (82.52%). Moreover, greater βC bioaccessibility (70.10%) were reported for tertiary emulsion compared to primary (30.24%), and secondary emulsion composed of lactoferrin/alginate (35.26%). The authors explained this difference by the increased amount of free fatty acids released during digestion (34.61% for primary, 61.03% for secondary, and 83.79% for tertiary emulsions) as the number of layers increased. The multi-layer emulsions presented good physical stability under stomach conditions until intestine where these different layers are completely separated from oil droplets that promote lipase action and lipid digestion (Gasa-Falcon et al. 2020). However, processing conditions, as well as food characteristics might influence their properties. Therefore, studying how different external conditions affect physical stability of these emulsions would afford useful information about where to use them.

Increase the bioavailability and bioaccessibility of lipophilic bioactive compounds is another goal of encapsulation process. In a study, βC was encapsulated separately in WPI, NaCas and soybean protein isolate (SPI) into nanoparticles through homogenization–evaporation method. The obtained results showed that βC encapsulated by WPI had the most favourable release properties. Release was low with pepsin but high with trypsin proposing that WPI might be the preferred protein delivery system to deliver βC to the intestine (Yi et al. 2015). Similarly, in another study higher in vitro bioaccessibility of βC from WPI-stabilized emulsions (58.50%) was reported in comparison to NaCas-stabilized (56.50%) and Tween 80-stabilized (41.30%) ones. In addition, it was found out that NaCas-stabilized βC emulsions display higher cellular uptake of βC (0.180 μg/mg protein) than the emulsions containing Tween 80 (0.146 μg/mg protein) or WPI (0.130 μg/mg protein) (Lu et al. 2017). Moreover, the bioaccessibility of βC was improved in the presence of L-α-phosphatidylcholine (PHC) and increased with increasing PHC concentration, ranging from 33.2% to 79.8% for a 1% and 4% PHC emulsion respectively (Verrijssen et al. 2015a). The same authors also showed that the effect of the degree of methylation (DM) of citrus pectin in emulsions on the in vitro bioaccessibility of βC and lipid digestion depends on the presence of PHC (Verrijssen et al. 2015b).

In efforts to use novel stabilizers from unconventional sources that may enhance bioaccessibility of emulsified βC, polysaccharides extracted from *Ulva fasciata* were used in a study. As the main finding, higher βC bioaccessibility (37.40%) was observed in comparison to that of beet pectin (31.60%), and gum arabic (24.90%) under in vitro oral-gastro-intestinal digestion conditions, which was due to the smaller droplet size and thereby higher surface area of *Ulva fasciata* polysaccharides-stabilized emulsions (Shao et al. 2017a). Additionally, new emulsifiers can be produced by modifying the properties of interfacial layers through covalent conjugation. For instance, ternary covalent conjugates of surface-active chlorogenic acid/lactoferrin/polydextrose were utilized as emulsifier in βC emulsions. Using these conjugates, βC bioaccessibility (5.58%) was improved through increasing the extent of βC transferred to mixed micelles as a result of decreased droplet aggregation rate under simulated gastrointestinal tract conditions compared to lactoferrin (2.64%), chlorogenic acid/lactoferrin conjugate (4.20%) or chlorogenic acid/lactoferrin/polydextrose physical mixture-stabilized (4.32%) emulsions (Liu et al. 2017). According to a very recent study, porous OSA-MS with different degree of substitution (DS) was used as emulsifier in βC-loaded emulsions. The pores and channels within porous starch provided more possibilities for OSA to modify starch, which resulted in higher bioaccessibility of βC as the DS of OSA-MS increased (Li et al. 2020a). Surfactant-dependent release rate of βC mainly results from interaction of the surfactant with bile extracts, pancreatin and lipase during in vitro digestion, which leads to different microstructural changes in emulsions (Hou et al. 2014).

Also, the bioaccessibility of βC is highly dependent on the carrier oil composition and its concentration. For example, in a study, the effects of carrier lipid physical state on in vitro digestive lipolysis and bioaccessibility of βC from emulsions containing compositionally equivalent solid or liquid cocoa butter were investigated. Under simulated gastric conditions, duodenal hydrolysis and βC bioaccessibility were lower in solid emulsions for the first 2 h. However, after 24 h, TAG of both emulsions (64.50% for liquid, and 60.80% for solid emulsions) were greatly digested. This study clarified that the solid state of carrier oil weakened the digestion rate but did not affect the maximum achievable TAG lipolysis. Moreover, it was reported that the extent of βC bioaccessibility were directly correlated with the extent of TAG lipolysis as both emulsions (liquid and solid) provided similar βC bioaccessibility after 24 h digestion (Hart et al. 2018).

In this context, it was shown that βC has much higher bioaccessibility in LCT nanoemulsions than in MCT nanoemulsions, which was attributed to the fact that the mixed micelles formed by long chain free fatty acids (FFA) have larger hydrophobic cores than those generated by medium chain FFA and therefore they are able to receive the relatively large linear non-polar βC molecules (Qian et al. 2012c; Han et al. 2019). This fact was confirmed in another study too, in which an increase in βC bioaccessibility for low fat nanoemulsions was observed, as the LCT content within the lipid phase increased, which was due to the increased solubilization capacity of mixed micelles formed by LCT. Whereas for high fat nanoemulsions, βC bioaccessibility displayed a complex dependence on LCT content, due to the opposing effects of lipid digestion and micelle solubilization (Salvia-Trujillo et al. 2013). Similarly, βC-loaded emulsions were stabilized using SGPIs with either LCT or MCT as a carrier oil and it was revealed that βC bioaccessibility in LCT emulsions was higher (65.50%) than in MCT emulsions (23.10%) (Han et al. 2019).

Finally, there is a strong correlation between the particle size of high pressure homogenized emulsions and the release and transport of bioactive compounds within in vitro digestion model. In vitro release of βC from emulsions can be promoted through fabrication of emulsions with smaller droplets with effective high-energy techniques like HPH or MF (Liu et al. 2012). In a study, βC was encapsulated in triglyceride emulsions stabilized by NaCas produced by MF. Subjecting the emulsions to in vitro digestion, a linear inverse relationship between initial droplet size and βC bioaccessibility was obtained (Fig. 2.10) (Yi et al. 2014). In another report, incorporation of mandarin fibre (up to 1 wt%) into βC nanoemulsions, homogenized with HSM and MF techniques, resulted in the improvement of βC bioaccessibility as the initial droplet size decreased (Gasa-Falcon et al. 2017).

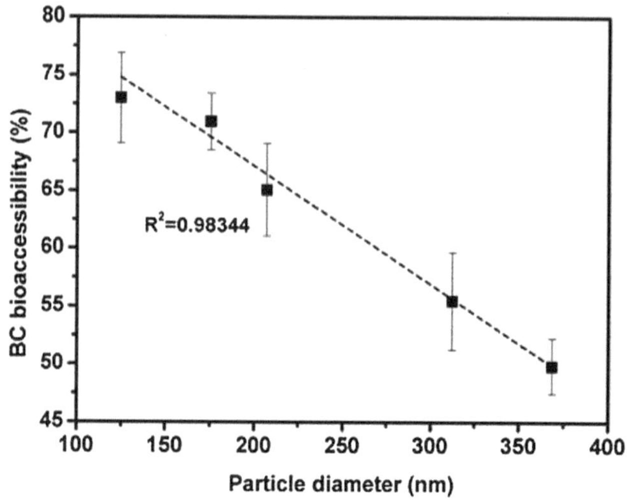

Fig. 2.10 The relationship between initial droplet size of emulsion and βC bioaccessibility. (Reproduced from the reference Yi et al. (2014) with permission from Elsevier)

2.3.2.2 Lycopene (LYC)

LYC, which is a long and linear molecule, is an unsaturated acyclic hydrocarbon carotenoid with 11 conjugated and 2 unconjugated double bonds (Rodriguez-Amaya 2016; Ranveer 2018). The well-known functions of lycopene are considered to originate from its highly strong antioxidant activity. It cannot be synthesized by humans and animals although plants and microbes such as fungi or bacteria can synthesize it (Liang et al. 2019). Besides its anti-inflammatory and radioprotective roles, LYC has protective effects against various cancer types including prostate, liver, ovarian, breast and colon cancer and cardiovascular diseases, Alzheimer's disease, diabetes, and beta-thalassemia (Fraser 2004; Kun et al. 2006; Islamian and Mehrali 2015; Costa-Rodrigues et al. 2018; Liang et al. 2019). Due to several conjugated bonds found in its structure, it is sensitive to oxidation reactions caused by heat, oxygen, light, and metals (Saini et al. 2020). Therefore, both stability and bioavailability of LYC can be enhanced by various encapsulation methods.

High-Energy Emulsification Approaches to Encapsulate LYC

Conventional and nano-sized LYC-loaded emulsions can be efficiently fabricated through high-energy emulsification techniques. Each homogenization equipment has its own advantages, and disadvantages. Therefore, necessary adjustments in either equipment conditions or emulsion properties should be properly done to ensure the maximum emulsification ability. A summary of the most relevant studies on the encapsulation of LYC in emulsion-based systems through using high-energy emulsification techniques is presented in Table 2.5.

Coarse and fine emulsions are commonly fabricated by using HSM technique. For instance, in a study, using only mechanical stirring, rice bran oil containing LYC and aqueous solution of casein was effectively homogenized. As a result, O/W emulsions were obtained which were used in the next step for fabrication of LYC-loaded microparticles (Jain et al. 2016). On the other side, coarse or fine emulsions can be homogenized by adjusting the rotation speed of HSM (10.000 rpm for coarse emulsions; 14.000–16.000–18.000 rpm for fine emulsions) as performed in a study. The authors found inverse relationships between droplet size, creaming index and the homogenization speed. On the other hand, they attributed the higher viscosity resulting from higher homogenization speed to an increase in the number of oil droplets and therefore, more droplet-droplet interactions that increase the resistance to flow. In addition to homogenization speed, droplet size and creaming index were directly proportional to LYC content, while inversely proportional to emulsifiers concentration. Finally, homogenization at 18.000 rpm, LYC content of 18.85 wt% and WPC + MD concentration of 37.16 wt% were suggested as the optimum conditions to obtain LYC emulsions (Salimi et al. 2015). Beyond mechanical homogenization, coarse emulsions can be exposed to USH technique to fabricate finer emulsions. In a study, coarse LYC emulsions were obtained through mechanical stirrer and then USH technique was applied to ensure an effective homogenization (Guo et al. 2014). On the contrary, in another study USH technique (amplitude of 80%, cycle of 0.8, 10 min) was employed for the first homogenization step before mechanical agitation (900 rpm–10 min) for fabrication of LYC nanodispersions (Anarjan and Jouyban 2017). In a study, the effects of light, pH, iron, and free radicals on oxidative stability of sodium dodecyl sulphate (SDS)-stabilized O/W LYC emulsions obtained through USH were investigated. Minor changes in LYC content under fluorescent light exposure at 15 °C were observed. On the other hand, a remarkable impact of pH on LYC degradation was observed with the greatest loss occurred in the emulsions at pH ≤4. The authors attributed this fact to the increased solubility of iron and its higher binding ability to droplet surfaces in low pH conditions. Moreover, addition of TBHQ at pH 7, decreased the LYC loss from 44% to 14–17% in the emulsions. As conclusion and to extend the stability of LYC-loaded emulsions the use of free radical scavengers (e.g. TBHQ) at higher pH and metal chelators like ethylenediaminetetraacetic acid (EDTA) at low-pH was suggested (Boon et al. 2009). However, in another study EDTA showed pro-oxidant activity in Tween 20-stabilized O/W LYC-loaded emulsions with low pH, while α-tocopherol, gallic acid and propyl gallate acted as strong antioxidants against LYC degradation (Bou et al. 2011).

Combination of two homogenization techniques has been employed more often to prepared LYC-loaded emulsions. In one of pioneering studies, LYC-loaded O/W emulsions were prepared using HSM and HPH techniques. The chemical and physical stability of the emulsions were studied by diluting them in skimmed milk, orange juice and water as control. As the main finding, LYC was particularly stable in orange juice and addition of α -tocopherol led to good stability of LYC in all food

Table 2.5 An overview of the most relevant studies on the encapsulation of LYC in emulsion-based systems via high-energy emulsification methods

Emulsification technique(s)	Emulsifier(s)[a]	Carrier oil	Emulsification conditions	Main findings	References
HSM	Denaturated WPC and MD	Soybean oil	10.000 rpm −5 min + 14.000/16.000/18.000 rpm −10 min	(a) Homogenization speed ↑, droplet size ↓ and viscosity ↑ (b) LYC content ↑, droplet size ↓	Salimi et al. (2015)
HSM	Casein and gum tragacanth	Double refined rice bran oil	6000 rpm −30 min	(a) Increase of rotation speed from 4000 to 8000 rpm, average particle size ↓ (b) Rotation speed <6000 rpm → Shear force was not enough for stable emulsion formation (c) Rotation speed at higher speeds → Lower entrapment efficiency due to emulsion breaking	Jain et al. (2016)
USH	Tween 20	MCT oil	2 min – 0.5 s pulses–70% amplitude	(a) Gallic acid + α-T → maximum antioxidant activity (b) Ascorbic acid + gallic acid + α-T → LYC loss ↑	Bou et al. (2011)
1. HSM + 2. HPH	Tween 20, CITREM or enzymatic hydrolysated soya lecithin	MCT oil	1. Homogenization conditions N/A 2. 65 MPa –number of passes N/A	Chemical stability of emulsified LYC depended on the type of food system	Ribeiro et al. (2003)
1. HSM + 2. HPH	Tween 20	MCT oil	1. Homogenization conditions N/A 2. 650 bar –number of passes N/A	(a) LYC degradation follows first-order kinetics (b) Temperature ↑, emulsified LYC loss ↑ (c) Absence of O_2 → LYC loss ↓	Ax et al. (2003)

1. HSM+ 2. HPH	DTAB, Brij® L23 or SDS	Corn oil, stripped corn oil or hexadecane	1. Rotation speed N/A –2 min 2. 4000 psi –4 passes	(a) LYC stability in decreasing order corn oil > hexadecane > tocopherol-stripped corn oil (b) Negatively charged interface of droplets in SDS-stabilized emulsions → Oxidation rates ↑	Boon et al. (2008)
1. HSM + 2. HPH	Sodium stearoyl glutamate	Orange wax: rice oil (9:1)	1. 12.000 rpm –30 s 2. 500 bar–3 passes	Concentration of LYC ↑, droplet size ↑	Okonogi and Riangjanapatee (2015)
1. HSM + 2. HPH	Tween 20	Linseed oil	1. 5000 rpm –10 min 2. 5, 10, 15, 30, 70 and 100 MPa–1–10 passes	(a) Trolox + BHT → Acted as antioxidants synergistically (b) 100 MPa-treated nanoemulsions → maximum LYC bioaccessibility	Sotomayor-Gerding et al. (2016)
1. HSM+ 2. HPH	OSA-MS	MCT oil	1. 18.000 rpm –4 min 2. 110 MPa –3 passes	LYC concentration ↑, droplet size ↓ and emulsion stability ↑	Li et al. (2018a)
1. HSM + 2. HPH	WPI	Different ratios of SCT and LCT oils	1. 1000 rpm –5 min 2. 50 MPa –2 passes	LCT ratio ↑, droplet size ↓, LYC bioaccessibility ↑	Raikos et al. (2019)
1. HSM+ 2. HPH	WPI	Orange oil, orange oil–tributyrin (SCT) or orange oil–corn oil (LCT)	1. Rotation speed N/A –2 min 2. 40 MPa–2 passes	(a) Addition of corn oil to orange oil-emulsions → physical stability ↑ (b) LCY retention in decreasing order 4 wt% orange oil > 2 wt% orange oil–2 wt% tributyrin >2 wt% orange oil–2 wt% corn oil (c) LYC bioaccessibility in decreasing order 100% orange oil >50% orange oil +50% SCT > 50% orange oil +50% corn oil	Meroni and Raikos (2018)

(continued)

Table 2.5 (continued)

Emulsification technique(s)	Emulsifier(s)[a]	Carrier oil	Emulsification conditions	Main findings	References
1. HSM + 2. HPH	Glucose-fructose syrup	Sunflower oil	1. 13.000 rpm –2 min 2. 60 MPa + 15 MPa – two-step homogenization	(a) Polarity of carotenoids influenced resistance of emulsions during storage. (b) Polarity ↓, resistance to viscoelastic changes ↑ (c) LYC (apolar carotenoid) → the highest stability	Domian and Szczepaniak (2020)
1. HSM+ 2. HPH	DBOS-MS	Soybean oil	1. N/A 2. 4/40 MPa –3 passes	(a) Chemical stability of LYC in the alginate bead > chemical stability of LYC in the emulsion (b) Bioaccessibility of LYC from emulsion > bioaccessibility of LYC from alginate bead	Jain et al. (2020)
1. HSM + 2. HPH	OSA-MS, WPI or NaCas	MCT oil	1. 10.000 rpm –5 min 2. 110 MPa –3 passes	(a) OSA-MS-stabilized emulsions → the smallest droplet size (b) WPI-stabilized emulsions → the lowest stability (c) Optimum surfactant amounts → 30 wt% for OSA-MS, 1 wt% for WPI and 2 wt% for NaCas	Wanyi et al. (2020)
1. HSM+ 2. HPH	LF	Sesame oil, linseed oil or walnut oil	1. 10.000 rpm –3 min 2. 10.000 psi–3 passes	(a) Sesame oil-based emulsion → highest LYC stability (b) Oil density ↑, LYC stability ↑ (c) Oil viscosity ↑, LYC stability ↓(d) LYC bioaccessibility from: sesame oil = linseed oil > walnut oil	Zhao et al. (2020)

1. HSM+ 2. MF	WPI or WPI/HMP	Corn oil	1. 5000 rpm– homogenization time N/A 2. 100 MPa –3 passes	Physicochemical stability → Double layer emulsion consisting of WPC/HMP > Single layer emulsion consisting of WPC	Shi et al. (2015)
1. HSM + 2. MF	WPI, NaCas, SPI or PPI	Stripped canola oil	1. 11.000 rpm–30 s 2. 800 bar –5 passes	(a) WPI or SPI-stabilized emulsions → unstable at refrigeration conditions for 14 days (b) PPI or NaCas as emulsifier → Physicochemical stabilization of LYC emulsions ↑ (c) SPI or WPI as emulsifier → Physical or chemical stabilization of LYC emulsions, not both	Ho et al. (2017)

[a]Abbreviations for emulsifiers: *CITREM* citric acid esters of mono- and diglycerides of fatty acids, *DBOS-MS* debranched octenyl succinic anhydride modified starch, *DTAB* dodecyltrimethylammonium bromide, *HMP* high-methylester-pectin, *LF* lactoferrin, *NaCas* sodium caseinate, *PPI* pea protein isolate, *SDS* sodium dodecyl sulphate, *SPI* soy protein isolate, *WPI* whey protein isolate

systems (Ribeiro et al. 2003). Tocopherols are oxidized prior to LYC, the preferential oxidation of tocopherols could be inhibiting the oxidation of lycopene. This fact was confirmed in another report as well (Boon et al. 2008).

The operating parameters of HPH technique (e.g. pressure and the number of homogenization) play an important role in controlling the characteristics of LYC nanodispersions, such as the particle size, PDI, zeta potential, etc. In a study, particle size decreased from 335 nm to 104, 97, and 92 nm when HPH (1 pass) was applied at 100, 300, and 500 bar, respectively. However, adjusting the pressure to more than 500 bar yielded emulsions with larger droplets due to over-processing effect which leads to enhance the re-coalescence rate of LYC particles (Boon et al. 2009). Number of HPH cycles is important too and when it increases, emulsification may be negatively affected due to mechanical or thermal destruction of LYC. For example, in a study, LYC nanoemulsion with highest stability and smallest size was obtained after 3 homogenization cycles. At higher homogenization cycles homogenization takes longer, which may mechanically or thermally destroy large numbers of LYC molecules (Kim et al. 2014).

Other variables including types of surfactant(s) and carrier oils and concentration of LYC and surfactant(s) should be optimized to provide maximum emulsification ability. In the former study, employing statistical analysis, LYC extract concentration was determined as the most important factor on stability, droplet size and emulsification value (Kim et al. 2014). On the other hand, it was shown that as LYC concentration increased from 0.1 to 0.5 wt%, droplet size tended to decrease, as a result of stronger packing of OSA-MS molecules on the interfacial membrane that was provided by more LYC molecules bound to oil-water interface (Li et al. 2018a). The advantages of OSA-MS and NaCas over WPI in terms of forming small droplets, stability and encapsulation efficiency of HSM and HPH-treated LYC nanoemulsions was demonstrated in another report (Wanyi et al. 2020). In a study, the effects of plant (SPI and pea) and dairy (WPI and NaCas) proteins on physical stability and retention of LYC in canola O/W emulsions (pH = 7.0, 10% oil) were compared and in general, SPI and WPI exhibited better physical and chemical stabilization rather than SC and pea protein isolate (Ho et al. 2017). Nonetheless, as well as surfactants, the polarity of carotenoids is an essential factor that influences the physical characteristics of emulsions. In a study, O/W emulsions containing carotenoids (2 wt%) with different polarity (LYC < βC < apocarotenal ≪ lutein) were prepared. The most polar carotenoid, lutein formed a very viscous emulsion structure after 3-month storage at 30 °C (Domian and Szczepaniak 2020). While, the emulsions containing the most apolar carotenoid, LYC, displayed the best resistance against viscoelastic changes during storage period due to better solubilization of apolar LYC in droplets.

Studies dealing with bioaccessibility and bioavailability of emulsified LYC should be carried out to evaluate the suitability of encapsulation systems for digestion of lipophilic bioactives. In this context, generally nano-sized emulsions are regarded as better vehicles compared to micro-sized ones. For example, the highest in vitro LYC bioaccessibility was detected in nanoemulsions with mean droplet diameters <100. Increased specific surface area promotes the access of lipases, co-lipases, bile salts, cholesterol and phospholipids and simultaneously increases the absorption rate of the bioactives by enhancing their solubilization (Ha et al. 2015). The effect of different carrier oils (sesame, linseed and walnut oils) on bioaccessibility of lactoferrin-stabilized LYC nanoemulsions was investigated in another study. As the main finding of this research, sesame oil with lower viscosity, higher density and lower unsaturation degree was suggested to be the best oil phase to ensure emulsion stability and LYC bioaccessibility, followed by linseed oil (Zhao et al. 2020). In addition, the impact of LCT to short chain triglyceride (SCT) ratio of the carrier on bioaccessibility of LYC was highlighted in another report, in which higher bioaccessibility of LYC was detected in O/W beverage emulsions that contained higher levels of LCT. This was due to capability of LCT in forming mixed micelles with a large hydrophobic core. Finally, the authors proposed an optimum LCT:SCT ratio of 75:25 to decrease phase separation induced by LCT as well as to obtain emulsions with smaller droplets and improved bioaccessibility (Raikos et al. 2019).

2.3.2.3 Lutein (LU)

LU belongs to xanthophyll subclass which is also known as oxygenated carotenoids (Ma and Lin 2010). It is widely found in egg yolk and green leafy vegetables including kale, spinach, broccoli, peas and Brussel sprouts. However, its red-orange colour is dominated by high levels of chlorophylls in these green vegetables (Johnson 2002; Nwachukwu et al. 2016). The well-known health benefits of LU are improvement of eye and skin health and immune system, prevention of cancer formation, cardiovascular diseases and degenerative diseases (Granado et al. 2003; Ma and Lin 2010). Lower bioavailability of LU is closely correlated with its lower water solubility due to highly lipophilic character. In addition, the high photooxidation sensitivity of LU requires a convenient process such as encapsulation via emulsion-based systems to overcome these problems.

High-Energy Emulsification Approaches to Encapsulate LU

HSM offers many advantages including adjustable rotation speed and easy accessibility. Regardless of emulsifier and carrier oil types, HSM is apparently the most common equipment used in preparation of LU-loaded emulsion systems. Table 2.6 summarizes the most relevant studies on the encapsulation of LU in emulsion-based systems through high-energy emulsification methods.

HSM technique, solely has been effectively used to fabricate LU-loaded emulsions in several studies (Lacatusu et al. 2013; Muhoza et al. 2016b; Murillo et al. 2016; Álvarez-Henao et al. 2018; Huang et al. 2019b; Wang et al. 2020; Su et al. 2020a, b; Li et al. 2020b). However, physicochemical stability of both emulsions and emulsified LU depend on various factors, mainly the composition of emulsion (type of surfactant, carrier oil, etc.). In addition to commonly used surfactants (e.g.

Table 2.6 An overview of the most relevant studies on the encapsulation of LU in emulsion-based systems via high-energy emulsification methods

Emulsification technique(s)	Emulsifier(s)[a]	Carrier oil	Emulsification conditions	Main findings	References
HSM	WPI, or WPI with CMC, HMP or LMP	MCT oil	12.000 rpm –5 min + 8000 rpm –3 min	(a) WPI → Maximum LU encapsulation efficiency at pH 3 and 4 (b) pH ↑, encapsulation efficiency ↓	Muhoza et al. (2016b)
HSM	OSQS	Corn oil	20.000 rpm –2 min	LU retention after a 31-day storage at RT → 55.38 %	Li et al. (2020b)
HSM	β-Lg/gum arabic composite particles	MCT oil	10.000 rpm –5 min	LU retention after a 12-week storage at 25 °C → 91.10 %	Su et al. (2020a)
HSM	β-Lg/EGCG composite nanoparticles	MCT oil	10.000 rpm –5 min	LU retention after a 30-day storage at 25 °C → 87.20%	Su et al. (2020b)
HSM	Xanthan and PGA	Coconut oil	18.000 rpm –2 min	Xanthan: PGA in the ratio of 4:6 → Emulsion stability ↑	Wang et al. (2020)
HPH	WPI	Corn oil	80 MPa – 4 passes	Cellular uptake of LU → Nanoemulsions > conventional emulsions	Teo et al. (2017)
USH	WPH, β-Lg or Tween 20 with/without lecithin	MCT oil	70% of maximum energy –5 min	WPH/lecithin → minimum droplet size and maximum LU uptake	Frede et al. (2014)
USH	WPI or PWP	MCT oil	40% of amplitude –5 min	(a) WPI-stabilized nanoemulsions → 4% of LU lost at 4 °C for 4 weeks (b) PWP-stabilized nanoemulsions → Unstable	Zhao et al. (2018)
1. HSM + 2. HPH	Whey protein with phosphatidylglycerol or PHC	Corn oil	1. Rotation speed N/A –2 min 2. 2000 psi –2 passes	(a) PHC-stabilized emulsions at 90 °C for 5 min → Collapse (b) Properties of phosphatidylglycerol-stabilized emulsions → Storage at 4 °C for 24 h = Thermal treatment at 90 °C for 5 min = Fresh emulsions	Losso et al. (2005)

1. HSM + 2. HPH	WPI, DTAB, fish gelatine, WPI/beet pectin or DTAB/beet pectin	MCT oil	1. Rotation speed N/A –2 min 2. 10.000 psi – 3 passes	(a) Primary fish gelatine LU-loaded emulsion → Unstable (b) Release of LU from WPI < Release of LU from DTAB	Beicht et al. (2013)
1. HSM + 2. USH	WPI with high methyl pectin or low methyl pectin	MCT oil	1. 12.000 rpm – 5 min 2. 40% maximum power – 6 min	(a) WPI-stabilized emulsions without pectin → Flocculation above 60 °C (b) WPI/pectin stabilized emulsions → Emulsion stability ↑ (c) Stabilizing ability → High methyl pectin > low methyl pectin	Muhoza et al. (2016a)
1. HSM + 2. HPH	Corn fibre gums or gum arabic	Corn oil	1. 13.000 rpm – 2 min 2. 50 MPa – 3 passes	LU bioaccessibility: (a) in corn fibre gums-stabilized emulsions = 24.90% & 32.40% (b) in gum arabic-stabilized emulsions = 27.70% (c) in corn oil = 13.80%	Feng et al. (2017)
1. HSM + 2. HPH	Bovine casein or caprine caseins with soybean lecithin	Corn oil	1. Conditions N/A 2. 12.000 psi –5 passes	(a) Chemical stability of LU in decreasing order caprine α_{s1}-II-casein/lecithin > caprine α_{s1}-I-casein/lecithin > bovine casein/lecithin (b) Thickness of interfacial layer affects chemical stability of LU	Mora-Gutierrez et al. (2018a)
1. HSM + 2. HPH	Bovine casein or caprine caseins with/without arabinogalactan	Corn oil	1. Conditions N/A 2. 12.000 psi –5 passes	(a) Chemical stability of LU in decreasing order caprine α_{S1}-II-casein > caprine α_{S1}-I-casein > bovine casein (b) Addition of arabinogalactan, chemical stability of LU against photooxidation ↑	Mora-Gutierrez et al. (2018b)

(continued)

Table 2.6 (continued)

Emulsification technique(s)	Emulsifier(s)[a]	Carrier oil	Emulsification conditions	Main findings	References
1. HSM + 2. HPH	Casein/dextran Maillard conjugates or casein/de xtran physical complexes	MCT oil or GSO	1. 20.000 rpm –2 min 2. 30.000 psi –5 passes	(a) Casein/dextran conjugates → Physical resistance of nanoemulsions ↑, colour stability ↓ (b) Casein/dextran physical complexes → Unstable at isoelectric point of casein (pH 4.6) (c) Resveratrol and GSO → LU degradation and colour fading ↓	Steiner et al. (2019)
1. HSM + 2. HPH	BSA, CA/BSA conjugate, BSA/dextran conjugate, or CA/BSA/dextran conjugate	Corn oil	1. 10.000 rpm –3 min 2. 800 bar –3 passes	(a) CA/BSA/dextran conjugate → Maximum LU bioaccessibility (b) Addition of VE → Physicochemical stability ↕ and LU bioaccessibility ↑	Yan et al. (2020)
1. HSM + 2. HPH	NaCas	MCT oils (MCT oil or coconut oil) or LCT oils (corn, olive or fish oil)	1. 10 Kr/min –2 min 2. 120 MPa –5 passes	MCT oils → overall digestibility and LU bioaccessibility↑ and mixed micelles size ↓	Yuan et al. (2018)
1. HSM + 2. HPH	NaCas	Corn oil	1. 10.000 rpm –2 min 2. 30-110 MPa – 1–12 passes	(a) Homogenization pressure ↑, droplet size ↓ (b) Number of homogenization cycles ↑, droplet size ↓ (c) Homogenization more than 5 cycles → droplet size ↕	Li et al. (2018b)
1. HSM + 2. HPH	NaCas	Different ratios of LCT and MCT oils	1. 10.000 rpm –2 min 2. 12.000 psi –5 passes	(a) MCT oil ratio ↑, LU bioaccessibility ↑ (b) Carotenoid bioaccessibility in LCT/MCT mixed oils before homogenization > Carotenoid bioaccessibility in LCT/MCT mixed oils after homogenization	Yao et al. (2019)

1. HSM + 2. MF	NaCas or NaCas/dextran Maillard conjugates	Corn oil	1. 10.000 rpm –2 min 2. 20.000 psi –3 passes	(a) Casein-stabilized emulsions → Flocculation at isoelectric point of casein (pH 4–5) (b) Casein/dextran conjugates-stabilized emulsions → stable between pH 3 and 7	Gumus et al. (2016)
HSM + 2. MF	NaCas	Corn oil	1. 10.000 rpm –2 min2. 12.000 psi –3 passes	(a) Droplet aggregation → pH 4–5 (b) pH → Physical stability of nanoemulsion (c) Storage temperature ↑, LU degradation and colour fading ↑	Davidov-Pardo et al. (2016)
1. HSM + 2. MF	WPI, WPI/flaxseed gum or WPI/flaxseed gum/chitosan	MCT oil	1. 10.000 rpm –3 min2. 50 MPa–3 passes	(a) Chitosan amount less than 1 wt% → Decreased stability of multilayer emulsion (b) Chemical stability → WPI/flaxseed gum/chitosan > WPI > WPI/flaxseed gum (c) EDTA and α-T addition → Lutein loss ↓	Xu et al. (2016)
1. HSM + 2. MF	QS, Tween 80, WPI or NaCas	Corn oil	1. 10.000 rpm –2 min2. 12.000 psi –3 passes	QS-stabilized nanoemulsions → Best stability against droplet aggregation, creaming and colour fading at 45 °C for 10 days	Weigel et al. (2018)
1. HSM + 2. MF	WPI	MCT oil	1. 19.000 rpm –3 min2. 50 MPa–3 passes	(a) Rosemary extract at a ratio of 0.05 wt% → the best LU protection (b) Addition of rosemary extract in aqueous phase was more effective than addition in oil phase.	Xu et al. (2020)

[a]Abbreviations for emulsifiers: *β-Lg* beta-lactoglobulin, *BSA* bovine serum albumin, *CA* chlorogenic acid, *CMC* carboxymethyl cellulose, *DTAB* dodecyltrimethylammonium bromide, *EGCG: EGCG* (−)-epigallocatechin-3-gallate, *HMP* high methyl pectin, *LMP* low methyl pectin, *OSQS* octeny lsuccinate quinoa starch, *PHC* phosphatidylcholine, *PGA* propylene glycol alginate, *PWP* polymerised WPI, *QS* Quillaja saponin, *WPH* whey protein hydrolysate

MD and WPI), some researchers have examined some new surfactant alternatives. For example, polar lipid fraction of oat was tested as emulsifier in HSM-treated LU emulsions. Emulsions were generated quite simply, but they were prone to rapid creaming even after a couple of days of storage at ambient conditions (Kaimainen et al. 2012). Recently, using xanthan- and propylene glycol alginate-derived carbohydrate-based emulsifiers highly stable LU-loaded emulsions were obtained by HSM. The authors suggested these emulsifiers as alternatives to protein-based ones, which their emulsification ability is greatly influenced by environmental conditions (Wang et al. 2020). Emulsions with fine droplets can be fabricated using only ultrasound technology. The coarse LU emulsions containing β-Lg, whey protein hydrolysate (WPH), Tween 20 or combinations of each of these surfactants with the lipid soluble emulsifier lecithin were subjected to USH to produce fine emulsions. As a result, nano-sized droplets ranging from 0.26 μm (in the case of WPH/lecithin) to 0.61 μm (in the case of Tween 20) were obtained. Furthermore, WPH/lecithin-stabilized emulsions exhibited the best LU protection against oxidation at 4 °C under dark conditions for 46 days (Frede et al. 2014). However, in another study using USH technique, stable LU nanoemulsions stabilized only by WPI as surfactant were developed. The LU content of the system was reduced by only 4% after 4 weeks storage at 4 °C (Zhao et al. 2018).

Coupling of two homogenization techniques is widely preferred to obtain fine LU-loaded emulsions. In one of first studies, HSM and HPH-treated LU-enriched O/W emulsions were effectively stabilized with phosphatidylglycerol. Based on the HPLC profile of LU, the obtained emulsions protected LU against thermal degradation during 5 min heat treatment at 90 °C (Losso et al. 2005). In a study, a combination of HSM and MF was used to encapsulate LU in corn oil-based, NaCas-stabilized emulsions. The authors stated that the physical stability of the emulsions was considerably influenced by pH (causing droplet aggregation at pH 4 and 5), while the chemical stability of LU was mainly affected by increasing temperature during storage time (Davidov-Pardo et al. 2016). The same combination of homogenization techniques was employed to prepare LU-enriched emulsions, and the effect of different emulsifiers (Tween 80 and quillaja saponin) and hydrophilic antioxidants (ascorbic acid, catechin, EDTA) and lipophilic antioxidants (ascorbic acid palmitate and α-tocopherol) on their physicochemical stability was evaluated. Due to great steric and electrostatic repulsion between saponin and oil droplets at neutral pH, quillaja saponin provided the most physically stable LU-enriched emulsions examined during a 10-day storage at 45 °C. On the other side, only ascorbic acid behaved as an antioxidant and significantly decreased the colour loss of quillaja saponin-stabilized LU emulsions. Furthermore, ineffectiveness of the metal chelator type antioxidant (EDTA) indicated that degradation of emulsified LU was not induced by cationic transition metals. Consequently, utilization of quillaja saponin as emulsifier and ascorbic acid as antioxidant in LU-enriched emulsions was proposed by the authors (Weigel et al. 2018). Contradictorily, oxidative stability of LU in emulsions stabilized with WPI-flaxseed gum (FG)-chitosan improved when EDTA or α-tocopherol was incorporated to the formulation (Xu et al. 2016). Recently, efficiency of rosemary extract (at optimal concentration of 0.05 wt%) for improvement of antioxidant capacity in WPI-stabilized LU emulsions was proven by the same authors (Xu et al. 2020).

An emulsified LU delivery system can be based on the solubility of LU in lipid media and its stabilization by a surfactant mixture of milk proteins (the caseins) and soybean lecithin. It was reported that depending on the type of casein that combines with lecithin, chemical stability of corn O/W emulsions containing LU can be improved. The combination of these surfactants forms a denser interfacial layer surrounding oil droplets and slows down the LU degradation (Mora-Gutierrez et al. 2018a). Higher binding affinity of some casein types to LU was contributed to the increased chemical stability of some emulsions. Furthermore, the same authors reported that addition of arabinogalactan to the emulsions improved the chemical stability of LU-casein complexes during storage under accelerated photo-oxidation conditions at 25 °C (Mora-Gutierrez et al. 2018b). Similarly, the efficiency of pectin addition to WPI-stabilized LU emulsions treated with the combination of HSM and USH techniques was determined. Results exhibited that the addition of second layer of high methyl or low methyl pectin enhanced the stability of the emulsion against the environmental stresses, which was due to the formation of steric barrier on the droplets (Muhoza et al. 2016a).

In addition to surfactants mixture, various conjugates have been formed to stabilize LU emulsions. For instance, MCT oil and grape seed oil (GSO)-based nanoemulsions containing LU were stabilized with Maillard conjugates or physical complexes of caseinate/dextran. As the main findings, MCT oil yielded smaller droplets than those prepared with GSO due to the lower viscosity of MCT compared with GSO. Maillard conjugates of caseinate/dextran (having stronger steric repulsion) produced finer nanoemulsions with improved pH- and salt-stability. Furthermore, incorporation of resveratrol into the interfacial layer around the lutein-enriched oil droplets, improved the colour stability of LU nanoemulsions as a result of its potent antioxidant activity. Similarly, GSO enhanced the colour and chemical stability of LU nanoemulsions owing to its unique composition rich in natural antioxidant compounds such as tocopherols, gallic acid, catechins, and proanthocyanidins (Steiner et al. 2019). Likewise, ternary conjugates of BSA/dextran with/without chlorogenic acid (CA) increased the physicochemical stability of high-pressure homogenized LU emulsions (Yan et al. 2020). The physicochemical stability of LU emulsions is also influenced by polar characteristics of LU, which induces its flocculation and increase viscosity in the emulsions stored at 30 °C for 3 months (Domian and Szczepaniak 2020). The findings of all aforementioned studies indicate that in food industry determining the optimum emulsion composition for a better functional quality of any food-grade emulsion is necessary.

Homogenization process parameters should be taken into consideration as well. Droplet size of any emulsion is the main index of homogenization conditions. In a study, droplet size of LU emulsions decreased from $\approx$ 643 nm to $\approx$ 221 nm as a result of higher magnitude of turbulent flow and shear forces applied on the droplets when homogenization pressure increased from 30 to 80 MPa. However, for pressures above 80 MPa, further reduction in droplet size was impeded due to the emulsifier content in the system. As the number of cycles increased from 1 to 5, mean droplet diameter decreased from ≈ 270 nm to ≈ 227 nm. Nonetheless, application of additional cycles (>5) did not produce smaller droplets because all droplets were already exposed uniformly to destructive forces produced by HPH (Li et al. 2018b).

Emulsion-based systems generally serve as convenient vehicles for enhanced bioaccessibility and bioavailability of lipid-soluble antioxidants such as LU (Vishwanathan et al. 2009; Kamil et al. 2016). However, bioaccessibility and bioavailability of emulsified LU depend on numerous factors. The importance of particle size on cellular uptake of emulsified LU was highlighted in a study, in which higher cellular uptake was observed in nano emulsions (with d ≈ 69 nm) than in conventional LU emulsions (with d ≈ 147 nm). The increased cellular uptake of LU from emulsions with smaller droplets has been attributed to (a) the passive transportation of nanoparticles through epithelium layer, (b) ability of smaller particles to penetrate better through cell membrane or (c) greater surface area to volume ratio of these particles (McClements and Xiao 2012; Teo et al. 2017). Thickness of interfacial layer, which depends on the type of surfactant used, also inversely affects the release rate of emulsified LU. A research revealed that WPI hinders the release of emulsified LU by forming more compact and denser interfacial membranes due to its globular structure, whereas dodecyltrimethylammonium bromide (DTAB) enhanced the release of LU by forming a thinner interfacial layer and solubilizing LU with its micelles in aqueous phase (Beicht et al. 2013). In addition to common surfactants, carbohydrates/proteins from unconventional sources or fabricated conjugates may serve as promising surfactants to obtain emulsions with improved LU bioaccessibility. For example, replacement of NaCas by Maillard conjugates of casein/dextran resulted in physically stable emulsions having LU bioaccessibility about 7.55% comparable to physically instable NaCas-stabilized ones (8.20%) (Gumus et al. 2016). Furthermore, corn fibre gums produced emulsions with excellent LU bioaccessibility (24.90 and 32.40%) similar to that of gum arabic-stabilized emulsion (27.70%) (Feng et al. 2017). Composition of carrier oil is another fundamental factor that should be considered to control the bioaccessibility of LU. For instance, it was found that LU and zeaxanthin in the meal containing butter had higher bioaccessibility and were better adsorbed in the rat plasma than that in the meal containing olive oil or fish oil. The authors reported that the size of mixed micelles originated in small intestine throughout in vivo digestion of lipids was significantly lower for saturated fatty acids-rich oils than for unsaturated fatty acids-rich oils (Gleize et al. 2013). Formation of small-sized mixed micelles (having higher specific surface area) promotes the release rate of emulsified LU. Moreover, smaller molecular size and more hydrophilic character of LU also boost its solubility and bioaccessibility when small-sized mixed micelles are formed (Yuan et al. 2018; Yao et al. 2019).

2.3.2.4 Astaxanthin (ASX)

ASX, a red-orange coloured carotenoid, is a member of xanthophllys group along with LU, zeaxanthin, and β-cryptoxanthin (Hussein et al. 2006; Thomas and Johnson 2018). Thanks to presence of oxygen in the form of both hydroxyl and carbonyl groups in chemical structure of astaxanthin, this xanthophyll has higher polarity and antioxidant activity compared to other xanthophylls (Higuera-Ciapara et al. 2006; Seabra and Pedrosa 2010). The antioxidant activity of ASX is 10 times higher than those of zeaxanthin, LU, canthaxanthin, and βC, whereas 100 times higher than that of α-tocopherol (Miki 1991). It has diverse health benefits such as anti-cancer and anti-inflammatory activities, prevention of cardiovascular diseases, diabetes, neurological disorders and skin aging, improvement of ocular and skeletal muscle health and enhancement of immune system as a result of its superior antioxidant capacity (Hussein et al. 2006; Higuera-Ciapara et al. 2006; Davinelli et al. 2018). Salmon, trout and marine invertebrates including krill, lobster, and shrimp are rich in ASX (Olaizola 2007). This carotenoid is responsible for the distinctive reddish colour of these marine animals (Hussein et al. 2006). However, double bonds found in the structure of ASX make it sensitive to oxidation reactions induced by particularly heat and light. Moreover, limited bioavailability of ASX due to its lipophilic nature may be improved by incorporation into lipid based formulations (Ambati et al. 2014).

High-Energy Emulsification Approaches to Encapsulate ASX

Combination of HSM with another high-energy emulsification technique (HPH, MF, etc.) has been widely utilized to form ASX-loaded emulsions. Table 2.7 summarizes the most relevant studies which aimed to encapsulate ASX in emulsion-based systems through high-energy emulsification approaches.

HPH was employed individually to form ASX-loaded O/W emulsions using NaCas or modified lecithin (ML). The utilization of NaCas and ML to improve chemical stability and bioaccessibility of ASX, respectively in O/W emulsions was recommended. ML-stabilized ASX emulsions were stable for 30 min against a wide range of pH (pH 2, 4, 6 and 8) due to negative charge of this surfactant coming from its anionic phosphate group. However, lower physical stability of NaCas-stabilized emulsions was attributed to their positive charges at pH levels near the isoelectric point of NaCas that reduced the zeta-potential of emulsions. Finally, the authors related the lower ASX bioaccessibility of NaCas-stabilized emulsions than that of ML-stabilized emulsions to the following reasons: (1) less FFA release from emulsions containing NaCas due to aggregation of oil droplets as a result of calcium bridging that supresses pancreatin penetration into the oil droplets; (2) formation of less mixed micelles from NaCas-stabilized emulsions owing to the inhibition of FFA release; (3) lower digestion of oil fraction in NaCas-stabilized emulsions, thus, more ASX retention in this undigested oil; (4) aggregation and precipitation of mixed micelles resulting from a possible interaction between caseinate and ASX during digestion process (Khalid et al. 2017).

Table 2.7 An overview of the most relevant studies on the encapsulation of ASX in emulsion-based systems via high-energy emulsification methods

Emulsification technique(s)	Emulsifier(s)[a]	Carrier oil	Emulsification conditions	Main findings	References
1. HSM + 2. ME	WPI and Tween 20	Palm oil (MCT)	1. 11.000 rpm–15 min 2. Pore size 0.8 µm – 5 to 15 bar –3 passes	(a) Pressure ↑ and dispersed phase fraction ↓, Droplet size ↓ (b) Loss of ASX after a 3-week storage → 30%	Ribeiro et al. (2005)
1. HSM + 2. USH	Tween 20, WPI, PWP, lecithin, WPI/ lecithin or PWP/ lecithin	MCT oil	1. 12.000 rpm –2 min 2. 40% amplitude –5 min	(a) WPI → maximum physical stability at 25 °C for 14 days and cellular uptake of ASX (b) WPI- and PWP-stabilized emulsions → ASX loss ↓	Shen et al. (2019)
1. HSM + 2. HPH	Tween 80 and lecithin	Pure palm olein	1. 3000, 6000 or 9000 rpm –5, 10, 15 or 20 min 2. 400, 600 or 800 bar – 1, 3, 5, 6 or 7 passes	(a) Rotation speed ↑, droplet size ↓ and PDI ↑ (b) Homogenization time ↑, droplet size ↓ and PDI ↑ (c) Homogenization pressure ↑, droplet size ↓ and PDI ↓ (d) Number of homogenization cycles ↑, droplet size ↓	Affandi et al. (2011)
1. HSM + 2. HPH	Tween 80 and lecithin	Pure palm olein	1. 9000 rpm–5 min (conventional emulsion) 2. 800 bar – 5 passes (nanoemulsion)	(a) Droplet size ↓, ASX bioavailability ↑ (b) ASX bioavailability: from nanoemulsion > from conventional emulsion	Affandi et al. (2012)
1. HSM + 2. HPH	Tween 20	Cold pressed linseed oil	1. 5000 rpm–10 min 2. 5, 10, 15, 30, 70 or 100 MPa – 1–10 passes	(a) Homogenization at 100 MPa → maximum ASX bioaccessibility (b) Homogenization pressure ↑, droplet size ↓	Sotomayor-Gerding et al. (2016)
1. HSM + 2. HPH	NaCas or ML	Soybean oil	1. 10.000 rpm –5 min 2. 100 MPa – 4 passes	(a) NaCas → Physicochemically stable for 30 days (b) ASX bioaccessibility → ML > NaCas	Khalid et al. (2017)

1. HSM + 2. HPH	Ginseng saponin	Soybean oil	1. 8000 rpm – 5 min 2. 20, 40, 60, 80 or 100 MPa–4 passes	(a) Emulsifier content ↑, droplet size ↓ (b) Homogenization pressure ↑, droplet size ↓ (c) Ginseng saponin → Instability at acidic conditions	Shu et al. (2018)
1. HSM + 2. HPH	Gypenosides or Tween 20	Refined soybean oil	1. 8000 rpm –5 min 2. 100 MPa – 4 passes	(a) ASX degradation after a 30-day storage at 5 or 25 °C → Tween 20 > Gypenosides (b) ASX bioaccessibility in: Tween 20 > in gypenosides	Chen et al. (2018)
1. HSM + 2. MF	NaCas	Corn oil	1. 10.000 rpm –2 min 2. 12.000 psi –5 passes	(a) pH ~ isoelectric point of NaCas → Droplet aggregation (b) Storage temperature ↑, chemical degradation of LU ↑ (c) Physically stable emulsions up to 70 °C	Liu et al. (2016b)
1. HSM + 2. MF	QS, QS/chitosan or QS/chitosan/pectin	Flaxseed oil	1. 10.000 rpm –2 min 2. 12.000 psi –3 passes	Multilayer emulsions → Aggregation stability of droplets and chemical stability of ASX ↑	Liu et al. (2019)

[a]Abbreviations for emulsifiers: *ML* modified lecithin, *NaCas* sodium caseinate, *QS* Quillaja saponin, *PWP* polymerized WPI, *SDS* sodium dodecyl sulphate, *WPI* whey protein isolate

In another report, stable ASX emulsions were produced by the combination of ME, a low-energy technique, with HSM a high-energy approach. To obtain more homogeneous droplets with a narrower size distribution, each O/W emulsion was passed three times through the membrane under pressures and disperse phase content of 5–15 bar and 10 wt% to 40 wt%, respectively. However, encapsulation was not enough to restrict the oxidative degradation of ASX which was almost 30% after 3 weeks (Ribeiro et al. 2005). The same research group in another study have shown that selection of surfactant(s) could bring about the expected properties from emulsification process such as increased bioavailability or extended shelf-life of emulsified molecule. Based on their results, a combination of hydrolysed WPI and Tween 20 improved the chemical stability of ASX and mixture of WPI and sucrose laurate enhanced the cellular uptake of ASX in MF-homogenized emulsions. The efficiency of α-tocopherol as a natural antioxidant for protection of ASX was demonstrated by them as well (Ribeiro et al. 2006). Like VE, EDTA and ascorbic acid have shown effective antioxidant activities to retard fluorescent light-induced oxidation of esterified ASX in the emulsions (Zhou et al. 2015). Another related study revealed that the highest ASX cellular uptake (10%) was provided by WPI, followed by polymerised whey protein (8.49%), WPI/lecithin (5.97%), polymerised whey protein/lecithin (5.05%), lecithin (3.37%) and Tween 20 (2.10%). Storing ASX-loaded emulsions at 4 °C was suggested by the authors for maximum ASX retention (Shen et al. 2019). Furthermore, some studies have shown that natural emulsifiers such as ginseng saponins and gypenosides could stabilize ASX-loaded O/W nanoemulsions produced via combination of HSM and HPH techniques. Ginseng saponins-stabilized ASX nanoemulsions were stable against thermal treatment (90 °C–30 min), while coalescence occurred under storage conditions with acidic (pH 3–6) medium and high salt content (>25 mM NaCl). On the other side, gypenosides yielded heat-resistant (120 °C– 60 min) ASX nanoemulsions, but acidic conditions and $CaCl_2$ addition (25–100 mM) resulted in physical instability of the emulsions (Shu et al. 2018; Chen et al. 2018).

As stated in previous sections, carrier oil is as important as emulsifier(s) in determining the bioavailability and bioaccessibility levels of an emulsion system. Higher ASX absorption was detected in olive oil-based emulsions than in corn oil-based emulsions. The authors explained the possible mechanisms causing different absorption rates of ASX as follow: (1) promotion of carotenoid oxidation in intestinal chyme due to higher polyunsaturated fatty acid (PUFA) contents and hence, lower carotenoid retention in emulsion to be absorbed through further steps of digestion system; (2) formation of larger mixed micelles from PUFA that penetrate slower through unstirred water layer adjacent to the enterocyte; (3) different phytosterol and stanol content of two oils that might interfere with the absorption of carotenoid (Clark et al. 2000).

Type and conditions of homogenization process have a direct influence on the physical properties of the resulting emulsion (e.g. droplet size). Higher ASX bioavailability was recorded in nanoemulsions than conventional emulsions due to higher surface areas of smaller oil droplets which improve their solubility and digestibility (Affandi et al. 2012). In a study, coarse ASX emulsions were formed by

HSM and then they converted into fine emulsions using HPH. Larger droplets (in the range 8.34–22.90 μm) were obtained in HSM-treated emulsions than in HPH-treated ones (in the range 122.70–724.7 nm). Increase of both rotation speed (up to 9000 rpm) and homogenization time (up to 20 min) contributed to the formation of finer droplets. However, this high energy input increased PDI values, as sufficient energy necessary to achieve the droplet size reduction process could not be provided only by HSM technique. Furthermore, pre-homogenization time in HSM was fixed at 5 min since heat generated during mechanical homogenization may impair the stability of ASX. In case of HPH-treated samples, as the homogenization pressure increased (up to 800 bar), droplet size and PDI values of ASX emulsions decreased due to higher shear and turbulence forces that produce smaller and homogenous droplets. Increasing the homogenization cycles up to certain extent (5 cycles) caused droplet size to be decreased as well (Affandi et al. 2011).

2.4 Conclusion

Emulsion-based systems have been widely used in order to enable the incorporation of lipid soluble molecules into food systems, to retard the oxidation reactions and to increase bioavailability and bioaccessibility in human body. These systems can be fabricated through low-energy and/or high-energy emulsification techniques. Among high-energy techniques, HSM is the most common one due to its simplicity and requirement for inexpensive, widely used and flexible equipment. Secondly, HPH and MF are commonly utilized for production of stable both conventional and nanoemulsions. However, USH technique is less frequently used in encapsulation of lipid-soluble antioxidants through emulsion-based systems.

Encapsulation of antioxidants in emulsion-based systems through high-energy methods is influenced by numerous factors that were explained in this chapter. Emulsification technique, types of emulsifier and/or carrier oil, and environmental conditions are the leading elements for emulsion's attributes. Similarly, properties of interfacial layer in emulsions can be easily manipulated by emulsifiers. Selection of suitable emulsification technique, emulsifier, and carrier oil should be certainly based on expectations from obtained emulsion. Improve the stability of lipid-soluble active molecules, incorporate these biologically active molecules with low water solubility in different food systems or increase the bioavailability/bioaccessibility are the underlying reasons for production of emulsions. Considering all these expectations, the most suitable process conditions and emulsion formulation should be determined. Moreover, it should be kept in mind that environmental conditions should be optimised to minimise degradation of emulsified compounds throughout storage period. Further studies are needed to find novel natural emulsifiers and cost-effective emulsification techniques that may yield emulsions with higher physico-chemical stability under extreme environmental conditions and improved bioaccessibility of emulsified compound.

References

Abbas S, Hayat K, Karangwa E et al (2013) An overview of ultrasound-assisted food-grade nanoemulsions. Food Eng Rev 5:139–157. https://doi.org/10.1007/s12393-013-9066-3

Affandi MMRMM, Julianto T, Majeed ABA (2011) Development and stability evaluation of Astaxanthin nanoemulsion. Asian J Pharm Clin Res 4:143–148

Affandi MMRMM, Julianto T, Majeed ABA (2012) Enhanced oral bioavailability of astaxanthin with droplet size reduction. Food Sci Technol Res 18:549–554. https://doi.org/10.3136/fstr.18.549

Akinosho HO, Wicker L (2015) Stability of β-carotene loaded emulsions vary by viscosity of hydroxypropyl methylcellulose dispersions. LWT - Food Sci Technol 63:582–589. https://doi.org/10.1016/j.lwt.2015.02.024

Álvarez-Henao MV, Saavedra N, Medina S et al (2018) Microencapsulation of lutein by spray-drying: characterization and stability analyses to promote its use as a functional ingredient. Food Chem 256:181–187. https://doi.org/10.1016/j.foodchem.2018.02.059

Ambati R, Phang S-M, Ravi S, Aswathanarayana R (2014) Astaxanthin: sources, extraction, stability, biological activities and its commercial applications—a review. Mar Drugs 12:128–152. https://doi.org/10.3390/md12010128

Anarjan N, Jouyban A (2017) Preparation of lycopene nanodispersions from tomato processing waste: effects of organic phase composition. Food Bioprod Process 103:104–113. https://doi.org/10.1016/j.fbp.2017.03.003

Assadpour E, Jafari SM (2019) An overview of specialized equipment for nanoencapsulation of food ingredients. In: Jafari SM (ed) Nanoencapsulation of food ingredients by specialized equipment. Academic, London, UK, pp 1–30

Ax K, Mayer-Miebach E, Link B et al (2003) Stability of lycopene in oil-in-water emulsions. Eng Life Sci 3:199–201. https://doi.org/10.1002/elsc.200390028

Ball GFM (2004) Vitamins: their role in the human body. Blackwell Publishing Ltd, Oxford, UK

Barbosa-Filho JM, Alencar AA, Nunes XP et al (2008) Sources of alpha-, beta-, gamma-, delta- and epsilon-carotenes: a twentieth century review. Rev Bras Farmacogn 18:135–154. https://doi.org/10.1590/S0102-695X2008000100023

Bauernfeind JC (1972) Carotenoid vitamin a precursors and analogs in foods and feeds. J Agric Food Chem 20:456–473. https://doi.org/10.1021/jf60181a003

Beicht J, Zeeb B, Gibis M et al (2013) Influence of layer thickness and composition of cross-linked multilayered oil-in-water emulsions on the release behavior of lutein. Food Funct 4:1457. https://doi.org/10.1039/c3fo60220f

Belitz H-DH-D, Grosch W, Schieberle P (2009) Food chemistry, 4th edn. Springer, Berlin Heidelberg, Berlin, Heidelberg

Binder HJ, Reuben A (2009) Nutrient digestion and absorption. In: Boron WF, Boulpaep EL (eds) Medical physiology: a cellular and molecular approach, 2nd edn. Saunders, Philadelphia, pp 949–979

Birringer M, Blumberg JB, Eggersdorfer M et al (2019) History of vitamin E research. In: Weber P, Birringer M, Blumberg JB et al (eds) Vitamin E in human health. Humana Press, Cham, pp 7–18

Bohn T (2008) Bioavailability of non-provitamin A carotenoids. Curr Nutr Food Sci 4:240–258. https://doi.org/10.2174/157340108786263685

Boon CS, Xu Z, Yue X et al (2008) Factors affecting lycopene oxidation in oil-in-water emulsions. J Agric Food Chem 56:1408–1414. https://doi.org/10.1021/jf072929+

Boon CS, McClements DJ, Weiss J, Decker EA (2009) Role of iron and hydroperoxides in the degradation of lycopene in oil-in-water emulsions. J Agric Food Chem 57:2993–2998. https://doi.org/10.1021/jf803747j

Boon CS, McClements DJ, Weiss J, Decker EA (2010) Factors influencing the chemical stability of carotenoids in foods. Crit Rev Food Sci Nutr 50:515–532. https://doi.org/10.1080/10408390802565889

Bou R, Boon C, Kweku A et al (2011) Effect of different antioxidants on lycopene degradation in oil-in-water emulsions. Eur J Lipid Sci Technol 113:724–729. https://doi.org/10.1002/ejlt.201000524
Burton G, Ingold K (1984) Beta-carotene: an unusual type of lipid antioxidant. Science (80-) 224:569–573. https://doi.org/10.1126/science.6710156
Canselier JP, Delmas H, Wilhelm AM, Abismaïl B (2002) Ultrasound emulsification—an overview. J Dispers Sci Technol 23:333–349. https://doi.org/10.1080/01932690208984209
Chen H, Zhong Q (2015) Thermal and UV stability of β-carotene dissolved in peppermint oil microemulsified by sunflower lecithin and Tween 20 blend. Food Chem 174:630–636. https://doi.org/10.1016/j.foodchem.2014.11.116
Chen J, Li F, Li Z et al (2017) Encapsulation of carotenoids in emulsion-based delivery systems: enhancement of β-carotene water-dispersibility and chemical stability. Food Hydrocoll 69:49–55. https://doi.org/10.1016/j.foodhyd.2017.01.024
Chen Z, Shu G, Taarji N et al (2018) Gypenosides as natural emulsifiers for oil-in-water nanoemulsions loaded with astaxanthin: insights of formulation, stability and release properties. Food Chem 261:322–328. https://doi.org/10.1016/j.foodchem.2018.04.054
Clark RM, Yao L, She L, Furr HC (2000) A comparison of lycopene and astaxanthin absorption from corn oil and olive oil emulsions. Lipids 35:803–806. https://doi.org/10.1007/s11745-000-0589-8
Cornacchia L, Roos YH (2011a) State of dispersed lipid carrier and interface composition as determinants of beta-carotene stability in oil-in-water emulsions. J Food Sci 76:C1211–C1218. https://doi.org/10.1111/j.1750-3841.2011.02376.x
Cornacchia L, Roos YH (2011b) Stability of β-carotene in protein-stabilized oil-in-water delivery systems. J Agric Food Chem 59:7013–7020. https://doi.org/10.1021/jf200841k
Costa-Rodrigues J, Pinho O, Monteiro PRR (2018) Can lycopene be considered an effective protection against cardiovascular disease? Food Chem 245:1148–1153. https://doi.org/10.1016/j.foodchem.2017.11.055
Davidov-Pardo G, Gumus CE, McClements DJ (2016) Lutein-enriched emulsion-based delivery systems: influence of pH and temperature on physical and chemical stability. Food Chem 196:821–827. https://doi.org/10.1016/j.foodchem.2015.10.018
Davinelli S, Nielsen M, Scapagnini G (2018) Astaxanthin in skin health, repair, and disease: a comprehensive review. Nutrients 10:522. https://doi.org/10.3390/nu10040522
Desmarchelier C, Borel P (2017) Overview of carotenoid bioavailability determinants: from dietary factors to host genetic variations. Trends Food Sci Technol 69:270–280. https://doi.org/10.1016/j.tifs.2017.03.002
Dickinson E (1994) Emulsions and droplet size control. In: Wedlock DJ (ed) Controlled particle, droplet and bubble formation. Butterworth-Heinemann Ltd, Oxford, UK, pp 191–216
Dima Ş, Dima C, Iordăchescu G (2015) Encapsulation of functional lipophilic food and drug biocomponents. Food Eng Rev 7:417–438. https://doi.org/10.1007/s12393-015-9115-1
Domian E, Szczepaniak M (2020) Rheological behavior of concentrated emulsions containing carotenoids with different polarity. J Food Eng 274. https://doi.org/10.1016/j.jfoodeng.2019.109827
Eitenmiller R, Lee J (2004) Vitamin E - Food chemistry, composition, and analysis. CRC Press, Boca Raton
Fahmi Wan Mohamad WA, McNaughton D, Buckow R, Augustin MA (2017) Stability and partitioning of β-carotene in whey protein emulsions during storage. Food Funct 8:3917–3925. https://doi.org/10.1039/C7FO01012E
Fan Y, Yi J, Zhang Y et al (2017) Physicochemical stability and in vitro bioaccessibility of β-carotene nanoemulsions stabilized with whey protein-dextran conjugates. Food Hydrocoll 63:256–264. https://doi.org/10.1016/j.foodhyd.2016.09.008
Feng H, Li C, Tan CP et al (2017) Physicochemical properties and in vitro bioaccessibility of lutein loaded emulsions stabilized by corn fiber gums. RSC Adv 7:38243–38250. https://doi.org/10.1039/c7ra04943a

Flores-Miranda GA, Valencia del Toro G, Yanez-Fernandez J (2015) Stability evaluation of B-carotene nanoemulsions prepared by homogenization-emulsification process using stearic acid as oil phase. Rev Mex Ing Quím 14:667–680

Fraser P (2004) The biosynthesis and nutritional uses of carotenoids. Prog Lipid Res 43:228–265. https://doi.org/10.1016/j.plipres.2003.10.002

Frede K, Henze A, Khalil M et al (2014) Stability and cellular uptake of lutein-loaded emulsions. J Funct Foods 8:118–127. https://doi.org/10.1016/j.jff.2014.03.011

Gandul-Rojas B, Minguez-Mosquera MI (1996) Chlorophyll and carotenoid composition in virgin olive oils from various Spanish olive varieties. J Sci Food Agric 72:31–39. https://doi.org/10.1002/(SICI)1097-0010(199609)72:1<31::AID-JSFA619>3.0.CO;2-5

Gangurde AB, Ali MT, Pawar JN, Amin PD (2017) Encapsulation of vitamin E acetate to convert oil to powder microcapsule using different starch derivatives. J Pharm Investig 47:559–574. https://doi.org/10.1007/s40005-016-0287-3

Gasa-Falcon A, Odriozola-Serrano I, Oms-Oliu G, Martín-Belloso O (2017) Influence of mandarin fiber addition on physico-chemical properties of nanoemulsions containing β-carotene under simulated gastrointestinal digestion conditions. LWT - Food Sci Technol 84:331–337. https://doi.org/10.1016/j.lwt.2017.05.070

Gasa-Falcon A, Acevedo-Fani A, Oms-Oliu G et al (2020) Development, physical stability and bioaccessibility of β-carotene-enriched tertiary emulsions. J Funct Foods 64:103615. https://doi.org/10.1016/j.jff.2019.103615

Gleize B, Tourniaire F, Depezay L et al (2013) Effect of type of TAG fatty acids on lutein and zeaxanthin bioavailability. Br J Nutr 110:1–10. https://doi.org/10.1017/S0007114512004813

Gómez-Mascaraque LG, López-Rubio A (2016) Protein-based emulsion electrosprayed micro- and submicroparticles for the encapsulation and stabilization of thermosensitive hydrophobic bioactives. J Colloid Interface Sci 465:259–270. https://doi.org/10.1016/j.jcis.2015.11.061

Gómez-Mascaraque LG, Perez-Masiá R, González-Barrio R et al (2017) Potential of microencapsulation through emulsion-electrospraying to improve the bioaccesibility of β-carotene. Food Hydrocoll 73:1–12. https://doi.org/10.1016/j.foodhyd.2017.06.019

Granado F, Olmedilla B, Blanco I (2003) Nutritional and clinical relevance of lutein in human health. Br J Nutr 90:487–502. https://doi.org/10.1079/BJN2003927

Gregory JF III (2017) Vitamins. In: Damodaran S, Parkin KL (eds) Fennema's food chemistry, 5th edn. CRC Press, Boca Raton, pp 543–626

Grolier P, Agoudavi S, Azais-Braesco V (1995) Comparative bioavailability of diet-, oil- and emulsion-based preparations of vitamin a and β-carotene in rat. Nutr Res 15:1507–1516. https://doi.org/10.1016/0271-5317(95)02023-O

Gu L, Su Y, Zhang M et al (2017) Protection of β-carotene from chemical degradation in emulsion-based delivery systems using antioxidant interfacial complexes: catechin-egg white protein conjugates. Food Res Int 96:84–93. https://doi.org/10.1016/j.foodres.2017.03.015

Guan Y, Wu J, Zhong Q (2016) Eugenol improves physical and chemical stabilities of nanoemulsions loaded with β-carotene. Food Chem 194:787–796. https://doi.org/10.1016/j.foodchem.2015.08.097

Gumus CE, Davidov-Pardo G, McClements DJ (2016) Lutein-enriched emulsion-based delivery systems: impact of Maillard conjugation on physicochemical stability and gastrointestinal fate. Food Hydrocoll 60:38–49. https://doi.org/10.1016/j.foodhyd.2016.03.021

Guo H, Huang Y, Qian J et al (2014) Optimization of technological parameters for preparation of lycopene microcapsules. J Food Sci Technol 51:1318–1325. https://doi.org/10.1007/s13197-011-0600-0

Gupta AK, Haberman HF, Pawlowski D et al (1985) Canthaxanthin. Int J Dermatol 24:528–532. https://doi.org/10.1111/j.1365-4362.1985.tb05554.x

Ha TVA, Kim S, Choi Y et al (2015) Antioxidant activity and bioaccessibility of size-different nanoemulsions for lycopene-enriched tomato extract. Food Chem 178:115–121. https://doi.org/10.1016/j.foodchem.2015.01.048

Han JR, Gu LP, Zhang RJ et al (2019) Bioaccessibility and cellular uptake of β-carotene in emulsion-based delivery systems using scallop (Patinopecten yessoensis) gonad protein isolates: effects of carrier oil. Food Funct 10:49–60. https://doi.org/10.1039/c8fo01390j

Han J-R, Shang W-H, Yan J-N et al (2020) Protection of β-carotene from chemical degradation in emulsion-based delivery systems using scallop (Patinopecten yessoensis) gonad protein isolates. Food Bioprocess Technol 13:680–692. https://doi.org/10.1007/s11947-020-02429-6

Hart SM, Lin X (Lois), Thilakarathna SH, Wright AJ (2018) Emulsion droplet crystallinity attenuates early in vitro digestive lipolysis and beta-carotene bioaccessibility. Food Chem 260:145–151. https://doi.org/10.1016/j.foodchem.2018.03.142

Hatanaka J, Chikamori H, Sato H et al (2010) Physicochemical and pharmacological characterization of α-tocopherol-loaded nano-emulsion system. Int J Pharm 396:188–193. https://doi.org/10.1016/j.ijpharm.2010.06.017

Hategekimana J, Masamba KG, Ma J, Zhong F (2015) Encapsulation of vitamin E: effect of physicochemical properties of wall material on retention and stability. Carbohydr Polym 124:172–179. https://doi.org/10.1016/j.carbpol.2015.01.060

Hejri A, Gharanjig K, Khosravi A, Hejazi M (2013) Effect of surfactants on kinetics of β-carotene photodegradation in emulsions. Chem Eng Commun 200:437–447. https://doi.org/10.1080/00986445.2012.712581

Hielscher T (2005) Ultrasonic production of nano-size dispersions and emulsions. In: ENS 2005, Paris, pp 138–143

Higuera-Ciapara I, Félix-Valenzuela L, Goycoolea FM (2006) Astaxanthin: a review of its chemistry and applications. Crit Rev Food Sci Nutr 46:185–196. https://doi.org/10.1080/10408690590957188

Ho KKHY, Schroën K, San Martín-González MF, Berton-Carabin CC (2017) Physicochemical stability of lycopene-loaded emulsions stabilized by plant or dairy proteins. Food Struct 12:34–42. https://doi.org/10.1016/j.foostr.2016.12.001

Hou Z, Gao Y, Yuan F et al (2010) Investigation into the physicochemical stability and rheological properties of β-carotene emulsion stabilized by soybean soluble polysaccharides and chitosan. J Agric Food Chem 58:8604–8611. https://doi.org/10.1021/jf1015686

Hou Z, Zhang M, Liu B et al (2012) Effect of chitosan molecular weight on the stability and rheological properties of β-carotene emulsions stabilized by soybean soluble polysaccharides. Food Hydrocoll 26:205–211. https://doi.org/10.1016/j.foodhyd.2011.05.013

Hou Z, Liu Y, Lei F, Gao Y (2014) Investigation into the in vitro release properties of β-carotene in emulsions stabilized by different emulsifiers. LWT - Food Sci Technol 59:867–873. https://doi.org/10.1016/j.lwt.2014.07.045

Huang E, Quek SY, Fu N et al (2019a) Co-encapsulation of coenzyme Q10 and vitamin E: a study of microcapsule formation and its relation to structure and functionalities using single droplet drying and micro-fluidic-jet spray drying. J Food Eng 247:45–55. https://doi.org/10.1016/j.jfoodeng.2018.11.017

Huang J, Bai F, Wu Y et al (2019b) Development and evaluation of lutein-loaded alginate microspheres with improved stability and antioxidant. J Sci Food Agric 99:5195–5201. https://doi.org/10.1002/jsfa.9766

Hussein G, Sankawa U, Goto H et al (2006) Astaxanthin, a carotenoid with potential in human health and nutrition. J Nat Prod 63:443–449. https://doi.org/10.1021/np050354+

Igielska-Kalwat J (2018) Studies on stability of emulsions containing carotenoids. Acta Biochim Pol 65:455–463. https://doi.org/10.18388/abp.2018_2589

Institute of Medicine (2000) Dietary reference intakes for vitamin C, vitamin E, selenium, and carotenoids. National Academy Press, Washington, D.C.

Institute of Medicine (2006) Dietary reference intakes: the essential guide to nutrient requirements. National Academies Press, Washington, D.C.

Islamian JP, Mehrali H (2015) Lycopene as a carotenoid provides radioprotectant and antioxidant effects by quenching radiation-induced free radical singlet oxygen: an overview. Cell J 16:386–391. https://doi.org/10.22074/cellj.2015.485

Jain A, Thakur D, Ghoshal G et al (2016) Formation and functional attributes of electrostatic complexes involving casein and anionic polysaccharides: an approach to enhance oral absorption of lycopene in rats in vivo. Int J Biol Macromol 93:746–756. https://doi.org/10.1016/j.ijbiomac.2016.08.071

Jain S, Winuprasith T, Suphantharika M (2020) Encapsulation of lycopene in emulsions and hydrogel beads using dual modified rice starch: characterization, stability analysis and release behaviour during in-vitro digestion. Food Hydrocoll 104:105730. https://doi.org/10.1016/j.foodhyd.2020.105730

Johnson EJ (2002) The role of carotenoids in human health. Nutr Clin Care 5:56–65. https://doi.org/10.1046/j.1523-5408.2002.00004.x

Kaimainen M, Ahvenainen S, Kaariste M et al (2012) Polar lipid fraction from oat (Avena sativa): characterization and use as an o/w emulsifier. Eur Food Res Technol 235:507–515. https://doi.org/10.1007/s00217-012-1780-1

Kamil A, Smith DE, Blumberg JB et al (2016) Bioavailability and biodistribution of nanodelivered lutein. Food Chem 192:915–923. https://doi.org/10.1016/j.foodchem.2015.07.106

Kentish S, Ashokkumar M (2011) The physical and chemical effects of ultrasound. In: Feng H, Barbosa-Cánovas GV, Weiss J (eds) Ultrasound technologies for food and bioprocessing. Springer Science+Business Media, New York, pp 1–12

Khalid N, Shu G, Holland BJ et al (2017) Formulation and characterization of O/W nanoemulsions encapsulating high concentration of astaxanthin. Food Res Int 102:364–371. https://doi.org/10.1016/j.foodres.2017.06.019

Kim SO, Ha TVA, Choi YJ, Ko S (2014) Optimization of homogenization-evaporation process for lycopene nanoemulsion production and its beverage applications. J Food Sci 79:N1604–N1610. https://doi.org/10.1111/1750-3841.12472

Kohler K, Schuchmann H (2015) Rotor–stator devices. In: Rayner M, Dejmek P (eds) Engineering aspects of food emulsification and homogenization. CRC Press, Boca Raton, pp 195–206

Komaiko JS, McClements DJ (2016) Formation of food-grade nanoemulsions using low-energy preparation methods: a review of available methods. Compr Rev Food Sci Food Saf 15:331–352. https://doi.org/10.1111/1541-4337.12189

Koroleva MY, Yurtov EV (2012) Nanoemulsions: the properties, methods of preparation and promising applications. Russ Chem Rev 81:21–43. https://doi.org/10.1070/RC2012v081n01ABEH004219

Kreps F, Vrbiková L, Schmidt Š (2014) Influence of industrial physical refining on tocopherol, chlorophyll and beta-carotene content in sunflower and rapeseed oil. Eur J Lipid Sci Technol 116:1572–1582. https://doi.org/10.1002/ejlt.201300460

Kun Y, Ssonko Lule U, Xiao-Lin D (2006) Lycopene: its properties and relationship to human health. Food Rev Int 22:309–333. https://doi.org/10.1080/87559120600864753

Lacatusu I, Mitrea E, Badea N et al (2013) Lipid nanoparticles based on omega-3 fatty acids as effective carriers for lutein delivery. Preparation and in vitro characterization studies. J Funct Foods 5:1260–1269. https://doi.org/10.1016/j.jff.2013.04.010

Leal-Calderon F, Schmitt V, Bibette J (2007) Emulsion science - basic principles, 2nd edn. Springer New York, New York

Lei F, Liu F, Yuan F, Gao Y (2014) Impact of chitosan–EGCG conjugates on physicochemical stability of β-carotene emulsion. Food Hydrocoll 39:163–170. https://doi.org/10.1016/j.foodhyd.2014.01.008

Li D, Li L, Xiao N et al (2018a) Physical properties of oil-in-water nanoemulsions stabilized by OSA-modified starch for the encapsulation of lycopene. Colloids Surf A Physicochem Eng Asp 552:59–66. https://doi.org/10.1016/j.colsurfa.2018.04.055

Li J, Guo R, Hu H et al (2018b) Preparation optimisation and storage stability of nanoemulsion-based lutein delivery systems. J Microencapsul 35:570–583. https://doi.org/10.1080/02652048.2018.1559245

Li H, Ma Y, Yu L et al (2020a) Construction of octenyl succinic anhydride modified porous starch for improving bioaccessibility of β-carotene in emulsions. RSC Adv 10:8480–8489. https://doi.org/10.1039/C9RA10079B

Li S, Zhang B, Li C et al (2020b) Pickering emulsion gel stabilized by octenylsuccinate quinoa starch granule as lutein carrier: role of the gel network. Food Chem 305:125476. https://doi.org/10.1016/j.foodchem.2019.125476

Liaaen-Jensen S (2004) Basic carotenoid chemistry. In: Krinsky NI, Mayne ST, Sies H (eds) Carotenoids in health and disease. Marcel Dekker Inc., New York, pp 1–30

Liang X, Ma C, Yan X et al (2019) Advances in research on bioactivity, metabolism, stability and delivery systems of lycopene. Trends Food Sci Technol 93:185–196. https://doi.org/10.1016/j.tifs.2019.08.019

Liu Y, Hou Z, Lei F et al (2012) Investigation into the bioaccessibility and microstructure changes of β-carotene emulsions during in vitro digestion. Innov Food Sci Emerg Technol 15:86–95. https://doi.org/10.1016/j.ifset.2012.04.002

Liu Y, Hou Z, Yang J, Gao Y (2015) Effects of antioxidants on the stability of β-carotene in O/W emulsions stabilized by gum arabic. J Food Sci Technol 52:3300–3311. https://doi.org/10.1007/s13197-014-1380-0

Liu L, Gao Y, McClements DJ, Decker EA (2016a) Role of continuous phase protein, (-)-epigallocatechin-3-gallate and carrier oil on β-carotene degradation in oil-in-water emulsions. Food Chem 210:242–248. https://doi.org/10.1016/j.foodchem.2016.04.075

Liu X, McClements DJ, Cao Y, Xiao H (2016b) Chemical and physical stability of astaxanthin-enriched emulsion-based delivery systems. Food Biophys 11:302–310. https://doi.org/10.1007/s11483-016-9443-6

Liu F, Ma C, Zhang R et al (2017) Controlling the potential gastrointestinal fate of β-carotene emulsions using interfacial engineering: impact of coating lipid droplets with polyphenol-protein-carbohydrate conjugate. Food Chem 221:395–403. https://doi.org/10.1016/j.foodchem.2016.10.057

Liu C, Tan Y, Xu Y et al (2019) Formation, characterization, and application of chitosan/pectin-stabilized multilayer emulsions as astaxanthin delivery systems. Int J Biol Macromol 140:985–997. https://doi.org/10.1016/j.ijbiomac.2019.08.071

Loh ZH, Samanta AK, Sia Heng PW (2014) Overview of milling techniques for improving the solubility of poorly water-soluble drugs. Asian J Pharm Sci 10:255–274. https://doi.org/10.1016/j.ajps.2014.12.006

Losso JN, Khachatryan A, Ogawa M et al (2005) Random centroid optimization of phosphatidylglycerol stabilized lutein-enriched oil-in-water emulsions at acidic pH. Food Chem 92:737–744. https://doi.org/10.1016/j.foodchem.2004.12.029

Lu W, Kelly A, Miao S (2017) Bioaccessibility and cellular uptake of β-carotene encapsulated in model O/W emulsions: influence of initial droplet size and emulsifiers. Nano 7:282. https://doi.org/10.3390/nano7090282

Luo X, Zhou Y, Bai L et al (2017) Fabrication of β-carotene nanoemulsion-based delivery systems using dual-channel microfluidization: physical and chemical stability. J Colloid Interface Sci 490:328–335. https://doi.org/10.1016/j.jcis.2016.11.057

Lv S, Gu J, Zhang R et al (2018) Vitamin E encapsulation in plant-based nanoemulsions fabricated using dual-channel microfluidization: formation, stability, and bioaccessibility. J Agric Food Chem 66:10532–10542. https://doi.org/10.1021/acs.jafc.8b03077

Lv S, Zhang Y, Tan H et al (2019) Vitamin E encapsulation within oil-in-water emulsions: impact of emulsifier type on physicochemical stability and bioaccessibility. J Agric Food Chem 67:1521–1529. https://doi.org/10.1021/acs.jafc.8b06347

Ma L, Lin X-M (2010) Effects of lutein and zeaxanthin on aspects of eye health. J Sci Food Agric 90:2–12. https://doi.org/10.1002/jsfa.3785

Maa YF, Hsu C (1996) Liquid-liquid emulsification by rotor/stator homogenization. J Control Release 38:219–228. https://doi.org/10.1016/0168-3659(95)00123-9

Maa Y-F, Hsu CC (1999) Performance of sonication and microfluidization for liquid–liquid emulsification. Pharm Dev Technol 4:233–240. https://doi.org/10.1081/PDT-100101357

Maiani G, Periago Castón MJ, Catasta G et al (2009) Carotenoids: actual knowledge on food sources, intakes, stability and bioavailability and their protective role in humans. Mol Nutr Food Res 53:S194–S218. https://doi.org/10.1002/mnfr.200800053

Manorama R, Rukmini C (1992) Crude palm oil as a source of beta-carotene. Nutr Res 12:S223–S232. https://doi.org/10.1016/S0271-5317(05)80466-X

Mao L, Xu D, Yang J et al (2009) Effects of small and large molecule emulsifiers on the characteristics of β-carotene nanoemulsions prepared by high pressure homogenization. Food Technol Biotechnol 47:336–342

Mao L, Yang J, Xu D et al (2010) Effects of homogenization models and emulsifiers on the physicochemical properties of β-carotene nanoemulsions. J Dispers Sci Technol 31:986–993. https://doi.org/10.1080/01932690903224482

Matus Z, Molnár P, Szabó LG (1993) Main carotenoids in pressed seeds (Cucurbitae semen) of oil pumpkin (Cucurbita pepo convar. pepo var. styriaca). Acta Pharm Hung 63:247–256

Mayer S, Weiss J, McClements DJ (2013) Behavior of vitamin E acetate delivery systems under simulated gastrointestinal conditions: lipid digestion and bioaccessibility of low-energy nanoemulsions. J Colloid Interface Sci 404:215–222. https://doi.org/10.1016/j.jcis.2013.04.048

McClements DJ (2015) Nanoparticle- and microparticle-based delivery systems. CRC Press, Boca Raton

McClements DJ (2016) Food emulsions - principles, practices, and techniques, 3rd edn. CRC Press, Boca Raton

McClements DJ, Li Y (2010) Structured emulsion-based delivery systems: controlling the digestion and release of lipophilic food components. Adv Colloid Interf Sci 159:213–228. https://doi.org/10.1016/j.cis.2010.06.010

McClements DJ, Rao J (2011) Food-grade nanoemulsions: formulation, fabrication, properties, performance, biological fate, and potential toxicity. Crit Rev Food Sci Nutr 51:285–330. https://doi.org/10.1080/10408398.2011.559558

McClements DJ, Xiao H (2012) Potential biological fate of ingested nanoemulsions: influence of particle characteristics. Food Funct 3:202–220. https://doi.org/10.1039/C1FO10193E

McClements DJ, Decker EA, Weiss J (2007) Emulsion-based delivery systems for lipophilic bioactive components. J Food Sci 72:R109–R124. https://doi.org/10.1111/j.1750-3841.2007.00507.x

McClements DJ, Saliva-Trujillo L, Zhang R et al (2016) Boosting the bioavailability of hydrophobic nutrients, vitamins, and nutraceuticals in natural products using excipient emulsions. Food Res Int 88:140–152. https://doi.org/10.1016/j.foodres.2015.11.017

Mehmood T (2015) Optimization of the canola oil based vitamin E nanoemulsions stabilized by food grade mixed surfactants using response surface methodology. Food Chem 183:1–7. https://doi.org/10.1016/j.foodchem.2015.03.021

Mehmood T, Ahmed A, Ahmad A et al (2018) Optimization of mixed surfactants-based β-carotene nanoemulsions using response surface methodology: an ultrasonic homogenization approach. Food Chem 253:179–184. https://doi.org/10.1016/j.foodchem.2018.01.136

Meng Q, Long P, Zhou J et al (2019) Improved absorption of β-carotene by encapsulation in an oil-in-water nanoemulsion containing tea polyphenols in the aqueous phase. Food Res Int 116:731–736. https://doi.org/10.1016/j.foodres.2018.09.004

Meroni E, Raikos V (2018) Formulating orange oil-in-water beverage emulsions for effective delivery of bioactives: improvements in chemical stability, antioxidant activity and gastrointestinal fate of lycopene using carrier oils. Food Res Int 106:439–445. https://doi.org/10.1016/j.foodres.2018.01.013

Miki W (1991) Biological functions and activities of animal carotenoids. Pure Appl Chem 63:141–146. https://doi.org/10.1351/pac199163010141

Monde AA, Michel F, Carbonneau M-A et al (2009) Comparative study of fatty acid composition, vitamin E and carotenoid contents of palm oils from four varieties of oil palm from Côte d'Ivoire. J Sci Food Agric 89:2535–2540. https://doi.org/10.1002/jsfa.3740

Monroy-Villagrana A, Alamilla-Beltrán L, Hernández-Sánchez H, Gutiérrez-López GF (2015) Hydrodynamic characterization of the formation of alpha-tocopherol nanoemulsions in a microfluidizer. In: Hernández-Sánchez H, Gutiérrez-López GF (eds) Food nanoscience and nanotechnology. Springer, Cham, pp 163–175

Mora-Gutierrez A, Attaie R, de González MN (2018a) Lutein-enriched emulsion-based delivery system: impact of casein-phospholipid emulsifiers on chemical stability. In: Zepka LQ, Jacob-Lopes E, De Rosso VV (eds) Progress in carotenoid research. InTechOpen, London, UK, pp 161–170

Mora-Gutierrez A, Attaie R, Núñez de González MT et al (2018b) Complexes of lutein with bovine and caprine caseins and their impact on lutein chemical stability in emulsion systems: effect of arabinogalactan. J Dairy Sci 101:18–27. https://doi.org/10.3168/jds.2017-13105

Mozafar A (2018) Plant vitamins: agronomic, physiological, and nutritional aspects. CRC Press, Boca Raton

Muhoza B, Karangwa E, Duhoranimana E et al (2016a) Influence of pectin on the stability of whey protein isolate stabilized emulsion for encapsulating lutein. Adv J Food Sci Technol 12:617–626. https://doi.org/10.19026/ajfst.12.3320

Muhoza B, Karangwa E, Zhang X, Xia S (2016b) Effect of pH, polysaccharide concentrations on properties of lutein-loaded oil whey protein isolate in water based emulsions. J Acad Ind Res 4:230–235

Müller L, Fröhlich K, Böhm V (2011) Comparative antioxidant activities of carotenoids measured by ferric reducing antioxidant power (FRAP), ABTS bleaching assay (αTEAC), DPPH assay and peroxyl radical scavenging assay. Food Chem 129:139–148. https://doi.org/10.1016/j.foodchem.2011.04.045

Murillo AG, Aguilar D, Norris GH et al (2016) Compared with powdered lutein, a lutein nanoemulsion increases plasma and liver lutein, protects against hepatic steatosis, and affects lipoprotein metabolism in Guinea pigs. J Nutr 146:1961–1969. https://doi.org/10.3945/jn.116.235374

Nagao A (2014) Bioavailability of dietary carotenoids: intestinal absorption and metabolism. Japan Agric Res Q 48:385–391. https://doi.org/10.6090/jarq.48.385

Nesterenko A, Alric I, Violleau F et al (2013) A new way of valorizing biomaterials: the use of sunflower protein for α-tocopherol microencapsulation. Food Res Int 53:115–124. https://doi.org/10.1016/j.foodres.2013.04.020

Nhung DTT, Bung PN, Ha NT, Phong TK (2010) Changes in lycopene and beta carotene contents in aril and oil of gac fruit during storage. Food Chem 121:326–331. https://doi.org/10.1016/j.foodchem.2009.12.032

Nik AM, Corredig M, Wright AJ (2011) Release of lipophilic molecules during in vitro digestion of soy protein-stabilized emulsions. Mol Nutr Food Res 55:S278–S289. https://doi.org/10.1002/mnfr.201000572

Nwachukwu ID, Udenigwe CC, Aluko RE (2016) Lutein and zeaxanthin: production technology, bioavailability, mechanisms of action, visual function, and health claim status. Trends Food Sci Technol 49:74–84. https://doi.org/10.1016/j.tifs.2015.12.005

Odriozola-Serrano I, Oms-Oliu G, Martín-Belloso O (2014) Nanoemulsion-based delivery systems to improve functionality of lipophilic components. Front Nutr 1:1–4. https://doi.org/10.3389/fnut.2014.00024

Okonogi S, Riangjanapatee P (2015) Physicochemical characterization of lycopene-loaded nanostructured lipid carrier formulations for topical administration. Int J Pharm 478:726–735. https://doi.org/10.1016/j.ijpharm.2014.12.002

Olaizola M (2007) The production and health benefits of astaxanthin. In: Barrow C, Shahidi F (eds) Marine nutraceuticals and functional foods. CRC Press, Boca Raton, pp 321–344

Olson DW, White CH, Richter RL (2004) Effect of pressure and fat content on particle sizes in microfluidized milk. J Dairy Sci 87:3217–3223. https://doi.org/10.3168/jds.S0022-0302(04)73457-8

Ong ASH, Packer L (1992) Lipid-soluble antioxidants: biochemistry and clinical applications. Birkhäuser Basel, Basel

Ozturk B, Argin S, Ozilgen M, McClements DJ (2014) Formation and stabilization of nanoemulsion-based vitamin E delivery systems using natural surfactants: quillaja saponin and lecithin. J Food Eng 142:57–63. https://doi.org/10.1016/j.jfoodeng.2014.06.015

Pacek AW, Hall S, Cooke M, Kowalski AJ (2013) Emulsification in rotor–stator mixers. In: Tadros TF (ed) Emulsion formation and stability. Wiley-VCH Verlag GmbH & Co. KGaA, Weinheim, pp 127–167

Packer L, Obermüller-Jevic UC (2002) Vitamin E: an introduction. In: Packer L, Traber MG, Kraemer K, Frei B (eds) Antioxidants: vitamins C and E for health. AOCS Press, Champaign, pp 133–151

Parry J, Su L, Luther M et al (2005) Fatty acid composition and antioxidant properties of cold-pressed marionberry, boysenberry, red raspberry, and blueberry seed oils. J Agric Food Chem 53:566–573. https://doi.org/10.1021/jf048615t

Parthasarathi S, Muthukumar SP, Anandharamakrishnan C (2016) The influence of droplet size on the stability,: in vivo digestion, and oral bioavailability of Vitamin E emulsions. Food Funct 7:2294–2302. https://doi.org/10.1039/c5fo01517k

Pattee HE, Purcell AE (1967) Carotenoid pigments of peanut oil. J Am Oil Chem Soc 44:328–330. https://doi.org/10.1007/BF02635627

Pénicaud C, Achir N, Dhuique-Mayer C et al (2011) Degradation of β-carotene during fruit and vegetable processing or storage: reaction mechanisms and kinetic aspects: a review. Fruits 66:417–440. https://doi.org/10.1051/fruits/2011058

Perry A, Rasmussen H, Johnson EJ (2009) Xanthophyll (lutein, zeaxanthin) content in fruits, vegetables and corn and egg products. J Food Compos Anal 22:9–15. https://doi.org/10.1016/j.jfca.2008.07.006

Ping BTY, Lian GEC (2005) Spectroscopic identification of geometrical isomers of A - and B -carotenes from palm oil. J Oil Palm Res 30:92–102

Qian C, Decker EA, Xiao H, McClements DJ (2012a) Inhibition of β-carotene degradation in oil-in-water nanoemulsions: influence of oil-soluble and water-soluble antioxidants. Food Chem 135:1036–1043. https://doi.org/10.1016/j.foodchem.2012.05.085

Qian C, Decker EA, Xiao H, McClements DJ (2012b) Physical and chemical stability of β-carotene-enriched nanoemulsions: influence of pH, ionic strength, temperature, and emulsifier type. Food Chem 132:1221–1229. https://doi.org/10.1016/j.foodchem.2011.11.091

Qian C, Decker EA, Xiao H, McClements DJ (2012c) Nanoemulsion delivery systems: influence of carrier oil on β-carotene bioaccessibility. Food Chem 135:1440–1447. https://doi.org/10.1016/j.foodchem.2012.06.047

Qian C, Decker EA, Xiao H, McClements DJ (2013) Impact of lipid nanoparticle physical state on particle aggregation and β-carotene degradation: potential limitations of solid lipid nanoparticles. Food Res Int 52:342–349. https://doi.org/10.1016/j.foodres.2013.03.035

Qiu D, Yang L, Shi Y-C (2015) Formation of vitamin E emulsion stabilized by octenylsuccinic starch: factors affecting particle size and oil load. J Food Sci 80:C680–C686. https://doi.org/10.1111/1750-3841.12841

Raikos V (2017) Encapsulation of vitamin E in edible orange oil-in-water emulsion beverages: influence of heating temperature on physicochemical stability during chilled storage. Food Hydrocoll 72:155–162. https://doi.org/10.1016/j.foodhyd.2017.05.027

Raikos V, Hayward N, Hayes H et al (2019) Optimising the ratio of long- to short-chain triglycerides of the lipid phase to enhance physical stability and bioaccessibility of lycopene-loaded beverage emulsions. Int J Food Sci Technol 54:1355–1362. https://doi.org/10.1111/ijfs.14024

Ranveer RC (2018) Lycopene: a natural red pigment. In: Grumezescu AM, Holban AM (eds) Natural and artificial flavoring agents and food dyes. Academic, London, UK, pp 427–456

Rao J, Decker EA, Xiao H, Mcclements DJ (2013) Nutraceutical nanoemulsions: influence of carrier oil composition (digestible versus indigestible oil) on β-carotene bioavailability. J Sci Food Agric 93:3175–3183. https://doi.org/10.1002/jsfa.6215

Rayner M (2015) Scales and forces in emulsification. In: Rayner M, Dejmek P (eds) Engineering aspects of food emulsification and homogenization. CRC Press, Boca Raton, pp 3–32

Read AE, Harvey RF, Naish JM (1981) The small intestine – malabsorption. In: Read AE, Harvey RF, Naish JM (eds) Basic gastroenterology, 3rd edn. John Wright & Sons Ltd, Bristol, UK, pp 280–310. https://doi.org/10.1016/B978-0-7236-0551-5.50020-2

Relkin P, Jung JM, Ollivon M (2009) Factors affecting vitamin degradation in oil-in-water nano-emulsions. J Therm Anal Calorim 98:13–18. https://doi.org/10.1007/s10973-009-0340-9

Ribeiro HS, Ax K, Schubert H (2003) Stability of lycopene emulsions in food systems. J Food Sci 68:2730–2734. https://doi.org/10.1111/j.1365-2621.2003.tb05796.x

Ribeiro HS, Rico LG, Badolato GG, Schubert H (2005) Production of O/W emulsions containing astaxanthin by repeated premix membrane emulsification. J Food Sci 70:E117–E123. https://doi.org/10.1111/j.1365-2621.2005.tb07083.x

Ribeiro HS, Guerrero JMM, Briviba K et al (2006) Cellular uptake of carotenoid-loaded oil-in-water emulsions in colon carcinoma cells in vitro. J Agric Food Chem 54:9366–9369. https://doi.org/10.1021/jf062409z

Ribeiro MC, Vilas Boas EV d B, Riul TR et al (2012) Influence of the extraction method and storage time on the physicochemical properties and carotenoid levels of pequi (Caryocar brasiliense Camb.) oil. Food Sci Technol 32:386–392. https://doi.org/10.1590/S0101-20612012005000053

Riedl J, Linseisen J, Hoffmann J, Wolfram G (1999) Some dietary fibers reduce the absorption of carotenoids in women. J Nutr 129:2170–2176. https://doi.org/10.1093/jn/129.12.2170

Rodrigues LM, Alcázar-Alay SC, Petenate AJ, Meireles MAA (2014) Bixin extraction from defatted annatto seeds. C R Chim 17:268–283. https://doi.org/10.1016/j.crci.2013.10.010

Rodriguez-Amaya DB (2016) Food carotenoids: chemistry, biology, and technology. Wiley, Chichester

Rodriguez-Amaya DB (2019) Natural food pigments and colorants. In: Mérillon J-M, Ramawat KG (eds) Bioactive molecules in food. Springer Nature Switzerland AG, Cham, pp 867–901

Rosdi MRH, Ariffin A, Ishak ZAM (2018) Optimizing homogenization parameters for improving ethylene vinyl acetate emulsion stability in pour point depressant application. J King Saud Univ - Eng Sci 30:105–115. https://doi.org/10.1016/j.jksues.2016.01.006

Saini RK, Nile SH, Park SW (2015) Carotenoids from fruits and vegetables: chemistry, analysis, occurrence, bioavailability and biological activities. Food Res Int 76:735–750. https://doi.org/10.1016/j.foodres.2015.07.047

Saini RK, Bekhit AE-D, Roohinejad S et al (2020) Chemical stability of lycopene in processed products: a review of the effects of processing methods and modern preservation strategies. J Agric Food Chem 68:712–726. https://doi.org/10.1021/acs.jafc.9b06669

Sajilata MG, Singhal RS, Kamat MY (2008) The carotenoid pigment zeaxanthin—a review. Compr Rev Food Sci Food Saf 7:29–49. https://doi.org/10.1111/j.1541-4337.2007.00028.x

Salimi A, Maghsoudlou Y, Jafari SM et al (2015) Preparation of lycopene emulsions by whey protein concentrate and maltodextrin and optimization by response surface methodology. J Dispers Sci Technol 36:274–283. https://doi.org/10.1080/01932691.2013.879833

Salvia-Trujillo L, Qian C, Martín-Belloso O, McClements DJ (2013) Modulating β-carotene bioaccessibility by controlling oil composition and concentration in edible nanoemulsions. Food Chem 139:878–884. https://doi.org/10.1016/j.foodchem.2013.02.024

Santana RC, Perrechil FA, Cunha RL (2013) High- and low-energy emulsifications for food applications: a focus on process parameters. Food Eng Rev 5:107–122. https://doi.org/10.1007/s12393-013-9065-4

Saravacos GD, Kostaropoulos AE (2016) Handbook of food processing equipment, 2nd edn. Springer International Publishing, Cham

Schultz S, Wagner G, Urban K, Ulrich J (2004) High-pressure homogenization as a process for emulsion formation. Chem Eng Technol 27:361–368. https://doi.org/10.1002/ceat.200406111

Schwartz SJ, Cooperstone JL, Cichon MJ et al (2017) Colorants. In: Damodaran S, Parkin KL (eds) Fennema's food chemistry, 5th edn. CRC Press, Boca Raton, pp 681–752

Seabra LMJ, Pedrosa LFC (2010) Astaxanthin: structural and functional aspects. Rev Nutr 23:1041–1050. https://doi.org/10.1590/S1415-52732010000600010

Shao P, Qiu Q, Xiao J et al (2017a) Chemical stability and in vitro release properties of β-carotene in emulsions stabilized by Ulva fasciata polysaccharide. Int J Biol Macromol 102:225–231. https://doi.org/10.1016/j.ijbiomac.2017.03.186

Shao P, Zhu Y, Jin W (2017b) Physical and chemical stabilities of β-carotene emulsions stabilized by Ulva fasciata polysaccharide. Food Hydrocoll 64:28–35. https://doi.org/10.1016/j.foodhyd.2016.10.023

Sharif HR, Goff HD, Majeed H et al (2017a) Physicochemical stability of β-carotene and α-tocopherol enriched nanoemulsions: influence of carrier oil, emulsifier and antioxidant. Colloids Surf A Physicochem Eng Asp 529:550–559. https://doi.org/10.1016/j.colsurfa.2017.05.076

Sharif HR, Sharif MK, Zhong F (2017b) Preparation, characterization and rheological properties of vitamin E enriched nanoemulsion. Pak J Food Sci 27:7–14

Shegokar R, Mitri K (2012) Carotenoid lutein: a promising candidate for pharmaceutical and nutraceutical applications. J Diet Suppl 9:183–210. https://doi.org/10.3109/19390211.2012.708716

Shen X, Fang T, Zheng J, Guo M (2019) Physicochemical properties and cellular uptake of astaxanthin-loaded emulsions. Molecules 24:727. https://doi.org/10.3390/molecules24040727

Shi J, Xue SJ, Wang B et al (2015) Optimization of formulation and influence of environmental stresses on stability of lycopene-microemulsion. LWT - Food Sci Technol 60:999–1008. https://doi.org/10.1016/j.lwt.2014.10.066

Shu G, Khalid N, Chen Z et al (2018) Formulation and characterization of astaxanthin-enriched nanoemulsions stabilized using ginseng saponins as natural emulsifiers. Food Chem 255:67–74. https://doi.org/10.1016/j.foodchem.2018.02.062

Sommerburg O, Keunen JEE, Bird AC, van Kuijk FJGM (1998) Fruits and vegetables that are sources for lutein and zeaxanthin: the macular pigment in human eyes. Br J Ophthalmol 82:907–910. https://doi.org/10.1136/bjo.82.8.907

Sotomayor-Gerding D, Oomah BD, Acevedo F et al (2016) High carotenoid bioaccessibility through linseed oil nanoemulsions with enhanced physical and oxidative stability. Food Chem 199:463–470. https://doi.org/10.1016/j.foodchem.2015.12.004

Soukoulis C, Bohn T (2018) A comprehensive overview on the micro- and nano-technological encapsulation advances for enhancing the chemical stability and bioavailability of carotenoids. Crit Rev Food Sci Nutr 58:1–36. https://doi.org/10.1080/10408398.2014.971353

Stahl W, Sies H (2003) Antioxidant activity of carotenoids. Mol Asp Med 24:345–351. https://doi.org/10.1016/S0098-2997(03)00030-X

Stang M, Schuchmann H, Schubert H (2001) Emulsification in high-pressure homogenizers. Eng Life Sci 1:151. https://doi.org/10.1002/1618-2863(200110)1:4<151::AID-ELSC151>3.0.CO;2-D

Steiner BM, Shukla V, McClements DJ et al (2019) Encapsulation of lutein in nanoemulsions stabilized by resveratrol and Maillard conjugates. J Food Sci 84:2421–2431. https://doi.org/10.1111/1750-3841.14751

Su J, Guo Q, Chen Y et al (2020a) Characterization and formation mechanism of lutein Pickering emulsion gels stabilized by β-lactoglobulin-gum arabic composite colloidal nanoparticles. Food Hydrocoll 98:105276. https://doi.org/10.1016/j.foodhyd.2019.105276

Su J, Guo Q, Chen Y et al (2020b) Utilization of β-lactoglobulin- (−)-epigallocatechin-3-gallate(EGCG) composite colloidal nanoparticles as stabilizers for lutein pickering emulsion. Food Hydrocoll 98:105293. https://doi.org/10.1016/j.foodhyd.2019.105293

Sweedman MC, Hasjim J, Schäfer C, Gilbert RG (2014) Structures of octenylsuccinylated starches: effects on emulsions containing β-carotene. Carbohydr Polym 112:85–93. https://doi.org/10.1016/j.carbpol.2014.05.067

Tadros T, Izquierdo P, Esquena J, Solans C (2004) Formation and stability of nano-emulsions. Adv Colloid Interf Sci 108–109:303–318. https://doi.org/10.1016/j.cis.2003.10.023

Teixeira MC, Severino P, Andreani T et al (2017) D-α-tocopherol nanoemulsions: size properties, rheological behavior, surface tension, osmolarity and cytotoxicity. Saudi Pharm J 25:231–235. https://doi.org/10.1016/j.jsps.2016.06.004

Teo A, Lee SJ, Goh KKT, Wolber FM (2017) Kinetic stability and cellular uptake of lutein in WPI-stabilised nanoemulsions and emulsions prepared by emulsification and solvent evaporation method. Food Chem 221:1269–1276. https://doi.org/10.1016/j.foodchem.2016.11.030

Thomas SE, Johnson EJ (2018) Xanthophylls. Adv Nutr 9:160–162. https://doi.org/10.1093/advances/nmx005

Toti E, Chen C-YO, Palmery M et al (2018) Non-provitamin A and provitamin A carotenoids as immunomodulators: recommended dietary allowance, therapeutic index, or personalized nutrition? Oxidative Med Cell Longev 2018:1–20. https://doi.org/10.1155/2018/4637861

Trentin A, De Lamo S, Güell C et al (2011) Protein-stabilized emulsions containing beta-carotene produced by premix membrane emulsification. J Food Eng 106:267–274. https://doi.org/10.1016/j.jfoodeng.2011.03.013

Tyssandier V, Lyan B, Borel P (2001) Main factors governing the transfer of carotenoids from emulsion lipid droplets to micelles. Biochim Biophys Acta Mol Cell Biol Lipids 1533:285–292. https://doi.org/10.1016/S1388-1981(01)00163-9

van het Hof KH, West CE, Weststrate JA, Hautvast JGAJ (2000) Dietary factors that affect the bioavailability of carotenoids. J Nutr 130:503–506. https://doi.org/10.1093/jn/130.3.503

Verrijssen TAJ, Smeets KHG, Christiaens S et al (2015a) Relation between in vitro lipid digestion and β-carotene bioaccessibility in β-carotene-enriched emulsions with different concentrations of l-α-phosphatidylcholine. Food Res Int 67:60–66. https://doi.org/10.1016/j.foodres.2014.10.024

Verrijssen TAJ, Verkempinck SHE, Christiaens S et al (2015b) The effect of pectin on in vitro β-carotene bioaccessibility and lipid digestion in low fat emulsions. Food Hydrocoll 49:73–81. https://doi.org/10.1016/j.foodhyd.2015.02.040

Vishwanathan R, Wilson TA, Nicolosi RJ (2009) Bioavailability of a nanoemulsion of lutein is greater than a lutein supplement. Nano Biomed Eng 1:57–73. https://doi.org/10.5101/nbe.v1i1.p38-49

Wackerbarth H, Stoll T, Gebken S et al (2009) Carotenoid–protein interaction as an approach for the formulation of functional food emulsions. Food Res Int 42:1254–1258. https://doi.org/10.1016/j.foodres.2009.04.002

Wang L, Gao Y, Li J et al (2016) Effect of resveratrol or ascorbic acid on the stability of α-tocopherol in O/W emulsions stabilized by whey protein isolate: simultaneous encapsulation of the vitamin and the protective antioxidant. Food Chem 196:466–474. https://doi.org/10.1016/j.foodchem.2015.09.071

Wang L, Li Y, Xiang D et al (2020) Stability of lutein in O/W emulsion prepared using xanthan and propylene glycol alginate. Int J Biol Macromol 152:371–379. https://doi.org/10.1016/j.ijbiomac.2020.02.162

Wanyi W, Lu L, Zehan H, Xinan X (2020) Comparison of emulsifying characteristics of different macromolecule emulsifiers and their effects on the physical properties of lycopene nanoemulsions. J Dispers Sci Technol 41:618–627. https://doi.org/10.1080/01932691.2019.1610421

Wei Z, Gao Y (2016a) Physicochemical properties of β-carotene bilayer emulsions coated by milk proteins and chitosan–EGCG conjugates. Food Hydrocoll 52:590–599. https://doi.org/10.1016/j.foodhyd.2015.08.002

Wei Z, Gao Y (2016b) Physicochemical properties of β-carotene emulsions stabilized by chitosan–chlorogenic acid complexes. LWT - Food Sci Technol 71:295–301. https://doi.org/10.1016/j.lwt.2016.04.007

Weigel F, Weiss J, Decker EA, McClements DJ (2018) Lutein-enriched emulsion-based delivery systems: influence of emulsifiers and antioxidants on physical and chemical stability. Food Chem 242:395–403. https://doi.org/10.1016/j.foodchem.2017.09.060

Winkler-Moser JK, Mehta BM (2015) Chemical composition of fat and oil products. In: Handbook of food chemistry. Springer, Berlin Heidelberg, Berlin, Heidelberg, pp 365–402

Wong DWS (2018) Mechanism and theory in food chemistry, 2nd edn. Springer International Publishing, Cham

Xu D, Wang X, Jiang J et al (2012) Impact of whey protein - beet pectin conjugation on the physicochemical stability of β-carotene emulsions. Food Hydrocoll 28:258–266. https://doi.org/10.1016/j.foodhyd.2012.01.002

Xu D, Wang X, Jiang J et al (2013a) Influence of pH, EDTA, α-tocopherol, and WPI oxidation on the degradation of β-carotene in WPI-stabilized oil-in-water emulsions. LWT - Food Sci Technol 54:236–241. https://doi.org/10.1016/j.lwt.2013.05.029

Xu D, Yuan F, Gao Y et al (2013b) Influence of pH, metal chelator, free radical scavenger and interfacial characteristics on the oxidative stability of β-carotene in conjugated whey protein–pectin stabilised emulsion. Food Chem 139:1098–1104. https://doi.org/10.1016/j.foodchem.2013.02.027

Xu D, Aihemaiti Z, Cao Y et al (2016) Physicochemical stability, microrheological properties and microstructure of lutein emulsions stabilized by multilayer membranes consisting of whey protein isolate, flaxseed gum and chitosan. Food Chem 202:156–164. https://doi.org/10.1016/j.foodchem.2016.01.052

Xu D, Hou Z, Liu G et al (2020) Influence of rosemary extract addition in different phases on the oxidation of lutein and WPI in WPI-stabilized lutein emulsions. J Food Qual 2020:1–10. https://doi.org/10.1155/2020/5894646

Yadav KS, Kale K (2019) High pressure homogenizer in pharmaceuticals: understanding its critical processing parameters and applications. J Pharm Innov. https://doi.org/10.1007/s12247-019-09413-4

Yan Y, Zhu Q, Diao C et al (2020) Enhanced physicochemical stability of lutein-enriched emulsions by polyphenol-protein-polysaccharide conjugates and fat-soluble antioxidant. Food Hydrocoll 101:105447. https://doi.org/10.1016/j.foodhyd.2019.105447

Yang Y, McClements DJ (2013a) Encapsulation of vitamin E in edible emulsions fabricated using a natural surfactant. Food Hydrocoll 30:712–720. https://doi.org/10.1016/j.foodhyd.2012.09.003

Yang Y, McClements DJ (2013b) Vitamin E bioaccessibility: influence of carrier oil type on digestion and release of emulsified α-tocopherol acetate. Food Chem 141:473–481. https://doi.org/10.1016/j.foodchem.2013.03.033

Yang Y, McClements DJ (2013c) Vitamin E and vitamin E acetate solubilization in mixed micelles: physicochemical basis of bioaccessibility. J Colloid Interface Sci 405:312–321. https://doi.org/10.1016/j.jcis.2013.05.018

Yang Y, Decker EA, Xiao H, McClements DJ (2015) Enhancing vitamin E bioaccessibility: factors impacting solubilization and hydrolysis of α-tocopherol acetate encapsulated in emulsion-based delivery systems. Food Funct 6:83–96. https://doi.org/10.1039/C4FO00725E

Yang Y, Xiao H, McClements DJ (2017) Impact of lipid phase on the bioavailability of vitamin E in emulsion-based delivery systems: relative importance of bioaccessibility, absorption, and transformation. J Agric Food Chem 65:3946–3955. https://doi.org/10.1021/acs.jafc.7b00955

Yao K, McClements DJ, Xiang J et al (2019) Improvement of carotenoid bioaccessibility from spinach by co-ingesting with excipient nanoemulsions: impact of the oil phase composition. Food Funct 10:5302–5311. https://doi.org/10.1039/C9FO01328H

Yeum K-J, Russell RM (2002) Carotenoid bioavailability and bioconversion. Annu Rev Nutr 22:483–504. https://doi.org/10.1146/annurev.nutr.22.010402.102834

Yi J, Li Y, Zhong F, Yokoyama W (2014) The physicochemical stability and invitro bioaccessibility of beta-carotene in oil-in-water sodium caseinate emulsions. Food Hydrocoll 35:19–27. https://doi.org/10.1016/j.foodhyd.2013.07.025

Yi J, Lam TI, Yokoyama W et al (2015) Beta-carotene encapsulated in food protein nanoparticles reduces peroxyl radical oxidation in Caco-2 cells. Food Hydrocoll 43:31–40. https://doi.org/10.1016/j.foodhyd.2014.04.028

Yi J, Fan Y, Zhang Y, Zhao L (2016) Characterization of catechin-α-lactalbumin conjugates and the improvement in β-carotene retention in an oil-in-water nanoemulsion. Food Chem 205:73–80. https://doi.org/10.1016/j.foodchem.2016.03.005

Yonekura L, Nagao A (2007) Intestinal absorption of dietary carotenoids. Mol Nutr Food Res 51:107–115. https://doi.org/10.1002/mnfr.200600145

Yuan Y, Gao Y, Zhao J, Mao L (2008) Characterization and stability evaluation of β-carotene nano-emulsions prepared by high pressure homogenization under various emulsifying conditions. Food Res Int 41:61–68. https://doi.org/10.1016/j.foodres.2007.09.006

Yuan X, Liu X, McClements DJ et al (2018) Enhancement of phytochemical bioaccessibility from plant-based foods using excipient emulsions: impact of lipid type on carotenoid solubilization from spinach. Food Funct 9:4352–4365. https://doi.org/10.1039/C8FO01118D

Zhang J, Zhang X, Wang X et al (2017) The influence of maltodextrin on the physicochemical properties and stabilization of beta-carotene emulsions. AAPS PharmSciTech 18:821–828. https://doi.org/10.1208/s12249-016-0572-5

Zhao C, Wei L, Yin B et al (2020) Encapsulation of lycopene within oil-in-water nanoemulsions using lactoferrin: impact of carrier oils on physicochemical stability and bioaccessibility. Int J Biol Macromol 1862:183135. https://doi.org/10.1016/j.ijbiomac.2020.03.063

Zhao C, Shen X, Guo M (2018) Stability of lutein encapsulated whey protein nano-emulsion during storage. PLoS One 13:e0192511. https://doi.org/10.1371/journal.pone.0192511

Zhou Y, Wang D, Dong P (2012) Vitamins. In: Wang D, Kan J, Liu L et al (eds) Food chemistry. Nova Science Publishers, Inc., New York, pp 191–221

Zhou Q, Xu J, Yang S et al (2015) The effect of various antioxidants on the degradation of O/W microemulsions containing esterified astaxanthins from Haematococcus pluvialis. J Oleo Sci 64:515–525. https://doi.org/10.5650/jos.ess14264

Chapter 3
Low-Energy Emulsification Methods for Encapsulation of Antioxidants

M. Ali Aboudzadeh and Shaghayegh Hamzehlou

3.1 Introduction

Emulsions are colloidal dispersions of minimum two immiscible liquids, which their properties depend not only on thermodynamic conditions (i.e., temperature, composition or pressure) but also on the formation methods, the order of addition and the nature of the ingredients. The structure of emulsions made up of droplets of the dispersed (or internal) phase in the continuous (or external) phase. Classic emulsions are divided into water-in-oil (W/O) or oil-in-water (O/W) systems, depending on which phase forms the disperse phase. Re-emulsion of classic emulsion systems within an aqueous or oil phase containing hydrophilic or lipophilic surfactant leads to the formation of water-in-oil-in-water (W/O/W) or oil-in-water-in-oil (O/W/O) emulsion systems, which are also called as double emulsions.

Microemulsions and nanoemulsions are the two main colloidal dispersions used in the food and pharmaceutical industries to encapsulate, protect and deliver bioactive components. Microemulsions are thermodynamically stable, which their formation is spontaneous and there is no need of supplying any external energy; although in practice some external energy may be needed to overcome kinetic

M. A. Aboudzadeh (✉)
Donostia International Physics Center (DIPC), Donostia-San Sebastian, Spain
e-mail: mohammadali.aboudzadeh@ehu.eus

S. Hamzehlou
POLYMAT and Kimika Aplikatua Saila, Kimika Fakultatea, University of the Basque Country UPV-EHU, Joxe Mari Korta Zentroa, Tolosa Hiribidea, Donostia-San Sebastian, Spain

M. A. Aboudzadeh (ed.), *Emulsion-based Encapsulation of Antioxidants*, Food Bioactive Ingredients, https://doi.org/10.1007/978-3-030-62052-3_3

energy barriers or mass transport limitations. On the other hand, nanoemulsions are not thermodynamically stable and external energy input is needed to overcome the free energy related to increasing the contact area between the oil and water phase. In general, microemulsions are formed at higher surfactant to oil ratio (SOR), while nanoemulsions are formed at lower SOR (McClements 2012).

Generally, emulsion droplet size lies in the micrometer range, a size range in which droplets are attracted by gravity forces. An important aspect in the preparation of emulsions from both fundamental and technological viewpoints is to obtain a desired droplet size and a narrow size distribution. Main factors that influence the size of emulsion systems include energy input and duration, the surfactant type and concentration, and the physicochemical properties of the oil and water phases.

The generating processes for emulsion-based system are divided into (i) high-energy and (ii) low-energy methods (Anton et al. 2008). In high-energy methods, required energy comes from mechanical forces applied to the system (such as shear, turbulence, or cavitation), which involve the use of specific devices such as rotor stator mixers, colloid mills, high-pressure homogenizers, sonicators or microfluidizers. During these processes only a very low amount (~0.1%) of the mechanical energy produced is used for emulsification (Tadros et al. 2004; McClements and Rao 2011). These technologies usually have several inconveniences. Firstly, they can have some limitations to control the dispersion of the final droplet size of the emulsion, and generally produce a big polydispersity. In addition, in these methods the consumption of energy is higher than what is required for droplet break-up, thus the product will be subjected to high shear stress and over-heating. This is very crucial for thermolabile or sensitive drugs and macromolecules such as proteins, enzymes, retinoids, peptides, and nucleic acids (Mishra et al. 2010). Industries are preferably using high-energy equipment that is capable of generating intense mechanical forces, which can disrupt and fuse the oil and water phases.

Low-energy approaches alter the inherent physicochemical features (such as pH or temperature) of the surfactants, co-surfactants and excipients in the formulation, bearing stable nano-sized emulsion droplets (Anton and Vandamme 2009). However, nowadays low-energy methods are not fully exploited at industrial use, even though at laboratory scale they show better performance than the high-energy approaches. Although studies have shown that low-energy methods are often more efficient in creating small droplet sizes than high-energy ones (Anton et al. 2008; Anton and Vandamme 2009; McClements and Rao 2011), still there is a need for a better insight on the types and amounts of ingredients required to form nano-sized emulsion droplets by low-energy methods and to build the most appropriate preparation method to apply for a particular application. The colloidal dispersions formed using these approaches can be either microemulsions or nano-emulsions depending on the SOR used.

In this chapter, first the basic concepts of low-energy emulsification methods, including mechanisms and influenced parameters are reviewed. Then, the advantages and disadvantages of each low-energy method are highlighted. Finally, a particular focus is given on the employment of these emulsification methods in the most relevant studies to encapsulate compounds having antioxidant properties.

3.2 Low-Energy Emulsification Methods

Low-energy methods, in contrast to the high-energy technique, are governed by the behavior of the systems and their intrinsic physicochemical properties. Low-energy processes mainly depend on the control of interfacial phenomenon at the boundary between oil–water phases and rely strongly on the nature of any surface active molecules present, e.g., their solubility and molecular geometry (Date et al. 2010). As the name suggests, in low-energy methods, only a low quantity of applied energy is applied to generate nano-sized emulsion droplets. For some particular commercial applications, the low-energy methods are better because they have simple implementation and it is not necessary to use expensive or sophisticated equipment for the formation of the colloidal objects, only simple stirring is required, which makes these methods more energy efficient and establish significant advantages of this technology (Sadurní et al. 2005; Solans and Solé 2012).

Based on recent review written by McClements et al., low-energy approaches have been classified into isothermal and thermal methods (Komaiko and McClements 2016). In this classification, in order to produce nano-sized droplets, isothermal approaches need a change in composition and thermal approaches need a change in temperature. Changing the temperature of large volumes of liquid is likely to be energy-intensive and therefore, the isothermal low-energy methods may be more appropriate for emulsion-based techniques for encapsulation of bioactive compounds in the food industry.

3.2.1 Isothermal Approaches

Low-energy isothermal approaches are those that do not require any specialized homogenization equipment or a change in temperature to generate nano-sized droplets. Isothermal methods allow us to prepare nano-sized droplets over a wide range of temperatures rather than at a fixed temperature close to the phase inversion temperature. In addition, in these methods, there is no necessity for temperature quenching after preparation, which could result in substantial energy savings and the capacity to encapsulate heat-sensitive bioactive compounds such as antioxidants.

Spontaneous emulsification (SE) and emulsion phase inversion (EPI) methods enter in the category of isothermal methods (Fernandez et al. 2004; Mayer et al.

2013; Saberi et al. 2013a). Membrane emulsification (ME) can be classified as an isothermal low-energy method too; due to the low-energy consummation of this technology, temperature does not increase during emulsification.

3.2.1.1 Spontaneous Emulsification (SE) Method

The spontaneous emulsification (SE), also known as self-emulsification, has been the center of attention since long time ago and have been used extensively in numerous and diverse applications (Solans et al. 2016).

The chemical composition of both the oil and aqueous phase and the role of surfactants are very important in this method. SE can occur as a result of many physicochemical mechanisms. The real SE takes place when two immiscible liquids in non-equilibrium conditions are put in contact and then emulsify without any external assist, neither thermal nor mechanical. Solvents can be used to promote this process in the presence or the absence of surfactants (Bouchemal et al. 2004; Ganachaud and Katz 2005). For this reason, SE also is known as emulsification by solvent diffusion (ESD) because it is mainly a diffusion-driven process.

Basically in SE approach, a homogeneous organic phase (oil + hydrophilic surfactant) is usually titrated into an aqueous phase (water or buffer solution) at a controlled rate and under nonstop magnetic stirring. SE may occur slowly due to kinetic barriers, while the ouzo effect initiates this process instantly (Espitia et al. 2019). The ouzo effect results from a significant supersaturation of the oil, facilitating the nucleation of oil droplets when combined with water. Consequently, instantaneous diffusion of the oil to the closest droplet occurs and decreases the supersaturation to avoid any further nucleation (Ganachaud and Katz 2005). Figure 3.1 shows a potential mechanism for SE method. In this mechanism, a bicontinuous microemulsion phase generates at the barrier where the organic and aqueous phases reach each other, subsequently the spontaneous formation of nano-sized oil droplets takes place when the bicontinuous microemulsion phase breaks up.

Fine droplet diameters ($d < 100$ nm) of dispersed phase coated with surfactant in the continuous phase will occur due to the high affinity of the surfactant to the oil or the aqueous phase that causes turbulence at the interface of both the dispersed phase and continuous phase as well as whether the surfactant is displaced toward the continuous phase. To produce very small emulsions, the turbulence at the interface of the two phases should be triggered and co-solvents such as ethanol, acetone, and propylene glycol can make a great help. The SE method has previously been used to form α -tocopherol nanoemulsions ($d = 170$ nm), however relatively high levels of organic solvent (acetone and ethanol) were utilized (Bouchemal et al. 2004), which may limit its practical application in the food industry.

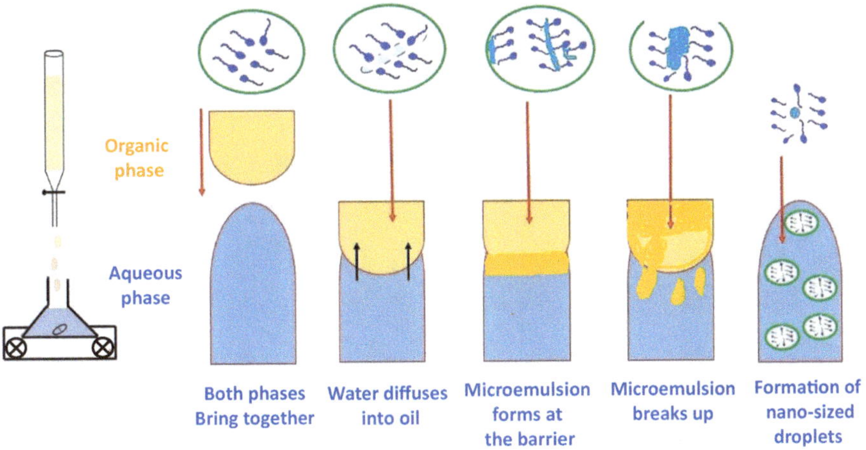

Fig. 3.1 Schematic sketch of potential mechanism for formation of nano-sized oil droplets through SE method. (Adapted from the reference (Asyikin et al. 2019) with permission)

Among low-energy methods, the SE presents some advantages to encapsulate and deliver bio-active compounds. In the pharmaceutical and food industries, it is not permitted to use high levels of surfactants and co-surfactants due to regulatory, cost, and sensory reasons. SE method allows to decrease the quantity of co-surfactants as well as to reduce the surfactant into the dispersed phase ratio (Bouchemal et al. 2004; Ahmed et al. 2012). Moreover, using SE, sensitive compounds are shielded against the severe conditions such as temperature and pressure.

3.2.1.2 Emulsion Phase Inversion (EPI) Method

The emulsion phase inversion (EPI) approach, or catastrophic phase inversion (CPI) as reported in some studies (Bilbao-Sáinz et al. 2010; Li et al. 2013), involves changing solution conditions in a surfactant–oil–water mixture to generate a phase inversion from a W/O to an O/W system (or oppositely), for example, changing pH or ionic strength (McClements 2013). In others words, a catastrophic-phase inversion happens when the ratio of the oil-to-water phases is altered while the surfactant properties remain constant. In order to avoid any confusion, it is worth to remind here that some scientists have mentioned this method as the phase inversion composition (PIC) (Koroleva and Yurtov 2012; Komaiko and McClements 2016).

The EPI method has a great potential for scale-up productions because of the ease of formation and relatively low energy costs. In EPI method unlike SE approach,

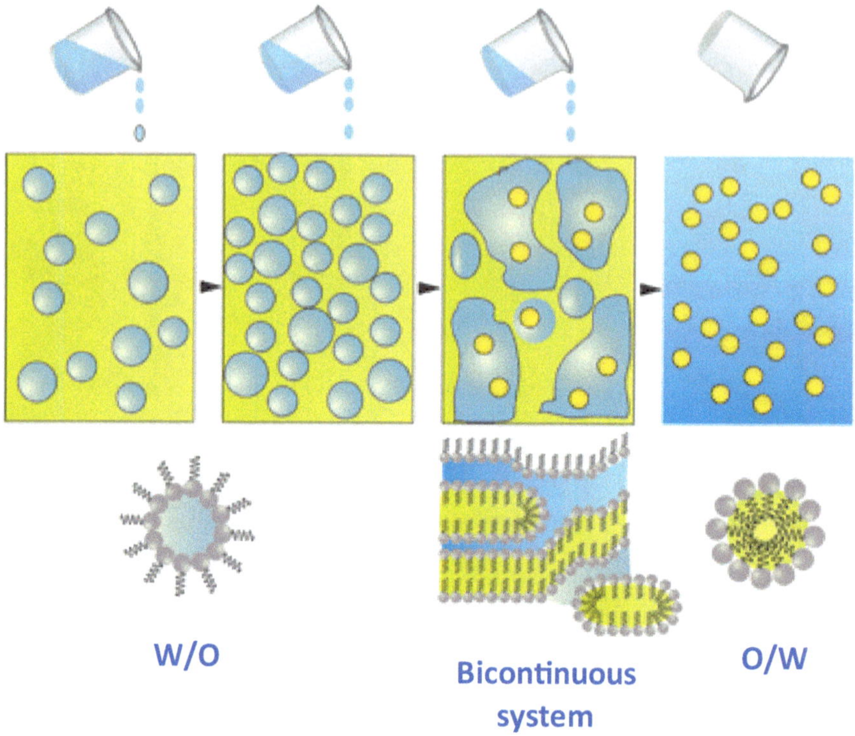

Fig. 3.2 Schematic sketch of proposed mechanism for formation of nano-sized oil droplets through EPI method. (Reproduced from the reference (McClements 2011) with permission)

an aqueous phase is added into a stirring organic phase, which usually consists of oil and surfactant (Fig. 3.2). However, there are some similarities between the proposed mechanism of the EPI method and of the SE. When the aqueous phase is initially added into the organic phase, a W/O emulsion is formed, as more water is titrated, a liquid crystalline (lamellar or cubic) or bi-continuous phase can be formed that can become very viscous and it may limits the proper agitation. In order to obtain small and uniform droplets, both phases need to cross this emulsification path, which is a critical intermediate step in creation of nano-sized emulsion droplets (Jahanzad et al. 2009). This bicontinuous phase can eventually catch one of the two phases into the other, allowing for inclusion of droplets into droplets (Perazzo et al. 2015). It has been reported that the value of the critical water concentration where phase inversion takes place, as well as the size of the oil droplets produced, depend on process parameters, such as rate of adding water, stirring velocity and surfactant concentration (Bilbao-Sáinz et al. 2010).

The EPI approach has many advantages, including low cost and the need for simple equipment. Nevertheless, the preparation time of this process was longer than that of SE technique due to the smaller driving forces of EPI method (Lippacher and Mu 2004).

3.2.1.3 Membrane Emulsification (ME) Method

Membrane emulsification (ME) technique can be classified as low-energy emulsification method as it needs 100 times lower energy comparing to high-energy methods. ME is a dispersion approach that uses a membrane to generate emulsion droplets. The two main configurations of ME are direct membrane emulsification (DME) and premix membrane emulsification (PME). In DME, the dispersed phase is pushed through the membrane pores into a stirring or cross-flowing continuous phase and small droplets are formed at the membrane/continuous phase interface. In PME, a coarse emulsion called premix is pushed through the membrane pores, reducing the droplet size and size distribution. Principles of each type of ME method, set-ups and involved parameters are fully discussed in Chap. 6.

As discussed previously, traditional low-energy emulsification methods depend on the physicochemical properties and therefore require the use of specific surfactants and/or co-surfactants at high concentration and they rely on the spontaneous formation of oil droplets, which may not be suitable for cosmetics or pharmaceutical applications. ME achieved advantage over traditional methods since with this technique it is possible to produce emulsion with quite a narrow droplet size distribution using relatively low shear and low energy input without temperature increase during the emulsification (Nakashima et al. 1991; Charcosset 2009). These strong points are clearly advantageous for applications demanding stable emulsions with shear or heat sensitive compounds, such as antioxidants.

3.2.2 Thermal Approaches

A change in temperature is needed in thermal methods in order to promote the generation of a nano-sized droplets. This temperature change hinders the occurrence of coalescence and thus, leads to the formation of stable droplets. The most important thermal method used to create nano-sized droplets is the phase inversion temperature (PIT) method which is particularly interesting since it is an organic solvent-free and low-energy method. These characteristics are potentially the most appropriate for application in the fields of pharmaceutical sciences, nano-medicine and cosmetics, to protect the encapsulated drug from degradation during processing.

3.2.2.1 Phase Inversion Temperature (PIT) Method

The PIT method was first defined by Shinoda and Saito as an alternative to high shear emulsification procedures (Shinoda and Saito 1968, 1969). In this method, temperature dependent solubility of nonionic surfactants (e.g. polyethoxylated surfactants) allows them to alter their affinities for water and oil as a function of the temperature and therefore to undergo a phase inversion. Water, oil and nonionic surfactants are mixed together at room temperature (RT) and stirred moderately. Then, the mixture is slowly heated up. Subsequently, the solubility of surfactant gradually shifts from the aqueous to the oily phase. Around or above the phase inversion temperature, the surfactant is completely solubilized in oil and thus the mixture undergoes a phase inversion, from an O/W to a W/O emulsion.

The reason behind this phase inversion is changes in the physicochemical properties of the surfactant with temperature. At low temperature, the monolayer of nonionic surfactant, such as polysorbates, is more hydrophilic and the surfactant head groups are well-hydrated, thus, it tends to be placed in the aqueous phase. In contrast, at high temperature, the surfactant becomes more lipophilic and its head groups are mostly dehydrated, hence it moves forward into the organic phase. At a particular intermediate temperature, around the PIT, the surfactant is uniformly distributed between the organic phase and the aqueous phase (McClements and Rao 2011; Komaiko and McClements 2016). For a fixed composition of water, oil and surfactant, a bicontinuous microemulsion is formed at this temperature, which consists of small hydrophobic domains. This is because at this temperature, the affinity of amphiphiles for each phase is similar and interfacial curvature is very low. It was proposed by Anton and Vandamme that the procedure of generation of nano-sized emulsion droplets by the PIT method is fairly similar to that of the SE method (Anton and Vandamme 2009). Afterwards, when the mixture is cooled down to the temperatures below the PIT, the surfactant molecules become more hydrophilic and they tend to migrate from the organic phase to the aqueous phase. This mechanism breaks the bicontinuous microemulsion phase and leads to the formation of small droplets of the organic component (Saberi et al. 2015a). In order to generate fine droplets, this cooling step must take place rapidly along with continuous stirring which is the principle behind the formation of nano-sized droplets using the PIT approach (Anton and Vandamme 2009), proved also in another study on the influence of cooling rate on droplet size (Saberi et al. 2015a).

3.3 Advantages and Disadvantages of Low-Energy Emulsification Methods

Over the last 30 years there has been a certain research interest in the development of low-energy techniques due to the advantages of these methods in terms of formulation, yield, potential industrial scale-up and non-aggressive characteristics (e.g.

against encapsulated delicate bioactive compounds). Moreover, low-energy emulsification does not need any specific devices or chemical reactions for the formation of the colloidal objects; the simplicity of the whole process as well as a high-energy yield, constitute significant advantages of this technology. These advantages come to sight when the process is compared with the general methods utilized to create nanoparticulate drug delivery systems, which are mainly high-energy lipid emulsification, commonly polymeric (Anton et al. 2008), or even inorganic chemistry via, for example, a sol–gel process defined as a 2-step polycondensation forming silica particles (Barbé et al. 2004).

Although, low-energy approaches are often more effective at producing small droplet sizes than high-energy approaches, they have some disadvantages. The main limitation of these methods is their dependence on the physicochemical properties of the system and on the type of oils and emulsifiers that can be used. For instance, it is currently not possible to use proteins or polysaccharides as emulsifiers in most of the low-energy methods used to generate nano-sized droplets. Instead, it is often necessary to utilize relatively high concentrations of synthetic surfactants by these approaches, which may limit their use for many food applications. In this context, the main limitations to the formulation of emulsion-based systems are related to the complexity of these systems and the use of food-grade ingredients.

Regarding the ME method, the main advantages are the low energy input leading to no temperature increase during emulsification and the low shear rate which gives better stability for shear sensitive actives. While the main disadvantage is the moderate output as the extrusion rate of the internal phase must be adequately low to prevent the formation of continuous jets flowing from the membrane pores (Kobayashi et al. 2003; Van Der Graaf et al. 2005). One-step membrane emulsification is the most efficient method for preparation of emulsions with the fraction of the internal phase up to 30 vol %. In this case, emulsions have narrow droplet size distributions. The polydispersity of droplets increases with an increase in the fraction of the internal phase.

Table 3.1 compares different methods involved in low-energy emulsification approach and highlights the advantages and disadvantages of each method.

3.4 Scale-Up of Low-Energy Emulsification Methods for Industrial Application

Low-energy emulsification processes are of interest from an economic perspective and as a potential matrix to encapsulate fragile molecules such as antioxidants. In addition, these methods are more energy-saving and therefore more attractive for large-scale production. Nevertheless, at present, these methods are not widely used in the food industry, and where they are used there is still a relatively poor understanding of the factors affecting their performance. Published reports about low-energy emulsification methods are generally made with a very small volume of the

Table 3.1 Comparison between different low-energy emulsification methods: by highlighting the advantages and disadvantages of each method

Method	Advantages	Limitations
SE	- Simple in practical implementation, as it just requires the addition of a surfactant/oil mixture into an aqueous phase with constant mixing at RT - No need to use temperature-sensitive surfactants - No need to apply high temperatures, which could lead to thermal degradation of sensitive components -No need for co-surfactants comparing with other low-energy methods (Solans et al. 2016)	- Suitable only for formulating emulsions with a low fraction of the internal phase which is ≈ 1% directly after dispersion and less than 5% after evaporation of the solvent (Ganachaud and Katz 2005; Carteau et al. 2008) - The co-solvent should be soluble in water in all proportions (Bouchemal et al. 2004)
EPI	- Possible to use a wide range of surfactants whose hydrophilic lipophilic balance (HLB) value is less dependent on temperature. - No need to use temperature gradients, which is important when encapsulating thermally unstable substances. -The temperature quench after preparation is not necessary in this method	- Need to use relatively high amounts of surfactant to create fine oil droplets - Sensitive to changes in the mixture composition, so the emulsification conditions should be rechecked when additional substances are added, e.g. when drug encapsulation is carried out. - Longer preparation time comparing to other low-energy methods due to the smaller driving forces of this method - Surfactants used in this method are usually limited to small molecules that can stabilize both W/O emulsions (over the short term) and O/W emulsions (over the long term)
ME	- No foaming, reduced coalescence phenomena (Drioli et al. 2005) - Narrow droplet size distributions - No moving parts, use of plastic equipment (absence of corrosion) - High compatibility and modularity, simple scale-up, easy control	- Membrane fouling (at the membrane surface and/in the membrane pores) - Low permeation rates related to narrow droplet size distribution and reduced productivity - Relatively low membrane lifetime; high replacement costs - Additional resistance to mass transfer created by the membrane (Charcosset 2009) - Expensive fabrication of perfectly monodisperse membranes from materials, which are not very favorable in food industry (e.g. nickel-alloy)
PIT	- Can be repeated many times by increasing and decreasing the temperatures - More suitable for forming solid lipid nanoparticles, as the lipid phase can be melted at high temperatures (> PIT), but crystallized at lower temperatures after the formation of nano-sized droplets.	- The formed droplets tend to be highly sensitive to droplet coalescence when the temperature of the system is raised - Limited to the non-ionic surfactants

ingredients, which is mostly prepared in a test tube or in a small container. However, if these methods are going to be used in industry, it is necessary to study how the scale-up should be done, and how it would influence the final properties of the product. For instance, in thermal approaches and specifically for the PIT method, although it is simple to rapidly cool a sample prepared in a test tube by immersion in ice, it seems more complicated to reach a very fast cooling of a big container of sample.

Among isothermal low-energy methods, SE and EPI methods are suitable for application in foods and beverages. The EPI method is considered to have a greater potential for a large-scale production than the PIT, because it is experimentally easier to add one component to a large volume of emulsion than to produce a sudden change in temperature. Moreover, EPI method is also preferred when dealing with components with temperature-stability problems. Sole et al. demonstrated a first approximation to the scale-up of the nanoemulsions preparation through the EPI method, which takes into account that preparation conditions must assure a proper transition of phases in different scales. In their study, the production of nanoemulsions was scaled-up from lab scale (100 mL) to medium scale (600 mL) (Solè et al. 2010). They analyzed that in both scales there should be a similarity in geometric form factors of both the stirrer and the container used. Also, mixing and addition conditions must be identical in both scales, in order to keep the same agitation level at any point and at any time into the container, which allows the appropriate transition of phases during emulsification. In a recent study, stable nanoemulsions of docosahexaenoic acid (DHA) and eicosapentaenoic acid (EPA) were successfully prepared by the EPI method and the authors demonstrated that EPI process is a promising approach to incorporate DHA and EPA into water-based systems to enhance their food fortification and large-scale production (Zhang et al. 2019).

On the other hand, SE method is used in the pharmaceutical industry to obtain O/W nano-sized droplets as carriers for lipophilic drugs in an aqueous media, with low-energy costs. In the literature, they are usually referred to as self-nanoemulsifying drug delivery systems (SNEDDS) (Wang et al. 2009; Ghai and Sinha 2012). Ease of scale-up is one of the most remarkable advantages that make SNEDDS exceptional comparing with other novel drug delivery systems, such as liposomes, solid dispersions and nanoparticles (Date et al. 2010). However, it is worth to remark here that SE is usually unsuitable for delivery of nutraceuticals for human applications because the high levels of surfactants required are inappropriate for food use. Conversely, they are applicable when only a low amount of oil in the final product is required, such as fortified waters and soft drinks, because the total amount of surfactant in the final product is then proportionately low, even though the SOR is high (Komaiko and McClements 2016). Moreover, it was shown that nanoemulsion formulated via SE method can find applications in the delivery of controlled amounts of drugs into the beverage of breeding animals (such as poultry, cattle, pigs) or be used for the controlled release of injectable poorly water-soluble drugs (Vandamme and Anton 2010).

Membrane processes have become major tools in the food processing industry over the last 25 years (Daufin et al. 2001). The ME process is also expected to gain an increasing interest in the food processing industry. Industrial scale-up of ME are quite rare in literature. Gijsbertsen-Abrahamse et al. presented an analysis of an industrial scale production of culinary cream, for which a microsieve membrane with a low porosity (≈ 1 m^2) was found the best suitable membrane (Gijsbertsen-Abrahamse et al. 2004). Due to the need of market for healthy foods, Moringa Milk Industry, one of the biggest milk industries in Japan, developed and commercialized a very low fat spread using ME technology (Nakashima et al. 2000). The product of this company has high stability during 6 months without the use of preservative, although the volume fraction of dispersed water-phase reaches up to 75%.

DME and PME are the two main configurations of ME. Naturally, DME is not recommended for high flow rates because the transmembrane flux has to be very low in order to obtain monodispersed emulsions. In contrast, in PME higher flow rates are more efficient for break-up of large droplets within the membrane due to the higher stress applied on the droplets inside the pores, which leads to a decrease in particles size and size distribution. Besides higher flowrate of PME comparing to DME, higher droplet concentrations are obtained and the experimental set-up is simpler and the process is easier to control and operate. Large-scale production (up to 500 mL) of monodisperse nanoemulsions (d ≈ 260 nm) was shown through PME using Shirasu Porous Glass (SPG) membranes with controlled size and very long stability (9 months at room temperature). The nanoemulsions were produced in only one cycle at moderate pressure, which can be appropriate for encapsulation of sensitive actives (Alliod et al. 2018). In another recent study, lemon oil was encapsulated (≈ 5%, O/W emulsion) by high-throughput PME system based on nickel microsieves. The ratio between the droplet size of the coarse emulsion and the hydraulic diameter of the micro-sieve kept about 1 or lower to refine emulsions at pressures ranging from 150 to 450 kPa. Under these conditions, high fluxes (400–800 $m^3 m^{-2} h^{-1}$) obtained during emulsification which shows the potential of microstructured microsieves to scale-up this technology (Kaade et al. 2019).

3.5 Low-Energy Emulsification Methods to Encapsulate Antioxidants: Formulations, Applications and Advances

Oxidation reactions are one of the main means of degradation of oils, fats and lipid-based edible materials, which lead to reduction in the nutritional value and sensory characteristics of food products. The oxidative deterioration of lipids (known as lipid peroxidation) proceeds by a free radical chain reaction mechanism. As with any radical reaction, the process involves three main steps: initiation, propagation, and termination. These steps are fully discussed in Chaps. 1 and 10, Sects. 1.2.2 and Sect. 10.2, respectively. The oxidation of biomolecules such as deoxyribonucleic acid (DNA) and proteins (structural and metabolic) and their biological implications have been related to several chronic and age-related disease, including cataract, cancer, neurodegenerative diseases, cardiovascular disorders and atherosclerosis (Guéraud et al. 2010; Vieira et al. 2017). Utilization of compounds with antioxidant properties is considered as one possible way to protect against oxidation. Antioxidant compounds are materials capable of delaying or preventing the development of any deterioration due to oxidation.

In order to incorporate antioxidants as supplement into food products, they must meet a number of standards. They should be economic, non-toxic, easy-to-blend, efficient at low concentrations, stable at least during the shelf-life of food product, capable of maintaining their activity and protect the finished product even on long-term storage without imparting a foreign color, odor or flavor to the food and finally capable of keeping their integrity during food processing (Barreira and Ferreira 2019). A potential approach to fulfil all these criteria is the employment of encapsulation systems, as advantageous platforms to entrap a bioactive compound in a core or a fill within a carrier (coating, membrane, capsule, matrix or shell) for enhancing its the delivery within living cells (Nedovic et al. 2011).

Among encapsulation approaches, colloidal emulsion-based systems are particularly interesting as they can easily be created from food-grade ingredients using relatively simple processing protocols. Emulsification is one of the favorable delivery systems applied to increase the solubility of phytochemicals, nutraceuticals and food additives.

Low-energy emulsification methods have the potential of being successfully implemented for the encapsulation of both hydrophilic (e.g. ascorbic acid and phenolics) and lipophilic antioxidant components (e.g. tocopherols and carotenoids). Schematic representation of different forms of emulsion-based encapsulation systems is illustrated in Fig. 3.3. In the case of lipophilic antioxidants, they are

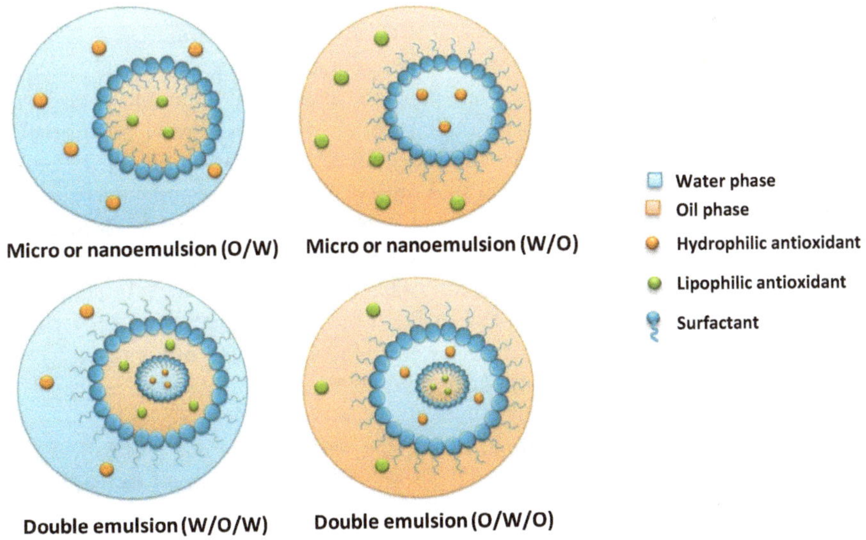

Fig. 3.3 Schematic representation of different forms of emulsion-based encapsulation systems. (Reproduced from the reference (Shishir et al. 2018) with permission)

solubilized within the oil core of the dispersed droplets (O/W micro or nanoemulsion), while in the case of hydrophilic ones they are solubilized within the water core of the dispersed droplets (W/O micro or nanoemulsion). Hydrophilic active antioxidants may be included in a primary aqueous phase, which may be dispersed in the oil phase, and subsequently emulsified, forming W/O/W double emulsions. Following, different studied examples of low-energy emulsification methods used for encapsulation of antioxidants are presented.

3.5.1 Encapsulation of Antioxidants via SE Method

Numerous studies have shown that SE, as an isothermal method, can be used for encapsulating of sensitive antioxidants in food and pharmaceutical industry, because no high temperatures are required. Moreover, this method allows emulsion-based encapsulation systems to be fabricated using simple stirring rather than expensive homogenization equipment. In this section, the most relevant studies are discussed.

In vitro physicochemical assays characterize most of essential oils as antioxidants (Bakkali et al. 2008). In one of the pioneer works, essential oil (carvacrol) was encapsulated through SE in O/W nanoemulsions consisting of medium chain triglyceride (MCT) and carvacrol (10 wt% total oil phase) and food-grade nonionic surfactants (Tween series). The smallest droplets ($d \approx 55$ nm) were formed when 7.5 wt% MCT, 2.5 wt% carvacrol and 20 wt% Tween 80 were used (Chang et al. 2013). The same authors developed transparent orange oil nanoemulsions with the same type of carrier oil and surfactant, in which the formulation containing 20 wt% Tween 80, 5 wt% orange oil and 5 wt% MCT remained transparent after thermal cycling (Chang and McClements 2014). The ability of Tween 80 (at SOR = 150%) to emulsify higher quantities of essential oils was confirmed in another study as well, in which SE was applied to encapsulate peppermint essential oil in microemulsions with droplet size of around 50 nm (Barzegar et al. 2018). Similarly, encapsulation of cinnamon oil by nanoemulsions using coconut oil as the carrier oil and Tween 80 as the surfactant has been reported. The formulated stable nanoemulsions exhibited approximate particle size of 100 nm and antimicrobial activity (Yildirim et al. 2017). The essential oil of cinnamon bark contains about 90% cinnamaldehyde. In this context, cinnamaldehyde has been encapsulated in spontaneously emulsified systems consisting of MCT and Tween 80. The encapsulation efficiency of the obtained nanoemulsions was 80% for 1 week and slow release of cinnamaldehyde could be expected. Phase separation occurred after 12 days of storage under 37 °C (Tian et al. 2015).

By means of self-emulsification process and using food-grade emulsifiers zein and sodium caseinate (NaCas) at mass ratio of 1:1, eugenol-loaded nanoemulsion was prepared which displayed a well-defined diameter of approximately 109–139 nm, a negative surface zeta-potential ranging between −28.5 mV and

−35.8 mV, as well as a spherical structure (Wang and Zhang 2017). In another study, limonene, a commonly used lipophilic flavoring agent with antioxidant properties, was loaded into emulsion-based delivery systems by SE. The smallest droplet diameters (25 nm) were produced when the system contained 15 or 20 wt% surfactant (Tween 60) and 6 wt% limonene (Saberi et al. 2015b). Related to this work, other nanoemulsions were developed via SE, using three different citrus oils as the oil phase (5 wt%) and Tween 80 (4 wt%) as the surfactant. Among the oils, Bergamot oil emulsions displayed small droplets with the best stability over 24 h, due to the relatively polar components (e.g. linalyl acetate) and water-insoluble constituents (e.g. γ-terpinene) of this oil (Zhao et al. 2018). In a very recent study, SE technique allowed a greener micelles formation in nanoemulsion system by slowly titrating the oil phase, containing *Eucalyptus globulus* (active compound), grape seed oil (carrier oil) and (Tween 40) hydrophilic surfactant into the aqueous phase, with subsequent stirring for 30 min to form a homogeneous solution. Surfactant at concentration 9.0 wt% promoted a thermodynamically stable micelles with spherical particles ($d = 17.13$ nm) and suitable viscosity (≈ 2.3 cP) and pH value (6.57) for transdermal purpose. Performing the in vivo analgesic activity assay on the optimized emulsion, the authors demonstrated the promising application of this finding as an efficient transdermal nanocarrier for the analgesic therapy alternative (Asyikin et al. 2019).

Antioxidant activity of Vitamin E (VE) has been shown in both biological and food systems (Traber and Packer 1995; Galli et al. 2017). The possibility of using SE method for encapsulation of VE was examined by multiple studies of Saberi et al. In these studies, the influence of system composition, preparation conditions, co-solvent type (e.g. glycerol, propylene glycol and ethanol) and presence of salt on the formation, stability and properties of VE-loaded mini- and nanoemulsions was investigated. It was found that in the nanoemulsions (without co-solvent) with 10 wt% oil (MCT + VE) and 10 wt% surfactant, the oil composition affected particle size, thereby minimum of mean droplet diameter obtained at 8 wt% VE and 2 wt% MCT. Surfactant type also had a considerable effect on particle size, with Tween 80 giving the smallest droplets ($d \approx 55$ nm) (Saberi et al. 2013a). The impact of two polar co-solvents (propylene glycol and ethanol) on characteristics of VE-loaded nanoemulsions (10 wt% VE + 10 wt% surfactant +80 wt% aqueous phase) formed by SE was investigated by these scientists as well. They discovered that type, concentration and initial location of co-solvent greatly influenced the droplet size and optical clarity of nanoemulsions. The smallest droplets ($d < 50$ nm) and highest transparency were observed when either 30 wt% propylene glycol or 20 wt% ethanol was present in the aqueous phase (Saberi et al. 2013b). Likewise, in the case of using glycerol, the oil phase composition can affect the distribution of particles, while reducing the amount of surfactant required. Utilizing glycerol in a system with a mixed oil phase (80 wt% vitamin E and 20 wt% MCT) led to the formation of smallest mean droplet diameters ($d < 50$ nm) at 40 and 50 wt% glycerol. The obtained particle size distribution was monomodal which was in contrast

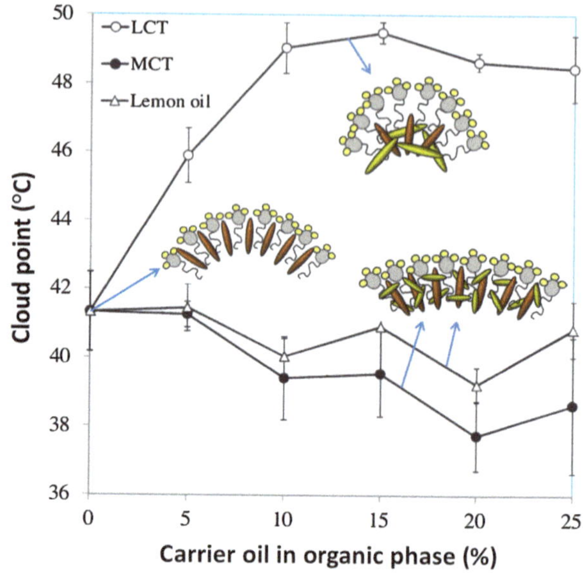

Fig. 3.4 Effect of oil type and concentration (wt%) on the cloud point of emulsion containing 10 wt% oil phase, 10 wt% surfactant phase (Tween 80), and 20 wt% co-solvent (ethanol). (Reproduced from the reference (Saberi et al. 2014a) with permission)

to emulsions formed with pure VE that had much broader distributions (Saberi et al. 2013c). Also, it was revealed that in VE emulsion containing 20 wt% ethanol, the carrier oil type had a considerable effect on the thermal stability of the emulsions: long-chain triglycerides (LCT) increased the cloud point, but MCT and lemon oil decreased it (Fig. 3.4). Contrarily, dilution of the aqueous phase of these emulsions (by adding water) caused an evident increase in the cloud point, which was attributed to the impact of ethanol on the optimum curvature of the surfactant monolayers (Saberi et al. 2014a).

In continuation of these researches and to study the temperature-dependent behavior of used non-ionic surfactants, the effect of heating–cooling conditions on the properties of VE-enriched emulsion-based delivery systems was examined. It was revealed that heating up to the temperatures well below the PIT (≤ 50 °C) and at around the PIT (≈ 75 °C) caused only a slight increase in the particle diameter, whereas heating to temperatures just below the PIT (50–70 °C), caused a significant increase in the particle diameter (Saberi et al. 2015a). This range often referred to as the droplet coalescence zone, where rapid oil droplet growth occurred due to changes in surfactant properties. Therefore, concerning the industrial application, it is important to keep emulsions away from this temperature range of high instability. Finally, the same scientists investigated the influence of salt on these emulsions. They found that salt type and concentration (0–1 N NaCl or 0–0.5 N $CaCl_2$) did not have a considerable impact on the initial droplet size of the emulsions, probably because the salts did not change the surfactant solubility at the tested concentrations. On the other hand, the isothermal and thermal stabilities of the emulsions were highly dependent on salt concentrations. The cloud point of the emulsions decreased with increasing salt level, which was associated to accelerated droplet coalescence in the presence of salts (Saberi et al. 2014b).

Antioxidant property of vitamin D is between the newest proposed non-calcemic roles of this compound. As it has a similar structure to cholesterol, thus, it has suggested that vitamin D may be classified as an antioxidant (Wiseman 1993). Similar to above-mentioned methods for oil-soluble VE, possibility of Vitamin D encapsulation in O/W nanoemulation via SE has been studied. As a result, nanoemulsions with small droplet diameters <200 nm were formed using a non-ionic surfactant (Tween 80) at a SOR $\geq$ 1:1 with constant stirring during the titration process. They were stable to droplet growth at ambient temperatures but unstable at high temperatures (> 80 °C). The thermal stability of the nanoemulsions was improved by using an anionic co-surfactant during the formulation (Guttoff et al. 2015).

The therapeutic advantages of the medicinal plants are generally associated to their antioxidant properties due to their large content of natural antioxidant compounds such as carotenoids, flavonoids and polyphenol (Mansour et al. 2011; Xu et al. 2017). In a study, capsanthin from paprika was encapsulated into stable and transparent nanoemulsions formed by SE. The smallest particle diameter ($\approx$ 30 nm) was obtained at 5 wt% oil phase (1 wt % capsanthin and 4 wt% MCT) containing mixed surfactants (10 wt%) of Tween 80 and Span 20 at weight ratio 3:1 (An et al. 2014). In other report, topical nanoemulsions containing the main flavonoids from liquid or dried plant extracts of *Achyrocline satureioides* were developed by SE. Formulations that contained plant extract up to 1 wt% of dry residue, remained monodispersed exhibiting a droplet size at the range of 200–300 nm (Bidone et al. 2014). In addition, the possibility of using SE method to encapsulate crocin (a carotenoid compound) was explored in W/O micro-emulsions (80 wt% olive oil) using nonionic surfactant polyglycerol polyricinoleate (PGPR). The smallest droplet size obtained when surfactant to water ratio was 100% (Mehrnia et al. 2016). In another study, a nanoemulsion delivery system composed of 10 wt% oil (grape seed oil plus orange oil), 10 wt% surfactant (Tween 80) and 80 wt% aqueous phase was formed to encapsulate resveratrol using SE. The optimum oil phase composition consisting of 50% grapeseed oil and 50% orange oil led to the formation of nanoemulsions containing small stable droplets (d $\approx$ 100 nm) (Davidov-Pardo and McClements 2015). In a recent report, capsaicin (the major ingredient of hot pepper) was loaded in olive oil O/W nanoemulsions prepared by SE. The smallest droplet size ($\approx$ 13 nm) and long-term stability ($\approx$ 8 months) in room and harsh temperature (i.e. 4 and 45 °C) was obtained by adjusting HLB value of system to 13.5 using Tween 80 and Span 80 surfactants (Ghiasi et al. 2019).

An overview of above-mentioned recent studies on O/W emulsion-based systems that have been developed to encapsulate compounds with antioxidants properties is presented in Table 3.2. Optimized formulations that led to the most stable system with the smallest particle diameter can be found in this table.

Table 3.2 An overview of the optimized formulations used in the most recent studies to encapsulate antioxidants in O/W emulsion systems via SE

Method	Oil (wt%) + Compound with antioxidant activity (wt%)	Surfactant(s) (wt%)	Co-solvent (wt%)	Preparation temp. (°C)	Smallest droplet diameter ≈ (nm)	Reference
SE	MCT (2) +VE (8)	Tween 80 (10)	–	RT	55	Saberi et al. (2013a)
	MCT (2) + VE (8)		Glycerol (50)		45	Saberi et al. (2013c)
	VE (10)					
	VE (10)		Propylene glycol (30)		47	Saberi et al. (2013b)
			Ethanol (20)		44	
	Lemon oil (4) + VE (6)		Ethanol (20)		40	Saberi et al. (2014a)
	Orange oil (5) + Grape seed oil (5) + Resveratrol (120 μg/ml)		–		100	Davidov-Pardo and McClements (2015)
	MCT (7.5) + Vitamin D (2.5)	Tween 80 (17.5)				Guttoff et al. (2015)
	MCT (7.5) + Carvacrol (2.5)	Tween 80 (20)			25	Chang et al. (2013)
	MCT (6) + Orange oil (4)	Tween 80 (20)			25	Chang and McClements (2014)
	Peppermint oil (20)	Tween 80 (30)			50	Barzegar et al. (2018)
	MCT (5) + Cinnamaldehyde (5)	Tween 80 (15)			30	Tian et al. (2015)
	Coconut oil (6) + Cinnamon oil (4)	Tween 80 (10)			100	Yildirim et al. (2017)
	Citrus oil (5)	Tween 80 (4)			10–30	Zhao et al. (2018)
	Grape seed oil (0.5) + Eucalyptus oil (2.5)	Tween 40 (9)			17	Asyikin et al. (2019)
	MCT (4) + Limonene (6)	Tween 60 (15 or 20)		45	25	Saberi et al. (2015b)
	MCT (4) + Capsanthin (1)	Tween 80 (10)		50	30	An et al. (2014)
	Olive oil (2) + Capsaicin (0.15)	Tween 80 (30.09) + Span 80 (4.91)	Ethanol (8.66)	40	13	Ghiasi et al. (2019)

At the end of this section, it is worth to remark that the potential of spontaneous or self-nanoemulsification (through SNEDDS) in oral drug delivery of antioxidants has been explored considerably in pharmaceutical applications. As an example, coenzyme Q10, an intracellular antioxidant, was formulated into SNEDDS to overcome its low bioavailability associated to hydrophobic nature of the drug. Hard fats such as Witepsol H35 (hydrogenated coco glycerides) was employed for the fabrication of SNEDDS owing to its excellent solubilization potential for coenzyme Q10 compared with the commonly used oils. As a result, stable SNEDDS with relatively good drug content (13 wt%) was prepared which emulsified easily with mean droplet size of 32.4 nm (Nepal et al. 2010). In another example, retinol acetate SNEDDS was prepared using different concentrations of soybean oil (solvent), Cremophor EL (surfactant) and Capmul MCM-C8 (co-surfactant). The optimum surfactant to co-surfactant ratio was found to be 2:1, which resulted in minimum particle size in the range of 51–103 nm (Taha et al. 2004). Also, lutein, a potent antioxidant, was formulated via SNEDDS in an optimal way containing 25 wt% oil (Phosal 53 MCT), 60 wt% surfactant (Labrasol) and 15 wt% co-surfactant (Transcutol-HP or Lutrol-E400). Furthermore, the physical mixture of the optimized SNEDDS and Aerosil 200 exhibited instantaneous dissolution of the drug within 5 minutes (Yoo et al. 2010).

3.5.2 Encapsulation of Antioxidants via EPI Method

EPI is also an isothermal method which has been used to encapsulate compounds. In most cases, food-grade non-ionic surfactants, e.g. Tweens, are regarded to be the best choice for formation of nano-sized emulsions by EPI method. In one of early studies, essential oils incorporated into isolated soybean protein (ISP) composite edible films through EPI. The essential oils chosen were cinnamon oil and oregano oil, both as potent antioxidants. In this nanoemulsion system consisting of Acetem 90–50 K and Tween 60, it was revealed that at a high surfactant level, increasing the stirring speed (700 to 1300 rpm) led to a decease in the particle size (190 to 120 nm) (Bilbao-Sáinz et al. 2010). In order to prevent rapid Ostwald ripening, these water-soluble essential oils need to be mixed with a carrier oil to form stable nanoemulsions, thus, only certain types of oil are suitable to be used in EPI method. This was confirmed in another study, in which a variety of oil types (most of them have considerable antioxidant activity) were encapsulated in O/W emulsions using Tween 80 as the surfactant. It was found that emulsions formed by flavor oils (e.g. orange oil and limonene), had smaller particle diameter than those that made of LCT oils (e.g. grape seed, peanut, sesame, olive and canola oils). This work also declared the major disadvantage of the EPI method; relatively high amounts of surfactant (SOR > 0.7) were used to produce small oil droplets (d < 160 nm) (Ostertag et al. 2012). In a related study, D-limonene nanoemulsions (using Tween 80 and LCT oil carriers) were prepared by EPI and it was found that the SOR value and the oil composition would significantly influence the turbidity and the mean particle diameter of emulsions. At a high concentration of surfactant (SOR = 1.5), O/W nanoemulsions could be obtained. Moreover, when the carrier oil contained less than 15 wt% olive oil, the nanoemulsions could be prepared (Li et al. 2013). The same scientists later observed that the particle diameter could increase slightly when nisin (a bioactive compound) was added to a nanoemulsion made through EPI with 4 wt% oil (D-limonene) and 6 wt% surfactant (Tween 80) (Zhang et al. 2014).

The potential of using the EPI method to form emulsion-based systems containing VE was investigated in some studies. For instance, in a study with 10 wt% oil (MCT + VE), it was found that oil composition, surfactant type and SOR affected particle size of nanoemulsions; there was a minimum in the mean droplet diameter (d = 41 nm) at 8 wt% VE, 2 wt% MCT and 20 wt% Tween 80. The authors concluded that EPI method is more effective at producing nanoemulsions at high SOR than high-energy methods (Mayer et al. 2013). Another related study reported the formation of VE-loaded nanoemulsion prepared with different carrier oils (LCT, MCT, and SCT oils) and Tween 80, which resulted in different droplet diameters. Regardless of oil type, all samples showed physical stability to heat shock (30–90 °C, 30 min), ionic strength (0–500 mM) and long-term storage (60 days). However, a significant difference in their VE degradation kinetic profiles in heat processing and long-term storage conditions was observed. SCT based nanoemulsions did not physically withstand temperatures above 25 °C, while VE in LCT nanoemulsions degraded at slow rate at both low and high temperatures (Hategekimana et al. 2015). The same authors investigated the thermal stability of VE-loaded emulsions made by EPI. It was found that emulsions made with 10 wt% oil composed of VE and tributyrin (a short chain triglyceride, SCT) and 10 wt% surfactant (Tween 80) were stable when exposed (for 30 minutes) to temperatures less than 75 °C, but unstable at temperatures greater than or equal to 75 °C. With increasing temperature, the water solubility of tributyrin increased which result in Ostwald ripening (Hategekimana and Zhong 2015).

Curcumin a natural antioxidant dye, was encapsulated by EPI method in nanoemulsions composed of soybean oil, Tween 80 and glycerol (the co-solvent). In this study, the most stable formulations were consisting of 20 wt% soybean oil, 10 wt% Tween 80 and 20 wt% glycerol and were generated with an anchor blade impeller conducting at 300 rpm; these conditions brought about 0.07% encapsulation of the curcumin. After 60 days, still 70% of the original amount of curcumin remained in the nanoemulsions (Borrin et al. 2016). Using the same protocol, the same authors reported the incorporation of curcumin-loaded nanoemulsions (produced via EPI)

Table 3.3 An overview of the optimized formulations used in the most recent studies to encapsulate antioxidants in emulsion systems formed via EPI

Method	Oil (wt%) + Compound with antioxidant activity (wt%)	Surfactant(s) (wt%)	Co-solvent (wt%)	Preparation temp. (°C)	Smallest droplet diameter ≈ (nm)	Reference
EPI (CPI)	Acetem (12) + regano oil (8)	Tween 60 (10–20)	–	RT	115–170	Bilbao-Sáinz et al. (2010)
	Acetem (12) + Cinnamon oil (8)					
	Orange oil (10)	Tween 80 (25)			< 300	Ostertag et al. (2012)
	Limonene (10)				< 400	
	MCT (2) + VE (8)	Tween 80 (20)			40	Mayer et al. (2013)
	Tributyrin (2) + VE (8)	Tween 80 (10)		Not stated	110	Hategekimana and Zhong (2015)
	Olive oil (≤ 0.4) + D-limonene (≤ 3.6)	Tween 80 (6)	Propylene glycol (30)		40	Li et al. (2013)
	Nisin (0.5–3) + D-limonene (4)				16–19	Zhang et al. (2014)
	Soybean oil (20) + Curcumin (0.07)	Tween 80 (20)	Glycerol (20)	RT	200	Borrin et al. (2016)

in pineapple ice creams, as a feasible substitute for artificial yellow dyes. This replacement only had a small impact on the physicochemical properties of the products (Borrin et al. 2018).

A summary of aforementioned studies on emulsion-based delivery systems that have been generated to encapsulate compounds with antioxidants activity is presented in Table 3.3. Optimized formulations that resulted in the formation of most stable system with the smallest particle size can be seen in this table.

3.5.3 Encapsulation of Antioxidants via PIT Method

An intrinsic limitation of PIT method for encapsulation of sensitive compounds is that it involves heating the nanoemulsions. The PIT of most surfactant-oil-water systems has been reported to be less than 90 °C (Anton and Vandamme 2009; Rao and McClements 2010). Heating may favor thermal degradation and loss of volatile active ingredients, which could decrease the bioactivity of the nanoemulsions produced (Komaiko and McClements 2016). In this context, it is preferred to utilize a lower PIT during the preparation of the nanoemulsions to prevent excessive thermal degradation.

Taking into account the biological properties of the essential oils, a number of studies have applied PIT method to generate essential oil nanoemulsions, and have studied some of the most important parameters that affect their formation and stability. In one of the first studies, PIT method was employed to prepare lemon oil-loaded colloidal dispersions using Tween 80 as nonionic surfactant. The impact of SOR on the formation and stability of fabricated emulsions was investigated. It was found that different systems could be formed by simple heat treatment (90 °C, 30 min) depending on the SOR: emulsions ($d > 200$ nm) at $SOR < 1$; nanoemulsions ($d < 200$ nm) at $1 < SOR > 2$ and microemulsions ($d < 20$ nm) at $SOR > 2$. The authors concluded that the employing this heating treatment seems to be more efficient than the application of mechanical energy (microfluidization, blending or sonication) at overcoming the energy barrier that exists at ambient temperature (Rao and Mcclements 2011). Lemon oil was also emulsified by Tween 20 and NaCas through PIT approach in another report. Turbidity and rheology results demonstrated a PIT between 80 and 90 °C. The nanoemulsions formulated with 2% w/v NaCas, 0.4–1.2% w/v Tween 20 and 1.5% w/v lemon oil had an approximate volume-area mean diameter of 100 nm and were stable during 15-day storage. This study approved the usage of NaCas as a partial replacement for synthetic surfactants in preparing flavor oils nanoemulsions (Su and Zhong 2016). Combination of surfactants was used to fabricate nanoemulsions in another study too, in which PIT method was applied to develop cinnamon bark oil (CBO)-loaded nanoemulsion that contained a combination of cationic (lauric arginate, LAE) and nonionic (Tween 80) surfactants. CBO was homogenized at 1 wt% in the aqueous phase with 3 wt% Tween 80 and 0.05–0.375% wt% LAE, along with heating at 90 °C for 30 min to obtain final emulsions. At $LAE \geq 0.125$ wt%, transparent emulsions with hydrodynamic diameter of ≈ 100 nm were observed to be stable during 30-day storage at 21 °C (Hilbig et al. 2016). Preparation of stable cinnamon oil-loaded nanoemulsions using PIT method was reported in another study as well, in which they possessed the lowest PIT (≈ 70 °C) and the smallest droplet diameter (101 nm) when using 40:60 wt% of cinnamon oil and medium chain triglyceride (MCT) in the total lipid phase. The nanoemulsions were stable for at least 31 days when stored at 4 °C or 25 °C (Chuesiang et al. 2018).

In efforts to encapsulate essential oils extracted from plant species, oregano (*Origanum vulgare*) oil was encapsulated in nanoemulsions using PIT method, where a combination of surfactants (Cremophor RH 40 + Span 80) was used to emulsify the lipid phase (oregano oil + sunflower oil). The value of PIT was calculated from the conductivity curves as 44 °C for the nanoemulsions produced with 5 g oregano oil/100 g, which were the most stable ones over the storage period of 45 days and had droplet diameter of 35–55 nm (Moraes-Lovison et al. 2017). Another essential oil, Spearmint oil (SMO), which can be obtained from Mentha spicata leaves, was formulated in nanoemulsions as carrier for targeting SMO to oral cancer cell. The most stable and transparent nanoemulsions for encapsulation of SMO and virgin coconut oil were fabricated via PIT method by using Cremophor® RH 40 surfactant. Transparent nanoemulsions (with droplet diameter <80 nm) were obtained at SMO to virgin coconut oil ratios in the limited range from 40:60 to 80:20 (Tubtimsri et al. 2018). Coconut oil was also used in another report for preparing stable and uniform curcuminoid-loaded nanoemulsions (through PIT method) again with using Cremophor® RH40 as the surfactant. The results revealed that at surfactant content of ≤10 wt%, an increase in curcuminoid concentration resulted in the migration of curcuminoids to the oil phase, confirmed by increasing the droplet size. The authors concluded that the optimum surfactant concentration and storage condition for curcuminoid-loaded nanoemulsions were 10 wt% and 4 °C, respectively (Jintapattanakit et al. 2018). In another study, stable nanoemulsions (with a droplet diameter < 100 nm) prepared by PIT approach, were used to encapsulate the essential oil from the leaves of *Cymbopogon densiflorus*. The formulation was composed of the nonionic surfactants (Span 80, 4 wt%; Croduret 50™, 6 wt%) with an oil phase consisted of soybean oil (5 wt%) and essential oil derived from leaves of *C. densiflorus* (5 wt%) and finally the aqueous phase (80 wt%). These nanoemulsions were able to reproduce the antioxidant activity at a concentration four times lower than that of the pure essential oil. They gave a 50% reduction of the initial levels of the free radicals DPPH (2,2-diphenyl-1-picrylhydrazyl) and ABTS (2,2′-azino-bis(3-ethylbenzothiazoline-6-sulphonic acid)), compared to the control nanoemulsion (Seibert et al. 2019). In a recent study, a black pepper essential oil nanoemulsion was formulated by PIT method. The emulsification process that gave the homogeneous and smallest droplets (17.9 nm) should be carried out with the ratio between Tween 80 and the oil phase (SOR) of 2.25:1, under the heating temperature of 75 °C (Vinh et al. 2020).

PIT method also was used recently to encapsulate vitamin D_3. Out of all formulations prepared, the high emulsion stability, high encapsulation efficiency (≈ 94%), small particle size (≈ 40 nm) and greater zeta potential (−18 mV) were achieved in nanoemulsion fabricated with 30% (v/v) of surfactant (Kolliphor HS 15), 20% (v/v) of carrier oil (caprylic-/capric triglyceride) and 50% (v/v) of water. Release of vitamin D_3 from this nanoemulsion was sustained over a significant period of time which is indicative of high bioavailability of the encapsulated vitamin D_3. At the end, the authors performed sensory evaluation and confirmed the suitability of the encapsulated vitamin D for the purpose of fortification for beverages (Maurya and Aggarwal 2019).

All above-mentioned studies on the application of PIT method to encapsulate compounds with antioxidants activity in emulsion-based systems are summarized in Table 3.4. Optimized formulations including the thermal treatment that led to the formation of most stable system with the smallest particle diameter can be found in this table.

Table 3.4 An overview of the optimized formulations used in the most recent studies to encapsulate antioxidants in emulsion systems formed via PIT

Method	Oil (conc.) + Compound with antioxidant activity (conc.)	Surfactant(s) (conc.)	Co-solvent (conc.)	Thermal treatment	Smallest droplet diameter ≈ (nm)	Reference
PIT	Lemon oil (10 wt%)	Tween 80 (7–13 wt%)	Propylene glycol (25–28 wt%)	Mixing all components for 30 min at RT, then heating the mixture to 90 °C at 0.8 °C min^{-1}, keeping the temperature at 90 °C for 30 min, followed by rapid cooling to RT	< 200	Rao and Mcclements (2011)
		Tween 80 (20 wt%)	Propylene glycol (23 wt%)		< 20	
	Lemon oil (1.5% w/v)	Tween 20 (0.4–1.2% w/v) + NaCas (2% w/v)	–	Heating coarse emulsions up to 98 °C for 0–3 h, followed by quick quenching in ice/water with hand shaking	100	Su and Zhong (2016)
	Cinnamon bark oil (1 wt%)	Tween 80 (3 wt%) + LAE (0.125–0.375 wt%)		Heating the homogenized emulsions at 90 °C for 30 min followed by quenching in ice/water	100	Hilbig et al. (2016)
	Sunflower oil (5 wt%) + Oregano oil (5 wt%)	Cremophor RH 40 (12 wt%) + Span 80 (8 wt%)		Heating all components to 20 °C above PIT, then fast cooling the mixture to RT at 10 °C/min, repeating the heating/cooling cycles	35–45	Moraes-Lovison et al. (2017)

(continued)

Table 3.4 (continued)

Method	Oil (conc.) + Compound with antioxidant activity (conc.)	Surfactant(s) (conc.)	Co-solvent (conc.)	Thermal treatment	Smallest droplet diameter ≈ (nm)	Reference
PIT	Coconut oil (2–6 wt%) + Spearmint oil (4–8 wt%)	Cremophor RH 40 (10 wt%)	–	Heating oils and surfactant mixture to 62 °C, then adding the mixture to hot water (65 °C), homogenizing the mixture and cooling to RT	50–70	Tubtimsri et al. (2018)
	Coconut oil (8.3 wt%) + Curcuminoid (≤ 1.67 mg/g)	Cremophor RH 40 (13–20 wt%)		Heating the mixture of all components to 90 °C followed by cooling down to 60 °C, repeating this cycle for 3 times. In the last cycle when T = 75 °C, rapid cooling the system by dilution with twice amount of cold water.	32–40	Jintapattanakit et al. (2018)
	MCT (6 wt%) + Cinnamon oil (4 wt%)	Tween 80 (10 wt%)		All mixing all components for 30 min at RT, then heating the mixture to 15 °C above PIT, next the temperature was decreased to PIT followed by a rapid cooling by adding cold water (4 °C) to the system with stirring form 3 min	101	Chuesiang et al. (2018)

(continued)

Table 3.4 (continued)

Method	Oil (conc.) + Compound with antioxidant activity (conc.)	Surfactant(s) (conc.)	Co-solvent (conc.)	Thermal treatment	Smallest droplet diameter ≈ (nm)	Reference
PIT	Soybean oil (5 wt%) + Oil of *C. densiflorus* leaves (5 wt%)	Span 80 (4 wt%) + Croduret 50 (6 wt%)	–	Heating oils/ surfactant mixture to 75 °C, then adding hot water (75 °C) to the mixture, keep stirring until the temperature reaches to RT	70	Seibert et al. (2019)
	Caprylic/capric triglyceride (CCTG) (20% v/v) + Vitamin D3 (2% w/v) + Leciva (2.5% w/v)	Kolliphor HS 15 (30% v/v)		Heating the mixture of all components to 85 °C, then cooling to 65 °C, repeating this cycle for 5 times. In the last cycle when T = 85 °C, titration of obtained clear brown mixture against hot water (65 °C) with stirring.	40	Maurya and Aggarwal (2019)
	Black pepper essential oil (4.3 wt%)	Tween 80 (9.7 wt%)		Slowly adding oil/surfactant mixture to water under stirring at 75 °C for 30 min, followed by fast cooling to 10 °C within 15 min	18	Vinh et al. (2020)

3.5.4 Encapsulation of Antioxidants via ME Method

Advantages of ME method for food applications include the low shear properties and the homogeneous size distribution that present high encapsulation efficiency and improved control of the release properties. Encapsulation of compounds with antioxidant activity through ME method has been reported in various articles, the most relevant ones are summarized in this section.

In one of the early studies, astaxanthin-loaded O/W emulsions were produced via PME approach. Pre-emulsions were generated by dispersing MCT (containing astaxanthin) in water. The oil droplets were stabilized with a combination of two surfactants, Tween 20 and whey protein isolate (WPI) in 1:1 ratio. To achieve more uniform droplets with a narrower size distribution over a wide range of dispersed phase fraction and pressures applied in the experiments, each O/W emulsion passed three times through the membrane under pressures of 5–15 bar and disperse phase fractions from 10 wt% to 40 wt%. However, encapsulation was not enough to completely inhibit the oxidative degradation of astaxanthin which was nearly 30% after 3 weeks (Ribeiro et al. 2005). In another study, O/W emulsions with encapsulated β-carotene in the oil phase (sunflower oil) were produced by PME using a polymeric membrane such as nitrocellulose mixed esters (MCE) and food grade proteins such as bovine serum albumin (BSA) and whey protein concentrate (WPC) as the main surfactant and nonionic Tween 20 as the co-surfactant. The combination of WPC or BSA with Tween 20, reduced the protein membrane fouling and at the same time decreased the mean droplet size to values close to the operating limit, i.e., the mean pore size (0.8 μm) of the membrane. In addition, it was shown that use of these proteins has shielding effect against degradation of β-carotene during storage at 35 °C compared to when using only Tween 20 (Trentin et al. 2011). The same research group reported the application of PME method to produce W/O/W double emulsions, containing procyanidin (having high antioxidant activity) stabilized with several WPI-polysaccharide soluble complexes (Berendsen et al. 2015a). Subsequently, in another report, they used spray drying to convert these W/O/W emulsions into powder form microcapsules. Among different emulsifiers which were used to stabilize the emulsion, WPI-CMC stabilized microcapsules showed the highest procyanidin content (5.3 g kg^{-1}) and gave the narrowest particle size distribution with the lowest particle size for both microcapsules and the corresponding emulsions after rehydration (7.7 and 9.9 μm, respectively) (Berendsen et al. 2015b). Another study also reported the efficiency of W/O/W double emulsions, made by ME, to encapsulate *trans*-resveratrol. In this work, using sodium carboxymethylcellulose (CMC) and Tween 20 as stabilizers, monodisperse food-grade emulsions with uniform droplets around 60 μm and span values in the range of ≈ 0.9 were provided (Matos et al. 2015).

Encapsulation of VE via ME method within nanoemulsions composed of MCT oil and surfactant mixture Tween80/Brij35 has been reported too. In this study, using a 0.9 μm SPG membrane, nanoemulsions with an average droplet size of 106 nm, a homogeneous size distribution (span factor = 0.30), a nearly high zeta-potential (−16.5 mV which is enough to inhibit droplet coalescence) and a high encapsulation efficiency (99.7%) were prepared. In addition, the obtained nano-emulsions exhibited good stability for at least 2 months (Laouini et al. 2012). The same research group, through ME technique, scaled-up the preparation of VE-loaded nanocapsules using optimal formulation obtained at the laboratory-scale. From laboratory-scale to pilot-scale, the important properties of the nanocapsules (e.g. droplet size and encapsulation efficiency) did not change significantly (Khayata et al. 2012). Later some of these authors, using the same protocol, reported the preparation of rosemary essential oil-loaded nanocapsules. The results exhibited that nanocapsules prepared at both scales were spherical with a mean droplet size around 220 nm (PDI < 0.25), negative zeta potential around -20 mV, good stability over time and high encapsulation efficiency (≈99%) for most of the rosemary essential oil components (Ephrem et al. 2014).

Biophenols are highly appreciated for their free radical scavenging and antioxidant activities (Obied et al. 2005). ME has been used for encapsulation of biophenols. For example, in a study catechol was used as the biophenol model, while a biophenols mixture extracted from olive mill waste was used as the real matrix. Using limonene as the continuous phase, 15 wt% PVA as the dispersed phase and 2 wt% Span 80 as the surfactant, W/O emulsions with droplet size approximately 2.3 times the membrane pore diameter, a span factor of 0.33, and high encapsulation efficiency (98% ± 1% and 92% ± 3%, for catechol and biophenols, respectively) were produced (Piacentini et al. 2016). One year later, the same authors reported the encapsulation of hydroxytyrosol, a compound with antioxidant properties from the olive biophenol family, in solid lipid particles using cold ME. The optimized process gave particle size (4 μm) nearly 3.6 times of the membrane por size and a span factor of 0.45. In comparison to the particles achieved by homogenization, the encapsulation efficiency of the particles obtained by ME was almost 2 times higher and also the energy consumption associated with ME process was two order of magnitude lower (Bazzarelli et al. 2017).

Tables 3.5 and 3.6 summarize all aforementioned reports on the encapsulation of compounds with antioxidant activity in O/W and in W/O/W double emulsion systems, respectively. Formulation parameters including the ME preparation method that resulted in the formation of most stable system with the smallest particle diameter can be found in these tables.

Table 3.5 An overview of the optimized conditions used in the most recent studies to encapsulate antioxidants in O/W emulsion systems formed via ME

Method	Oil (conc.) + Compound with antioxidant activity (conc.)	Surfactant(s) (conc.)	Membrane type/ material	Pore size (μm)	ME preparation method	Droplets diameter (Polydispersity, Span)	Reference
ME	MCT (10–40 wt%) + Astaxanthin (9 mmol/L)	Tween 20 (0.5 wt%) + WPI (0.5 wt%)	Polyamid 6, 6	0.8	Dispersing hot oil phase into the cold aqueous phase containing surfactants by a rotor-stator-system, then each emulsion was passed the membrane 3 times under pressures of 5 to 15 bar	(≈ 1)	Ribeiro et al. (2005)
	Sunflower oil (10% v/v) + β-carotene (0.3 wt%)	WPC (1 or 2 wt%) +Tween 20 (2 wt%)	Nitrocellulose mixed ester	0.8	Preparing a coarse emulsion by adding the oil phase to the water phase and mixing in a rotor–stator system, then passing the emulsion 4–5 times through the membrane forced with N_2 (500 or 900 kPa)	1.28–1.69 μm (< 1)	Trentin et al. (2011)
		BSA (1 or 2 wt%) +Tween 20 (2 wt%)				(7)	
	MCT (17.75 wt%) + VE (5 wt%)	Tween 80 (1.12 wt%) + Brij35 (1.12 wt%)	Tubular SPG	0.9	Placing oil phase in the pressurized vessel (2.4 bar), then pumping water phase containing surfactants via the membrane module (33 ml/s). When the water phase arrived to the outlet of the membrane device, the valve of the pressurized vessel was opened, so the oil phase permeated via the membrane pores into the water phase	106 nm (0.30)	Laouini et al. (2012)

(continued)

Table 3.5 (continued)

Method	Oil (conc.) + Compound with antioxidant activity (conc.)	Surfactant(s) (conc.)	Membrane type/ material	Pore size (μm)	ME preparation method	Droplets diameter (Polydispersity, Span)	Reference
ME	Polycaprolactone (125 mg) + VE (100 mg)	Tween 80 (50 mg)	Tubular SPG	0.9	Placing oil phase in the pressurized vessel (0.5 bar), while the water phase was circulating tangentially to the membrane surface (700 ml min^{-1}). At time t = 0, the valves connecting the pressurized vessel to N_2 and to the filtrate side of the membrane module were opened, which resulted in formation of nanoparticles.	172 nm (0.28)	Khayata et al. (2012)
	Polycaprolactone (750 mg) + Rosemary oil (2400 mg)	Tween 80 (1200 mg) + Span 20 (300 mg)				230 nm (< 0.25)	Ephrem et al. (2014)
	Cocoa butter (≤ 40 wt%) + Hydroxytyrosol (55 mg L^{-1})	Brij 78 (4 wt%)		1.1	Using pulsed back-and-forward ME, water phase was circulating (0.57 L min^{-1}) into lumen of the membrane, while the lipid phase was pressed through the membrane pores (under 0.25 to 0.55 bar). ME went on until the dispersed phase concentration was 30 or 40 wt%.	≈ 4 μm (≈ 0.4)	Bazzarelli et al. (2017)
	Lemon oil (5 wt%)	Tween 20 (2 wt%)	Microstructured nickel sieves	4.5 × 291.2 or 17.7 × 20.6	Placing coarse emulsion in the pressure tank and pushing it through the membrane module using N_2 (150 to 450 kPa). The micro-sieve was placed at the bottom of module. The process was repeated 5 times, keeping the sieve in the holder	≈ 0.5 μm (1.2–1.7)	Kaade et al. (2019)

Table 3.6 An overview of the optimized conditions used in recent studies to encapsulate antioxidants in W/O/W double emulsions formed via ME

Method	Oil phase (Conc.)	Antioxidant compound in water phase (Conc.)	Hydrophilic surfactant(s) (Conc.)	Membrane type/ material	Pore size (μm)	ME preparation method	Final droplet diameter (Polydispersity, Span)	Reference
ME	Sunflower oil (14 wt%) + PGPR (0.6 wt%)	Vitaflavan source of procyanidin (0.6 wt%)	WPI – polysaccharide complexes (2 wt%)	SPG	10	Loading W/O/W coarse emulsions into the membrane with N_2 pressures (150–800 kPa). Repeating the process 3 times	8–12 μm (≈ 1)	Berendsen et al. (2015a)
	Sunflower oil (9 wt%) + PGPR (0.36 wt%)	Vitaflavan source of procyanidin (0.38 wt%)	WPI-CMC (2 wt%) + Maltodextrin (27 wt%)				7.7 μm (≈ 1)	Berendsen et al. (2015b)
	Miglyol 812 (16% v/v) + PGPR (5% w/v)	*trans*-resveratrol (50 mg/L)	Tween 20 (2% w/v) + CMC (0.5% w/v)			Feeding primary W_1/O emulsion via the membrane under pressure 20–80 kPa, while pumping the external water phase at a flow rate of 150 L/h. Three replicates of each experiment were done.	60 μm (≈ 0.9)	Matos et al. (2015)

3.6 Summary and Future Perspectives

Antioxidants are compounds that inhibit oxidation reactions. They are used during food processing as food additives to increase the functional properties of the food by scavenging the free radicals. Emulsion-based systems are of great interest as delivery systems for bioactive compounds because their small size leads to high physical stability, good optical clarity, and high bioavailability, features that are important for their application in many foods and drugs. Formulated with cost-saving low-energy methods, these emulsions can find application in encapsulation of shear and heat sensitive antioxidants.

In this chapter, first we reviewed different low-energy approaches (i.e. SE, EPI, PIT and ME) for the preparation of emulsion-based systems; characteristics, feasibility and limitations of each method in terms of physical properties, stability and microstructure of obtained colloidal system were discussed. Second, we presented several examples that demonstrate the potential of low-energy emulsification methods for encapsulation of compounds with antioxidant properties.

Despite all studies carried out in this field, there is little understanding of the possible industrial pertinence of many of low-energy approaches as the physics of emulsion formation through these methods is still semi-empirical and rational scale-up protocols have not been extensively explored. On the other hand, there are miscellaneous regulatory features that should be considered to allow the broad applications of these systems. First of all, most of the ingredients used in formulation of emulsion-based systems in low-energy approaches are unsuitable for widespread utilization within the food industry such as synthetic oils, surfactants or organic solvents. In this context, low-energy emulsification approaches may not be proper for developing food and drug products that contain relatively high contents of oil, because naturally high levels of synthetic surfactants (such as Tweens and Spans) would be needed in the final product. Contrarily, they are appropriate for applications that only a low amount of oil in the final product is required such as in fortified waters. As potential alternatives, natural food-grade ingredients such as proteins, phospholipids, polysaccharides, flavor oils and triglyceride oils should be utilized in their formulations as these ingredients are legally approved, label-friendly and economically practicable. Secondly, it is essential to optimize the bioactivity and bioavailability of the encapsulated antioxidant for scaled-up production, as there are certain safety factors associated with the influence of particle size, charge and composition on the absorption and digestion of encapsulated compound. Therefore, additional investigations should direct on the biological events and risks related with the use of emulsion-based delivery systems in food and drug products to guarantee the safety of the consumers.

References

Ahmed K, Li Y, McClements DJ, Xiao H (2012) Nanoemulsion- and emulsion-based delivery systems for curcumin: encapsulation and release properties. Food Chem 132:799–807. https://doi.org/10.1016/j.foodchem.2011.11.039

Alliod O, Valour JP, Urbaniak S et al (2018) Preparation of oil-in-water nanoemulsions at large-scale using premix membrane emulsification and Shirasu Porous Glass (SPG) membranes. Colloids Surf A Physicochem Eng Asp 557:76–84. https://doi.org/10.1016/j.colsurfa.2018.04.045

An Y, Yan X, Li B, Li Y (2014) Microencapsulation of capsanthin by self-emulsifying nanoemulsions and stability evaluation. Eur Food Res Technol 239:1077–1085. https://doi.org/10.1007/s00217-014-2328-3

Anton N, Vandamme TF (2009) The universality of low-energy nano-emulsification. Int J Pharm 377:142–147. https://doi.org/10.1016/j.ijpharm.2009.05.014

Anton N, Benoit J, Saulnier P (2008) Design and production of nanoparticles formulated from nano-emulsion templates — a review. J Control Release 128:185–199. https://doi.org/10.1016/j.jconrel.2008.02.007

Asyikin Z, Aziz A, Nasir HM et al (2019) Enrichment of Eucalyptus oil nanoemulsion by micellar nanotechnology: transdermal analgesic activity using hot plate test in rats' assay. Sci Rep:1–16. https://doi.org/10.1038/s41598-019-50134-y

Bakkali F, Averbeck S, Averbeck D, Idaomar M (2008) Biological effects of essential oils – a review. Food Chem Toxicol 46:446–475. https://doi.org/10.1016/j.fct.2007.09.106

Barbé C, Bartlett J, Kong L et al (2004) Silica particles: a novel drug-delivery system. Adv Mater 16:1959–1966. https://doi.org/10.1002/adma.200400771

Barreira JCM, Ferreira ICFR (2019) Artificial antioxidants. In: Melton L, Shahidi F, Varelis P (eds) Encyclopedia of food chemistry. Elsevier, pp 283–290

Barzegar H, Mehrnia MA, Nasehi B, Alipour M (2018) Fabrication of peppermint essential oil nanoemulsions by spontaneous method: effect of preparing conditions on droplet size. Flavour Fragr J:1–6. https://doi.org/10.1002/ffj.3455

Bazzarelli F, Piacentini E, Giorno L (2017) Biophenols-loaded solid lipid particles (SLPs) development by membrane emulsification. J Membr Sci 541:587–594. https://doi.org/10.1016/j.memsci.2017.07.029

Berendsen R, Güell C, Ferrando M (2015a) A procyanidin-rich extract encapsulated in water-in-oil-in-water emulsions produced by premix membrane emulsification. Food Hydrocoll 43:636–648. https://doi.org/10.1016/j.foodhyd.2014.07.023

Berendsen R, Güell C, Ferrando M (2015b) Spray dried double emulsions containing procyanidin-rich extracts produced by premix membrane emulsification: effect of interfacial composition. Food Chem 178:251–258. https://doi.org/10.1016/j.foodchem.2015.01.093

Bidone J, Zorzi GK, Carvalho ELS et al (2014) Incorporation of Achyrocline satureioides (Lam.) DC extracts into topical nanoemulsions obtained by means of spontaneous emulsification procedure. Ind Crop Prod 62:421–429. https://doi.org/10.1016/j.indcrop.2014.08.042

Bilbao-Sáinz C, Avena-Bustillos RJ, Wood DF et al (2010) Nanoemulsions prepared by a low-energy emulsification method applied to edible films. J Agric Food Chem 58:11932–11938. https://doi.org/10.1021/jf102341r

Borrin TR, Georges EL, Moraes ICF, Pinho SC (2016) Curcumin-loaded nanoemulsions produced by the emulsion inversion point (EIP) method: An evaluation of process parameters and physico-chemical stability. J Food Eng 169:1–9. https://doi.org/10.1016/j.jfoodeng.2015.08.012

Borrin TR, Georges EL, Brito-Oliveira TC et al (2018) Technological and sensory evaluation of pineapple ice creams incorporating curcumin-loaded nanoemulsions obtained by the emulsion inversion point method. Int J Dairy Technol 71:491–500. https://doi.org/10.1111/1471-0307.12451

Bouchemal K, Briançon S, Perrier E, Fessi H (2004) Nano-emulsion formulation using spontaneous emulsification: solvent, oil and surfactant optimisation. Int J Pharm 280:241–251. https://doi.org/10.1016/j.ijpharm.2004.05.016

Carteau D, Bassani D, Pianet I (2008) The "ouzo effect": following the spontaneous emulsification of trans-anethole in water by NMR. C R Chim 11:493–498. https://doi.org/10.1016/j.crci.2007.11.003

Chang Y, McClements DJ (2014) Optimization of orange oil nanoemulsion formation by isothermal low-energy methods: in fl uence of the oil phase, surfactant, and temperature. J Agric Food Chem 62:2306–2312. https://doi.org/10.1021/jf500160y

Chang Y, McLandsborough L, McClements DJ (2013) Physicochemical properties and antimicrobial efficacy of carvacrol nanoemulsions formed by spontaneous emulsification. J Agric Food Chem 61:8906–8913. https://doi.org/10.1021/jf402147p

Charcosset C (2009) Preparation of emulsions and particles by membrane emulsification for the food processing industry. J Food Eng 92:241–249. https://doi.org/10.1016/j.jfoodeng.2008.11.017

Chuesiang P, Siripatrawan U, Sanguandeekul R et al (2018) Optimization of cinnamon oil nanoemulsions using phase inversion temperature method: impact of oil phase composition and surfactant concentration. J Colloid Interface Sci 514:208–216. https://doi.org/10.1016/j.jcis.2017.11.084

Date AA, Desai N, Dixit R, Nagarsenker M (2010) Self-nanoemulsifying drug delivery systems: formulation insights, applications and advances. Nanomedicine 5:1595–1616. https://doi.org/10.2217/nnm.10.126

Daufin G, Escudier JP, Carrére H et al (2001) Recent and emerging applications of membrane processes in the food and dairy industry. Food Bioprod Process Trans Inst Chem Eng Part C 79:89–102. https://doi.org/10.1016/S0960-3085(01)70244-1

Davidov-Pardo G, McClements DJ (2015) Nutraceutical delivery systems: resveratrol encapsulation in grape seed oil nanoemulsions formed by spontaneous emulsification. Food Chem 167:205–212. https://doi.org/10.1016/j.foodchem.2014.06.082

Drioli E, Curcio E, Di Profio G (2005) State of the art and recent progresses in membrane contactors. Chem Eng Res Des 83:223–233. https://doi.org/10.1205/cherd.04203

Ephrem E, Greige-Gerges H, Fessi H, Charcosset C (2014) Optimisation of rosemary oil encapsulation in polycaprolactone and scale-up of the process. J Microencapsul 31:746–753. https://doi.org/10.3109/02652048.2014.918669

Espitia PJP, Fuenmayor CA, Otoni CG (2019) Nanoemulsions: synthesis, characterization, and application in bio-based active food packaging. Compr Rev Food Sci Food Saf 18:264–285. https://doi.org/10.1111/1541-4337.12405

Fernandez P, Rieger J, Angelika K (2004) Nano-emulsion formation by emulsion phase inversion. Colloids Surf A Physicochem Eng Asp 251:53–58. https://doi.org/10.1016/j.colsurfa.2004.09.029

Galli F, Azzi A, Birringer M et al (2017) Vitamin E: emerging aspects and new directions. Free Radic Biol Med 102:16–36. https://doi.org/10.1016/j.freeradbiomed.2016.09.017

Ganachaud F, Katz JL (2005) Nanoparticles and nanocapsules created using the ouzo effect: spontaneous emulsification as an alternative to ultrasonic and high-shear devices. ChemPhysChem 6:209–216. https://doi.org/10.1002/cphc.200400527

Ghai D, Sinha VR (2012) Nanoemulsions as self-emulsified drug delivery carriers for enhanced permeability of the poorly water-soluble selective β 1 -adrenoreceptor blocker Talinolol. Nanomed Nanotechnol Biol Med 8:618–626. https://doi.org/10.1016/j.nano.2011.08.015

Ghiasi Z, Esmaeli F, Aghajani M, Ghazi-khansari M (2019) Enhancing analgesic and anti-inflammatory effects of capsaicin when loaded into olive oil nanoemulsion : An in vivo study. Int J Pharm 559:341–347. https://doi.org/10.1016/j.ijpharm.2019.01.043

Gijsbertsen-Abrahamse AJ, Van Der Padt A, Boom RM (2004) Status of cross-flow membrane emulsification and outlook for industrial application. J Membr Sci 230:149–159. https://doi.org/10.1016/j.memsci.2003.11.006

Guéraud F, Atalay M, Bresgen N et al (2010) Chemistry and biochemistry of lipid peroxidation products. Free Radic Res 44:1098–1124. https://doi.org/10.3109/10715762.2010.498477

Guttoff M, Saberi AH, Mcclements DJ (2015) Formation of vitamin D nanoemulsion-based delivery systems by spontaneous emulsification : factors affecting particle size and stability. Food Chem 171:117–122. https://doi.org/10.1016/j.foodchem.2014.08.087

Hategekimana J, Zhong F (2015) Degradation of vitamin E in nanoemulsions during storage as affected by temperature, light and darkness. Int J Food Eng 11:199–206. https://doi.org/10.1515/ijfe-2014-0256

Hategekimana J, Chamba MVM, Shoemaker CF et al (2015) Vitamin E nanoemulsions by emulsion phase inversion: effect of environmental stress and long-term storage on stability and degradation in different carrier oil types. Colloids Surf A Physicochem Eng Asp 483:70–80. https://doi.org/10.1016/j.colsurfa.2015.03.020

Hilbig J, Ma Q, Davidson PM et al (2016) Physical and antimicrobial properties of cinnamon bark oil co-nanoemulsified by lauric arginate and tween 80. Int J Food Microbiol 233:52–59. https://doi.org/10.1016/j.ijfoodmicro.2016.06.016

Jahanzad F, Crombie G, Innes R, Sajjadi S (2009) Catastrophic phase inversion via formation of multiple emulsions: a prerequisite for formation of fine emulsions. Chem Eng Res Des 87:492–498. https://doi.org/10.1016/j.cherd.2008.11.015

Jintapattanakit A, Hasan HM, Junyaprasert VB (2018) Vegetable oil-based nanoemulsions containing curcuminoids: formation optimization by phase inversion temperature method. J Drug Deliv Sci Technol 44:289–297. https://doi.org/10.1016/j.jddst.2017.12.018

Kaade W, Ferrando M, Khanmohammed A et al (2019) Low-energy high-throughput emulsification with nickel micro-sieves for essential oils encapsulation. J Food Eng 263:326–336. https://doi.org/10.1016/j.jfoodeng.2019.07.006

Khayata N, Abdelwahed W, Chehna MF et al (2012) Preparation of vitamin e loaded nanocapsules by the nanoprecipitation method: from laboratory scale to large scale using a membrane contactor. Int J Pharm 423:419–427. https://doi.org/10.1016/j.ijpharm.2011.12.016

Kobayashi I, Nakajima M, Mukataka S (2003) Preparation characteristics of oil-in-water emulsions using differently charged surfactants in straight-through microchannel emulsification. Colloids Surf A Physicochem Eng Asp 229:33–41. https://doi.org/10.1016/j.colsurfa.2003.08.005

Komaiko JS, McClements DJ (2016) Formation of food-grade nanoemulsions using low-energy preparation methods: a review of available methods. Compr Rev Food Sci Food Saf 15:331–352. https://doi.org/10.1111/1541-4337.12189

Koroleva MY, Yurtov EV (2012) Nanoemulsions: the properties, methods of preparation and promising applications. Russ Chem Rev 81:21–43. https://doi.org/10.1070/rc2012v081n01abeh004219

Laouini A, Fessi H, Charcosset C (2012) Membrane emulsification: a promising alternative for vitamin E encapsulation within nano-emulsion. J Membr Sci 423–424:85–96. https://doi.org/10.1016/j.memsci.2012.07.031

Li Y, Zhang Z, Yuan Q et al (2013) Process optimization and stability of d-limonene nanoemulsions prepared by catastrophic phase inversion method. J Food Eng 119:419–424. https://doi.org/10.1016/j.jfoodeng.2013.06.001

Lippacher A, Mu RH (2004) Liquid and semisolid SLN dispersions for topical application: rheological characterization. Eur J Pharm Biopharm 58:561–567. https://doi.org/10.1016/j.ejpb.2004.04.009

Mansour RB, Gargouri B, Bouaziz M et al (2011) Antioxidant activity of ethanolic extract of inflorescence of Ormenis Africana in vitro and in cell cultures. Lipids Health Dis 10:78. https://doi.org/10.1186/1476-511X-10-78

Matos M, Gutiérrez G, Iglesias O et al (2015) Enhancing encapsulation efficiency of food-grade double emulsions containing resveratrol or vitamin B12 by membrane emulsification. J Food Eng 166:212–220. https://doi.org/10.1016/j.jfoodeng.2015.06.002
Maurya VK, Aggarwal M (2019) A phase inversion based nanoemulsion fabrication process to encapsulate vitamin D3 for food applications. J Steroid Biochem Mol Biol 190:88–98. https://doi.org/10.1016/j.jsbmb.2019.03.021
Mayer S, Weiss J, McClements DJ (2013) Vitamin E-enriched nanoemulsions formed by emulsion phase inversion: factors influencing droplet size and stability. J Colloid Interface Sci 402:122–130. https://doi.org/10.1016/j.jcis.2013.04.016
McClements DJ (2011) Edible nanoemulsions: fabrication, properties, and functional performance. Soft Matter 7:2297–2316. https://doi.org/10.1039/c0sm00549e
McClements DJ (2012) Nanoemulsions versus microemulsions: terminology, differences, and similarities. Soft Matter 8:1719–1729. https://doi.org/10.1039/C2SM06903B
McClements DJ (2013) Nanoemulsion-based oral delivery systems for lipophilic bioactive components: nutraceuticals and pharmaceuticals. Ther Deliv 4:841–857. https://doi.org/10.4155/tde.13.46
McClements DJ, Rao J (2011) Food-grade nanoemulsions: formulation, fabrication, properties, performance, biological fate, and potential toxicity. Crit Rev Food Sci Nutr 51:285–330. https://doi.org/10.1080/10408398.2011.559558
Mehrnia M, Jafari S, Makhmal-zadeh BS (2016) Crocin loaded nano-emulsions: factors affecting emulsion properties in spontaneous emulsification. Int J Biol Macromol 84:261–267. https://doi.org/10.1016/j.ijbiomac.2015.12.029
Mishra B, Patel BB, Tiwari S (2010) Colloidal nanocarriers : a review on formulation technology , types and applications toward targeted drug delivery. Nanomed Nanotechnol Biol Med 6:9–24. https://doi.org/10.1016/j.nano.2009.04.008
Moraes-Lovison M, Marostegan LFP, Peres MS et al (2017) Nanoemulsions encapsulating oregano essential oil: production, stability, antibacterial activity and incorporation in chicken pâté. LWT – Food Sci Technol 77:233–240. https://doi.org/10.1016/j.lwt.2016.11.061
Nakashima T, Shimizu M, Kukizaki M (1991) Membrane emulsification by microporous glass. Inorg Membr ICIM2-91 62:513–516. https://doi.org/10.4028/www.scientific.net/kem.61-62.513
Nakashima T, Shimizu M, Kukizaki M (2000) Particle control of emulsion by membrane emulsification and its applications. Adv Drug Deliv Rev 45:47–56. https://doi.org/10.1016/S0169-409X(00)00099-5
Nedovic V, Kalusevic A, Manojlovic V et al (2011) An overview of encapsulation technologies for food applications. Procedia Food Sci 1:1806–1815. https://doi.org/10.1016/j.profoo.2011.09.265
Nepal PR, Han HK, Choi HK (2010) Preparation and in vitro-in vivo evaluation of Witepsol® H35 based self-nanoemulsifying drug delivery systems (SNEDDS) of coenzyme Q10. Eur J Pharm Sci 39:224–232. https://doi.org/10.1016/j.ejps.2009.12.004
Obied HK, Allen MS, Bedgood DR et al (2005) Bioactivity and analysis of biophenols recovered from olive mill waste. J Agric Food Chem 53:823–837. https://doi.org/10.1021/jf048569x
Ostertag F, Weiss J, McClements DJ (2012) Low-energy formation of edible nanoemulsions: factors influencing droplet size produced by emulsion phase inversion. J Colloid Interface Sci 388:95–102. https://doi.org/10.1016/j.jcis.2012.07.089
Perazzo A, Preziosi V, Guido S (2015) Phase inversion emulsification: current understanding and applications. Adv Colloid Interf Sci 222:581–599. https://doi.org/10.1016/j.cis.2015.01.001
Piacentini E, Poerio T, Bazzarelli F, Giorno L (2016) Microencapsulation by membrane emulsification of biophenols recovered from olive mill wastewaters. Membranes (Basel) 6. https://doi.org/10.3390/membranes6020025

Rao J, McClements DJ (2010) Stabilization of phase inversion temperature nanoemulsions by surfactant displacement. J Agric Food Chem 58:7059–7066. https://doi.org/10.1021/jf100990r
Rao J, Mcclements DJ (2011) Formation of flavor oil microemulsions, nanoemulsions and emulsions: influence of composition and preparation method. J Agric Food Chem 59(9):5026–5035
Ribeiro HS, Rico LG, Badolato GG, Schubert H (2005) Production of O/W repeated, containing Astaxanthin by premix membrane emulsification. J Food Sci 70:E117–E123. https://doi.org/10.1111/j.1365-2621.2005.tb07083.x
Saberi AH, Fang Y, McClements DJ (2013a) Fabrication of vitamin E-enriched nanoemulsions: factors affecting particle size using spontaneous emulsification. J Colloid Interface Sci 391:95–102. https://doi.org/10.1016/j.jcis.2012.08.069
Saberi AH, Fang Y, McClements DJ (2013b) Fabrication of vitamin E-enriched nanoemulsions by spontaneous emulsification: effect of propylene glycol and ethanol on formation, stability, and properties. Food Res Int 54:812–820. https://doi.org/10.1016/j.foodres.2013.08.028
Saberi AH, Fang Y, McClements DJ (2013c) Effect of glycerol on formation, stability, and properties of vitamin-E enriched nanoemulsions produced using spontaneous emulsification. J Colloid Interface Sci 411:105–113. https://doi.org/10.1016/j.jcis.2013.08.041
Saberi AH, Fang Y, McClements DJ (2014a) Stabilization of vitamin E-enriched mini-emulsions: influence of organic and aqueous phase compositions. Colloids Surf A Physicochem Eng Asp 449:65–73. https://doi.org/10.1016/j.colsurfa.2014.02.042
Saberi AH, Fang Y, McClements DJ (2014b) Effect of salts on formation and stability of vitamin E-enriched mini-emulsions produced by spontaneous emulsification. J Agric Food Chem 62:11246–11253. https://doi.org/10.1021/jf503862u
Saberi AH, Fang Y, McClements DJ (2015a) Thermal reversibility of vitamin E-enriched emulsion-based delivery systems produced using spontaneous emulsification. Food Chem 185:254–260. https://doi.org/10.1016/j.foodchem.2015.03.080
Saberi AH, Fang Y, McClements DJ (2015b) Formation of thermally reversible optically transparent emulsion-based delivery systems using spontaneous emulsification. Soft Matter 11:9321–9329. https://doi.org/10.1039/c5sm02221e
Sadurní N, Solans C, Azemar N, García-Celma MJ (2005) Studies on the formation of O/W nano-emulsions, by low-energy emulsification methods, suitable for pharmaceutical applications. Eur J Pharm Sci 26:438–445. https://doi.org/10.1016/j.ejps.2005.08.001
Seibert JB, Rodrigues IV, Carneiro SP et al (2019) Seasonality study of essential oil from leaves of Cymbopogon densiflorus and nanoemulsion development with antioxidant activity. Flavour Fragr J 34:5–14. https://doi.org/10.1002/ffj.3472
Shinoda K, Saito H (1968) The effect of temperature on the phase equilibria and the types of dispersions of the ternary system composed of water, cyclohexane, and nonionic surfactant. J Colloid Interface Sci 26:70–74. https://doi.org/10.1016/0021-9797(68)90273-7
Shinoda K, Saito H (1969) The stability of O/W type emulsions as functions of temperature and the HLB of emulsifiers: the emulsification by PIT-method. J Colloid Interface Sci 30:258–263. https://doi.org/10.1016/S0021-9797(69)80012-3
Shishir MRI, Xie L, Sun C et al (2018) Advances in micro and nano-encapsulation of bioactive compounds using biopolymer and lipid-based transporters. Trends Food Sci Technol 78:34–60. https://doi.org/10.1016/j.tifs.2018.05.018
Solans C, Solé I (2012) Nano-emulsions: formation by low-energy methods. Curr Opin Colloid Interface Sci 17:246–254. https://doi.org/10.1016/j.cocis.2012.07.003
Solans C, Morales D, Homs M (2016) Spontaneous emulsification. Curr Opin Colloid Interface Sci 22:88–93. https://doi.org/10.1016/j.cocis.2016.03.002
Solè I, Pey CM, Maestro A et al (2010) Nano-emulsions prepared by the phase inversion composition method: preparation variables and scale up. J Colloid Interface Sci 344:417–423. https://doi.org/10.1016/j.jcis.2009.11.046

Su D, Zhong Q (2016) Lemon oil nanoemulsions fabricated with sodium caseinate and Tween 20 using phase inversion temperature method. J Food Eng 171:214–221. https://doi.org/10.1016/j.jfoodeng.2015.10.040

Tadros T, Izquierdo P, Esquena J, Solans C (2004) Formation and stability of nano-emulsions. Adv Colloid Interf Sci 108–109:303–318. https://doi.org/10.1016/j.cis.2003.10.023

Taha EI, Al-Saidan S, Samy AM, Khan MA (2004) Preparation and in vitro characterization of self-nanoemulsified drug delivery system (SNEDDS) of all-trans-retinol acetate. Int J Pharm 285:109–119. https://doi.org/10.1016/j.ijpharm.2004.03.034

Tian W-L, Lei L, Zhang Q, Li Y (2015) Physical stability and antimicrobial activity of encapsulated cinnamaldehyde by self-emulsifying nanoemulsion. J Food Process Eng 39:462–471. https://doi.org/10.1111/jfpe.12237

Traber MG, Packer L (1995) Vitamin E: beyond antioxidant function. Am J Clin Nutr 62:1501S–1509S. https://doi.org/10.1093/ajcn/62.6.1501S

Trentin A, De Lamo S, Güell C et al (2011) Protein-stabilized emulsions containing beta-carotene produced by premix membrane emulsification. J Food Eng 106:267–274. https://doi.org/10.1016/j.jfoodeng.2011.03.013

Tubtimsri S, Limmatvapirat C, Limsirichaikul S et al (2018) Fabrication and characterization of spearmint oil loaded nanoemulsions as cytotoxic agents against oral cancer cell. Asian J Pharm Sci 13:425–437. https://doi.org/10.1016/j.ajps.2018.02.003

Van Der Graaf S, Schroën CGPH, Boom RM (2005) Preparation of double emulsions by membrane emulsification – a review. J Membr Sci 251:7–15. https://doi.org/10.1016/j.memsci.2004.12.013

Vandamme TF, Anton N (2010) Low-energy nanoemulsification to design veterinary controlled drug delivery devices. Int J Nanomedicine 5:867–873. https://doi.org/10.2147/IJN.S13273

Vieira SA, Zhang G, Decker EA (2017) Biological implications of lipid oxidation products. JAOCS J Am Oil Chem Soc 94:339–351. https://doi.org/10.1007/s11746-017-2958-2

Vinh TDT, Hien LTM, Dao DTA (2020) Formulation of black pepper (Piper nigrum L.) essential oil nano-emulsion via phase inversion temperature method. Food Sci Nutr:1741–1752. https://doi.org/10.1002/fsn3.1422

Wang L, Zhang Y (2017) Eugenol Nanoemulsion stabilized with Zein and sodium Caseinate by self-assembly. J Agric Food Chem 65:2990–2998. https://doi.org/10.1021/acs.jafc.7b00194

Wang L, Dong J, Chen J et al (2009) Design and optimization of a new self-nanoemulsifying drug delivery system. J Colloid Interface Sci 330:443–448. https://doi.org/10.1016/j.jcis.2008.10.077

Wiseman H (1993) Vitamin D is a membrane antioxidant ability to inhibit iron-dependent lipid peroxidation in liposomes compared to cholesterol, ergosterol and tamoxifen and relevance to anticancer action. FEBS Lett 326:285–288. https://doi.org/10.1016/0014-5793(93)81809-E

Xu D, Li Y, Meng X et al (2017) Natural antioxidants in foods and medicinal plants: extraction , assessment and resources. Int J Mol Sci 18:1–32. https://doi.org/10.3390/ijms18010096

Yildirim ST, Oztop MH, Soyer Y (2017) Cinnamon oil nanoemulsions by spontaneous emulsification: formulation, characterization and antimicrobial activity. LWT – Food Sci Technol 80:122–128. https://doi.org/10.1016/j.lwt.2017.05.041

Yoo JH, Shanmugam S, Thapa P et al (2010) Novel self-nanoemulsifying drug delivery system for enhanced solubility and dissolution of lutein. Arch Pharm Res 33:417–426. https://doi.org/10.1007/s12272-010-0311-5

Zhang Z, Vriesekoop F, Yuan Q, Liang H (2014) Effects of nisin on the antimicrobial activity of d-limonene and its nanoemulsion. Food Chem 150:307–312. https://doi.org/10.1016/j.foodchem.2013.10.160

Zhang L, Zhang F, Fan Z et al (2019) DHA and EPA nanoemulsions prepared by the low-energy emulsification method: process factors influencing droplet size and physicochemical stability. Food Res Int 121:359–366. https://doi.org/10.1016/j.foodres.2019.03.059

Zhao S, Tian G, Zhao C et al (2018) The stability of three different citrus oil-in-water emulsions fabricated by spontaneous emulsification. Food Chem 269:577–587. https://doi.org/10.1016/j.foodchem.2018.07.062

Chapter 4
Nanoemulsions as Carriers for Natural Antioxidants: Formulation Development and Optimisation

Ines Nikolić, Ana Gledović, Slobodanka Tamburić, Tamara Major, and Snežana Savić

4.1 Introduction

Antioxidants derived from natural sources are gaining increased popularity in the past few decades as an alternative to synthetic molecules, which could exhibit side effects or low bioactivity, while imposing ecological burden and negative perception by the consumers (Lou et al. 2017; Raut and Karuppayil 2014; Aburjai and Natseh 2003). Phytotherapy and cosmetic use of botanical raw materials dates back to ancient times (Weber et al. 2009), but the renewal of interest in botanical actives is markedly accelerated due to recent advancements in analytical methods for their characterisation and bioactivity assessment, as well as the use of advanced carrier systems to optimise their performance (Majeed et al. 2015).

It is well-known that human body gets exposed to various external factors that generate free radicals, mostly reactive oxygen species (ROS) for example superoxide anion (O_2^-), peroxyl radical (RO_2), hydroxyl radical (OH), nitric oxide radical (NO), hydrogen peroxide (H_2O_2). There are endogenous factors that serve to neutralise these ROS, but sometimes, due to increased and chronic exposure to reactive

Ines Nikolić and Ana Gledović contributed equally.

I. Nikolić · A. Gledović · S. Savić (✉)
Department of Pharmaceutical Technology and Cosmetology,
University of Belgrade – Faculty of Pharmacy, Belgrade, Serbia
e-mail: snezana.savic@pharmacy.bg.ac.rs

S. Tamburić
Cosmetic Science Research Group, London College of Fashion, University of the Arts London, London, UK

T. Major
Center for Stem Cell Research and Drug Development, University of Belgrade - Faculty of Pharmacy, Belgrade, Serbia

M. A. Aboudzadeh (ed.), *Emulsion-based Encapsulation of Antioxidants*, Food Bioactive Ingredients, https://doi.org/10.1007/978-3-030-62052-3_4

molecules, the skin antioxidant capacity becomes overwhelmed and it experiences oxidative stress. This condition could lead to the ROS-induced damage of lipids, proteins and DNA, causing (photo)aging, reduction of immune response or cancer (Naidoo and Birch-Machin 2017; Reis Mansur et al. 2016; Chen et al. 2012). Therefore, a well-planned intake of antioxidants through balanced diet, topical application and supplementation, in line with healthy lifestyle, should boost body's ability to prevent and fight many pathological conditions (Lobo et al. 2010; Masaki 2010).

In addition to the health aspects, encapsulation of natural antioxidants and their application in the material and packaging industry is constantly increasing, aiming to develop novel "green" materials and packaging solutions that will improve safety, quality and shelf-life of the final product by minimising the amount of externally added additives (Han et al. 2018; Prakash et al. 2018).

An important class of natural lipophilic substances with high bioactivity are plant/fruit seed oils containing essential polyunsaturated fatty acids and antioxidant molecules such as carotenoids, tocopherols, tocotrienols and polyphenols (e.g. olive, grape seed oil, sunflower, passion fruit, pomegranate, wheat germ, blackberry and red raspberry seed oils) (Michalak and Dadasiewicz 2018; Pereira et al. 2016; Bushmann et al. 2004). Some of these oils, such as olive oil, have been used for centuries in human nutrition and skincare preparations (Aburjai and Natseh 2003). Another important group of lipophilic bioactives are plant essential oils, complex mixtures of up to 70 various volatile and aromatic compounds, which are known for their antimicrobial, antioxidant and anticarcinogenic activity. Each essential oil is characterised by two or three principal compounds (usually phenolic constituents, flavonoids and terpenoids) which can act alone or synergistically, ensuring the important biological effects (Majeed et al. 2015). Essential oils prepared from clove, oregano, thyme, sage, rosemary and many other herbs and spices are known as potent antioxidants (Raut and Karuppayil 2014; Viuda-Martos et al. 2010; Dorman et al. 2000). Therefore, it is no wonder that there is an intensive focus on plant bioactives among pharmaceutical, cosmetic, agricultural and food industries. It is worth mentioning that the exact mechanism of action and safety profiles of many natural antioxidants are still unknown, because of the large number of naturally-derived plant extracts, their intrinsic composition variations and the fact that they are sometimes not precisely characterised (Gledovic et al. 2020). Isolated compounds from essential oils (e.g. thymol, carvacrol, eucalyptol, eugenol) and other extracted compounds from plant sources (e.g. curcumin, resveratrol, lycopene, β-carotene) with defined composition are more convenient for formulation development compared to the multicomponent extracts, because they are standardised and more concentrated (Weber et al. 2009).

Although many of these natural oils and isolated molecules are generally recognised as safe (GRAS) for humans, they are prone to heat degradation, oxidation, polymerisation and hydrolysis in aqueous formulations, which is a limiting factor for their application. Therefore, adequate formulations are necessary to preserve their bioactivity and safety profile. In response to that need, many researches have been carried out showing the feasibility of nanoemulsion carriers for natural antioxidants (Ahmadi and Jafarizadeh-Malmiri 2020; Lou et al. 2017; Pereira et al. 2016).

Nanoemulsions present innovative colloidal delivery systems with very small droplet sizes (usually up to 300 nm) consisting of the organic phase (natural and/or synthetic oils and lipophilic actives), water and surfactants (natural and/or synthetic), and optionally some hydrophilic and/or rheological additives, prepared via appropriate low-energy or high-energy methods. Depending on the obtained droplet sizes, nanoemulsions are transparent, translucent or milky white fluids, with typical blue-shining appearance. Small droplet size of nanoemulsions increases their stability towards gravity-induced phenomena (creaming and sedimentation), leading to improved shelf-life compared to standard macroemulsions. In addition, an increased surface area between oil and water phases leads to increased solubility and bioavailability of the lipophilic actives that are entrapped into the internal phase of oil-in-water nanoemulsions (O/W nanoemulsions), while the surfactant layer protects bioactives from deterioration. Theoretically, controlled release and/or targeted delivery are possible when lipophilic antioxidants are incorporated into O/W nanoemulsions. Most importantly, improved stability can be achieved for the natural antioxidants, which are usually sensitive to heat, air and light exposure (Ahmadi and Jafarizadeh-Malmiri 2020; Pavoni et al. 2020; Pereira et al. 2016). It is worth mentioning that hydroglycolic plant extracts can also be added to O/W nanoemulsions, which can sometimes lead to synergistic antioxidant effects with the lipophilic and hydrophilic molecules in the nanoemulsion carrier and improved stability of such multicomponent products (Gledovic et al. 2020). Taken all together, natural antioxidants can be powerful adjuvants in combination with other active ingredients, which makes them very promising materials for future research and application.

With all of the above in mind, the aim of this chapter is to present a comprehensive review of the literature related to the formation, properties and formulation optimisation of nanoemulsions with various lipophilic and hydrophilic plant extracts and isolated compounds with proven antioxidant activity. Special focus will be placed on the multidisciplinary role of natural antioxidants in nanoemulsion carriers. This will be presented with respect to the structural specificities of nanoemulsions and the interactions between natural antioxidants and their nanoemulsion carrier, encompassing in vitro antioxidant and cytotoxicity assays.

4.2 Encapsulation of Antioxidants in Nanoemulsions: Methods and Basic Principles

Based on the energy-efficiency and underlying mechanisms responsible for nanoemulsion formation, nanoemulsions can be produced via different methods, generally classified as low-energy and high-energy emulsification methods (Fig. 4.1). In the following subsections, the main advantages and disadvantages of each method for preparation of nanoemulsions containing natural antioxidants will be briefly

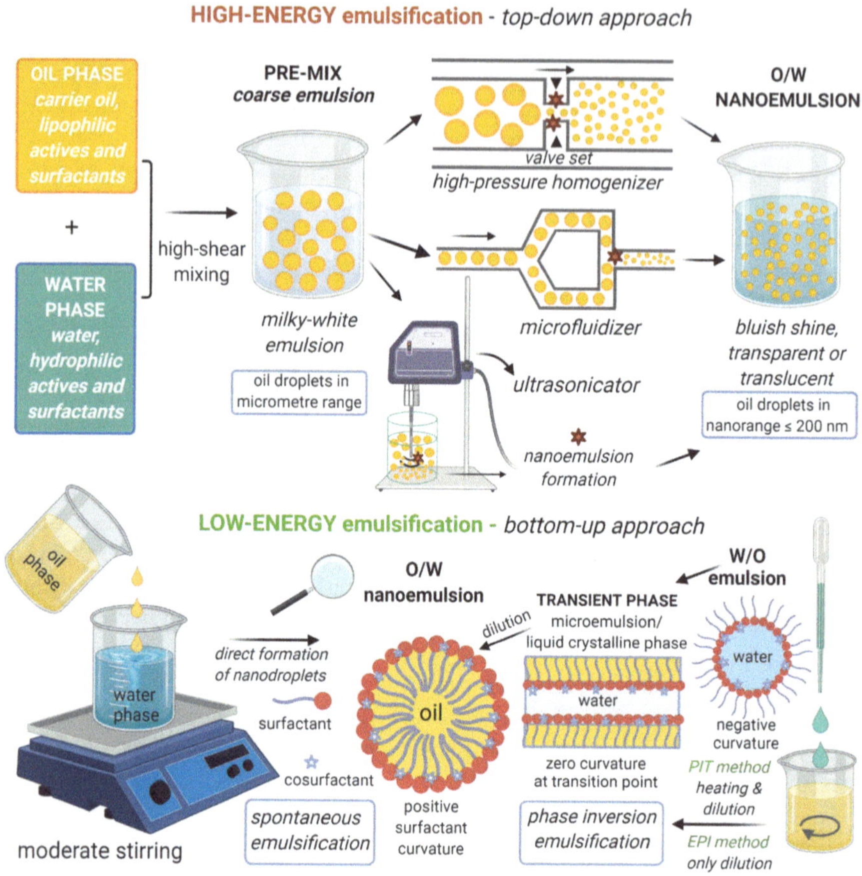

Fig. 4.1 Schematic representation of O/W nanoemulsion formation through high-energy and low-energy emulsification methods

described. Original research papers reflecting the complexity of formulation development and production optimisation of nanoemulsion carriers with natural antioxidants will be discussed as a guide for the researchers in this field (Table 4.1).

4.2.1 *High-Energy Emulsification Methods*

High-energy methods represent a top-down approach to nanoemulsion formation. These methods are based on the usage of mechanical devices, for instance high-pressure homogeniser (HPH), ultrasonic homogeniser (USH), microfluidiser (MF) and high-speed mixer (HSM) to produce intense disruptive forces such as collision, compression and cavitation, which can break the micrometre emulsion droplets into the smaller nanosized dimensions. The input energy density in these high-energy processes is very high (the order of 10^8–10^{10} W kg^{-1}), but only a small amount (around 0.1%) of this energy is used for emulsification and a large amount of energy is dissipated as friction due to the presence of high shear rates, which make them cost-inefficient (Azmi et al. 2019; Gupta et al. 2016). This dissipated energy is converted into heat, which raises the nanoemulsion temperature. Therefore, a cooling system is necessary to prevent heat damage of the sensitive ingredients such as vitamins, proteins and enzymes (Espitia et al. 2018; Chong et al. 2018). However, the main advantages of the high-energy methods in nanoemulsions preparation are the possibility to use lower amounts of surfactants and smaller surfactant-to-oil ratios (SOR), which is especially important in food or certain pharmaceutical applications, where high surfactant concentration is a limiting factor (Gothani and Prasert 2014; Solans and Sole 2012). The two most widely used high-energy emulsification methods on the laboratory scale are HPH and USH, while for the industrial application HPH and MF are more suitable (Azmi et al. 2019; Gupta et al. 2016).

4.2.1.1 High-Pressure Homogeniser (HPH)

HPH is a device originally used in the food and beverage industry. It is the most straightforward approach for nanoemulsification due to its versatility and suitability in laboratory settings and easy scale-up (Espitia et al. 2018). Three main process parameters influence HPH emulsification: pressure, number of cycles and temperature. Conventional HPHs employ pressures between 50 and 100 MPa, however, operating pressures of about 350 to 400 MPa can also be used. HPH is based on the forced passage of pre-emulsion/pre-mix (prepared via HSM) through a specially designed narrow valve, causing a sudden pressure drop across HPH to reach a few thousand bars. Multiple passes are employed until uniform nanodroplets of desired sizes are formed as a result of the combination of intensive disruptive forces, such as shear stress, cavitation and turbulent flow conditions (Salem and Ezzat 2018). In general, droplet sizes decrease with an increased number of passes, until a minimum value is reached. When the droplet size plateau is reached, further homogenisation may lead to coalescence and increase in droplet dimension due to the over-processing phenomenon (Azmi et al. 2019). Viscosities of the oil and water phases are important formulation parameters, as well as sufficient amount of surfactant necessary to quickly and effectively cover the surfaces of the newly formed droplets (Gupta et al. 2016).

4.2.1.2 Ultrasonic Homogeniser (USH)

An USH involves inserting the sonicator probe into a sample. The probe creates sound waves of high frequency (> 20 kHz), causing shock waves which result in turbulence due to cavitation in the surrounding liquid. The mechanical vibrations lead to the formation of liquid jets at high speed, while the collapse of the micro-bubbles generates intense disruptive forces that lead to droplet disruption and the formation of nanodroplets. It is recommended to prepare a pre-emulsion before USH (Rinaldi et al. 2017). The droplet sizes significantly decrease when the intensity of ultrasonic waves, sonication time, power level and surfactant concentration increase, while over-processing can cause droplet coalescence and instability (Azmi et al. 2019). The amount of surfactant and the viscosity of the oil and aqueous phases are among important parameters in nanoemulsion formation, therefore should be precisely optimised. The USH method has several advantages, such as simple manipulation using the lower-cost equipment, easy operation, cleaning and servicing (Rebolleda et al. 2015). The main disadvantage of USH is the production of small batches on a lab scale. Moreover, USH can lead to protein denaturation, polysaccharide depolymerisation and lipid oxidation during homogenisation, which could limit its application in industrial settings (Salem and Ezzat 2018).

4.2.1.3 Microfluidiser (MF)

In MF device, the liquid mixture is passed through the interaction chamber comprising microchannels under a high pressure, resulting in the breakage of micrometre-sized droplets into nanodroplets. Similar to other high-energy methods, the pre-emulsion is prepared first and then it is passed through the interaction chamber consisting of two separate flow channels. Due to the specific design of the channels, the two streams of the pre-emulsion collide at high velocity, generating a very high shearing action, which results in droplet breakage and formation of nanoemulsion. The desired droplet size and distribution are achieved through multiple passages with sufficient amount of surfactant. MF is operating based on similar principles as HPH. Therefore, these two homogenisation methods are usually interchangeable, and both are applicable at a laboratory and industrial scale (Salem and Ezzat 2018; Salvia-Trujillo et al. 2013).

High-speed mixer (HSM) such as rotor-stator devices as a single technique is not suitable to generate nanoemulsions with small droplet sizes and narrow polydispersity. However, it is widely used to form pre-emulsions as a primary step, before applying other high-energy emulsification techniques (Salem and Ezzat 2018). Alternatively, they can be combined with low-energy emulsification methods, as a final homogenisation step, once the nanoemulsion is formed (Teo et al. 2010).

4.2.2 Low-Energy Emulsification Methods

Low-energy methods represent a bottom-up approach, which is based on the tendency of certain surfactants to self-assemble and produce nanodroplets due to the release of chemical energy when selected ingredients are mixed in a specific way. The requirements for the formation of nanoemulsion through low-energy methods are dominantly related to the surfactant-oil-water (SOW) composition, which is limited to only certain types of surfactants, oils and additives (Solans and Sole 2012). In other words, the low-energy emulsification methods are based on the control of interfacial phenomena at the boundary between organic and aqueous phases and depend upon the intrinsic properties (e.g. solubility and molecular geometry) of any surface-active molecules present (e.g. surfactants and co-solvents such as polyols) (Nikolic et al. 2018; Chang and McClements 2014). Since these methods require significantly lower energy density input (10^3–10^5 W kg^{-1}), nanoemulsification can be performed applying simple equipment (magnetic stirrer or vortex mixer) to mix the components and release the chemical energy responsible for the nanoemulsion formation (Espitia et al. 2018; Gothani and Prasert 2014). Many of these low-energy methods can be performed at room temperature, which is a significant benefit for the thermosensitive ingredients.

Classification of low-energy emulsification methods is sometimes unclear, giving the fact that many mechanisms of low-energy nanoemulsion formation are rarely fully understood. The most straightforward distinction could be done if any phase inversion from the spontaneous surfactant curvature is produced during the nanoemulsion formation or not (Solans and Sole 2012). In spontaneous emulsification (SE) method (sometimes marked as self-emulsification), the organic phase (surfactants and oil) is being added to the water phase stepwise under continuous mixing, and ultrafine nanoemulsion droplets are formed as a result of rapid diffusion of surfactant to the continuous aqueous phase, without any change in surfactant curvature (Nikolic et al. 2018; Chang and McClements 2014). However, when the aqueous phase is added to the organic phase, the process is classified as emulsion phase inversion (EPI) method (sometimes also referred to as the phase inversion composition or phase inversion concentration – PIC method). When the change of spontaneous surfactant curvature is caused by the change in temperature, the approach is named as phase inversion temperature (PIT). Many possible transient phases can occur at the phase transition point, such as liquid crystalline phase (cubic or lamellar), microemulsion (ME), oil-in-water-in-oil (O/W/O) multiple emulsions. They are all characterised by low interfacial tension, thus enabling the formation of nanodroplets (Azmi et al. 2019; Gupta et al. 2016).

In the past decade the low-energy methods gained higher popularity over conventional high-energy techniques due to several reasons: low-energy consumption, inexpensive equipment, simple manipulation and fast preparation. Most importantly, they are particularly suitable for the production of nanoemulsions with shear- and thermo-sensitive natural ingredients, such as essential or plant seed oils and fruit extracts (Chang and McClements 2014; Solans and Sole 2012). The main disadvantages of these methods are using relatively high concentration of surfactants (usually 10 to 20 wt%) and/or the addition of co-solvents (polyols), which is necessary to form small nanodroplets (< 100 nm). Also, there is a lack of published data regarding the scale-up and the industrial application, which is unfortunate since these methods seem to hold great promise for future applications (Azmi et al. 2019; Salem and Ezzat 2018; Gupta et al. 2016). Therefore, the key requirement for the widespread use of low-energy emulsification methods is the fundamental understanding of how to control the transient phases responsible for nanoemulsion formation, so that the nanoemulsions can be readily reproduced with the desired characteristics in the industrial settings.

4.3 Formulation Optimisation of Nanoemulsions with Natural Antioxidants

Natural antioxidants are challenging ingredients because of their complex composition and instability to environmental stress (heath, light, air, changes of pH values or ionic strength). Therefore, it can be a very difficult task to create nanoemulsions with ultra-fine droplets and long-term physicochemical stability (Zhong et al. 2017; Bajerski et al. 2016). The properties and the ratio of the oil phase, water phase and surfactants, as well as the production method, contribute significantly to the final characteristics and performance of nanoemulsions (Azmi et al. 2019; Gupta et al. 2016; Gothani and Prasert 2014). Therefore, it is necessary to meticulously choose all ingredients and carefully adjust the production procedure to ensure the stability of these sensitive bioactive molecules in a particular nanoemulsion carrier.

An overview of recently published papers is presented in Table 4.1. Current trends in formulation development and some important unresolved issues regarding the encapsulation of natural antioxidants in nanoemulsion-based carriers are mentioned in this table.

4.3.1 Stability of Antioxidant-Loaded Nanoemulsions

There are several reasons for the lack of physical and/or chemical stability in nanoemulsion systems. The primary reason is the choice of surfactants. The most widely employed surfactants, regardless of the nanoemulsion production method, are hydrophilic low molecular weight non-ionic surfactants: Tween 80 (polyoxyethylene 20 sorbitan monooleate) and Tween 20 (polyoxyethylene 20 sorbitan monolaurate). The Tweens are known for their suitability in food, pharmaceutical and cosmetic industry and their versatility and excellent compatibility with various natural and synthetic oils and other ingredients, as well as having a good safety profile. However, these surfactants exhibit a decrease of hydrophilic-lipophilic balance (HLB) value at elevated temperatures that changes their affinity for water and oil phases, consequently leading to physical instability of nanoemulsions (e.g. increase in droplet sizes and polydispersity index (PDI) values) (Chuesiang et al. 2018). The monolayer made of a single surfactant may not be sufficient to prevent the influence of pro-oxidants, air and light on the nanoemulsion lipid core, leading to chemical instability of the incorporated bioactives. Thus, it is always advisable to employ a blend of surfactants (Nikolic et al. 2018; Mao et al. 2009). Lecithin is one of the recommended surfactants compatible with Tween 80. Its positioning at the oil-water interface increases Z-potential and provides electrostatic stabilisation of nanoemulsion. Additionally, lecithin has the potential to block the permeation of peroxyl radicals across the oil/water interface and to decrease the rate of oxidation of bioactive encapsulates (Pan et al. 2013). An increase in surfactant concentration leads to a tighter packing of surfactant molecules at the oil-water boundary, providing an efficient physical barrier to oxidative species. Alternatively, some findings support the opinion that the excess of surfactant forms micelles in the aqueous phase which can encapsulate pro-oxidative species (McClements and Decker 2000). However, when using higher amounts of surfactants, the formed micelles can sometimes facilitate the migration of smaller oil droplets to the larger ones (Ostwald ripening), or an excess of surfactant may alter the properties of the interfacial layer, thereby enhancing coalescence (Chuesiang et al. 2018). Lipophilic non-ionic surfactants Span 80 (Sorbitan monooleate) and Span 20 (Sorbitan monolaurate) are also commonly used to decrease the HLB value of the Tween-Span mixture and to match it with the HLB of natural oils, leading to a decrease in droplet size and an increase in physical and/or chemical stability (Chong et al. 2018; Zhong et al. 2017; Rocha-Filho et al. 2014). However, looking at the available data presented in Table 4.1, it can be concluded that lecithin and Spans, without additional stabilisers, were not particularly effective at inhibiting chemical instability at elevated temperatures.

To conclude, impaired physical and/or chemical stability at elevated temperatures is apparent in most nanoemulsion systems with delicate compounds such as carotenoids (β-carotene and astaxanthin), tocopherols (α-tocopherol/α-TOC and tocopheryl acetate/VE acetate), curcumin, ɣ-oryzanol and plant seed oils (Pereira et al. 2016; Zhong et al. 2017; Hategekimana et al. 2015). In order to preserve the initial optimal nanoemulsion properties (e.g. small droplet sizes and high entrapment efficiency of the isolated compounds), the general recommendation is to store the antioxidant-loaded nanoemulsions at temperatures lower than 25 °C, protected from light and air. However, in a real-life scenario, it is not always possible to retain such ideal storage conditions. Thus, researchers are putting more effort in improving the stability of antioxidant-loaded nanoemulsions under more challenging conditions, such as elevated temperature, UV-irradiation or oxygen exposure (Kaur et al. 2017; Kim et al. 2011; Bernardi et al. 2011). There have been several promising solutions reported to improve the nanoemulsion stability at elevated temperatures, which are presented as follows.

4.3.1.1 Introduction of High Molecular Weight Surfactants

In a study, the influence of different surfactants and production parameters was investigated on the physicochemical properties of β-carotene nanoemulsions developed via HPH. The resulting nanoemulsions stabilised with Tween 20 and decaglycerol monolaurate (DML) had significantly smaller droplet sizes, but they were less stable compared with the ones stabilised with octenyl succinate anhydride modified starch (OSA-MS) and whey protein isolate (WPI). It was found that WPI was the only emulsifier able to protect β-carotene effectively from degradation, ensuring that 72% of β-carotene remained after 12 days of storage at 55 °C. It was pointed out that the low molecular weight surfactants (Tween 20 and DML) generally do not provide highly cohesive or viscous surface layers around the nanodroplets, and generate low Z-potential values (~ − 5 mV), which can explain the obtained poor stability. However, the large molecule emulsifiers (OSA-MS and WPI) can form mechanically strong interfacial layers and cause steric hindrance to prevent droplet coalescence. Droplets in WPI emulsions showed more negative zeta potential values (~ − 17 mV), while OSA-MS solution showed a relatively high viscosity, which might improve their stabilising properties. It has also been discussed that WPI can act as an antioxidant since the main constituents of WPI (β-lactoglobulin and α-lactalbumin) both contain cysteyl residues, disulphide bonds and thiol functional groups, which can inhibit lipid oxidation by scavenging free radicals. Interestingly, the combination of Tween 20 and WPI did not result in a synergistic protective effect. Instead, the β-carotene degradation rates were just between those in the nanoemulsions stabilised by WPI and Tween 20 alone. Droplet sizes of β-carotene nanoemulsions could be further reduced by increasing the homogenisation pressure to 140 MPa. However, the processing conditions of 80 MPa and 5 cycles were found to be optimal to avoid possible formation of free radicals during the HPH process and temperature rise, which would lead to thermal and oxidative degradation of β-carotene (Mao et al. 2009).

4.3.1.2 Introduction of Lipophilic and/or Hydrophilic Antioxidants

In another study, the physical and chemical stability of β-carotene enriched nanoemulsions were tested using different types of surfactants (Tween 20 or β-lactoglobulin) and different types of water-soluble (EDTA or ascorbic acid) and oil-soluble antioxidants (coenzyme Q10 or VE acetate). Nanoemulsions were prepared with 10 wt% oil phase (0.5 wt% β-carotene in corn oil) and 90 wt% aqueous phase (2% β-lactoglobulin in buffer solution), via MF method (3 passes, at 9000 psi) with temperature control, to avoid degradation of β-carotene and other antioxidants. It was found that the addition of antioxidants to the nanoemulsions did not have a major influence on the droplet size (around 90 nm) and long-term physical stability. A possible explanation for this phenomenon is that the antioxidants were primarily present in the oil phase; they were not particularly surface-active and did not compete with the globular protein at the oil-water interface. However, β-carotene degradation at 55 °C and consequent colour fading of the product could be effectively supressed by adding water-soluble or oil-soluble antioxidants to the nanoemulsions, with the following order of effectiveness: EDTA> ascorbic acid> coenzyme Q10 > VE acetate. The effectiveness of EDTA in inhibiting colour loss was attributed to its ability to strongly chelate and inactivate transition metals (such as iron) that normally promote carotenoid oxidation. Among the oil-soluble antioxidants, coenzyme Q10 was shown to provide better protection against colour fading than α-TOC, which might be due to its ability to regenerate other antioxidants present in the system (such as tocopherols in the corn oil used as a carrier oil). Similar to the previously described study (Mao et al. 2009), protein-stabilised nanoemulsions were found to exhibit better β-carotene stability and less colour fading than non-ionic surfactant-stabilised nanoemulsions, which was attributed to the antioxidant effects of globular proteins. Interestingly, the synergistic effect between the hydrophilic (EDTA) and lipophilic antioxidant (VE acetate) was not observed in this study (Qian et al. 2012).

In another interesting study, astaxanthin-loaded nanoemulsions were prepared via the HPH using a mixture of natural surfactants (glyceryl/citrate/lactate/linoleate/oleate) or hydrogenated lecithins with different coantioxidants. The glyceryl ester stabilised-nanoemulsions had uniform droplet sizes after three passes at 1000 bar, which then became independent of the number of additional cycles. However, the hydrogenated lecithin nanoemulsions had much larger droplet sizes, which were not constant with the increasing number of cycles. Therefore, the glyceryl ester was superior to the lecithin in terms of forming semi-transparent nanoemulsions with uniform droplet distribution, even at the high concentration of astaxanthin (5.5%). Among tested coantioxidants, only α-TOC or hydroxy dimethoxy benzyl malonate (HDBM) were suitable to keep the monomodal distribution and to partially inhibit the colour change after UV-irradiation of the tested nanoemulsions. The astaxanthin content remained stable (>90%) after exposure to potentially destabilising physical stresses (− 5, 5, 25 and 45 °C, 12 h on each temperature, for 4 weeks), indicating the robustness of the astaxanthin-loaded nanoemulsions. The improved stability of glycerol-stabilised nanoemulsions was confirmed by SEM micrographs, in which a tight multilayer structure with a lamellar form was observed, whereby astaxanthin was completely incorporated in the hydrophobic core of nanodroplets (Kim et al. 2011).

Several other studies concluded that VE acetate can act as an additional surfactant due to its ability to position at the oil-water interface. It was observed that it can decrease droplet size of the nanoemulsions and improve their physical stability (Gledovic et al. 2020; Rocha-Filho et al. 2014; Teo et al. 2010). For example, in the latter study, VE acetate and Pluronic F-68 were found to co-stabilize the formulations and nanoemulsions with droplet size ~94 nm, physically stable for 4 weeks at >45 °C were obtained. The optimal formulation contained 24 wt% Tween 80, 2.4 wt% Pluronic F-68, 10 wt% palm oil esters (POEs), 10 wt% α-TOC and 53.6 wt% deionised water (Teo et al. 2010).

4.3.2 The Multifunctional Role of Natural (Essential and Seed) Oils in Nanoemulsions

The choice of oil phase is another crucial factor in nanoemulsion formation and stability. The incorporation of natural oils instead/in addition to the traditionally used carrier oil medium-chain triglycerides (MCT) is currently the focus of research (Zhong et al. 2017; Hategekimana et al. 2015; Bernardi et al. 2011). The incorporation of essential oils in O/W nanoemulsions converts them into an aqueous-based product suitable for oral and topical usage since they cannot be used undiluted. Given the fact that essential oils are liquids composed of small aromatic and volatile molecules that can act as an additional surfactants (e.g. alcohols and esters), they play an important role in nanoemulsion formation (Pavoni et al. 2020; Rocha-Filho et al. 2014). The formulation development of nanoemulsions loaded with essential oils is usually done empirically when a part of the carrier oil is replaced by essential oil. At certain oil phase composition, the droplet size tends to decrease and stable nanoemulsions can be formed. Since essential oils have a complex composition, a detailed characterisation is needed for the standard molecules isolated from them, in order to discover which components are involved in co-stabilising action and which ones remain entrapped in the nanoemulsion droplet core.

The low-energy methods that can be performed at room temperature (EPI and SE) are the preferred choice for the production of essential oil-loaded nanoemulsions. In the study involving the formation of orange oil nanoemulsions by SE, it was found that the surfactant type and concentration and oil phase composition (orange oil/MCT ratio) had notable effects on nanoemulsion formation and stability. Transparent nanoemulsions could be formed under certain conditions: 20 wt% surfactant (Tween 40, 60, or 80) and 10 wt% oil phase (4–6 wt% orange oil and 6–4 wt% MCT). Surfactant type and oil phase composition also affected thermal stability of the nanoemulsions. The system that retained ultrafine nanodroplets of ≈ 25 nm even after thermal cycling (from 20 to 90 °C and back to 20 °C) was the nanoemulsion prepared with 20 wt% Tween 80, 5 wt% orange oil and 5 wt% MCT, which is a clear indication of the role of the oil phase composition on the nanoemulsion stability (Chang and McClements 2014). In a similar study, citrus essential oil was incorporated in Tween 80-stabilised nanoemulsions prepared via SE method. It was also observed that the smallest droplets (≈ 73 nm) were obtained using relatively high amount of Tween 80 (20 wt%) at the optimal ratio of essential oil to MCT of 1:1 (Lou et al. 2017).

PIT method was also found suitable for the production of nanoemulsions with lavender essential oil, employing a passion fruit oil (PFO) as a carrier oil with natural origin. The minimum surfactant concentration necessary for the formation of nanoemulsions was 5.0 wt%. The addition of lavender essential oil (LO) to the system consisting of PFO and mixed surfactants with HLB = 10 (PEG-30 castor oil and sorbitan monooleate) reduced the droplet size compared to nanoemulsions without LO, due to its courfactant properties. LO-loaded nanoemulsions could be formed at different PFO:LO ratios e.g. 5:1, 5:2 and 5:5 (giving droplet sizes of ≈

105 nm, 54 and 38 nm, respectively), but the optimal ratio was selected as 5:5, a similar finding aforementioned studies (Chang and McClements 2014 and Lou et al. 2017). The chromatographic analysis of LO-loaded nanoemulsions indicated no change to the LO main constituents at the temperatures up to 25 °C, but the degradation of linalyl acetate and increased concentration of linalool after thermal cycling indicated some chemical instability at higher temperatures (Rocha-Filho et al. 2014).

Sometimes, the essential oil nanoemulsions are prepared without using any carrier oil, which in this case, it seems that the high-energy methods are more appropriate, as in low-energy approaches the amount of surfactant to form nanoemulsions is insufficient. For example, lemongrass oil (LEO)-loaded nanoemulsions were prepared by MF method with LEO: Tween 80 ratio of 1:1, and sodium-alginate 1 vol% was used as a costabiliser. The mechanism of action of food hydrocolloids such as sodium alginate is related to their ability to adsorb to the interfacial layer, causing possible interactions and competition with the main surfactant. Also, they can modify the viscosity in the aqueous continuous phase which can decrease the rate of creaming and coalescence (Mao et al. 2009). The nanoemulsion with very small average droplet sizes (≈7.35 nm, PDI ≈ 0.34) were prepared after three passes through a MF device working at 150 MPa and the temperature in the reaction chamber was kept to <20 °C, to prevent the evaporation of LEO (Salvia-Trujillo et al. 2013). Basil oil nanoemulsions were prepared by HSM (up to 17,000 rpm) with temperature control, containing basil oil 7.5 wt%, Tween 80 2 wt% and Span 80 2 wt% as surfactants, and water. These nanoemulsions showed good physicochemical stability after 90 days of storage in refrigerator (4 °C), retaining 87% of estragole as the basil oil main constituent in the nanoemulsion (Da Silva Gundel et al. 2018).

As it can be seen from the literature review (Table 4.1), the usage of natural plant seed oils (e.g. sunflower, corn, olive, wheat bran and berry fruit seed oils) as carriers and actives is a growing trend in formulation development (Pereira et al. 2016; Rebolleda et al. 2015; Bernardi et al. 2011). These oils are a rich source of fatty acid triglycerides, thus they can substitute the commonly used MCT or long-chain triglycerides (LCT). However, the minor constituents of these oils (i.e. tocopherols, tocotrienols, carotenoids, phenolic compounds, phytosterols) which play a big role in their bioactivity, could vary depending on the oil production procedure, climate, soil and plant material (Gledovic et al. 2020; Chong et al. 2018). Some of these molecules such as tocopherols and phenolics are also potential surface-active ingredients. Therefore, they can significantly impact the nanoemulsion formation and stability.

It is well known that there are different grades of the same natural raw materials available on the market, but not much is known about the impact of such fine differences on the formation of nanoemulsions via low-energy methods, given the fact that these methods are very sensitive to any change in composition (Chang and McClements 2014; Solans and Sole 2012). Aiming to elaborate on this topic, Gledovic et al. investigated the impact of different red raspberry seed oils (ROs) on nanoemulsion formation via the EPI method at room temperature. The oils used were representatives of several groups, i.e. cold-pressed oils: non-organic, refined (RO1) vs. organic, unrefined (RO2) and CO_2-extracted unrefined oils: non-organic (RO3) vs. organic (RO4). A very important finding was that all ROs can form nanoemulsion in a simple ternary system composed of Tween 80/RO/water by moderate mixing. The minimal surfactant to emulsion ratio (SER) was 10, and the minimal SOR value was 1.0 (50:50 ratio). However, the obtained nanoemulsions had significantly different droplet sizes and preliminary stability, which can be ascribed to the fine differences in their composition (saturated vs. unsaturated fatty acid profiles and the content of tocopherols and carotenoids). Raman spectroscopy confirmed these chemical differences among various ROs as well as their respective nanoemulsions, and it also approved the interactions among nanoemulsion components. It detected interaction among RO2, Tween 80 and glycerol in an aqueous environment which resulted in the formation of nanoemulsions with the smallest droplets (size: $\approx$125 nm, PDI < 0.1), compared to the nanoemulsions prepared with other oils ($\approx$ 144 to 157 nm, PDI < 0.14) and poorer stability. Therefore, one RO could not be directly exchanged for the other (e.g. the cold-pressed with CO_2-extracted oil) without a significant impact on nanoemulsion properties. In addition, the textural analysis confirmed that cubic liquid crystalline gel phase was a necessary step in the preparation of RO-loaded nanoemulsions via the EPI method, and that all ingredients that disturb this step are unfavourable (e.g. polyol concentration above 15 wt% or fruit hydro-glycolic antioxidant extracts above 10 wt%, relative to the water phase). Similarly to the other previously mentioned studies (Kim et al. 2011; Qian et al. 2012), this study also revealed that lipophilic (VE acetate) and/or hydrophilic antioxidants (in this case hydro-glycolic fruit extracts from red raspberry fruit – RE or French oak fruit – FE) improve the stability of the nanoemulsions. Although they were physically stable at all tested temperatures (5, 25 or 40 °C, for 45 days), the only formulation which could inhibit the oil oxidation and rancid odour was the nanoemulsion containing FE extract in the water phase (4 wt%) (Gledovic et al. 2020).

4.3.3 *Conclusions and Future Guidelines for Formulation Development*

Based on the presented findings, it is clear that formulation optimisation of nanoemulsions with natural antioxidants is a complex task and many formulation parameters and production variables have to be taken into account to obtain a stable product. Central Composite Design (CCD) is one of the statistical analysis techniques of Response Surface Methodology (RSM), which is sometimes used as a tool in the formulation development (Chong et al. 2018; Alzorqi et al. 2016). This methodology was found helpful in assessing the interactive effect of the independent variables (e.g. surfactant concentration, SOR, water content, costabiliser content) and production variables (homogenisation pressure, number of passes, ultrasonication, irradiation time and power) on the dependent variables related to nanoemulsion physical characteristics (e.g. droplet size, PDI and viscosity).

Future research should be focused on the multicomponent systems (e.g. systems with mixed surfactants, coantioxidants and rheological additives, Fig. 4.2) to mimic the more realistic scenario of a product ready for the market. The group of natural emulsifiers are markedly underexplored, especially in combination with the low-energy methods; however, recent reports on the usage of WPI, OSA-MS, sodium stearoyl lactate (SSL) and glycerol esters (glyceryl citrate/lactate/linoleate/oleate) in combination with HPH and USH methods are promising (Kim et al. 2011; Qian et al. 2012; Mao et al. 2009). It should be noted that the literature data regarding the long-term physicochemical stability of nanoemulsions (e.g. 6–12 months), even at the room temperature, are practically non-existent. In vitro data regarding cytotoxicity and in vivo data on human volunteers are also insufficient, having in mind the number of natural antioxidants available in the market. The advancements in these research areas should lead to more reproducible scaling-up procedures and industrial applications as a step towards the creation of stable, safe and efficient nanoemulsions.

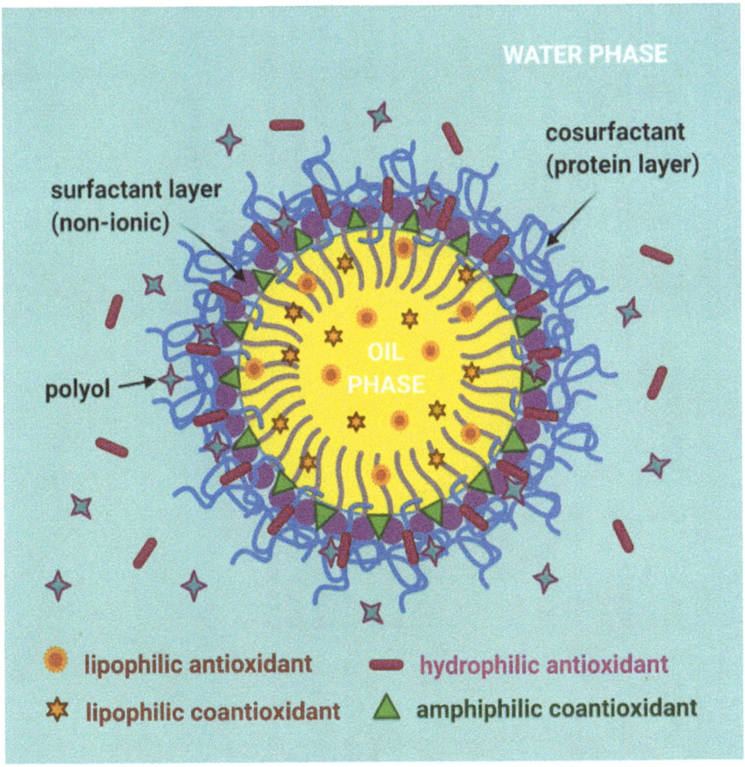

Fig. 4.2 Prototype of a stable O/W nanoemulsion with natural antioxidants

Table 4.1 Nanoemulsions with natural antioxidants – formulation optimisation via high-energy and low-energy methods

Antioxidant/ Carrier oil	Surfactants	Water phase	Production method	Main findings (droplet size distribution, stability)	Application/ Reference
Astaxanthin/ Liquid paraffin, cholesterol, ceramide, ethanol, coantioxidant	Glyceryl citrate/ lactate/ linoleate/ oleate or hydrogenated lecithin	Water Glycerin Tween 60	**HPH** 1. Pre-mix prepared at 70–75 °C, 3000–4000 rpm, 10 min. 2. Cooling to 50 °C, then homogenised with astaxanthin, at 3000–4000 rpm, 5 min, then cooling to RT 3. HPH at 1000 bar.	Nanoemulsionss prepared with mixed glycerol esters (3 wt%) and up to 5.5 wt% astaxanthin had smaller droplets and PDI, and they were more stable then lecithin-based nanoemulsions. The obtained droplet size was: 160–190 nm, narrow size distribution. Astaxanthin nanomulsions were stabilised by a multilamellar surfactant layer, as observed by FF-SEM. Coantioxidants (α-TOC or HDBM) protected astaxanthin from heat and light degradation.	Cosmetics/ Pharmaceutics Kim et al. (2011)
β-carotene suspension (30 wt% of β-carotene in sunflower oil)/ MCT	<u>Small molecular weight:</u> Tween 20 DML <u>High molecular weight:</u> OSA-MS WPI	50 mM Aqueous phosphate buffer solution of pH 7 Preservative: Sodium azide 0.01%	**HPH** 1. β-carotene suspension dissolved in MCT (0.03% β-carotene), at 140 °C for a few seconds. 2. Pre-mix: surfactants (1 wt%) dissolved in buffer solution added to the oil phase under continuous mixing at 5000 rpm. 3. HPH at 20, 80 or 140 MPa, for 3 cycles, then cooling to room temperature.	Optimal pressure was 80 MPa, although at 140 MPa nanoemulsions had smaller droplets. Size: Tween 20–132.0 ± 2.5 nm; DML −132.5 ± 2.9 nm; WPI - 183.3 ± 3.6 nm; WPI+ tween 20–140 ± 2.3 nm; OSS - 212.2 ± 2.3; PDI for all nanoemulsions was <0.22 (at 80 MPa). Nanoemulsions stabilised with Tween-20 and DML had smaller droplet sizes, but poorer stability, compared with the OSA-MS and WPI-stabilised nanoemulsions. WPI was able to protect β-carotene from degradation, whereas OSA-MS was not when nanoemulsions were stored at 55 °C for 12 days. Mixture of Tween-20 + WPI improved nanoemulsions physical stability but it did not improve the stability of β-carotene in the nanoemulsions.	Food/ Pharmaceutics/ Cosmetics Mao et al. (2009)

β-carotene/ corn oil	β-lactoglobulin or Tween 20	10 mM Aqueous phosphate buffer solution of pH 7, Preservative: Sodium azide 0.01%	**MF** 1. β-carotene 0.5 wt% dispersed in corn oil with mild heating (<5 min, 50–60 °C), then stirring at RT ~ 1 h, to fully dissolve. 2. β-lactoglobulin 2 wt% or Tween 20 1.5% dissolved in the aqueous phase. 3. Pre-mix: 10 wt% oil phase was mixed with 90 wt% (w/w) aqueous phase at RT, with a high-speed blender, 2 min. 4. MF: 3 cycles at 9000 psi.	Size: initially around 90 nm for all β-lactoglobulin nanoemulsions, after 15 days storage at 55 °C, there was little change (<7%) in droplet size, but the distribution became bimodal, probably due to flocculation. Therefore, antioxidants did not influence the nanoemulsion physical stability, i.e. droplet size and distribution. The rate of β-carotene degradation decreased upon antioxidant addition with the following order of efficacy: EDTA> ascorbic acid> Q10 > VE acetate. No synergism was observed between ETDA and α-TOC. Nanoemulsions stabilised by β-lactoglobulin were more stable to colour fading than those stabilised by Tween 20.	Food/ Pharmaceutics/ Cosmetics Qian et al. (2012)
β-D-glucan from *Ganoderma Lucidum)*/Palm olein, refined	Polyoxyl 40 Hydrogenated Castor Oil (hydrophilic surfactant)/ Oleoyl macrogol-6 glycerides (lipophilic surfactant) HLB of the mix: 10	Water, deionised	**USH** 1. Water phase: hydrophilic surfactant and β-glucan dispersed in water at 40 °C. 2. Lipophilic surfactant dissolved with stirring in the oil phase at 40 °C. 3. Pre-mix: oil phase was added to the water phase dropwise, at 40 °C, and then mixed with at 12000 rpm, 20 min at RT. 4. Ultrasonication at 30 °C, max 1000 W, at 20 kHz, with cooling.	Nanoemulsions were optimised with experimental design: CCD+ RSM. The optimal composition of formulated nanoemulsions (water: 85%, oil/surfactant ratio: 3), as well as the ultrasonic emulsification conditions, (power: 700 W, irradiation time: 300 s) enhanced the nanoemulsions' physical properties by producing lower droplet diameter (263 nm), narrower PDI (0.244) and lower viscosity (1.85 cP). β-D-glucan nanoemulsions exhibited higher stability at lower concentration (1%) compared to the highest concentration (85%) and showed higher antioxidant activity than the free β-glucan (FRAP and DPPH assays).	Cosmetics/ Pharmaceutics Alzorqi et al. (2016)

(continued)

Table 4.1 (continued)

Antioxidant/ Carrier oil	Surfactants	Water phase	Production method	Main findings (droplet size distribution, stability)	Application/ Reference
Curcumin/ VE Coantioxidant: Benzyl isothiocyanate-BITC	Tween 80/ SSL	Water, deionised ethanol	**USH** 1. Pre-mix: the organic phase was added to the water phase (containing SSL) in a drop-wise manner with continuous stirring. 2. Ultrasonication for 10 min and then ethanol was added and probe-sonicated for an additional 3 min (20% amplitude, 60 W).	The optimal ratio of water: surfactant: oil was (94:4.5:1.5) with (SSL/Tween-80) = 0.023, and ethanol (2 wt%), to yield transparent nanoemulsions. Size: pure nanoemulsion: 38 ± 3 nm, curcumin nanoemulsion: 49 ± 3 nm, curcumin + BITC nanoemulsion: 53 ± 2 nm. All nanoemulsions exhibited good storage stability up to 90 days at 4, 25 and 37 °C. Nanoemulsions protected curcumin from UV light i.e. only 7% curcumin degraded in nanoemulsion compared to 78% in water/ ethanol system after 120 min. The IC_{50} value for pure nanoemulsion, curcumin nanoemulsion, BITC nanoemulsion and curcumin + BICT nanoemulsion was: 85.46 μM, 104.24 μM, 58.73 μM and 75.35 μM, respectively (DPPH assay). Curcumin + BITC nanoemulsion shows a synergistic antioxidant effect.	Pharmaceutics/ Cosmetics/ Food Kaur et al. (2017)
Curcumin/ MCT	Tween 80/ Lecithin (soybean) HLB at 9:1 ratio ~ 14	Water, ultrapure	**SE** 1. Lecithin was dissolved in the MCT, Tween 80 was added, and mixed for 30 min. 2. Curcumin was dissolved in the oil phase. 3. The oil and surfactant blend was added dropwise to water under constant stirring at 1000 rpm, with a magnetic stirrer.	The optimal formulation components: Tween 80: Lecithin (9:1, 10 wt%), MCT 10% and ultrapure water 80 wt% + curcumin (1, 2 and 3 mg/mL). Physicochemical stability was demonstrated during 3 months at RT (mean droplet size: 111.3–146.8 nm; PDI < 0.2; pH: 4.73–5.73). Potent antioxidant activity of curcumin in nanoemulsions was confirmed via DPPH (IC_{50} = 0.1187 mg/mL) and FRAP (1.19 ± 0.02 mmol/g), with no alterations after incorporation in the formulation.	Pharmaceutics/ Cosmetics Nikolic et al. (2018)

VE acetate/ POEs	Tween 80/ Poloxamer 188 (Pluronic 68) Optimal ratio: 40:1	Water, deionised	**EPI** 1. The surfactant mixture was first dissolved into a mix of POEs and VE acetate, stirring at 150 rpm. 2. Deionised water was added dropwise while stirring at 150 rpm, after that it was homogenised at 250–350 rpm for 4 hours. 3. Applying HSM at 10000 rpm, 5 min.	The optimal formulation contained: 10 wt% POEs, 10 wt% VE acetate, 24 wt% Tween 80, 2.4% Pluronic F-68 and 53.6% deionised water. This formulation is considered to be the best as a nanocosmeceutical product due to the small droplet size (94.21 nm), low occurrence of Ostwald ripening and stable at different storing temperatures (5, 25 and 45 °C) for 4 weeks. In conclusion: VE and Pluronic improved nanoemulsion stability at elevated temperatures, and VE decreased droplet sizes <100 nm (at 8 or 10 wt%).	Cosmetics/ Pharmaceutics Teo et al. (2010)
α-TOC / Olive oil (as a source of LCT) MCT Short-chain triglyceride - SCT	Tween 80	Water, Citric buffer of pH 3.0	**EPI** 1. The organic phase containing the carrier oil, tween 80 and α-TOC was stirred for 30 min at 500 rpm with a magnetic stirrer. 2. The water phase (citric buffer, pH 3.0) was titrated into the oil phase at a flow rate of 20 drops per 10 s, with constant stirring at 500 rpm.	The optimal formulation contained: 8 wt% α -TOC, and 2 wt% carrier oil, SER 1, SOR 1. Droplet sizes of nanoemulsions prepared with different carrier oils: 82.6 ± 0.7, 113.4 ± 1.4 and 87.7 ± 2.1 nm) using SCT, MCT and LCT as carrier oils, respectively. α-TOC-loaded nanoemulsions with SCT, MCT and LCT showed physical stability to heat shock (30–90 °C, 30 min), ionic strength (0–500 mM), pH (2.0–8.5) and long term storage (60 days, under light and darkness, 4, 25, 40 °C), but there was significant α-TOC degradation in heat processed and long-term storage samples.	Pharmaceutics/ Cosmetics Hategekimana et al.(2015)

(continued)

Table 4.1 (continued)

Antioxidant/ Carrier oil	Surfactants	Water phase	Production method	Main findings (droplet size distribution, stability)	Application/ Reference
γ- Oryzanol/ MCT Fish oil (as a source of LCT)	Tween 80/ Span 20 Optimal ratio 3:1 HLB mix: 13.4	Water, deionised Citric acid 1% Sodium benzoate 0.1%	**SE** 1. γ-oryzanol (1% wt) was dissolved in the oils phase, then the surfactant mix was added and stirred for a minimum of 20 min at RT to obtain the organic phase. 2. The organic phase was added to the water phase under stirring (850 rpm), with a titrating speed of 60 drops per minute. 3. Homogenisation: extra 5 min, at 850 rpm.	Optimal formulation contained: 1 wt% γ-oryzanol, 10 wt % oil phase (MCT to fish oil =7:3) and 10 wt% surfactant mix. Nanoemulsion droplet size: ~ 157 nm, PDI < 0.14. γ-oryzanol nanoemulsions were physically stable at a broad pH range (2–7), high salt levels (≤0.8 mol L^{-1}), high sugar concentrations (≤16%), and heating temperatures below 50 °C. The oxidative stability of the nanoemulsions compromised at temperatures above 37 °C, as confirmed by elevated peroxide and p-anisidine values, therefore storage at temperatures <23 °C are recommended.	Food/ Pharmaceutics/ Cosmetics Zhong et al. (2017)
Cinnamon essential oil/ MCT	Tween 80	Water, deionised	**PIT** 1. Pre-mix: mixing cinnamon oil and MCT for 3 min, then Tween 80 and deionised water were added and mixed for 30 min. 2. Pre-mix is heated to temperature 15 °C above the PIT. 3. Two-step cooling process: cooling to the PIT to form the microemulsion phase. Secondly, rapid cooling by adding cold water (4 °C) to the system with stirring 3 min.	The cinnamon oil-to-carrier oil ratio in the lipid phase impacted the PIT temperature, initial mean droplet diameter, droplet size distribution, and stability of the cinnamon oil nanoemulsions. Optimal Cinnamon oil: MCT ratio was found to be 4: 6 (10 wt% oil phase). Cinnamon oil nanoemulsions prepared using a PIT with cooling-dilution method had droplet diameters ~100 nm (at 10 wt% Tween 80) and 20 nm (at 20 wt% Tween80), and they were stable during storage at low temperature (4 °C) or ambient temperature (25 °C) for at least 31 days.	Food/ Pharmaceutics Chuesiang et al. (2018)

Orange peel essential oil/ MCT	Tween 20, Tween 40, Tween 60, Tween 80, Tween 85 Span 20	5 mM Aqueous citrate buffer of pH 3.5	**SE** Nanoemulsions were prepared by titration of a mixture of orange oil, carrier oil - MCT, surfactant (Tween) into an aqueous solution (5 mM citrate buffer at pH 3.5) with continuous stirring. The oil/emulsion ratio content was kept constant (10 wt %), while the SER varied (2.5–20 wt %).	Transparent nanoemulsions could be formed under certain conditions: 20 wt % surfactant (Tween 40, 60, or 80) and 10 wt% oil phase (4–6 wt% orange oil +6–4 wt% MCT). Surfactant type and oil-phase composition also affected the thermal stability of the nanoemulsions. Most of the nanoemulsions broke down after thermal cycling (from 20 to 90 °C and back to 20 °C). Only one formulation remained transparent (droplet size ~25 nm) after thermal cycling, which was the one prepared with: 20 wt% Tween 80, 5 wt% orange oil, and 5 wt% MCT.	Food/ Pharmaceutics Chang and McClements (2014)
Basil essential oil/ MCT	Tween 80 Span 80	Water, deionised	**HSM** 1. Organic phase (basil oil + lipophilic surfactant Span 80 and water phase (water + hydrophilic surfactant Tween 80) were mixed separately with a magnetic stirrer. 2. HSM: adding organic phase to water phase at 10000 rpm then increasing speed to 17,000 rpm and mixing for 30 min with cooling.	The optimal formulation contained: basil oil 7.5 wt%, Span 80 2 wt%, Tween 80 2 wt%, water 88.5 wt%, with a droplet size of ~119 nm, PDI ~ 0.16. The obtained nanoemulsions were physically stable at 4, 25 up to 45 days, while at 45 °C the pH value decreased due to hydrolysis of fatty esters in the aqueous surrounding. Therefore, storage at 4 °C is recommended, in which case, 87% of estragole remained after 90 days. Moreover, basil-loaded nanoemulsions maintained antioxidant activity in NE carrier (DPPH test) with reduced cytotoxicity.	Food/ Pharmaceutics Da Silva Gundel et al. (2018)

(continued)

Table 4.1 (continued)

Antioxidant/ Carrier oil	Surfactants	Water phase	Production method	Main findings (droplet size distribution, stability)	Application/ Reference
Citrus medica essential oil/ MCT	Tween 80	5 mM Aqueous citrate buffer of pH 6.0	**SE** 1. The organic phase (essential oil, surfactant and MCT were mixed, at 500 rpm with a magnetic stirrer. 2. The water phase (aqueous citrate buffer) was previously prepared. 3. The organic phase was titrated into the water phase at 2 ml/min, under continuous mixing at 500 rpm.	Optimal formulation: Citrus essential oil 5 wt%, MCT 5%, Tween 80 20 wt% and water phase 70 wt%. Citrus nanoemulsions had droplet size ~73 nm and were stable 30 days at RT. The nanoemulsification significantly increased the antioxidant, antibacterial and antibiofilm activity of essential oil. DPPH assay - citrus oil: 44.3% vs. citrus nanoemulsion: 72.4%; OH radical scavenging - citrus oil: 26.1% vs. citrus nanoemulsion: 58.7% Iron reducing power - citrus oil: 0.106 vs. citrus nanoemulsion: 0.218 at 0.48 mg/ml.	Food/ Pharmaceutics/ Cosmetics Lou et al. (2017)
LEO	Tween 80	Water, deionised Sodium alginate	**MF** 1. Water phase: dissolving sodium alginate (1 wt%) in hot water at 70 °C under stirring. 2. Pre-mix emulsion was made by mixing the water phase and LEO (1 vol%) as a lipid phase plus Tween 80 (1 vol% (as a surfactant, with an HSM, at 3400 rpm for 2 min. 3. MF: at 50, 100 or 150 MPa, for 1, 2, 3, 4, 5 and 10 cycles, with temperature constantly cooled to <20 °C	The average droplet size, viscosity and whiteness index of nanoemulsions decreased by increasing the processing pressure and the cycles through the interaction chamber of the microfluidizer device. After homogenization with MF, the interfacial electrical charge of droplets ranged between −36.66 and − 51.95 mV, irrespectively of the pressure applied and the number of cycles. Therefore, LEO-alginate nanoemulsions obtained with MF are more stable than pre-mix. The optimal formulation was obtained after 3 cycles at 150 MPa, with very small droplet sizes: ~7 nm, PDI ~ 0.34.	Food/ Pharmaceutics/ Cosmetics Salvia-Trujilo et al. (2013)

LO/ PFO	Tween 80 Span 80 PEG-30, PEG-40, PEG-60 Castor oil	Water deionised	**PIT** 1. The water phase and oily phase + surfactants blend were heated separately at 75 ± 3 °C. 2. The water phase was added to the oily phase (PFO with or without LO) under 600 rpm 3. Cooling 25 ± 3 °C under stirring.	The minimum surfactant concentration necessary for the formation of nanoemulsions was 5.0 wt%. LO caused the reduction in droplet sizes in mixed LO + PFO oil phases due to its co-stabilizing properties. LO-loaded nanoemulsions could be formed at several PFO: LO ratios: 5:1, 5:2 and 5:5 (droplet sizes ~105 nm, 54, and 38 nm, respectively). There were no observed changes in the LO main constituents in LO-loaded nanoemulsions at temperature up to 25 °C, but the degradation of linalyl acetate was observed after thermal stress.	Cosmetics/ Pharmaceutics Rocha-Filho et al. (2014)
RO/ Isostearyl isostearate (ISIS) Coantioxidant: VE acetate	Tween 80	Water, deionised Glycerol or Hydro-glycolic antioxidant fruit extracts: RE or FE	**EPI** 1. The organic phase (RO and Tween 80, with or without ISIS and/or VE acetate) and the water phase (water, glycerol/hydro-glycolic extracts) were mixed separately at 1300 rpm. 2. The water phase was added gradually, to the organic phase, with hand-mixing with glass laboratory sticks until the gel phase is crossed, and then vortex mixing at 1300 rpm continued until the nanoemulsion was formed and the sample was homogenised 2 min at 1300 rpm.	It was found the oil type had a major impact on nanoemulsion formation and stability. The organic, cold-pressed, unrefined oil RO2 gave the optimal nanoemulsions with the smallest droplets and PDI and overall stability, regarding all oil phase (VE acetate, ISIS) and water phase variations (size: 125 to 135 nm; PDI < 0.1). The synergistic free radical scavenging effect was pronounced in nanoemulsions with combined lipophilic (in RO2) and hydrophilic antioxidants (in FE) with very high DPPH and ABTS results (>90% inhibition), and good stability at 40 °C. All raw materials and low-energy nanoemulsions showed satisfactory safety profiles in the MTT test on MRC-5 cells, while the anti-proliferative effect was more pronounced on HeLa cells when using nanoemulsions than neat ingredients.	Cosmetics/ Pharmaceutics Gledovic et al. (2020)

(continued)

Table 4.1 (continued)

Antioxidant/ Carrier oil	Surfactants	Water phase	Production method	Main findings (droplet size distribution, stability)	Application/ Reference
Red palm oil (RPO)	Tween 80/ Span 80	Water, deionised Glycerol Preservative: Citric acid 0.08%	**HPH** 1. The organic phase (RPO and Span 80) was added to the water phase (water, glycerol, Tween 80, citric acid) mixed using high shear at 6000 rpm for 10 minutes. 2. Pre-mix then were passed through high-pressure homogeniser (up to 7 cycles, at 500–900 bar).	The formulation was optimised using CCD coupled to RSM and it contained: 6.09 wt% mixed surfactant (Tween 80/Span 80 (63:37, wt)), 20 wt% glycerol as a cosolvent via homogenisation pressure (500 bar). The optimised RPO-based nanoemulsion had droplet size and PDI were 119.49 nm and 0.286, respectively, which was in agreement with the RSM predicted values, and nanoemulsions were preliminary stable for 35 days at RT.	Cosmetics/ Pharmaceutics Chong et al. (2018)
Rice bran oil (RBO) Coantioxidant: butyl hydroxytoluene	Span 80 PEG-30 Castor oil Surfactant mix HLB: 8.0	Water, deionised Preservative: 0.5%	**PIT** Water and organic phases were heated separately at 75 °C, the water phase was added into the organic phase (RBO and surfactants) with continuous stirring at 600 rpm. After nanoemulsion formation, the mixture was cooled to 25 °C while stirring.	The optimal nanoemulsion was composed of 10% RBO, 10% mix, 0.05% antioxidant and 0.50% preservatives in water (size: 69 ± 17 nm). Nanoemulsions were physically stable at 4, 25 and 40 °C for 90 days. In vivo studies showed that nanoemulsions have hydrating properties on healthy volunteers and psoriasis patients and maintained normal skin pH. In vitro HET-CAM test revealed the non-irritant nature of RBO-loaded nanoemulsions as opposed to the slight irritant nature of the neat surfactant.	Cosmetics/ Pharmaceutics Bernardi et al. (2011)

Wheat bran oil (WBO)	Tween 80 Tween 20 Span 80 Diacetyl tartaric acid ester of mono- and diglycerides (DATEM)	Water, deionised	**USH** 1. Pre-mix preparation: WBO and surfactants were mixed separately before water was added and homogenised with a high-speed blender at 29000 pm. 2. High-intensity ultrasonication: at 500 W, 20 kHz, at 20% amplitude and in pulses of 5 s (5 s ultrasound and 5 s pause) to avoid heating of the sample.	The optimal nanoemulsion was obtained via CCD and RSM when 1 wt% of WBO and 7.3 wt% of a surfactant mixture of Span 80: Tween 80 (37.4:62.6) were emulsified in water by high-intensity ultrasonication for 50 s after pre-emulsification with a high-speed blender for 5 min (~39 nm, PDI ~ 0.25). Nanoemulsions showed good stability when stored at 4 °C during 60 days. Antioxidant activity of WBO nanoemulsions and pure oil were confirmed in vitro with several tests (DPPH, FRAP and ABTS) and formulation also inhibited mushroom tyrosinase activity (skin whitening action).	Cosmetics/ Pharmaceutics Rebolleda et al. (2015)

4.4 Screening the Antioxidant Activity – An Overview of the Well-Established Methods and Future Perspectives

An important feature of hydrophilic nanoemulsions in pharmaceutical, food and cosmetic industry is their ability to effectively solubilise/encapsulate lipophilic active and/or functional components such as vitamins and nutraceuticals, flavours, colouring agents, antioxidants, preservatives. Due to their properties, nanoemulsions provide improved stability and handling, facilitated incorporation of the specific component within a product, increased (bio)availability and efficacy, with good visual appearance of the final product (McClements and Rao 2011).

The concept of antioxidant capacity first originated from chemistry and it was later adapted to biology, medicine, epidemiology and nutrition, describing the ability of redox molecules to scavenge free radicals (Floegel et al. 2011). For the determination of antioxidant ability of different antioxidant molecules per se, as well as the ability of antioxidant-loaded nanoemulsions, different methods have been introduced.

A standardised method for antioxidant activity should, ideally, meet the following requirements (Prior et al. 2005):

- It utilises a biologically relevant radical source;
- It is simple;
- It has defined endpoint and chemical mechanism;
- Required chemicals and instrumentation are readily available;
- It has good within-run and between-day reproducibility;
- It should be adaptable for assay of both hydrophilic and lipophilic antioxidants and radicals from different sources;
- It should be as close as possible to the real application;
- It is adaptable to "high-throughput" analysis for routine quality control analyses.

In line with the assessment of other nanoemulsion properties, the antioxidant activity assessment should not be based on a single antioxidant test. It is always advised to perform several in vitro antioxidant evaluation procedures. Moreover, due to significant differences underlying each method, the obtained results cannot be easily compared (Alam et al. 2013). Therefore, the results should always be critically analysed before reaching any conclusion.

In general, antioxidant can deactivate free radicals by hydrogen atom transfer (HAT) or single electron transfer (SET). Consequently, methods developed for antioxidant activity assessment can be categorised as HAT-based or SET-based methods. HAT-based methods measure the ability of an antioxidant to quench free radicals by hydrogen donation, whereas SET-based methods detect the ability of a potential antioxidant to transfer one electron and reduce any compound. The two mechanisms may occur in parallel, but the dominating one will be determined based on the antioxidant structure and properties (Prior et al. 2005). In addition, it is noteworthy that nanomaterials may possess inherent antioxidant ability (regardless of the presence of an antioxidant molecule) due to their possibility to trap or adsorb free radicals, thus preventing them from continuing oxidative reactions. Therefore, such property should also be appropriately checked (Valgimigli et al. 2018).

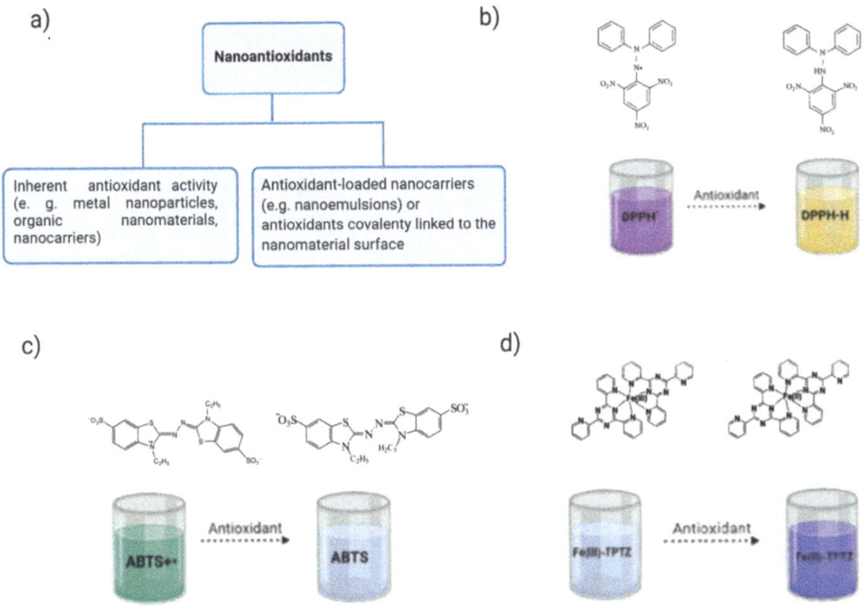

Fig. 4.3 (**a**) Nanoantioxidants: structural properties determine the antioxidant activity; (**b**) Schematic representation of the DPPH radical scavenging assay; (**c**) Schematic representation of the ABTS radical scavenging assay; (**d**) Schematic representation of the FRAP assay

The most popular assays of radical trapping share a simple principle: the change in absorbance or fluorescence of an indicator solution (free radical/oxidising agent) is measured upon addition of an antioxidant (Fig. 4.3). The measurement is generally performed after a certain time in order to allow equilibrium to be established (Li and Pratt 2015). Even though translation of the results obtained after colourimetric antioxidant assays to physiological context may be difficult, they are fast and easy to perform, presenting good screening tools. In this section, several commonly applied methods for antioxidant determination will be described underlining their advantages and disadvantages with respect to antioxidant-loaded nanoemulsion testing. The most important findings are summarised in Table 4.2.

4.4.1 Spectrophotometric Methods

4.4.1.1 DPPH Radical Scavenging Assay

This method is widely used for the determination of free radical scavenging activity. DPPH• (2,2-diphenyl-1-picrylhydrazyl) represents a stable free radical. It is soluble in organic solvents (e.g. ethanol, methanol), rendering intensively purple-coloured solutions (λ_{max} ~ 515–520 nm). An antioxidant with proton-donating ability can

Table 4.2 Selection of commonly applied tests for antioxidant activity, accompanied with useful references providing methodological entries

Antioxidant assay	Principle/Detection method	Advantages	Disadvantages	Useful references[a]
DPPH radical scavenging assay	Mixed HAT- and SET-based method. UV-Vis Spectrophotometric determination.	Affordable, simple, fast and reproducible method. Applicable to both hydrophilic and lipophilic antioxidants. Apart from spectroscopy measurements, DPPH radical can be determined applying EPR-spectroscopy, as well.	DPPH radical may react with reductants having no antioxidant activity. For instance, DPPH• can be completely reduced by H_2O_2, which cannot be considered an antioxidant. Results highly depend on the reaction time, so obtained data can be compared only in case of identical experimental setting. It cannot determine the reaction kinetics. The reaction milieu (organic solvents) breaks the structure of lipid-based carriers (e.g. nanoemulsions).	Gledovic et al. (2020) (nanoemulsions prepared with RO and/or hydrophilic antioxidant fruit extracts) Nikolic et al. (2018) (curcumin and curcumin-loaded nanoemulsions) Rinaldi et al. (2017) (nanoemulsions prepared with neem oil) Zugic et al. (2016) (different *Usnea barbata* extracts)
ABTS radical scavenging assay	SET-based mechanism UV-Vis Spectrophotometric determination	It is usually performed in aqueous environment (PBS buffer), so that hydrophilic nanoemulsions containing antioxidants can retain their structure. As various solvents can be used (ethanol, methanol, DMSO), this method is adaptable to hydrophilic and lipophilic antioxidants.	ABTS radical has to be generated at the beginning of the reaction. Time consuming. Results are dependent on the reaction time.	Gledovic et al. (2020) (nanoemulsions prpepared with RO and/or hydrophilic antioxidant fruit extracts) Rinaldi et al. (2017) (nanoemulsions prepared with neem oil) Rebolleda et al. (2015) (WBO-loaded nanoemulsions)

FRAP assay	SET-based mechanism UV-Vis spectrophotometric determination	It is performed in aqueous environment, adaptable to hydrophilic and lipophilic antioxidants, as well as their carriers (such as hydrophilic nanoemulsions).	Time consuming – the FRAP reagent should be prepared prior to the reaction. Temperature control is required. Detects antioxidant activity only for the antioxidants acting via SET mechanism.	Nikolic et al. (2018) (curcumin and curcumin-loaded nanoemulsions) Rebolleda et al. (2015) (WBO-loaded nanoemulsions)
EPR spectroscopy	Determination of the EPR spectra of a free radical	It can be performed both in aqueous environment and in organic solvents. Interference due to the overlap of the absorption bands of the probe or of the reaction products with the nanomaterial that may hinder spectroscopic measurements ca be avoided. It is possible to evaluate the kinetics of the process. It is possible to detect inherent antioxidant activity of the nanocarrier.	Experiments are costly due to sophisticated equipment (EPR spectrometer)	Nikolic et al. (2020) (curcumin and curcumin-loaded nanoemulsions) Mitsou et al. (2019) (Hydroxytyrosol-loaded microemulsions) Sanna et al. (2019) (myrtle hydroalcoholic extracts) Aboudzadeh et al. (2018) (VE-loaded microemulsions) Chatzidaki et al. (2015) (microemulaina loaded with various phenolic antioxidants) Polovka et al. (2003) (green, black and mixed fruit tea samples)
Electrochemical method (CV)	Analyses changes in the CV recorded upon successive addition of sample containing antioxidant molecules that originate from its chemical reaction with $O2^{\bullet -}$.	Gives insight into free radical scavenging activity of a tested compound in reaction with a biologically compatible ROS, such as superoxide anion ($O2^{\bullet -}$). It can be performed both in aqueous environment and in organic solvents. It is inexpensive and easy to reproduce.	The results of different studies can be compared only if the test is performed in an identical way.	Benedetti et al. (2012) (nanoemulsions loaded with olive oil, or caffeic acid in sunflower oil). Janosevic Lezaic et al. (2014) (polyaniline tannate solid microspheres).

[a]Tested antioxidants or antioxidant-loaded carriers are indicated in the parentheses

react with DPPH•, forming DPPH-H, which fades out the purple colour of DPPH• solution. The degree of discolouration depends on the potency and concentration of the antioxidant, but also on the reaction time (duration of the experiment). During the experiment, prepared mixtures of the radical with different concentrations of the antioxidant should be protected from light and continuously shaken. The inhibition percentage (%) of radical is calculated applying the following formula (Eq. 4.1):

$$\% \ of \ inhibition = \frac{A_{blank\ sample} - A_{test\ sample}}{A_{blank\ sample}} * 100 \tag{4.1}$$

After plotting the inhibition percentage versus antioxidant concentration, results are usually expressed as IC_{50} (inhibition concentration of 50%), defined as the concentration of the potential antioxidant able to decrease 50% of the initial absorbance of the radical (Nile et al. 2012). However, taken alone, this parameter does not provide comparable results with those obtained applying the same assay, but for different reaction time. It can be used only to compare the activity of different antioxidants evaluated in the identical experimental setting (Amorati and Valgimigli 2015). In addition, there are some disadvantages when lipid-based antioxidant carriers (e.g. nanoemulsions) are tested through this methodology, since it is performed in an organic solvent. By dissolving the lipid-based components of the carrier in the organic solvent, the antioxidant would be released from the formulation immediately upon the contact with the solvent. Consequently, it cannot be claimed that measured effect comes from the loaded formulation, but from the antioxidant released upon disruption of the carrier's structure. However, even with that disadvantage, this method can be used as a stability assessment tool for antioxidants incorporated in a carrier (Nikolic et al. 2018).

4.4.1.2 ABTS Radical Scavenging Assay

ABTS scavenging assay is also a spectrophotometric test. In this assay, ABTS (2,2′-azinobis-(3-ethylbenzothiazoline-6-sulphonic acid)) is oxidised to its radical cation, $ABTS^{\bullet +}$, which has intensive blue-green colour (Gledovic et al. 2020). Originally, this assay used metmyoglobin and H_2O_2 to generate ferrylmyoglobin, which then reacted with ABTS, forming $ABTS^{\bullet +}$ (Miller et al. 1993). In the following years, various oxidising agents have been used to generate the radical (potassium persulfate – $K_2S_2O_8$ (Re et al. 1999), manganese dioxide – MnO_2 (Miller et al. 1996) or oxidising enzymes (Cano et al. 2013). The $ABTS^+$ has strong absorption at 734 nm. Upon reaction with an antioxidant, there is a decrease in the colour intensity, which correlates with the antioxidant ability of the tested compound. The antioxidant activity is defined as the amount of $ABTS^{\bullet +}$ quenched after a fixed time, usually 6–30 minutes, which has to be optimised (Amorati and Valgimigli 2015). Since this test is usually performed in an aqueous environment (such as phosphate-buffered saline, pH 7.4), when the hydrophilic nanoemulsions loaded with antioxidants are tested, their structure can be preserved (Gledovic et al. 2020, Rebolleda et al. 2015). In the comparative study of the antioxidant activity of a variety of fresh fruits, vegetables and beverages consumed as a source of antioxidants, it was observed that the antioxidant capacity detected by ABTS assay was significantly higher compared to that by DPPH assay. It was concluded that the high-pigmented and hydrophilic antioxidants were better reflected by ABTS assay (in PBS buffer - aqueous environment) than DPPH assay (in methanol) which suggests that ABTS assay may be more useful than DPPH assay for detecting antioxidant capacity in a variety of foods (Floegel et al. 2011). Another advantage of the ABTS test is the fact that it can also be performed in other solvents e.g. ethanol, methanol and dimethyl sulfoxide (DMSO) (Rinaldi et al. 2017). Therefore, the choice of solvent is one of the crucial parameters in ABTS test and it can be adjusted to the solubility of the tested antioxidants.

4.4.1.3 FRAP Assay (Ferric Reducing Antioxidant Power)

This is a frequently used assay, based on the reduction of Fe^{3+} ion. Prior to the analysis, FRAP reagent, consisting of acetate buffer (pH 3.6), TPTZ (2,4,6-tripyridyl-s-triazine) solution in HCl and $FeCl_3x6H_2O$ solution in purified water, should be formed (Nile et al. 2012). The antioxidant should react with the reagent for 30 minutes under controlled temperature conditions (37 °C). During the reaction, Fe^{2+} ion is generated, rendering blue solution and absorbance can be detected at 595 nm. The results are expressed as FRAP value – mmol of Fe^{2+} per gram of dry matter. The test is useful for the assessment of hydrophilic and lipophilic antioxidants, as well as antioxidant-loaded carriers, such as hydrophilic nanoemulsions (Nikolic et al. 2018). However, as the mechanism involved is SET-based, the assay may not detect the activity of antioxidants, acting only via HAT (Pisoschi et al. 2016).

4.4.2 Non-spectrophotometric Methods

4.4.2.1 Electron Paramagnetic Resonance Spectroscopy

Antioxidant activity evaluation through electron paramagnetic resonance (EPR) spectroscopy represents a valuable method and an alternative to spectrophotometric approaches, avoiding some interferences that may occur due to the overlap of the absorption bands of the probe or of the reaction products with that of the nanomaterial (Valgimigli et al. 2018). The method itself involves the absorption of microwave energy by paramagnetic species (molecules with unpaired electron – such as free radicals) produced during transition of spin states in the presence of an external magnetic field. Therefore, it can be used in the antioxidant activity assessment (Nawab et al. 2017). Unlike other usually applied methods, EPR spectroscopy enables detection of the inherent antioxidant activity of a nanocarrier (e.g.

nanoemulsions). EPR antioxidant activity tests can be performed in aqueous environment, enabling experiments with hydrophilic lipid-based systems, such as O/W nanoemulsions. One of the widely used free radicals in this kind of experiments is Tempol (4-hydroxy-2,2,6,6 - tetramethyl-l-piperidinyloxy) – a stable nitroxide radical, characterised by a well-defined EPR spectrum consisting of three peaks. In the presence of compounds with antioxidant activity, the EPR spectrum of Tempol is reduced after the scavenging reaction with the antioxidant. The % of inhibition of the EPR spectrum is calculated from the following equation (Eq. 4.2):

$$\%\ of\ inhibition = \frac{A_0 - A_s}{A_0} * 100 \tag{4.2}$$

where A_0 represents the integral intensity of the EPR spectrum of a control sample (blank), and A_s is the integral intensity of the EPR spectrum in the presence of the test sample.

Many research groups have successfully applied EPR to perform antioxidant activity evaluation of various types of antioxidants or antioxidant-loaded nano-dispersed systems (Mitsou et al. 2019; Sanna et al. 2019; Aboudzadeh et al. 2018; Chatzidaki et al. 2015). However, the main disadvantage is associated to the costs of the equipment.

4.4.2.2 Electrochemical Method Based on Cyclic Voltammetry (CV)

Another promising non-spectrophotometric method which can be used to assess the antioxidant activity of natural antioxidants before and after nanoemulsification is electrochemical method based on CV. Unlike the standardly used DPPH and ABTS radicals, which do not occur in nature, the oxygen-derived free radicals are readily produced in biological systems, causing damage to lipids, proteins and DNA, as a main cause of many diseases. Therefore, it can be of great interest to directly test the radical scavenging activity of a potential antioxidant with more biologically compatible ROS. Prior to electrochemical measurements, the investigated antioxidant compounds are usually dissolved in DMSO, but it has been reported that O/W nanoemulsions, or standard antioxidants diluted in sunflower oil, can be analysed without the use of organic solvents, e.g. in their original state, which is a very favourable experimental setting (Benedetti et al. 2012).

In the electrochemical approach, CV is used to electrocatalytically reduce oxygen to O_2•-, which further reacts with a radical scavenger. The information regarding a radical scavenging activity can be obtained analysing the evolution of the electrochemical response upon successive addition of a radical scavenger. The antioxidant activity ranking is established by evaluating the decrease of charge under anodic wave upon addition of antioxidant (Q), relative to the charge observed in O_2-saturated solution without antioxidants (Q_0), where O_2• – scavenging activity is quantified as an absolute value of a slope of a Q/Q_0 vs. concentration of the sample (Janosevic Lezaic et al. 2014; Dimitric Markovic et al. 2012). Two possible mechanisms for the reaction of polyphenols with O_2• − (proton-transfer and H-transfer i.e. radical-transfer), can be investigated based on the analysis of their cyclic voltammograms. For instance, in the presence of polyphenols (which are main antioxidant actives in many natural extracts and oils), there is no increase of cathodic peak current upon sample addition into O_2-saturated solution, and the appearance of cathodic pre-peaks (or pre-waves) is indicative of a prevailing H-transfer (radical-transfer) mechanism. Although electrochemical analysis is a convenient method that can be performed in an aqueous environment (Benedetti et al. 2012), it should be noted that a direct comparison of radical scavenging activities of two different compounds or systems can be made only if the electrochemical assay is performed under identical conditions (Janosevic Lezaic et al. 2014; Dimitric Markovic et al. 2012).

4.4.3 Conclusion

Due to the growing interest in the substances of natural sources with antioxidant effects, the area of their application has been significantly enlarged. A great diversity of antioxidant test methods is described in the literature, but the three presented methods (DPPH, ABTS and FRAP) remain well-established since they are applicable to various isolated compounds and antioxidant-loaded emulsions. However, the comparison of the results obtained in different studies remains a difficult task, due to large variations in experimental setting, solvent type and test concentration range. In order to generate valid conclusions, the selection of appropriate methods and their concomitant use is an imperative for a reliable assessment of antioxidant activity of the molecule per se, but also of the antioxidant-loaded carrier. Non-spectrophotometric methods are a valuable addition to the standard protocols, since they can be performed with nanoemulsions in their original state (in the aqueous environment). They should be used more often in future work.

4.5 In Vitro Safety and Efficacy Screening of Nanomaterials Using Cell Cultures: Focus on Nanoemulsions

Despite evident increase in the use of nanomaterials in medicine and consumer products, there is a general lack of standardised protocols for their characterisation, especially in terms of safety aspects and biological interactions (Rischitor et al. 2016). Even though in vitro test methods with cell cultures represent substantial tools for both mechanistic toxicity studies in fundamental research and for toxicity screening purposes, the existing established protocols usually apply to pure chemicals/test compounds, whereas they are not completely applicable to complex nanoformulations, such as nanoemulsions (Gioria et al. 2018).

Due to the observed cell-specific effects, the selection of optimal cell line for toxicity screening should be done based on the intended route of administration, target tissue and estimated exposure, preferably applying more than one test (Aslantürk 2018; Joris et al. 2016). However, many studies indicate that, after systemic exposure, most of the nanoparticulate systems are eliminated from the body through liver and kidneys, making these organs suitable for toxicity tests. Therefore, there are some available protocols for cytotoxicity assessment applying porcine kidney (LLC-PK1) and human cancerous liver cells (Hep G2) (Potter and Stern 2011). As common methodologies, 3-(4,5-dimethyl-2-thiazolyl)-2,5-diphenyl-2H-tetrazolium bromide (MTT) reduction and lactate dehydrogenase (LDH) release are performed. To complement the findings, and to reveal the exact mechanism of the obtained effects, genotoxicity assessment is also advised applying, for example, COMET assay (Krug and Wick 2011).

Being a highly interdisciplinary field, working with nanosystems in a biomedical context requires a strong understanding of colloidal behaviour and familiarity with specific characterisation methods, biology, and bio-nano interactions. When reporting results, some seemingly simple and trivial experimental details may be overlooked, giving misleading interpretations (Moore et al. 2019). Reliable experimental protocols are essential for proper processing of the results. Much work has been done in order to investigate the effect of particle physicochemical properties on particle-cell interactions, targeting, cellular uptake and toxicity (Kinnear et al. 2017; Zhao et al. 2011). However, issues of reproducibility still persist, which are, at least partly, caused by lack of reporting, non-standardised characterisation of nanomaterials and variations in biological assays (Faria et al. 2018).

Due to their diversity, and sometimes inherent complexity, there are numerous obstacles that may be involved in the in vitro screening of nanomaterials. A thorough characterisation of the tested nanomaterial and its behaviour both before and during toxicity assessment is a prerequisite to properly address these issues (Bouwmeester et al. 2011). With this in mind, in the following section, we present the methods used for in vitro nanomaterial safety and efficacy assessment, underlining some useful considerations aiming to prevent commonly occurring pitfalls. Some specific aspects related to the nanoemulsions are also addressed.

4.5.1 Commonly Used In Vitro Cell Viability Assays

As previously mentioned, MTT and LDH represent routinely performed in vitro assays for cytotoxicity assessment. Despite some limitations, according to ISO 10993–5:2009 (2009) and ASTM E2526–08 (2013), they are considered as standard methods for biological evaluation of nanomaterials.

MTT assay represents the most commonly used colourimetric test, providing insights into the proliferation and viability of cultured cells. The test itself is based on the reduction activity of metabolically active cells (Aslantürk 2018). Namely, MTT is a water-soluble yellow die, which can, following reduction by dehydrogenases present in metabolically active cells, be turned into a water-insoluble violet-blue formazan product (Fig. 4.4). Formazan deposits are extracted and calorimetrically assessed (Stockert et al. 2018). The intensity of violet-blue colour indicates the extent of metabolic activity, which is related to the cell viability. Even though there are findings suggesting that MTT reduction may occur elsewhere, it mainly takes place in mitochondria. Therefore, this test is considered as an efficient assay for mitochondrial function (Potter and Stern 2011).

On the other hand, LDH represents a cytosolic enzyme which can be released upon cell damage. The extent of the LDH release can be correlated with cell membrane integrity (Fig. 4.5).

Therefore, MTT and LDH assay are complementary methods for in vitro cytotoxicity assessment. The principle of this assay lays in the ability of LDH to oxidise lactate to pyruvate. Pyruvate can then react with a tetrazolium salt, forming water-soluble formazan, which can also be detected spectrophotometrically (Kaja et al. 2017).

Even though these methods are commonly applied, there are some identified pitfalls due to the colourimetric detection and the possibility of nanomaterial

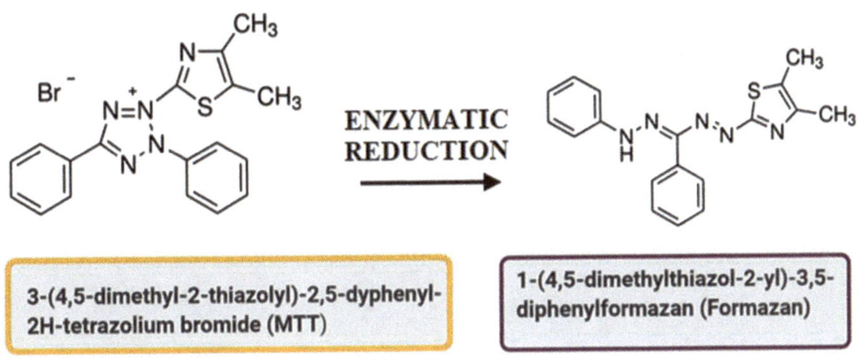

Fig. 4.4 Principle of the MTT assay: reduction of the MTT reagent and formazan formation, which can be spectrophotometrically assessed

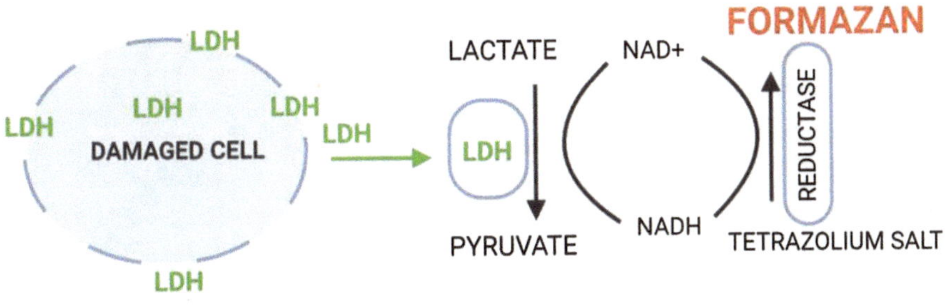

Fig. 4.5 Scheme of the LDH assay: two-step reaction and formation of coloured formazan

interferences with the test reagents (Gioria et al. 2018). In addition to the obtained results, to avoid any misleading interpretation, careful visual examination of the treated cells should be an important step in any assessment (Guadagnini et al. 2015).

In general, standardised protocols should be followed, but there are some specific considerations that ought to be addressed when testing nanoparticulated material. Hence, they are discussed in more details.

4.5.2 *Adsorption of Proteins: Formation of a Protein Corona*

Apart from their apparent difference from small molecules, some properties of the nanomaterials may change significantly during the assay, which can further complicate the data interpretation. For instance, a well-known phenomenon is the formation of a protein corona (an adsorbed layer of various proteins) around the nanomaterial upon its contact with the serum or cell culture medium (Fig. 4.6). Such occurrence may affect the nanomaterial-cell interaction by either facilitating or making the internalisation harder (Monopoli et al. 2012; Walkey and Chan 2012; Lynch et al. 2009). It has been shown that components of the biological surrounding (proteins and lipids) can affect recognition and processing of the nanomaterials by the cells. In other words, biological outcomes, at least partially, are dependent on the identity of the protein corona (its structure and protein orientation) and its residence time on the nanodroplet surface (Albanese et al. 2014).

Any nanomaterial exposed to a physiological environment interacts with proteins. However, the level of these interactions, as well as the properties of the formed

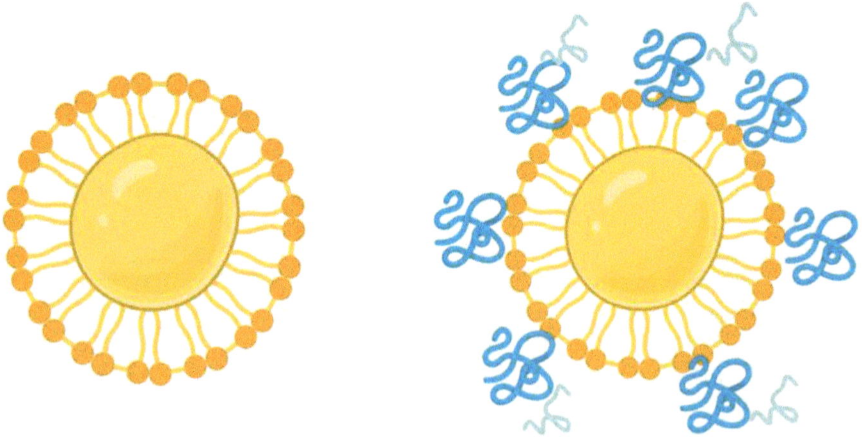

Fig. 4.6 Schematic representation of a nanoemulsion droplet per se (left), and in physiological environment/cell culture medium (right): adsorption of proteins at the interface (formation of protein corona) is visible

protein corona, depend on the so called "synthetic" identity of the nanomaterial, meaning that size, composition, topography and curvature of the nanomaterial surface are the most important parameters governing interactions with proteins, which further determines the extent of cellular uptake (Walkey and Chan 2012). Despite considerable research effort, the exact relationship between nanomaterial properties and protein adsorption has not yet been completely elucidated. What can be generalised is that nanomaterials with hydrophobic or charged surfaces tend to bind more proteins, whereas highly curved nanomaterial surfaces tend to decrease interactions with proteins (Aggarwal et al. 2009).

Literature provides an overview of techniques that are appropriate for the estimation of some properties of the protein corona. They can be broadly divided into in situ (the ones that allow direct measurements of the nanomaterial in the biological medium) and ex situ (the ones that require nanomaterial isolation). Both approaches have their advantages and disadvantages. Even though the in situ techniques are more relevant, they are limited in terms of the amount of information they could provide. On the other hand, nanomaterial isolation for ex situ techniques inevitably causes some structural changes in the protein corona and the loss of the loosely attached proteins (Weber et al. 2019). Nevertheless, the structural assessment of the protein corona encompasses its thickness, density, protein identity and affinity towards the cells (Walkey and Chan 2012).

One of the commonly applied techniques for the size and size distribution assessment of nanomaterials is dynamic light scattering, which also finds its place in the estimation of the protein corona thickness. Protein corona evaluation may be easily performed by comparing the size in water (the usual dispersant) to the size obtained using cell culture medium or full serum as a dispersing medium. Applying this technique, Walczyk et al. (2010) have obtained reproducible results with surface-carboxylated polystyrene particles, showing that protein corona in full serum is formed in a relatively stable manner after 1 h. Obtained results were consistent with those extracted through TEM analysis, as an example of ex situ technique. Other corona parameters and selection of appropriate techniques for its assessment are summarised in Table 4.3. As isolation techniques for ex situ analysis, differential centrifugation and size exclusion chromatography are typically applied (Walkey and Chan 2012).

In addition to this, in silico simulations of protein–nanomaterial interactions have been attracting attention recently. So far, such approach has not been sufficiently powerful to take into consideration all the peculiarities and to predict the complexity of interactions in physiological environment. However, it is expected that the improvement of the computational methodologies will enable them to reach the appropriate level of accuracy (Di Felice and Corni 2011).

In spite of all difficulties, the characterisation and analysis of protein adsorption to the nanomaterial represents a step towards understanding the true nature of its biological effects (Weber et al. 2019). It provides important insights into the cellular uptake mechanism, cytotoxicity, inflammation potential and other biological effects caused by the nanomaterials (Saptarshi et al. 2013).

Table 4.3 Selected methods for protein corona assessment

Corona parameters	In situ techniques	Ex situ techniques
Thickness	Dynamic light scattering Fluorescence correlation spectrometry	Differential centrifugal sedimentation Size exclusion chromatography Transmission electron microscopy
Density of the adsorbed proteins		Colourimetric protein assays
Identity and quantity of the adsorbed proteins		Poly(acrylamide) gel electrophoresis Liquid chromatography tandem mass spectrometry
Protein conformation		Circular dichroism Fluorescence quenching
Affinity		Size exclusion chromatography Surface plasmon resonance Isothermal titration calorimetry

4.5.3 Relation Between Nanomaterial Physicochemical Properties and Dosimetry

Because of the small size and specific stabilisers, nanomaterials are practically not affected by the gravitational force, representing stable dispersions. Such property presents a potential difficulty for in vitro assays, where cells adhering to the bottom of the well may not be exposed to the sufficient amount of the tested nanomaterial (Lison et al. 2008).

In order to avoid any misleading results from in vitro tests, determination of the dose of that can effectively get in contact with the cells during the specific assay is also an important aspect to be considered (Rischitor et al. 2016). Usually, when chemicals are being tested in in vitro studies, the initial concentration added to the culture medium would be considered as the effective concentration (due to diffusion and expected homogeneity). However, such approach should not be taken with nanomaterials. In contrast to chemical compounds, nanomaterials interact with culture medium, and depending on their properties (size and size distribution, shape, mass density and solubility) utilise different transport mechanisms to reach the cell monolayer (Cohen et al. 2014; Cho et al. 2011; Teeguarden et al. 2007). Furthermore, apart from aforementioned protein corona, other dynamic modifications of the nanomaterial properties may occur in the cell culture medium, leading to potential agglomeration and aggregation. Consequently, the particle size of the test nanocarrier could be changed, directly influencing its transport towards the cell monolayer and the cellular uptake (internalisation) of the tested material (Cohen et al. 2014). Having in mind the importance of this specific aspect, much research has been done so far dealing with dosimetry (Rischitor et al. 2016).

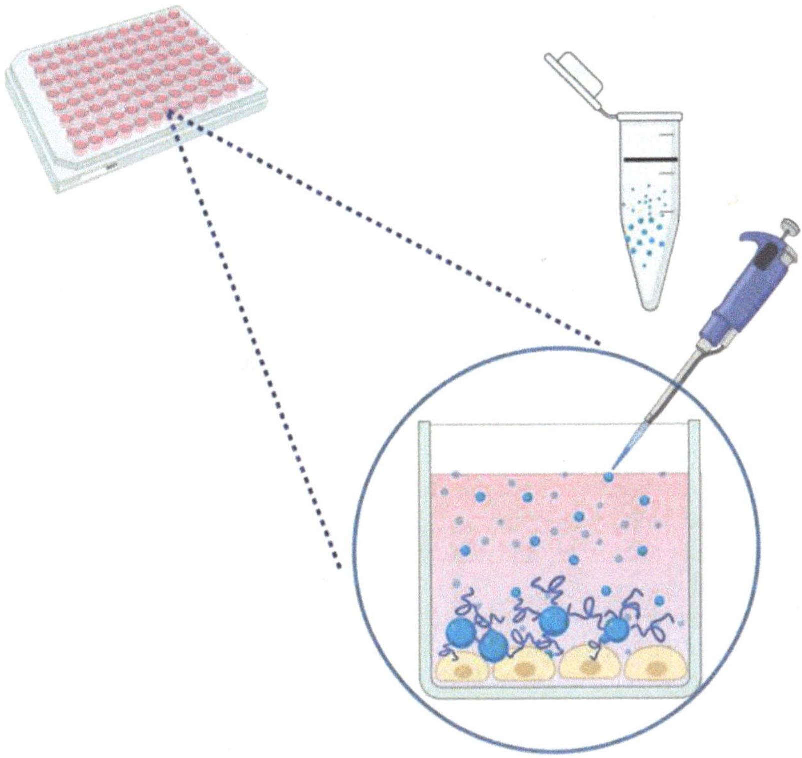

Fig. 4.7 Deposition of the nanomaterial onto the cell monolayer

Studies of cellular uptake are closely related to the transport of the nanomaterial to the cell monolayer (Fig. 4.7). *Cellular dose* – the number of nanounits that reach the cell membrane and that can eventually be uptaken, depends on the transport of the nanomaterial and its affinity to the cells. Rischitor et al. (2016) presented an interesting work, calculating the amount of deposited and/or uptaken gold nanoparticles by measuring the UV-Vis spectra of the cell culture medium supernatant from the wells at several time points and comparing it to the spectra obtained for the supernatant from the well without cells. The authors found that the fraction of nanoparticles deposited on the cellular layer is dependent on their size and density, as these parameters govern their transport towards the cell monolayer. Also, the duration of exposure is an important experimental condition that should be controlled.

It is worth noting that initial nanomaterial concentration used in the in vitro assay should be carefully selected because an excess of nanostructured material can, in some cases, be misinterpreted as toxicity, even though a decrease in cell viability is actually caused by mechanical obstruction due to the overload (Wittmaack 2011).

Interestingly, Moore and et al. (2019) demonstrated that even the manner of administration of the nanoparticles to the well could significantly influence the observed interactions. They evaluated concentrated (bolus) administration, concentrated administration followed by pipette mixing, and premixing to the desired concentration with culture medium prior to the addition. Observed differences in cellular response were attributed to the different ability of nanoparticles to deposit onto cell monolayer.

4.5.4 Specific Observations Related to the Nanoemulsions

Nanoemulsions, as soft matter nanosystems, are usually applied as carries for medicinal compounds, nutrients, food additives and cosmetic ingredients (McClements and Jafari 2018; Nikolic et al. 2018). It is well known that certain components, such as nanoemulsion stabilisers, may induce toxicity to cultured cells due to their solubilising effects (Vater et al. 2019), even though they are declared as biocompatible. Therefore, attention should be paid to the selected concentration range for toxicity testing. Whenever possible, for any tested nanomaterial, cell exposure to the test sample should realistically reflect intended application (Krug and Wick 2011).

It has been reported that some antioxidants possess additional properties, such as anticancer and/or antigenotoxic activities, which should be assessed applying cell cultures (Gledovic et al. 2020; Nikolic et al. 2020; Nikolic et al. 2018). In order to estimate their efficacy when applied in the form of nanoemulsion, it is crucial to eliminate any potential effect of the empty carrier. It addition, it is sometimes possible to discover a beneficial contribution of this specific delivery system, comparing the outcomes of the "free" antioxidant to the one solubilised in the formulation. Therefore, it is advised that the placebo naoemulsion formulation is tested first, in order to obtain a specific concentration range with no observed effects. Further on, based on the findings related to the "free" compound, the loaded formulation can be diluted to the desired concentration for efficacy assessment (Nikolic et al. 2020; Theochari et al. 2017). Moreover, since nanoemulsions are designed as carriers for the active ingredient (e.g. antioxidants), in some cases, there is no need for the uptake of the whole nanoemulsion droplet. It is expected that nanoemulsions should provide stability for the antioxidant, enabling its efficacy at the moment of application by facilitating internalisation of the compound into the cell (Pavoni et al. 2020). Therefore, appropriate release studies should be conducted with a view to estimate the amount of available active ingredient.

4.5.5 Novel Approaches for In Vitro Testing: Microfluidic Technology

In order to overcome some of the mentioned limitations of the classical in vitro biological assessment, considerable attention is currently being devoted to the cell-based microfluidic model systems, an emerging biomimetic screening tools. In such setting, cells are not static, but exposed to the constant flow of the culture medium, which is an environment with physiological relevance (Jie and Lin 2018; Perestrelo et al. 2015). Moreover, microfluidic cell platforms enable culturing and screening of a range of miniaturised 3D organ and tissue models, representing elegant and reliable tissue/organ-on-a-chip platforms (Trietsch et al. 2013). Described technology could bridge the existing gap between the in vitro experimental setting and in vivo application, while reducing experimental costs (smaller amounts of samples and reagents are needed), with perspectives that are beyond the proof-of-concept stage (Esch et al. 2015).

4.6 Conclusion and Closing Remarks

For meaningful evaluation of nanomaterials it is necessary to perform their complete physicochemical assessment (including size, morphology, surface charge determination, loading and release profile assessment) in addition to the characterisation of interactions between the nanomaterial and the culture medium, considering all potential experimental issues. Such comprehensive approach represents an important step towards the next stage, a more relevant in vitro screening. There is a need for further standardisation and development of robust methods for nanomaterial characterisation, which should lead to a reliable and physiologically closer assessment of their safety and efficacy aspects.

Based on the findings presented in this chapter, it is clear that the preparation of nanoemulsions as carriers for natural antioxidants is a complex task. Numerous factors should be considered in order to achieve the formulation with good chemical stability, desired release profile and high antioxidant efficacy. Those factors include right selection of oils and stabilisers, appropriate preparation method and processing conditions, often followed by the adjustment of characterisation protocols. The theoretical concepts and experimental results presented in this chapter, alongside relevant literature sources, should provide a useful guide to formulating innovative solutions for tailor-made nanoemulsions with natural antioxidants.

References

Aboudzadeh MA, Mehravar E, Fernandez M, Lezama L, Tomovska R (2018) Low-energy encapsulation of α-tocopherol using fully food grade oil-in-water microemulsions. ACS Omega 3:10999–11008. https://doi.org/10.1021/acsomega.8b01272

Aburjai T, Natseh FM (2003) Plants used in cosmetics. Phytother Res 17:987–1000. https://doi.org/10.1002/ptr.1363

Aggarwal P, Hall JB, McLeland CB, Dobrovolskaia MA, McNeil SE (2009) Nanoparticle interaction with plasma proteins as it relates to particle biodistribution, biocompatibility and therapeutic efficacy. Adv Drug Deliv Rev 61:428–437. https://doi.org/10.1016/j.addr.2009.03.009

Ahmadi O, Jafarizadeh-Malmiri H (2020) Green approach in food nanotechnology based on subcritical water: effects of thyme oil and saponin on characteristics of the prepared oil in water nanoemulsions. Food Sci Biotechnol 29:783–792. https://doi.org/10.1007/s10068-019-00727-0

Alam N, Bristi NC, Rafiquzzaman M (2013) Review on in vivo and in vitro methods evaluation of antioxidant activity. Saudi Pharm J 21:143–152. https://doi.org/10.1016/j.jsps.2012.05.002

Albanese A, Walkey CD, Olsen JB, Guo H, Emili A, Chan WCW (2014) Secreted biomolecules alter the biological identity and cellular interactions of nanoparticles. ACS Nano 8:5515–5526. https://doi.org/10.1021/nn4061012

Alzorqi I, Ketabchi MR, Sudheer S, Manickam S (2016) Optimization of ultrasound induced emulsification on the formulation of palm-olein based nanoemulsions for the incorporation of antioxidant b-D-glucan polysaccharides. Ultrason Sonochem 31:71–84. https://doi.org/10.1016/j.ultsonch.2015.12.004

Amorati R, Valgimigli L (2015) Advantages and limitations of common testing methods for antioxidants. Free Radic Res 49:633–649. https://doi.org/10.3109/10715762.2014.996146

Aslantürk ÖS (2018) In vitro cytotoxicity and cell viability assays: principles, advantages, and disadvantages. In: Larramendy ML, Soloneski S (eds) Genotoxicity: a predictable risk to our actual world. InTechOpen, London, pp 1–17. https://doi.org/10.5772/intechopen.71923

ASTM (2013) Standard test method for evaluation of cytotoxicity of nanoparticulate materials in porcine kidney cells and human hepatocarcinoma cells (E2526–08). https://doi.org/10.1520/E2526-08R13

Azmi NAN, Elgharbawy AAM, Motlagh SR, Samsudin N, Salleh HM (2019) Nanoemulsions: factory for food, pharmaceutical and cosmetics. PRO 7:617. https://doi.org/10.3390/pr7090617

Bajerski L, Michels LR, Marques Colome L, Bender EA, Freddo RJ, Bruxel F, Haas SE (2016) The use of Brazilian vegetable oils in nanoemulsions: an update on preparation and biological applications. Braz J Pharm 52:347–363. https://doi.org/10.1590/s1984-82502016000300001

Benedetti S, Cosio MS, Scampicchio M, Mannino S (2012) Nanoemulsions for the determination of the antioxidant capacity of oils by an electrochemical method. Electroanalysis 24:1356–1361. https://doi.org/10.1002/elan.201200063

Bernardi DS, Pereira TA, Maciel NR, Bortoloto J, Viera GS, Oliveira GC, Rocha-Filho PA (2011) Formation and stability of oil-in-water nanoemulsions containing rice bran oil: in vitro and in vivo assessments. J Nanobiotechnology 9:44. https://doi.org/10.1186/1477-3155-9-44

Bouwmeester H, Lynch I, Marvin HJ, Dawson KA, Berges M, Braguer D, Byrne HJ, Casey A, Chambers G, Clift MJ, Elia G, Fernandes TF, Fjellsbø LB, Hatto P, Juillerat L, Klein C, Kreyling WG, Nickel C, Riediker M, Stone V (2011) Minimal analytical characterization of engineered nanomaterials needed for hazard assessment in biological matrices. Nanotoxicology 5:1–11. https://doi.org/10.3109/17435391003775266

Bushmann BS, Phillips B, Isbell T, Ou B, Crane JM, Knapp SJ (2004) Chemical composition of Caneberry (Rubus spp.) seeds and oils and their antioxidant potential. J Agric Food Chem 52:7982–7987. https://doi.org/10.1021/jf049149a

Cano A, Alcaraz O, Acosta M, Arnao MB (2013) On-line antioxidant activity determination: comparison of hydrophilic and lipophilic antioxidant activity using the ABTS assay. Redox Rep 7:103–109. https://doi.org/10.1179/135100002125000334

Chang YY, McClements DJ (2014) Optimization of Orange oil Nanoemulsion formation by isothermal low-energy methods: influence of the oil phase, surfactant, and temperature. J Agric Food Chem 62:2306–2312. https://doi.org/10.1021/jf500160y

Chatzidaki MD, Mitsou E, Yaghmur A, Xenakis A, Papadimitriou V (2015) Formulation and characterization of food-grade microemulsions as carriers of natural phenolic antioxidants. Colloids Surf A Physicochem Eng Asp 483:130–136. https://doi.org/10.1016/j.colsurfa.2015.03.060

Chen L, Hu JY, Steven Q, Wang SQ (2012) The role of antioxidants in photoprotection: a critical review. J Am Acad Dermatol 67:1013–1024. https://doi.org/10.1016/j.jaad.2012.02.009

Cho EC, Zhang Q, Xia Y (2011) The effect of sedimentation and diffusion on cellular uptake of gold nanoparticles. Nat Nano 6:385–391. https://doi.org/10.1038/nnano.2011.58

Chong WT, Tan CP, Cheah YK, Lajis AFB, Habi Mat Dian NL, Kanagaratnam S, Ming Lai O-M (2018) Optimization of process parameters in preparation of tocotrienol-rich red palm oil-based nanoemulsion stabilized by Tween80- Span 80 using response surface methodology. PLoS One 13(8):e0202771. https://doi.org/10.1371/journal.pone.0202771

Chuesiang P, Siripatrawan U, Sanguandeekul R, McLandsborough L, McClements DJ (2018) Optimization of cinnamon oil nanoemulsions using phase inversion temperature method: impact of oil phase composition and surfactant concentration. J Colloid Interface Sci 514:208–216. https://doi.org/10.1016/j.jcis.2017.11.084

Cohen J, Teeguarden J, Demokritou P (2014) An integrated approach for the in vitro dosimetry of engineered nanomaterials. Part Fibre Toxicol 11:20. https://doi.org/10.1186/1743-8977-11-20

Da Silva Gundel S, Callegaro Velho M, Kerche Diefenthaler M, Reis Favarin F, Marquezan Copetti P, de Oliveira Fogaça A, Klein B, Wagner R, Gundel A, Rorato Sagrillo M, Ferreira Ourique A (2018) Basil oil-nanoemulsions: development, cytotoxicity and evaluation of antioxidant and antimicrobial potential. J Drug Deliv Sci Tec 46:378–383. https://doi.org/10.1016/j.jddst.2018.05.038

Di Felice R, Corni S (2011) Simulation of peptide–surface recognition. J Phys Chem Letter 2:1510–1519. https://doi.org/10.1021/jz200297k

Dimitric-Markovic JM, Markovic ZS, Pasti IA, Brdaric TP, Popovic-Bijelic A, Mojovic M (2012) A joint application of spectroscopic, electrochemical and theoretical approaches in evaluation of the radical scavenging activity of 3-OH flavones and their iron complexes towards different radical species. Dalton Trans 41:7295. https://doi.org/10.1039/C2DT30220A

Dorman HJD, Surai P, Deans SG (2000) In vitro antioxidant activity of a number of plant essential oils and phytoconstituents. J Essent Oil Res 12(2):241–248. https://doi.org/10.1080/10412905.2000.9699508

Esch EW, Bahinski A, Huh D (2015) Organs-on-chips at the frontiers of drug discovery. Nat Rev Drug Discov 14:248–260. https://doi.org/10.1038/nrd4539

Espitia PJP, Fuenmayor CA, Otoni CG (2018) Nanoemulsions: synthesis, characterization, and application in bio-based active food packaging. Compr Rev Food Sci Food Saf 18:264–285. https://doi.org/10.1111/1541-4337.12405

Faria M, Björnmalm M, Thurecht KJ, Kent SJ, Parton RG, Kavallaris M, Johnston APR, Gooding JJ, Corrie SR, Boyd BJ, Thordarson P, Whittaker AK, Stevens MM, Prestidge CA, Porter CJH, Park WJ, Davis TP, Crampin EJ, Caruso F (2018) Minimum information reporting in bio–nano experimental literature. Nature Nanotech 13:777–785. https://doi.org/10.1038/s41565-018-0246-4

Floegel A, Kim DO, Chung SJ, Koo SI, Chun OK (2011) Comparison of ABTS/DPPH assays to measure antioxidant capacity in popular antioxidant-rich US foods. J Food Compos Anal 24:1043–1048. https://doi.org/10.1016/j.jfca.2011.01.008

Gioria S, Caputo F, Urbán P, Maguire CM, Bremer-Hoffmann S, Prina-Mello A, Calzolai L, Mehn D (2018) Are existing standard methods suitable for the evaluation of nanomedicines: some case studies. Nanomedicine 13:539–554. https://doi.org/10.2217/nnm-2017-0338

Gledovic A, Janosevic Lezaic A, Krstonosic V, Djokovic J, Nikolic I, Bajuk-Bogdanovic D, Antic Stankovic J, Randjelovic D, Savic SM, Filipovic M, Tamburic S, Savic SD (2020) Low-energy nanoemulsions as carriers for red raspberry seed oil: formulation approach based on Raman spectroscopy and textural analysis, physicochemical properties, stability and in vitro antioxidant/ biological activity. PLoS One 15:e0230993. https://doi.org/10.1371/journal.pone.0230993

Gothani and Prasert (2014) Nano-emulsions; emulsification using low energy methods. Japan J Food Eng 15:119–130. https://doi.org/10.11301/jsfe.15.119

Guadagnini R, Halamoda Kenzaoui B, Walker L, Pojana G, Magdolenova Z, Bilanicova D, Saunders M, Juillerat-Jeanneret L, Marcomini A, Huk A, Dusinska M, Fjellsbo LM, Marano F, Boland S (2015) Toxicity screenings of nanomaterials: challenges due to interference with assay processes and components of classic in vitro tests. Nanotoxicology 9:13–24. https://doi.org/10.3109/17435390.2013.829590

Gupta A, Eral HB, Hatton TA, Doyle PS (2016) Nanoemulsions: formation, properties and applications. Soft Matter 12:2826–2841. https://doi.org/10.1039/C5SM02958A

Han JW, Ruiz-Garcia L, Qian JP, Yang XT (2018) Food packaging: a comprehensive review and future trends. Compr Rev Food Sci Food Safety 17:860–877. https://doi.org/10.1111/1541-4337.12343

Hategekimana J, Chamba MVM, Shoemaker CF, Majeed H, Zhong F (2015) Vitamin E nanoemulsions by emulsion phase inversion: effect of environmental stress and long-term storage on stability and degradation in different carrier oil types. Colloids Surf A Physicochem Eng Asp 483:70–80. https://doi.org/10.3390/molecules24234242

ISO (2009) Biological evaluation of medical devices (ISO 10993:2009). ISO 2009:10993–10995

Janosevic Lezaic A, Pasti I, Vukomanovic M, Ciric-Marjanovic G (2014) Polyaniline tannate - synthesis, characterization and electrochemical assessment of superoxide anion radical scavenging activity. Electrochim Acta 142:92–100. https://doi.org/10.1016/j.electacta.2014.07.073

Jie M, Lin JM (2018) Microfluidic cell culture systems for drug research. In: Lin JM (ed) Cell analysis on microfluidics. Springer, Singapore, pp 339–370. https://doi.org/10.1007/978-981-10-5394-8

Joris F, Valdepérez D, Pelaz B, Soenen SJ, Manshian BB, Parak WJ, De Smedt SC, Raemdonck K (2016) The impact of species and cell type on the nanosafety profile of iron oxide nanoparticles in neural cells. J Nanobiotechnol 14:1–13. https://doi.org/10.1186/s12951-016-0220-y

Kaja S, Payne AJ, Naumchuk Y, Koulen P (2017) Quantification of lactate dehydrogenase for cell viability testing using cell lines and primary cultured astrocytes. Cur Protoc Toxicol 72:2–26. https://doi.org/10.1002/cptx.21

Kaur K, Kaur J, Kumar R, Mehta SK (2017) Formulation and physiochemical study of a-tocopherol based oil in water nanoemulsion stabilized with nontoxic, biodegradable surfactant: sodium stearoyl lactate. Ultrason Sonochem 38:570–578. https://doi.org/10.1016/j.ultsonch.2016.08.026

Kim DM, Hyun SS, Yun P, Lee CH, Byun SY (2011) Identification of an emulsifier and conditions for preparing stable nanoemulsions containing the antioxidant astaxanthin. Int J Cosmet Sci 34:64–73. https://doi.org/10.1111/j.1468-2494.2011.00682.x

Kinnear C, Moore TL, Rodriguez-Lorenzo L, Rothen-Rutishauser B, Petri-Fink A (2017) Form follows function: nanoparticle shape and its implications for nanomedicine. Chem Rev 117:11476–11521. https://doi.org/10.1021/acs.chemrev.7b00194

Krug HF, Wick P (2011) Nanotoxicology: an interdisciplinary challenge. Angew Chem Int Ed 50:1260–1278. https://doi.org/10.1002/anie.201001037

Li B, Pratt DA (2015) Methods for determining the efficacy of radical-trapping antioxidants. Free Radic Biol Med 82:187–202. https://doi.org/10.1016/j.freeradbiomed.2015.01.020

Lison D, Thomassen LC, Rabolli V, Gonzalez L, Napierska D, Seo JW, Kirsch-Volders M, Hoet P, Kirschhock CAE, Martens JA (2008) Nominal and effective dosimetry of silica nanoparticles in cytotoxicity assays. Toxicol Sci 104:155–162. https://doi.org/10.1093/toxsci/kfn072

Lobo V, Patil A, Phatak A, Chandra N (2010) Free radicals, antioxidants and functional foods: impact on human health. Pharmacogn Rev 4:118. https://doi.org/10.4103/0973-7847.70902

Lou Z, Chen J, Yu F, Wang H, Kou X, Ma C, Zhu S (2017) The antioxidant, antibacterial, antibiofilm activity of essential oil from Citrus medica L. var. sarcodactylis and its nanoemulsion. LWT - Food Sci Technol 80:371–377. https://doi.org/10.1016/j.lwt.2017.02.037

Lynch I, Salvati A, interactions DKA P-n (2009) What does the cell see? Nat Nano. 4:546–547. https://doi.org/10.1038/nnano.2009.248

Majeed H, Bian YY, Ali B, Jamil A, Usman Majeed U, Khan QF, Iqbal KJ, Shoemaker CF, Fang Z (2015) Essential oil encapsulations: uses, procedures, and trends. RSC Adv 5:58449. https://doi.org/10.1039/C5RA06556A

Mao L, Xu D, Yang J, Yuan F, Gao Y, Zhao J (2009) Effects of small and large molecule emulsifiers on the characteristics of b-carotene Nanoemulsions prepared by high pressure homogenization. Food Technol Biotechnol 47:336–342

Masaki H (2010) Role of antioxidants in the skin: anti-aging effects. J Dermatol Sci 58:85–90. https://doi.org/10.1016/j.jdermsci.2010.03.003

McClements DJ, Decker EA (2000) Lipid oxidation in oil-in-water emulsions: impact of molecular environment on chemical reactions in heterogeneous food systems. JFS 65:1270–1282. https://doi.org/10.1111/j.1365-2621.2000.tb10596.x

McClements DJ, Jafari SM (2018) General aspects of nanoemulsions and their formulation. In: Jafari SM, McClements DJ (eds) Nanoemulsions: formulation, applications, and characterization. Academic, San Diego, pp 3–20. https://doi.org/10.1016/B978-0-12-811838-2.00001-1

McClements DJ, Rao J (2011) Food-grade nanoemulsions: formulation, fabrication, properties, performance, biological fate, and potential toxicity. Crit Rev Foof Sci Nutr 51:285–330. https://doi.org/10.1080/10408398.2011.559558

Michalak M, Dadasiewicz AK (2018) Oils from fruit seeds and their dietetic and cosmetic significance. Herba Pol 64:63–70. https://doi.org/10.2478/hepo-2018-0026

Miller NJ, Diplock AT, Rice-Evans C, Davies MJ, Gopinathan V, Milner A (1993) A novel method for measuring antioxidant capacity and its application to monitoring the antioxidant status in premature neonates. Clin Sci 84:407–412. https://doi.org/10.1042/cs0840407

Miller NJ, Sampson J, Candeias LP, Bramley PM, Rice-Evans CA (1996) Antioxidant activities of carotenes and xanthophylls. FEBS Lett 384:240–242. https://doi.org/10.1016/0014-5793(96)00323-7

Mitsou E, Dupin A, Sassi AH, Monteil J, Sotiroudis GT, Leal-Calderon F, Xenakis A (2019) Hydroxytyrosol encapsulated in biocompatible water-in-oil microemulsions: how the structure affects in vitro absorption. Colloids Surf B Biointerfaces 184:110482. https://doi.org/10.1016/j.colsurfb.2019.110482

Monopoli MP, Åberg C, Salvati A, Dawson KA (2012) Biomolecular coronas provide the biological identity of nanosized materials. Nat Nano 7:779–786. https://doi.org/10.1038/nnano.2012.207

Moore TL, Urban DA, Rodriguez-Lorenzo L, Milosevic A, Crippa F, Spuch-Calvar M, Balog S, Rothen-Rutishauser B, Lattuada M, Petri-Fink A (2019) Nanoparticle administration method in cell culture alters particle-cell interaction. Sci Rep 9:1–9. https://doi.org/10.1038/s41598-018-36954-4

Naidoo K, Birch-Machin M (2017) Oxidative stress and ageing: the influence of environmental pollution, sunlight and diet on skin. Cosmetics 4:4. https://doi.org/10.3390/cosmetics4010004

Nawab A, Nichols A, Klug R, Shapiro JI, Sodhi K (2017) Spin trapping: a review for the study of obesity related oxidative stress and Na+/K+-ATPase. J Clin Cell Immunol 8(3):505. https://doi.org/10.4172/2155-9899.1000505

Nikolic I, Lunter DJ, Randjelovic D, Zugic A, Tadic V, Markovic B, Cekic N, Zivkovic L, Topalovic D, Spremo-Potparevic B, Daniels R, Savic S (2018) Curcumin-loaded low-energy nanoemulsions as a prototype of multifunctional vehicles for different administration routes: physicochemical and in vitro peculiarities important for dermal application. Int J Pharm 550:333–346. https://doi.org/10.1016/j.ijpharm.2018.08.060

Nikolic I, Mitsou E, Damjanovic A, Papadimitriou V, Antic-Stankovic J, Stanojevic B, Xenakis A, Savic S (2020) Curcumin-loaded low-energy nanoemulsions: linking EPR spectroscopy-analysed microstructure and antioxidant potential with in vitro evaluated biological activity. J Mol Liq 301:112479. https://doi.org/10.1016/j.molliq.2020.112479

Nile SH, Khobragade NC, Park WS (2012) Optimized and comparative antioxidant assays and its applications in herbal and synthetic drug analysis as antioxidants. Mini Rev Med Chem 12(10):1007–1014. https://doi.org/10.2174/138955712802762310

Pan Y, Tikekar RV, Nitin N (2013) Effect of antioxidant properties of lecithin emulsifier on oxidative stability of encapsulated bioactive compounds. Int J Pharm 450:129–137. https://doi.org/10.1016/j.ijpharm.2013.04.038

Pavoni L, Perinelli DR, Bonacucina G, Cespi M, Palmieri GF (2020) An overview of micro- and nanoemulsions as vehicles for essential oils: formulation, preparation and stability. Nanomaterials 10:135. https://doi.org/10.3390/nano10010135

Pereira TA, Guerreiro CM, Maruno M, Ferrari M, Rocha-Filho PA (2016) Exotic vegetable oils for cosmetic O/W nanoemulsions: in vivo evaluation. Molecules 21:248. https://doi.org/10.3390/molecules21030248

Perestrelo AR, Águas AC, Rainer A, Forte G (2015) Microfluidic organ/body-on-a-chip devices at the convergence of biology and microengineering. Sensors 15:31142–31170. https://doi.org/10.3390/s151229848

Pisoschi AM, Pop A, Cimpeanu C, Predoi G (2016) Antioxidant capacity determination in plants and plant-derived products: a review. Oxidative Med Cell Longev 2016:9130976. https://doi.org/10.1155/2016/9130976

Polovka M, Brezová V, Staško A (2003) Antioxidant properties of tea investigated by EPR spectroscopy. Biophys Chem 106:39–56. https://doi.org/10.1016/S0301-4622(03)00159-5

Potter TM, Stern ST (2011) Evaluation of cytotoxicity of nanoparticulate materials in porcine kidney cells and human hepatocarcinoma cells. In: McNeil SE (ed) Characterization of nanoparticles intended for drug delivery. Human Press, New Jersey, pp 157–165. https://doi.org/10.1007/978-1-60327-198-1_16

Prakash B, Kujur A, Yadav A, Kumar A, Singh PP, Dubey NK (2018) Nanoencapsulation: an efficient technology to boost the antimicrobial potential of plant essential oils in food system. Food Control 89:1–11. https://doi.org/10.1016/j.foodcont.2018.01.018

Prior RL, Wu X, Schaich K (2005) Standardized methods for the determination of antioxidant capacity and phenolics in foods and dietary supplements. J Agric Food Chem 53:4290–4302. https://doi.org/10.1021/jf0502698

Qian C, Decker EA, Xiao H, McClements DJ (2012) Inhibition of b-carotene degradation in oil-in-water nanoemulsions: influence of oil-soluble and water-soluble antioxidants. Food Chem 135:1036–1043. https://doi.org/10.1016/j.foodchem.2012.05.085

Raut JS, Karuppayil SM (2014) A status review on the medicinal properties of essential oils. Ind Crop Prod 62:250–264. https://doi.org/10.1016/j.indcrop.2014.05.055

Re R, Pellegrini N, Proteggente A, Pannala A, Yang M, Rice-Evans C (1999) Antioxidant activity applying and improved ABTS radical cation decolorization assay. Free Radic Biol Med 26:1231–1237. https://doi.org/10.1016/S0891-5849(98)00315-3

Rebolleda S, Sanz MT, Benito JM, Beltran S, Escudero I, González San-José ML (2015) Formulation and characterisation of wheat bran oil-in-water nanoemulsions. Food Chem 167:16–23. https://doi.org/10.1016/j.foodchem.2014.06.097

Reis Mansur MCPP, Leitão SG, Cerqueira-Coutinho C, Vermelho AB, Silva RS, Presgrave OAF, Leitão AAC, Leitão GG, Ricci-Júnior E, Santos EP (2016) In vitro and in vivo evaluation of efficacy and safety of photoprotective formulations containing antioxidant extracts. Rev Bras Farmacogn 26:251–258. https://doi.org/10.1016/j.bjp.2015.11.006

Rinaldi F, Hanieh PN, Longhi C, Carradori S, Secci D, Zengin G, Ammendolia MG, Mattia E, Del Favero E, Marianecci C, Carafa M (2017) Neem oil nanoemulsions: characterisation and antioxidant activity. J Enzym Inhib Med Chem 32:1265–1273. https://doi.org/10.1080/14756366.2017.1378190

Rischitor G, Parracino M, La Spina R, Urbán P, Ojea-Jiménez I, Bellido E, Valesia A, Gioria S, Campomaccio R, Kinsner-Ovaskainen A, Gilliland D, Rossi F, Colpo P (2016) Quantification of the cellular dose and characterization of nanoparticle transport during in vitro testing. Part Fibre Toxicol 13:47. https://doi.org/10.1186/s12989-016-0157-1

Rocha-Filho PA, Camargo MFP, Ferrari M, Maruno M (2014) Influence of Lavander essential oil addition on passion fruit oil nanoemulsions: stability and in vivo study. J Nanomed Nanotechnol 5:2. https://doi.org/10.4172/2157-7439.1000198

Salem MA, Ezzat SM (2018) Nanoemulsions in food industry. In: Milani JM (ed) Some new aspects of colloidal systems in foods. InechOpen, London, pp 31–51. https://doi.org/10.5772/intechopen.79447

Salvia-Trujillo L, Rojas-Grau MA, Soliva-Fortuny R, Martín-Belloso O (2013) Effect of processing parameters on physicochemical characteristics of microfuidized lemongrass essential oil-alginate nanoemulsions. Food Hydrocoll 30:401–407. https://doi.org/10.1016/j.foodhyd.2012.07.004

Sanna D, Mulas M, Molinu MG, Fadda A (2019) Oxidative stability of plant hydroalcoholic extracts assessed by EPR spin trapping under forced ageing conditions: a myrtle case study. Food Chem 271:753–761. https://doi.org/10.1016/j.foodchem.2018.07.156

Saptarshi SR, Duschl A, Lopata AL (2013) Interaction of nanoparticles with proteins: relation to bio-reactivity of the nanoparticle. J Nanobiotechnol 11:26. https://doi.org/10.1186/1477-3155-11-26

Solans C, Sole I (2012) Nano-emulsions: formation by low-energy methods. Curr Opin Colloid Interface Sci 17:246–254. https://doi.org/10.1016/j.cocis.2012.07.003

Stockert JC, Horobin RW, Colombo LL, Blázquez-Castro A (2018) Tetrazolium salts and formazan products in cell biology: viability assessment, fluorescence imaging, and labeling perspectives. Acta Histochem 120:159–167. https://doi.org/10.1016/j.acthis.2018.02.005

Teeguarden JG, Hinderliter PM, Orr G, Thrall BD, Pounds JG (2007) Particokinetics in vitro: dosimetry considerations for in vitro nanoparticle toxicity assessments. Toxicol Sci 95:300–312. https://doi.org/10.1093/toxsci/kfl165

Teo BSX, Basri M, Zakaria MRS, Salleh AB, Rahman RNZRA, Rahman MBA (2010) A potential tocopherol acetate loaded palm oil esters-in-water nanoemulsions for nanocosmeceuticals. J Nanobiotechnol 8:4. https://doi.org/10.1186/1477-3155-8-4

Theochari I, Goulielmaki M, Danino D, Papadimitriou V, Pintzas A, Xenakis A (2017) Drug nanocarriers for cancer chemotherapy based on microemulsions: the case of Vemurafenib analog PLX4720. Colloid Surf B Biointerfaces 154:350–356. https://doi.org/10.1016/j.colsurfb.2017.03.032

Trietsch SJ, Israëls GD, Joore J, Hankemeier T, Vulto P (2013) Microfluidic titer plate for stratified 3D cell culture. Lab Chip 13:3548–3554. https://doi.org/10.1039/c3lc50210d

Valgimigli L, Baschieri A, Amorati R (2018) Antioxidant activity of nanomaterials. J Mater Chem B 6:2036–2051. https://doi.org/10.1039/C8TB00107C

Vater C, Adamovic A, Ruttensteiner L, Steiner K, Tajpara P, Klang V, Elbe-Burger A, Wirth M, Valenta C (2019) Cytotoxicity of lecithin-based nanoemulsions on human skin cells and ex vivo skin permeation: comparison to conventional surfactant types. Int J Pharm 566:383–390. https://doi.org/10.1016/j.ijpharm.2019.05.078

Viuda-Martos M, Navajas YR, Sánchez Zapata E, Fernandez-Lopez J, Perez-Alvarez JA (2010) Antioxidant activity of essential oils of five spice plants widely used in a Mediterranean diet. Flavour Fragr J 25:13–19. https://doi.org/10.1002/ffj.1951

Walczyk D, Bombelli FB, Monopoli MP, Lynch I, Dawson KA (2010) What the cell "sees" in bionanoscience. J Am Chem Soc 132:5761–5768. https://doi.org/10.1021/ja910675v

Walkey CD, Chan WC (2012) Understanding and controlling the interaction of nanomaterials with proteins in a physiological environment. Chem Soc Rev 41:2780–2799. https://doi.org/10.1039/c1cs15233e

Weber C, Morsbach S, Landfester K (2019) Possibilities and limitations of different separation techniques for the analysis of the protein corona. Angew Chem Int Ed 58:12787–12794. https://doi.org/10.1002/anie.201902323

Weber SU, Lodge JK, Saliou C, Packer L (2009) Antioxidants. In: Barel AO, Paye M, Maibach H (eds) Handbook of cosmetic science and technology, vol 36. Informa Healthcare, London, pp 301–310. https://doi.org/10.1111/j.1524-4725.2009.01444.x

Wittmaack K (2011) Excessive delivery of nanostructured matter to submersed cells caused by rapid gravitational settling. ACS Nano 5:3766–3778. https://doi.org/10.1021/nn200112u

Zhao F, Zhao Y, Liu Y, Chang X, Chen C, Zhao Y (2011) Cellular uptake, intracellular trafficking, and cytotoxicity of nanomaterials. Small 7:1322–1337. https://doi.org/10.1002/smll.201100001

Zhong J, Liu X, Wang Y, Qin X, Li Z (2017) γ-Oryzanol nanoemulsions produced by a low-energy emulsification method: an evaluation of process parameters and physicochemical stability. Food Funct 8(6):2202–2211. https://doi.org/10.1039/C7FO00023E

Zugic A, Jeremic I, Isakovic A, Arsic I, Savic S, Tadic V (2016) Evaluation of anticancer and antioxidant activity of a commercially available CO_2 supercritical extract of Old Man's Beard (Usnea barbata). PLoS One 11 (1): 1–13. https://doi.org/10.1371/journal.pone.0146342

Chapter 5
Microemulsions as Antioxidant Carriers

Anna Froelich and Tomasz Osmałek

5.1 Introduction

Microemulsions have been known and investigated for more than 70 years since the introduction of pioneering works of Hoar & Schulman (Hoar and Schulman 1943) and Winsor (Winsor 1954). Even though these systems seem to be well known and thoroughly described, it is noteworthy that still they are gaining scientific attention regarding numerous interesting properties, which make them suitable for the application in many different areas. Among numerous research and industrial fields investigating the potential of microemulsions petroleum recovery (Nazar et al. 2011), nanoparticle synthesis (Wolf and Feldmann 2016), drug delivery (Callender et al. 2017), extraction processes (Ghouas et al. 2016) and novel cleaning technologies should be mentioned. Specific properties of microemulsions such as thermodynamic stability, extremely low interfacial tension and high interfacial surface, make their macroscopic features different from coarse emulsions and nanoemulsions. Moreover, because of the presence of surfactants (and sometimes co-surfactants), microemulsions display good solubilizing properties, which may be utilized in formulation and delivery of the substances with poor water solubility. The latter phenomenon is a frequently encountered problem in the studies related to the delivery of drugs and natural compounds, including antioxidants.

A. Froelich (✉) · T. Osmałek
Poznan University of Medical Sciences, Chair and Department of Pharmaceutical Technology, Poznań, Poland
e-mail: froelich@ump.edu.pl

M. A. Aboudzadeh (ed.), *Emulsion-based Encapsulation of Antioxidants*, Food Bioactive Ingredients, https://doi.org/10.1007/978-3-030-62052-3_5

5.2 Structure and Classification

Microemulsions are systems composed of polar and nonpolar phases stabilized with one or more surfactants. In order to decrease the interfacial tension to extremely low values, which is typical for these systems, usually a co-surfactant is required. Co-surfactant is a low molecular weight compound revealing weak surface activity. It is usually miscible with both oil and water phase which means it is localized not only in the interfacial layer but might be also present in both phases. Due to this property, it may be assumed that it can decrease the polarity difference between them and make the interfacial layer more flexible that contributes to the increased stability of the dispersion comparing to coarse emulsions and nanoemulsions. High flexibility of the interfacial layer and low interfacial tension induce high structural variability for microemulsions, which may occur not only as droplet systems but also as bicontinuous structures with hydrophilic and lipophilic domains intertwining each other (Fig. 5.1). The dispersed phase particles might adopt a shape different from spherical one, e.g. elongated ellipsoids (Lawrence and Rees 2012). One of

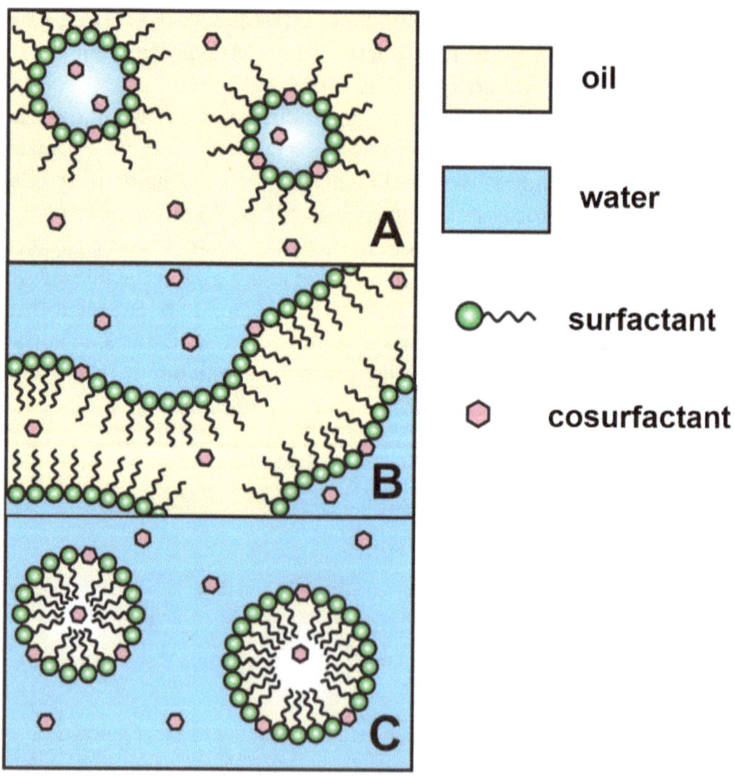

Fig. 5.1 Microemulsion structural types: (**a**) water in oil (W/O), (**b**) bicontinuous, (**c**) oil in water (O/W)

the most important factors determining the type of microemulsion is a quantitative relationship between polar and non-polar phases. The systems containing similar amounts of both phases tend to reveal bicontinuous structure while in other situations the prevailing phase tends to be a continuous (external) one. Another factor playing an important role in the formation of microemulsion is the characteristics of the interfacial layer, which is related mostly to the chemical and structural properties of the surfactant. One of the most popular parameters commonly used in the description of surfactant is hydrophilic-lipophilic balance (HLB) showing the ratio between polar and non-polar moieties in the surfactant molecule. The strong hydrophilic characteristic of surface-active agents is translated to high HLB numbers which usually leads to formation of O/W systems, while W/O microemulsions are formed by less polar surfactants with low HLB values. Another tool used in the description of microemulsion formation process is critical packing parameter (CPP, Eq. 5.1) reflecting the molecular geometry of the surfactant.

$$CPP = V / Al \tag{5.1}$$

where V is the partial molar volume of the nonpolar part of the surfactant molecule, A is the optimal polar head moiety area and l is the length of the hydrophobic chain (Lawrence and Rees 2012). In general, CPP adopts values smaller than one for the surfactants with more bulky polar head groups. In such cases, O/W microemulsions will be formed. Surfactants with large hydrophobic groups and CPP higher than one reveal the tendency to form W/O systems. When CPP = 1, lamellar systems and worm-like micelles might be formed. It is important to notice that the properties and curvature of the interfacial layer can be modified by the addition of other components, which can interact with surfactant molecules and participate in the layer formation.

It is important to notice that macroscopically microemulsions are perceived as transparent fluids displaying low viscosity. The definition formulated by Danielsson and Lindman in 1981 (Danielsson and Lindman 1981) allowed to differentiate between microemulsions and systems that are macroscopically similar but reveal different structure, such as liquid crystals, nanoemulsions and different micellar structures resulting from surfactant molecules association. According to the restrictions introduced by Danielsson and Lindman, microemulsions must be optically isotropic and thermodynamically stable liquids consisting of water, oil and amphiphilic component. Nevertheless, the term 'microemulsion' is still regarded as confusing and it is frequently used to describe for example submicron emulsions which might be kinetically stable but do not display thermodynamic stability (Anton and Vandamme 2011; McClements 2012). The most important systems similar to microemulsions are summarized in Table 5.1.

An important consequence of thermodynamic stability of microemulsions is their spontaneous formation without any significant amount of required energy. Taking into consideration the technological or industrial aspects of microemulsion manufacturing, the mentioned phenomenon is an enormous advantage because it excludes the necessity to apply expensive and energy-consuming high-shear

Table 5.1 The differences between microemulsions (ME) and similar systems

System	*System features*						
	Liquid	Oil	Water	Amphiphile	Thermodynamic stability	Isotropy	Droplet/micelle diameter (Rao and McClements 2011)
Microemulsion	✓	✓	✓	✓	✓	✓	<100 nm
Nanoemulsion	✓	✓	✓	✓	×	✓	<200 nm
Coarse emulsion	✓	✓	✓	✓	×	✓	>200 nm
Micellar solution	✓	×	✓	✓	✓	✓	<100 nm
Liquid crystal/lamellar phase	✓/×	✓	✓	✓	✓	×	No droplets

homogenizers or microfluidizing devices. The thermodynamic effects associated with microemulsion formation can be explained with Eq. 5.2 (Lawrence and Rees 2012):

$$\Delta G_f = \gamma \Delta A - T\Delta S \tag{5.2}$$

where ΔG_f is the free energy of microemulsion formation, γ is the surface tension, ΔA is the change of the interfacial area observed during the process, T is the temperature and ΔS is a change of entropy. Each spontaneous process is characterized by negative value of free energy. During microemulsion formation two significant changes in the parameters occur, i.e. increase of interfacial area and increase of entropy. Both of them are a result of the dispersion into numerous small droplets or domains. From the Eq. 5.1 it can be concluded that both described phenomena have an opposite effect on the free energy of the process. Therefore, to overcome the energetically unfavorable effects related to the increase of the interfacial surface, the interfacial tension value must be small. As it was mentioned above, in microemulsion this parameter assumes to be extremely low (close to zero), which explains spontaneous formation of these systems. Moreover, the described effects explain the necessity to employ co-surfactants to obtain microemulsions. Most of the commonly applied surfactants do not reveal the ability to decrease the surface tension at oil and water interface to extremely low values, so an additional element enhancing their action is required.

Considering microemulsion as a multicomponent system, it is important to note that it is not formed at any quantitative composition of oil, water, surfactant and co-surfactant. The same components may form structurally different systems, i.e. coarse emulsion, microemulsion and lamellar structure, depending on the concentrations of particular substances (Mouri et al. 2014). In order to establish the area of microemulsion occurrence, usually pseudoternary phase diagrams employing Gibbs triangle are used (Fanun 2008). However, graphical presentation of phase equilibria in the system containing four components with the use of three-dimensional space

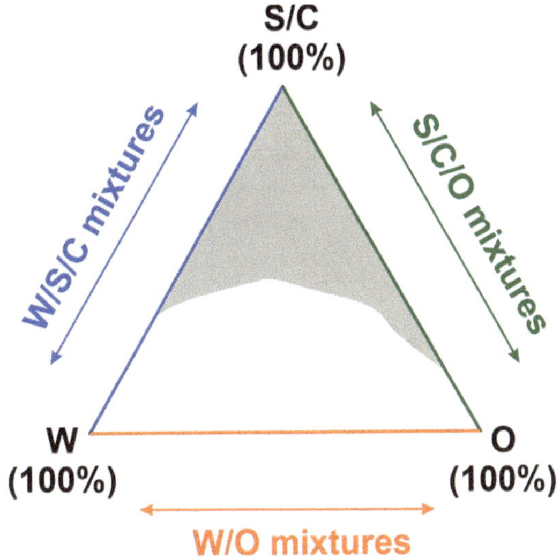

Fig. 5.2 Hypothetical pseudoternary plot for the system containing water (W), oil (O), surfactant (S) and co-surfactant (C). Gray area corresponds to transparent liquids identified as microemulsions, white corresponds to non-transparent coarse emulsions

requires some simplification. For this purpose, the mass ratio of two components is fixed and they are treated as one. As presented in the hypothetical pseudoternary phase diagram (Fig. 5.2), the corners of triangle are a graphical representation of three components (oil, water, surfactant/co-surfactant mixture at a fixed ratio). The points located in the sidelines correspond to the mixtures containing two components at different ratios. The closer to the particular corner the point is located, the higher the amount of this component in the mixture. The points present inside the plot contain all three components at different concentrations. The area corresponding to microemulsion occurrence is usually estimated with the use of titration experiment performed at fixed temperature and pressure conditions. In the first step, a set of binary mixtures at different ratios are prepared. Next, the mixtures are titrated with the third component. The systems which are low viscosity transparent liquids are classified as microemulsions (Djordjevic et al. 2004). The transparency loss indicates the transformation from microemulsion to coarse emulsion, while the increase of viscosity and gel formation can be interpreted as lamellar phase formation.

5.3 Applications

Microemulsions reveal numerous advantages making them useful for various practical applications in different scientific and industrial areas. It is important to note that they were used as household cleaning products long before the first scientific reports regarding their structure and properties were published. The first microemulsion-based products were used as washing and polishing liquids, usually containing natural oils and waxes and the first industrial formulations involving microemulsions included cutting oils and lubricants. With the increasing popularity of nonionic surfactants in 1940s and 50s, microemulsions gained much attention also in other areas e.g. food, agrochemicals and paints industry (Solans and Kunieda 1997). Currently, the most important applications comprise tertiary oil recovery, drug delivery and analytical chemistry. It is also noteworthy that microemulsions still are acquiring a lot of scientific attention as extraction media or nanoreactor media for synthesis or templates for nanoparticles formulation.

5.3.1 Drug Delivery

One of the most important practical approaches involving microemulsions is the design and development of novel pharmaceutical formulations. It is important to note that in order to observe the therapeutic effect, the active pharmaceutical ingredient (API) must be absorbed either from gastrointestinal tract or from skin or mucous membrane surface. The absorption is possible only if either the drug is dissolved in an appropriate carrier or it can be dissolved directly at the absorption site. It is estimated that about 70% of newly synthesized APIs reveal low water solubility, which can be a potential problem in terms of therapeutic efficacy. Therefore, the design of a carrier allowing for the dissolution of the drug and enabling its efficient delivery is an extremely important part of pharmaceutical product development. Microemulsions are known for their high solubilization capacity both in the case of polar and nonpolar substances (Constantinides 1995). This phenomenon is a result of specific composition of microemulsions that usually contain relatively high amount of surfactant and co-surfactants, which can act as co-solvents for poorly soluble substances. Moreover, numerous studies indicate that microemulsions can be successfully employed as drug carriers improving bioavailability of active pharmaceutical ingredient. This effect was observed in various formulations, including oral, dermal ophthalmic and many other ones. In oral drug delivery particularly self-microemulsifying drug delivery systems (SMEDDS) should be mentioned (Kohli et al. 2010; Kuentz 2011). These systems are in fact in situ microemulsion-forming liquid mixtures of oil, surfactant and co-surfactant. After oral administration, the system is mixed with the fluid present in the gastrointestinal tract and because of its peristaltic movements, microemulsion is formed. It is noteworthy that these lipid-based formulations were successfully introduced to the pharmaceutical market and allowed for the significant improvements in the case of drugs with poor and variable bioavailability, and displayed significant differences in therapeutic effects observed inter- and intra-individually. Moreover, it was noted that incorporation of the drug in lipid-based formulation allowed for reduction of so called "food effects" related to the significant impact of the ingested food on the bioavailability of the drug administered orally (Pouton 2000; Perlman et al. 2008; Kohli et al. 2010). The mentioned phenomenon is observed mostly for strongly lipophilic drugs, which are absorbed more efficiently in the presence of dietary lipids, while in the fasting state the absorption is impaired.

Microemulsions are also considered as very efficient delivery systems for the drugs administered to the skin surface. Numerous studies reported usefulness of these systems in terms of overcoming the barrier function of *stratum corneum*, the external layer of the skin (Sintov and Shapiro 2004; Kogan and Garti 2006; Lopes et al. 2010; Gannu et al. 2010; Lopes 2014; Pillai et al. 2015; Aliberti et al. 2017). The exact mechanism underlying this phenomenon has not been precisely explained yet (Santos et al. 2008). However, there are several hypotheses that can describe it. One of them indicates high ability to improve the solubility of different drugs, which enables incorporation of higher amounts of active ingredients (He et al. 2010; Xing et al. 2016). As a result, higher concentration gradient between the formulation and skin is obtained, which allows for more efficient drug delivery. Another important feature of microemulsions frequently mentioned as a factor improving drug permeation through the skin is their composition. Surfactants and co-surfactants present in the formulation usually act as permeation promotors, temporarily disrupting lipid organization in *stratum corneum*. In the case of droplet-like O/W systems, the internal phase containing the drug may act as a reservoir. In such formulations, the active ingredient is transported continuously from the internal phase to the external one and the gradient between the formulation and the skin is maintained at the same level. Another factor important in terms of bioavailability of drugs incorporated in microemulsions is an extremely low interfacial tension that allows for good contact between the formulation and the skin. It is noteworthy that the droplet size in O/W and W/O systems seems to be of little importance in terms of skin permeation (Kreilgaard 2002).

Transparency and low viscosity of microemulsion make them interesting potential drug delivery systems for ocular delivery. Due to the aforementioned features, they can be spread easily at the surface of the eye without impairing vision. Moreover, they offer the possibility to transform into liquid crystalline system of increased viscosity upon the contact with tear fluid, which might provide prolonged drug release. This phenomenon seems to be particularly advantageous in ocular drug delivery because of specific physiology of the human eye and quick removal of the active ingredient with tear fluid that decreases the efficacy of conventional ocular drops instilled in the conjunctival sac. However, ocular drug delivery with the use of microemulsion-based systems might be challenging because of eye sensitivity and limitations related to the use of surfactants and co-surfactants. Similar problems may be encountered in the case of the drugs administered via parenteral route.

5.3.2 Petroleum Industry

Microemulsions have been investigated in the area of petroleum industry as well. Naturally occurring crude oil impregnates rocks and it is estimated that only 10–20% of the available resource can be effectively extracted with conventionally applied methods, such as drilling and pumping. In order to retrieve the remaining amount, some additional techniques must be applied. In secondary methods, water or steam is used to push out the oil from its source and direct it to the wellbore. In enhanced oil recovery methods known also as tertiary techniques, oil is obtained with the use of various materials injected into the reservoir. In the process involving microemulsion, surfactant solution with some additives is used in order to decrease the interfacial tension between water and oil to reduce capillary forces. In this way, the oil can be displaced from reservoir rock. Extremely low interfacial tension observed in microemulsions, as well as their low viscosity, are regarded as useful features in terms of tertiary oil recovery (Santanna et al. 2009; Nazar et al. 2011; Bera and Mandal 2015).

5.3.3 Microemulsion Electrokinetic Chromatography (MEEKC)

Microemulsion electrokinetic capillary chromatography (MEEKC) is a separation technique that employs electrical current. The separation process takes place in a capillary filled with O/W microemulsion. Microemulsion droplets are stabilized by an anionic surfactant, which is usually sodium dodecylsulfate, and n-butanol as a co-surfactant. The droplets play the same role as the stationary phase in different chromatography techniques. As a mobile phase, aqueous buffer is applied. The technique can be used to separate various range of solutes, including charged and uncharged ones. The basic principle of the described method is presented in Fig. 5.3. Microemulsion droplets are negatively charged and reveal electrophoretic mobility, however, after applying an electrical current to the capillary, the buffer generate high electroosmotic flow (EOF) which overcomes the electrostatic force driving microemulsion droplets towards anode.

The partitioning of the solutes is based both on their electrokinetic mobility and hydrophobicity, allowing for efficient separation of both neutral and charged

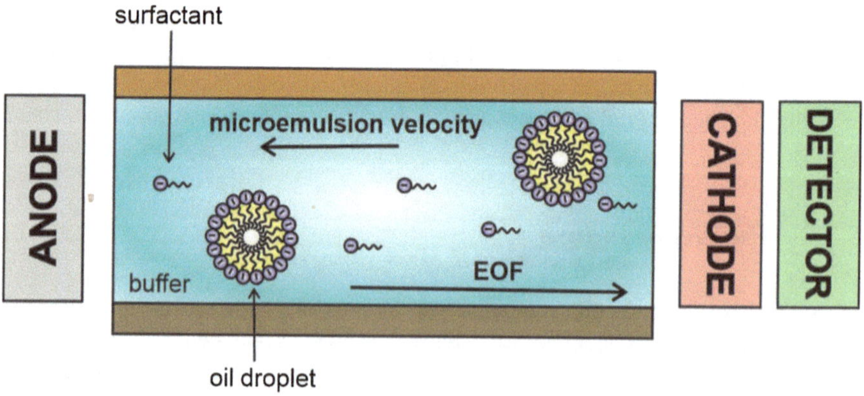

Fig. 5.3 The principle of MEEKC (Altria 2000); EOF – electroosmotic flow

molecules displaying different water affinity (Altria 2000; Altria et al. 2003). When a neutral compound is injected into the capillary, it will be partitioned between microemulsion droplets and buffer. Strongly hydrophobic analytes will reveal the tendency to dissolve in microemulsion droplets and migrate through the capillary with the same velocity. In such cases, long retention time is observed. Hydrophilic solutes are present mostly in the aqueous phase, revealing different retention time depending on their electrophoretic mobility and interaction with negatively charged droplets. It is noteworthy that negatively charged analytes will be repulsed by the droplets, while in the case of positively charged compounds ion pairing is observed. However, in both types of analytes, the diffusion into the hydrophobic core is prevented and the analyzed substance is present mostly at the hydrophilic site (Klampfl 2003).

5.3.4 Nanoparticle Formation

Microemulsions are frequently utilized as reaction media or templates for the preparation of various types of nanoparticles. In the case of inorganic nanoparticles, W/O microemulsion droplets may serve as containers for reagents. Mixing two different microemulsion systems containing separated substrates leads to droplets collisions and precipitation of the product. The synthesis of inorganic particles with the use of O/W microemulsions has also been described in the literature and the precursor might be either an ionic salt dissolved in water phase or an organometallic salt dissolved in oil phase (Magno et al. 2011; Housaindokht and Nakhaei Pour 2012; Kong et al. 2012; Sanchez-Dominguez et al. 2012). Microemulsions can be employed in the production of solid lipid nanoparticles (Müller et al. 2000; Shah et al. 2014; Chirio et al. 2019). In such cases, O/W systems obtained at the temperature above melting point of the lipid component are used. Finally, liquid droplets are transformed into solid lipid nanoparticles as a result of microemulsion dispersion in cold aqueous medium (Mehnert 2001).

Moreover, microemulsions can be used as reaction media for polymerization reactions of various types (Pavel 2004). The process performed for hydrophobic monomers incorporated in the internal phase of O/W microemulsions leads to the formation of polymer particles of various diameters, depending usually on the reaction conditions (He et al. 2003; Chen et al. 2010). In another approach, water-soluble monomers dissolved in W/O microemulsions are used (McAllister et al. 2002; Deng et al. 2003). The results of polymerization reaction performed with the use of bicontinuous systems (Peinado et al. 2006; Wang et al. 2017) were also reported and they depended mostly on the type of the applied monomers and the composition of microemulsions.

5.4 Antioxidants in Microemulsions

Taking into account the versatile and unique properties of microemulsions, including their excellent ability to dissolve substances of different polarities, they are frequently investigated as carrier systems for various active compounds. Very often microemulsions are considered as vehicles used in the development of novel pharmaceutical delivery systems (Talegaonkar et al. 2008) or cosmetics (Boonme 2007), but also as components of other biomedical devices (Malik et al. 2012). Most of the mentioned applications include liquid and semi-solid formulations for topical use (Froelich et al. 2017; Nastiti et al. 2017), as well as various oral dosage forms (Gibaud and Attivi 2012). For many years, another major area for microemulsions use has been the food industry.

Both in medical and food applications microemulsions are broadly investigated as carriers for antioxidant agents. It has been proven that in many cases, incorporation of such compounds into microemulsions may result in not only higher solubility and stability, but is followed by improved bioavailability, both after external application and ingestion. This in turn allows to obtain better effectiveness with lower concentrations of the actives.

The abundance of experimental data in literature clearly indicates multiple beneficial effects related to microemulsion-based systems for the delivery of antioxidants. The main targets posed to microemulsions include an improvement of skin anti-aging efficacy or anti-cancer activity, which are obtained due to the stability increase, deeper percutaneous penetration and stronger antioxidant effects. In terms of other administration routes, microemulsions can efficiently promote nose-to-brain transport. Several studies indicate that active ingredients incorporated in microemulsion-based media applied to the nasal mucous membrane can quickly reach the brain tissue and achieve higher concentrations compared to the conventional formulations (Shah et al. 2016). The described approach can also be useful to avoid systemic side effects frequently encountered after oral administration. The observed beneficial effects are related mostly to the direct transport from olfactory region in nasal cavity to central nervous system. In this way, the active ingredient bypasses the blood-brain barrier, the most important factor limiting the effectiveness of brain-targeted drug delivery (Alam et al. 2010; Lu et al. 2014).

In the case of natural products, when active ingredients are obtained from plant sources, microemulsions can also be used as excellent extraction media. Moreover, in some cases the protective action of microemulsions in preventing antioxidants from degradation was proven.

In the following paragraphs, the literature reports regarding the most frequently investigated antioxidant agents have been reviewed and discussed. It is noteworthy that all of the mentioned substances reveal poor solubility in water, which is challenging in terms of both technological process and absorption from gastrointestinal tract. The application of microemulsion as carrier or extraction medium is a simple and efficient method to overcome the potential difficulties related to low solubility and low absorption of active ingredients.

5.4.1 Carotenoids

Carotenoids are phytochemical dyes occurring abundantly in plant kingdom. Numerous scientific studies show the evidence for their preventive role in several chronic diseases related to the occurrence of an oxidative stress, like cancer and cardiovascular disorders (D'Odorico et al. 2000; Rao and Rao 2007; Riccioni 2009; Karppi et al. 2012; Jomova and Valko 2013; Kasperczyk et al. 2014). All carotenoids are lipophilic which might contribute to the formulation and processing difficulties encountered during extraction and obtaining stable carotenoid-loaded product.

β-Carotene is a one of the most frequently occurring plant carotenoids which displays antioxidant properties. It is present in large amounts in fresh fruits and vegetables, like carrots. Numerous scientific studies indicate its advantageous properties as a food component (Johnson and Krinsky 2009; Fiedor and Burda 2014; Woodside et al. 2015). These properties include chemopreventive potential, as well as the possibility to improve the efficiency of immune response and reduce the risk of coronary heart disease. Taking into consideration its beneficial properties and high concentration in plant sources, it is utilized as an alternative for synthetic food dyes. Depending on the localization and a physical form of β-carotene occurring in plants, the extraction might be more or less complicated. It was noted that the use of conventional methods involving organic solvents is usually related to the presence of toxic residues, which poses an important problem whenever food or pharmaceutical products are designed. Roohinejad et al. proposed the method for carotene extraction using O/W microemulsion as safe, efficient and cheap alternative to the conventional techniques employing relatively toxic organic solvents. In this method, in the first step pulsed electric field was applied to the carrot pomace to increase the efficiency of carotene extraction. Next, carrot pomace was freeze-dried to avoid water content changes in microemulsion used for extraction. Finally, the dried product was subjected to the extraction process using different times and temperatures and the impact of different variables was evaluated with the use of statistical methods. The obtained stable and transparent β-carotene-loaded microemulsion can be used as promising vehicle for fortification of transparent food, beverage and pharmaceutical products (Roohinejad et al. 2014b). The same authors in another report, prepared microemulsions containing β-carotene from carrot pomace and investigated their cytotoxicity toward Caco-2 cells. The microemulsions contained glycerol monocaprylocaprate (Capmul MCM), Tween® 80 and phosphate buffer. The idea was to check whether the carriers could protect the cells from damage in the presence of H_2O_2. It turned out that β-carotene concentration of 0.0313% gave bioprotection to the cells. The increase in cytotoxicity at the concentrations higher than 0.0313% was interpreted as the result of increasing Tween® 80 precipitation on the cell culture monolayer (Roohinejad et al. 2014a).

Chen & Zhong investigated the protective potential of microemulsion carrier with incorporated β-carotene against oxidation of the active ingredient. It was shown that the most important factor improving stability of the product was lecithin content which might be related both to the properties of lecithin and to the increased viscosity of the system (Chen and Zhong 2015). Moreover, the increased content of lecithin (as a surfactant) contributed to better antioxidant performance of peppermint oil, which was applied as an oil phase. The analyzed formulations showed significantly increased thermal stability and resistance to ultraviolet (UV) radiation.

Chaari et al. presented the study related to carotenoids obtained from halophilic *Archaea* microorganisms. The aim of the project was to solubilize lipophilic compounds in micro- and nanoemulsions and to compare the stability and antioxidative potential of the obtained systems. It was shown that both investigated dispersions had similar radical-scavenging effects. However, microemulsion revealed better physical stability, which was obvious considering the thermodynamic and kinetic stabilities of both systems (Chaari et al. 2018). Another study focused on the stability of β-carotene and lemon oil-loaded microemulsion, in which the basic idea was to incorporate lipophilic flavors and coloring agents to transparent beverages with the use of microemulsion systems. The beverages obtained with the described microemulsions were physically stable and no phase separation was observed after 30 days. However, the color was not stable during storage which indicated oxidation of β-carotene (Calligaris et al. 2019).

Another extensively investigated carotenoid compound is lycopene, red compound present in tomatoes, watermelon and papaya. As an antioxidant, it reveals higher activity than β-carotene (Kelkel et al. 2011). It is noteworthy that the compound is poorly soluble in water and most food-grade oils, which is a cause of poor absorption of lycopene from the gastrointestinal tract. The mentioned effect contributed also to low bioavailability after oral administration. Therefore, the studies involving microemulsions as carriers usually aim at the solubilization of lycopene and improvement of biopharmaceutical properties (Spernath et al. 2002). Moreover, similarly to other carotenoids, alternative methods for lycopene extraction from plant sources are investigated.

According to Lopes et al., microemulsions can be used for topical supplementation of skin with antioxidants. The authors prepared two types of emulsions containing lycopene (0.05%, w/w). The formulations were composed of BRIJ-propylene glycol mixture (2:1, w/w) and different oil phases: mono/diglycerides of capric and caprylic acids or triglycerides of the same fatty acids. The aim was to increase lycopene penetration through the skin layers in order to obtain better protection against UV radiation. The experiments conducted on porcine ear skin showed that microemulsion-based solutions of the active compound revealed 6 to 3.6-fold greater penetration than a control solution (oil/propylene glycol). Moreover, significant effects of the increase of lycopene concentration in the skin were observed after addition of ascorbic acid to the formulations. The obtained microemulsions showed very low irritation tendency against skin cells (Lopes et al. 2010).

Pepe et al. presented another skin-related study performed with the use of lycopene-loaded microemulsion. In order to increase the antioxidant potential of the investigated topical formulation, they used ascorbic acid as another active ingredient. As an oil phase, monoglycerides of different chain lengths were applied. It was shown that the length of the lipophilic chain affected both water solubilization capacity and skin permeation process. It was found that the shorter the monoglyceride hydrophobic chain, the higher water solubilization capacity of the obtained microemulsion. Regardless of the oil type, all lycopene-loaded microemulsions could significantly enhance the delivery of the active ingredient to viable skin layers and *stratum corneum*. However, the weakest effect was observed for the system with the lipid of intermediate chain length. The same system did not significantly enhance the delivery of ascorbic acid compared to the plain solution used as a reference. The enhancement effects observed for the monoglycerides with short and long chain for ascorbic acid were similar, while in the case of lycopene the systems with short chain monoglyceride performed better. It is noteworthy that lycopene was retained in the skin but did not cross it, which may indicate possible difficulties with transdermal delivery of this antioxidant (Pepe et al. 2012).

Guo et al. investigated microemulsion as a carrier improving bioavailability and brain targeting efficiency of orally administered lycopene. The pharmacokinetics and tissue distribution of lycopene were tested with the use of animal model with lycopene dissolved in olive oil as a reference formulation. It was shown that the microemulsion-based system significantly increased absorption of the active ingredient from gastrointestinal tract. Moreover, after oral administration of microemulsion-based medium, higher amounts of lycopene were found in the brain tissue when compared to the simple solution. Several different hypotheses explaining the possible impact of microemulsion on blood-brain barrier were presented. The obtained results are particularly important for the research exploring novel therapeutic approaches in neurodegenerative disorders (Guo et al. 2019).

Application of microemulsion technique to extract lycopene from tomato pomace (considered as a waste) was investigated in some studies. The obtained results allowed simple and safe recovery of lycopene from tomato pomace (and possibly from tomato industrial wastes). The recovery value reached up to 35% in the case of the best formulation by using saponin as a natural surfactant and combined ultrasonic and enzymatic pretreatment (Amiri-Rigi and Abbasi 2016). The same authors studied the feasibility of microemulsion technique in enhancing the solubility of lycopene, which could lead to an efficient extraction from tomato pomace. The obtained lycopene-loaded microemulsions were analyzed for the stability during several technological processes, such as pasteurization, sterilization, freezing and UV-irradiation. It was shown that the investigated systems were not affected by the applied procedures as long as the temperature and time were controlled (Amiri-Rigi and Abbasi 2017). The same research group presented the results obtained for the food-grade microemulsion composed of different proportions of olive oil, lecithin, 1-propanol and water applied as an extraction medium for retrieving lycopen from tomato pomace. After four cycles of extraction with the optimized microemulsion, the maximum extraction efficiency of 88% was obtained. It is noteworthy that the applied medium contained mostly biocompatible components and can be used as an interesting alternative for toxic organic solvents (Amiri-Rigi and Abbasi 2019).

5.4.2 Curcumin

Curcumin is a yellow polyphenol compound occurring in *Curcuma longa* rhizome, which is popularly used as a spice and food colorant. It is well known mostly for its antioxidant and anticancer properties (Parvathy et al. 2009) but reveals also beneficial effects in inflammatory (Mobasheri et al. 2012), neurological (Cole et al. 2007) and infectious (Padmanaban and Rangarajan 2016) conditions. Even though the antioxidant activity of curcumin has been proven in numerous scientific studies, the exact mechanism of this phenomenon remains unclear and several different hypotheses have been proposed, including free radical scavenging effects and preventive effects in lipid peroxidation (Galano et al. 2009). One of the most important difficulties encountered in the curcumin delivery is its low water solubility affecting its bioavailability. In order to achieve therapeutic plasma concentrations, extremely high oral doses of curcumin are required. The ingestion of high amounts of the active ingredient is not comfortable for the patients and also results in an increased risk of side effects (Yen et al. 2010). Another important disadvantage of curcumin is its susceptibility to photo-degradation and hydrolysis in alkaline conditions (Tønnesen et al. 2002). Microemulsions applied as carriers dissolve high amounts of this active ingredient, as well as improve its bioavailability. Bergonzi et al. designed and evaluated microemulsion as a medium for dissolution and oral administration of curcumin. The solubility of the active ingredient depended on the composition of micremulsion, however, in the case of all investigated formulations significant improvement with respect to aqueous environment was observed. The obtained microemulsions were physically stable and remained transparent for two months. Moreover, the dilution tests indicate that they would not transform into coarse emulsions after oral administration (Bergonzi et al. 2014).

Hu et al. also designed and evaluated curcumin-loaded microemulsions for potential oral administration. Based on the solubility screening, the optimal formulation was composed of Capryol™ 90 (oil phase) Cremophor® RH40 (surfactant), Transcutol® P (co-surfactant) and water. The plasma concentration levels observed with the use of animal model indicate that microemulsions can be utilized as drug delivery systems for the active ingredients revealing poor water solubility and poor bioavailability. The obtained results were significantly better when compared to oral suspension. The authors suggest that the presented system might be potentially useful in the production of nutraceuticals and functional food (Hu et al. 2012). Similar bioavailability improvement was reported by Xiao et al. based on the study performed on the mixture of three different curcuminoids incorporated in microemulsion (Xiao et al. 2013).

In addition, microemulsions have been investigated as systems for delivering curcumin through other routes (nasal, skin, etc.). For example, Liu et al. showed that microemulsions can be used as carriers for antioxidants in photodynamic treatment of localized bacterial inflammations. The authors prepared microemulsions using distilled water, geraniol oil, propylene glycol, and Tween® 80. Curcumin (4000 ppm) was used as a photoactive agent. The overall idea was to increase the solubility of curcumin due to improve the photodynamic effects toward *Pseudomonas aeruginosa* colonies. Blue light-emitting diode (455 nm) was used for activation of curcumin and further generation of singlet oxygen. It was clearly shown that microemulsions increased photoinactivation of *P. aeruginosa* both in the planktonic and biofilm form. Moreover, ex vivo experiments on neonatal porcine ear skin revealed that microemulsions decreased the penetration of curcumin in comparison to aqueous solutions. The effect was significant regarding potential application on patients' skin and limited possible damage to healthy tissue (Liu et al. 2016).

Mandal et al. designed mucoadhesive microemulsion for intranasal delivery of curcumin. The aim was to increase the brain uptake of the compound. According to the drug (curcumin) solubility, Capmul® MCM, a mono-diglyceride of caprylic and capric acids, was selected as the oil phase and polycarbophil as mucoadhesive polymer was added to the formulations as well. The in vivo experiments on male albino rats showed that following intranasal administration, brain concentrations of the compound were higher than after an intravenous injection. Moreover, no damage of the nasal mucosa was observed (Mandal et al. 2016).

Ghosh et al. presented an interesting study on the physiochemical properties of curcumin-loaded microemulsions containing gold and silver nanoparticles with the aim of checking the possible impact of metal nanoparticles on photochemical properties and antioxidant performance of curcumin. The study revealed that curcumin formed conjugates with silver nanoparticles and consequently the excited state lifetime of the active ingredient increased. In the case of the system with gold nanoparticles, the opposite effect was observed. The excited state lifetime of curcumin was decreased which was explained with possible nanometal surface energy transfer (NSET). The observed phenomenon was a result of the overlap between emission and absorption spectra of curcumin and gold nanoparticles, respectively. In the investigated system, curcumin acted as the fluorophore donor while gold was the acceptor. The results obtained for both systems can be useful in bioimaging applications (Ghosh et al. 2020).

5.4.3 *Flavonoids*

Flavonoids are wide group of phytochemical compounds occurring abundantly in plants. Considering their chemical structure, they are categorized into several different classes including flavonols, anthocyanins, isoflavones, flavanones, proanthocyanidins and flavones (Brodowska 2017). It was shown that their presence in daily diet is beneficial in many different health aspects, for example they are efficient scavengers for reactive oxygen species (Terao 2009). This effect is important in prevention of cancer and complication of many chronic diseases, like diabetes and cardiovascular problems (Le Marchand 2002). Flavonoids reveal good thermal stability making them resistant to the conditions applied during industrial processing. The most important problem encountered in flavonoids delivery is their poor and highly variable bioavailability depending on their molecular weight, glycosylation and esterification (Thilakarathna and Rupasinghe 2013). It was found that small molecular weight polyphenols, like caffeic acid, can be easily absorbed from the gastrointestinal tract, while large proanthocyanidins, which are in fact flavonoid polymers, reveal poor bioavailability. Another important factor is the type of sugar moiety bonded to the aglycone. It was found that glucoside derivative of quercetin revealed better intestinal absorption than the aglycone and corresponding rhamnoside derivatives (Scalbert et al. 2002). Incorporation of flavonoids in different carriers, such as liposomes (Kerdudo et al. 2014), nanoparticles (Roussaki et al. 2014) or microemulsions is one of the most commonly applied methods to increase the bioavailability of different classes of these compounds. It is important to notice that the applied carrier might have an impact directly on the antioxidant activity of the incorporated agent. Fan et al. revealed that propolis flavone had higher in vivo antioxidant activity in microemulsion than when administrating alone (Fan et al. 2014). Among the most frequently investigated antioxidant flavonoids incorporated in microemulsions quercetin, resveratrol and its derivatives, hesperitin and catechins should be mentioned.

5.4.3.1 Quercetin

Quercetin is a very popular plant flavonol. It can be found in numerous different edible plants, however, the highest amounts of quercetin are observed in onion. It occurs in a form of different glycosides, including the most popular rutin (3-rhamnoglucoside), galactosides, arabinosides and glucosides present in onion (Erlund 2004). Quercetin glycosides reveal hydrophilic properties, therefore, initially it was expected that only non-glycosylated form could be absorbed from gastrointestinal tract. However, the studies comparing different forms of quercetin indicate that the presence of sugar moiety significantly improves bioavailability which might be related either to deglycosylation or carrier-mediated transport (Boots et al. 2008). Poor water solubility and instability of quercetin seem to be the most important problems in its oral and topical delivery. Therefore, several different technologies have been investigated in order to overcome these difficulties, including incorporation in microemulsion (Nagula and Wairkar 2019). Skin penetration of quercetin from microemulsions was investigated by Kitagawa et al. They prepared quercetin-loaded microemulsions composed of isopropyl myristate, 150 mM NaCl solution, Tween® 80 and ethanol. It was observed that intradermal delivery of the active compound from microemulsion vehicle was more efficient than from unmixed organic solvents. The authors supported the theory that the effect was related to microemulsion breakage upon contact with skin followed by quercetin release from the interface region. Another possibility is that microemulsion partly mixes with *stratum corneum* lipids and therefore enhances the permeation of the drug. The authors stated that continuous and spontaneous fluctuations of microemulsions interfaces contributed to high drug mobility and thus enhanced the drug diffusion process (Kitagawa et al. 2009). Quercetin-loaded microemulsions for topical use were also prepared by Kajbafvala et al. It was observed that the skin retention decreased with increasing surfactant/co-surfactant ratio, which was related to increasing droplet size. The authors have proven that in vitro quercetin release directly depended on surfactant concentration (Kajbafvala et al. 2018).

In another work, the influence of the storage conditions (temperature, humidity, light) on physical, chemical and functional stability of the quercetin-loaded microemulsions was investigated. As a result, they were physically stable; however, some chemical and functional changes occurred indicating the necessity of special storage conditions. Promising results were obtained when the samples were exposed to UV-B radiation and it was shown that the microemulsions can be used as protective agents against radiation skin damage (Vicentini et al. 2011).

Another study devoted to design quercetin-loaded microemulsion to improve solubility and bioavailability of the active ingredient was presented by Gao et al. To obtain the final microemulsion composition, simplex lattice experiment design was employed. Quercetin absorption after oral administration was evaluated with the use of rat model with micellar solution applied as a reference. The recorded differences were statistically significant and indicated faster absorption from the intestine and prolonged presence in the plasma in the case of microemulsion-based carrier. The observed effect was assigned to absorption enhancing effect exerted by lipid on the lymphatic route (Gao et al. 2009).

Censi et al. indicated the difficulties frequently encountered in obtaining therapeutic concentrations of quercetin in systemic circulation after transdermal delivery. They mentioned that problems with poor absorption of the active ingredient could be overcome with the use of properly selected permeation enhancers, like Transcutol® P. It is noteworthy that in microemulsions, this component might act as a solubilizer for the active component and co-surfactant, which enables the formation of microemulsion due to its ability to decrease interfacial tension. Moreover, it is widely known as a valuable skin permeation enhancer. Quercetin-loaded microemulsion in this study was evaluated using excised pig skin model. The obtained results were compared to those recorded for corresponding Transcutol® P solution, oil/water/surfactant solution and oil/Transcutol® P/surfactant solutions and it was found out that solubilizing agent had a significant impact on quercetin absorption. The highest permeation rate was observed for simple Transcutol® P solution while the system without it (i.e. oil/water/surfactant solution), had the lowest rate. Quercetin-based microemulsion revealed better properties than the formulations without solubilizer and without water. It is also important to note that in the case of the systems that improve permeation through the skin, lower amounts of quercetin were accumulated in skin (Censi et al. 2012).

Microemulsions can be considered as carriers to improve the stability of quercetin. Lv et al. obtained essential oil-based systems to improve water solubility of quercetin and to reduce its sensitivity to light and pH instability. The active ingredient was significantly more resistant to alkaline pH and UV radiation (Lv et al. 2017).

5.4.3.2 Resveratrol

Resveratrol is a polyphenolic component of grapes, berries and other plants. It is commonly known as a compound related to so called "French paradox" describing low prevalence of cardiovascular disease among French population, despite saturated lipid-rich diet. The explanation of this phenomenon was based on the data on high consumption of red wine, which is a good source of resveratrol. Its antioxidant properties are related not only to its free radicals scavenging activity but also to an ability to induce the expression of specific enzymes revealing antioxidant activity (Smoliga et al. 2011). Similar to other flavonoids, resveratrol reveals low bioavailability, even though its absorption from gastrointestinal tract reaches about 75% (Pangeni et al. 2014). It was shown that this phenomenon is related mostly to its rapid transformation to sulfate and glucuronide metabolites (Walle et al. 2004). Another important disadvantage of resveratrol is its poor solubility in water. Therefore, numerous technological approaches aiming at reduction of the mentioned effects have been investigated. Most of the studies proposed its incorporation into liposomes, micro- and nanocapsules and nanoparticles as efficient platforms to overcome the described difficulties (Augustin et al. 2013). Microemulsions due to their ability to improve dermal absorption of active ingredients are considered as potential carriers in transdermal resveratrol delivery. Sucrose fatty acid ester microemulsions were designed by Yutani et al. with the aim of promoting skin delivery of resveratrol. In vitro intradermal and transdermal experiments on Yukatan micropig skin revealed that sucrose oleate had the best performance among the other sucrose esters (laurate, myristate, palmitate, stearate). The authors stated that resveratrol skin incorporation efficiency was approximately inversely proportional to its concentration (Yutani et al. 2016). In another study, Das et al. investigated microemulsion-based gels containing tea tree oil and medium chain glyceride as carriers for resveratrol (Das et al. 2020). The obtained systems were tested for drug release and permeation with the use of polysulfone and Strat-M™ synthetic membranes, respectively. Strat-M™ is usually applied as an alternative to human or animal skin ex vivo model. The obtained results indicate that the active ingredient permeated through the membrane in relatively small amounts and in a prolonged manner, which is favorable in cosmetic applications.

For oral administration route, self-emulsifying drug delivery systems (SEDDS) are considered as potential carriers for resveratrol delivery. For instance, Chen et al. incorporated resveratrol into a SEDDS with purpose of improving its absorption after oral administration. The formulations were composed of ethyl oleate, Tween® 80 and polyethylene glycol (PEG) 400 as the oil, surfactant and co-surfactant, respectively. The drug concentration was 5%. The formulation revealed stronger antioxidant activity and was less toxic to cells than free resveratrol. The results were very promising in terms of food supplementation with the active ingredient (Chen et al. 2015).

In a study performed by Bolko et al., resveratrol was incorporated in an innovative mixed lipid phase SMEDDS with the aim of enhancing its solubility. The mixed lipid phase was composed of long-chained triglyceride plus medium chain mono- and diglycerides and it showed the best self-emulsifying ability in terms of self-emulsifying time as well as droplet size and monodispersity of microemulsions achieved upon SMEDDS dilution with aqueous phase. This formulation also displayed higher drug release rate in comparison to the corresponding system containing single lipid in an oil phase (Bolko et al. 2014). The same research group using Caco-2 model showed that the applied formulation had high drug loading capacity and the incorporated active ingredient was released rapidly and did not impair the viability of in vitro cell cultures. Moreover, it was shown that the applied excipients might reduce the efflux of resveratrol metabolites which can contribute also to the reduction of intestinal metabolism and improvement of bioavailability (Seljak et al. 2014). Finally, in another report about incorporation of resveratrol in SMEDDS, it was shown that the solubility of the active ingredient in self-microemulsifying carrier was about 1000 times higher than its solubility in water. On the other hand, the authors obtained satisfactory results in in vitro drug release tests, showing that the process was faster in the case of encapsulated antioxidant than for resveratrol powder. Moreover, the concentrations of the drug observed in the receptor fluids were not depended on pH value of media (Tang et al. 2019).

5.4.3.3 Other Flavonoids

Apart from the two most extensively studied antioxidant flavonoids (quercetin and resveratrol), there are several studies describing other compounds with related structure. For example, hesperetin-loaded microemulsion for topical use was designed by Tsai et al., which revealed better in vitro permeation in comparison to the aqueous and isopropyl myristate suspension dosage form of hesperetin. Studies on the influence of co-surfactant on the drug permeation capacity showed that propylene glycol yielded the highest permeation rate, followed by ethanol, glycerol and PEG 400. The permeation also depended on HLB of co-surfactants. The analyzed microemulsion demostrated very good whitening effects on skin (Tsai et al. 2010).

Solubility of apigenin, a bioactive flavonoid with various pharmacological activities was improved by incorporating apigenin/hydroxypropyl-β-cyclodextrin complex into microemulsions. No co-surfactant was used for preparation of the investigated system. In vitro drug release profile obtained with the use of dialysis technique indicated zero-order kinetic process at the whole range of measurement. The authors concluded that the aqueous solubility of apigenin remarkably increased in the complex with Tween® 80 based O/W microemulsions, via solubilizing in the palisade layer, the inner oil core, and outer water phase (Zhao et al. 2016). The same authors presented another related study in which apigenin was encapsulated in complex system composed of O/W microemulsion thickened with gellan gum. The apigenin release was studied under different pH conditions. The observed results indicated two different release mechanisms depending on the pH of receptor media. At pH = 1.2 corresponding to the stomach environment, the active ingredient was released in a diffusion-controlled manner, while at pH = 7.4 similar to the duodenal conditions, the process was controlled by erosion (Zhao and Wang 2019).

5.4.4 Vitamin E

The term "vitamin E" in fact comprises a group of tocopherols and tocotrienols. Both forms are structurally similar but tocotrienols contain three double bonds in the side chain instead of two. Among all naturally occurring and synthetic tocopherols and tocotrienols, α-tocopherol reveals the highest biological activity. The most important function of vitamin E is its free radical-scavenging ability protecting biological macromolecules from damage. Vitamin E is lipophilic and occupies cell membranes and protects membrane lipids from oxidation (Dutta and Dutta 2003; Traber and Atkinson 2007). As a hydrophobic compound, it is poorly soluble in water and its absorption from gastrointestinal tract is closely related to lipid absorption (Rigotti 2007). Similar to other hydrophobic active ingredients, efficient uptake and delivery of vitamin E can be challenging. Therefore, in the studies focusing on the delivery of this compound, usually lipid-based carriers, including microemulsions, are applied. Another beneficial effect of vitamin E incorporation as an excipient is reduction of gastrointestinal side effects that can be observed after oral administration of microemulsions (Gibaud and Attivi 2012).

With the purpose of improving the bioavailability of vitamin E, it was encapsulated in microemulsion system and its antinociceptive, antioxidant, antidepressant- and anxiolytic-like activities in mice were evaluated. It was shown that α-tocopherol incorporated in microemulsion could protect lipids from peroxidation more efficiently comparing to free vitamin E. The observed results suggest that oxidative stress may be involved in the mechanism of some neurological disorders and vitamin E-loaded microemulsion has tremendous potential for the treatment of these conditions (Wilhelm et al. 2018). Carvalho et al. presented a study involving microemulsions as carriers for dermal delivery of vitamin E and other compounds. Microemulsions were applied in order to overcome technological challenges related to obtaining stable and effective formulation with poorly water-soluble drugs. It was shown that depending on the lipophilicity of the incorporated active ingredient, the applied approach could allow for more efficient dermal topical delivery of the drug (Carvalho et al. 2017).

Co-delivery of two synergistically acting antioxidant agents is useful for topical skin protection and treatment. For example, in a study presented by Cichewicz et al., microemulsions were used to promote the concomitant delivery of α-tocopherol and lipoic acid into viable skin layers. Microemulsions with different water content and droplet charge were investigated. It was shown that the structural features of the carrier affected the cutaneous delivery of α-tocopherol. However, the same factors had no impact on lipoic acid. The authors concluded that combination of these two antioxidants could potentially protect skin against damage associated with the generation of reactive oxygen species (Cichewicz et al. 2013). Similarly, Praça et al. investigated vitamins A and E-loaded W/O microemulsions for potential dermal application. The obtained system was physically stable and in vivo treatments showed reduced dermal expression of tumor necrosis factor alpha (TNF)-α by 1.3-fold ($p < 0.01$), when compared to unloaded microemulsion treatment group (Praça et al. 2020).

5.5 Summary

Antioxidants are wide and diverse group of compounds revealing different physicochemical properties. Nevertheless, there are few features, which seem to be commonly described as challenges in the studies focusing on antioxidant technological and delivery issues. Most of the described compounds reveal poor water solubility that causes difficulties both in extraction from plant sources and in the delivery as active pharmaceutical ingredients. It is noteworthy that in the case of such compounds bioavailability is usually low after oral administration, which also limits the possibility to achieve the therapeutic effect. Therefore, the selection of proper carrier enabling dissolution of active ingredient may be crucial for achieving the main goal of applying antioxidants. On the other hand, the efficient extraction process also requires the selection of proper medium revealing good solubilizing properties. Microemulsions that contain several different solubilizing agents seem to be perfect carriers and extraction liquids for the compounds revealing low solubility in water. Moreover, numerous studies confirm their ability to increase the absorption of the therapeutic agent from skin or gastrointestinal tract, which contributes to better efficacy of the applied formulation too. In the case of antioxidants which are often susceptible to degradation, a proper carrier may also improve the stability of the main ingredient and decrease its sensitivity to light and oxygen, as was proven for quercetin-loaded microemulsions (Lv et al. 2017).

Despite numerous advantages of microemulsions in antioxidant technology, it must be kept in mind that they are not free from drawbacks. One of the most important ones is the presence of surfactants and co-surfactants, which may act as irritants both applied topically and in gastrointestinal tract. The side effects related to skin irritation, impairing the barrier function of *stratum corneum* and dehydration were already described in the literature (Lehmann et al. 2001). Similar issues have been observed for self-microemulsifying systems administered orally (Gursoy and Benita 2004). It is also noteworthy that the components responsible for skin and mucous membrane irritation are necessary for microemulsion formation and usually occur in such systems at high concentrations, which increase the risk of side effects occurrence.

References

Alam MI, Beg S, Samad A et al (2010) Strategy for effective brain drug delivery. Eur J Pharm Sci 40:385–403. https://doi.org/10.1016/j.ejps.2010.05.003

Aliberti ALM, de Queiroz AC, Praça FSG et al (2017) Ketoprofen microemulsion for improved skin delivery and in vivo anti-inflammatory effect. AAPS PharmSciTech 18:2783–2791. https://doi.org/10.1208/s12249-017-0749-6

Altria KD (2000) Background theory and applications of microemulsion electrokinetic chromatography. J Chromatogr A 892:171–186. https://doi.org/10.1016/S0021-9673(00)00088-1

Altria KD, Mahuzier P-E, Clark BJ (2003) Background and operating parameters in microemulsion electrokinetic chromatography. Electrophoresis 24:315–324. https://doi.org/10.1002/elps.200390041

Amiri-Rigi A, Abbasi S (2016) Microemulsion-based lycopene extraction: effect of surfactants, co-surfactants and pretreatments. Food Chem 197:1002–1007. https://doi.org/10.1016/j.foodchem.2015.11.077

Amiri-Rigi A, Abbasi S (2017) Stability assessment of lycopene microemulsion prepared using tomato industrial waste against various processing conditions. J Sci Food Agric 97:4922–4928. https://doi.org/10.1002/jsfa.8368

Amiri-Rigi A, Abbasi S (2019) Extraction of lycopene using a lecithin-based olive oil microemulsion. Food Chem 272:568–573. https://doi.org/10.1016/j.foodchem.2018.08.080

Anton N, Vandamme TF (2011) Nano-emulsions and micro-emulsions: clarifications of the critical differences. Pharm Res 28:978–985. https://doi.org/10.1007/s11095-010-0309-1

Augustin MA, Sanguansri L, Lockett T (2013) Nano- and micro-encapsulated systems for enhancing the delivery of resveratrol. Ann N Y Acad Sci 1290:107–112. https://doi.org/10.1111/nyas.12130

Bera A, Mandal A (2015) Microemulsions: a novel approach to enhanced oil recovery: a review. J Pet Explor Prod Technol 5:255–268. https://doi.org/10.1007/s13202-014-0139-5

Bergonzi MC, Hamdouch R, Mazzacuva F et al (2014) Optimization, characterization and invitro evaluation of curcumin microemulsions. LWT – Food Sci Technol 59:148–155. https://doi.org/10.1016/j.lwt.2014.06.009

Bolko K, Zvonar A, Gašperlin M (2014) Mixed lipid phase SMEDDS as an innovative approach to enhance resveratrol solubility. Drug Dev Ind Pharm 40:102–109. https://doi.org/10.3109/03639045.2012.749888

Boonme P (2007) Applications of microemulsions in cosmetics. J Cosmet Dermatol 6:223–228. https://doi.org/10.1111/j.1473-2165.2007.00337.x

Boots AW, Haenen GRMM, Bast A (2008) Health effects of quercetin: from antioxidant to nutraceutical. Eur J Pharmacol 585:325–337. https://doi.org/10.1016/j.ejphar.2008.03.008

Brodowska K (2017) Natural flavonoids: classification, potential role, and application of flavonoid analogues. Eur J Biol Res 7:108–123. https://doi.org/10.5281/zenodo.545778

Callender SP, Mathews JA, Kobernyk K, Wettig SD (2017) Microemulsion utility in pharmaceuticals: implications for multi-drug delivery. Int J Pharm 526:425–442. https://doi.org/10.1016/j.ijpharm.2017.05.005

Calligaris S, Manzocco L, Valoppi F et al (2019) Microemulsions as delivery systems of lemon oil and β-carotene into beverages: stability test under different light conditions. J Sci Food Agric 99:7016–7020. https://doi.org/10.1002/jsfa.9973

Carvalho VF, de Lemos DP, Vieira CS et al (2017) Potential of non-aqueous microemulsions to improve the delivery of lipophilic drugs to the skin. AAPS PharmSciTech 18:1739–1749. https://doi.org/10.1208/s12249-016-0643-7

Censi R, Martena V, Hoti E et al (2012) Permeation and skin retention of quercetin from microemulsions containing Transcutol ® P. Drug Dev Ind Pharm 38:1128–1133. https://doi.org/10.3109/03639045.2011.641564

Chaari M, Theochari I, Papadimitriou V et al (2018) Encapsulation of carotenoids extracted from halophilic Archaea in oil-in-water (O/W) micro- and nano-emulsions. Colloids Surf B Biointerfaces 161:219–227. https://doi.org/10.1016/j.colsurfb.2017.10.042

Chen H, Zhong Q (2015) Thermal and UV stability of β-carotene dissolved in peppermint oil microemulsified by sunflower lecithin and Tween 20 blend. Food Chem 174:630–636. https://doi.org/10.1016/j.foodchem.2014.11.116

Chen W, Liu X, Liu Y et al (2010) Synthesis of PMMA and PMMA/PS nanoparticles by microemulsion polymerization with a new vapor monomer feeding system. Colloids Surf A Physicochem Eng Asp 364:145–150. https://doi.org/10.1016/j.colsurfa.2010.05.010

Chen Y, Zhang H, Yang J, Sun H (2015) Improved antioxidant capacity of optimization of a self-microemulsifying drug delivery system for resveratrol. Molecules 20:21167–21177. https://doi.org/10.3390/molecules201219750

Chirio D, Peira E, Dianzani C et al (2019) Development of solid lipid nanoparticles by cold dilution of microemulsions: curcumin loading, preliminary in vitro studies, and biodistribution. Nano 9:247. https://doi.org/10.3390/nano9020230

Cichewicz A, Pacleb C, Connors A et al (2013) Cutaneous delivery of α-tocopherol and lipoic acid using microemulsions: influence of composition and charge. J Pharm Pharmacol 65:817–826. https://doi.org/10.1111/jphp.12045
Cole GM, Teter B, Frautschy SA (2007) Neuroprotective effects of curcumin. In: Advances in experimental medicine and biology. Springer, New York, pp 197–212
Constantinides PP (1995) Lipid microemulsions for improving drug dissolution and oral absorption: physical and biopharmaceutical aspects. Pharm Res Off J Am Assoc Pharm Sci 12:1561–1572. https://doi.org/10.1023/A:1016268311867
D'Odorico A, Martines D, Kiechl S et al (2000) High plasma levels of α- and β-carotene are associated with a lower risk of atherosclerosisResults from the Bruneck study. Atherosclerosis 153:231–239. https://doi.org/10.1016/S0021-9150(00)00403-2
Danielsson I, Lindman B (1981) The definition of microemulsion. Colloids Surf 3:391–392. https://doi.org/10.1016/0166-6622(81)80064-9
Das S, Lee SH, Chow PS, Macbeath C (2020) Microemulsion composed of combination of skin beneficial oils as vehicle: development of resveratrol-loaded microemulsion based formulations for skin care applications. Colloids Surf B Biointerfaces 194:111161. https://doi.org/10.1016/j.colsurfb.2020.111161
Deng Y, Wang L, Yang W et al (2003) Preparation of magnetic polymeric particles via inverse microemulsion polymerization process. J Magn Magn Mater 257:69–78. https://doi.org/10.1016/S0304-8853(02)00987-3
Djordjevic L, Primorac M, Stupar M, Krajisnik D (2004) Characterization of caprylocaproyl macrogolglycerides based microemulsion drug delivery vehicles for an amphiphilic drug. Int J Pharm 271:11–19. https://doi.org/10.1016/j.ijpharm.2003.10.037
Dutta A, Dutta SK (2003) Vitamin e and its role in the prevention of atherosclerosis and carcinogenesis: a review. J Am Coll Nutr 22:258–268. https://doi.org/10.1080/07315724.2003.10719302
Erlund I (2004) Review of the flavonoids quercetin, hesperetin, and naringenin. Dietary sources, bioactivities, bioavailability, and epidemiology. Nutr Res 24:851–874. https://doi.org/10.1016/j.nutres.2004.07.005
Fan Y, Ma L, Zhang W et al (2014) The design of propolis flavone microemulsion and its effect on enhancing the immunity and antioxidant activity in mice. Int J Biol Macromol 65:200–207. https://doi.org/10.1016/j.ijbiomac.2014.01.041
Fanun M (ed) (2008) Microemulsions. Properties and applications, 1st edn. CRC Press, Boca Raton
Fiedor J, Burda K (2014) Potential role of carotenoids as antioxidants in human health and disease. Nutrients 6:466–488. https://doi.org/10.3390/nu6020466
Froelich A, Osmałek T, Snela A et al (2017) Novel microemulsion-based gels for topical delivery of indomethacin: formulation, physicochemical properties and in vitro drug release studies. J Colloid Interface Sci 507:323–336. https://doi.org/10.1016/j.jcis.2017.08.011
Galano A, Álvarez-Diduk R, Ramírez-Silva MT et al (2009) Role of the reacting free radicals on the antioxidant mechanism of curcumin. Chem Phys 363:13–23. https://doi.org/10.1016/j.chemphys.2009.07.003
Gannu R, Palem CR, Yamsani VV et al (2010) Enhanced bioavailability of lacidipine via microemulsion based transdermal gels: formulation optimization, ex vivo and in vivo characterization. Int J Pharm 388:231–241. https://doi.org/10.1016/j.ijpharm.2009.12.050
Gao Y, Wang Y, Ma Y et al (2009) Formulation optimization and in situ absorption in rat intestinal tract of quercetin-loaded microemulsion. Colloids Surf B Biointerfaces 71:306–314. https://doi.org/10.1016/j.colsurfb.2009.03.005
Ghosh M, Kundu S, Pyne A, Sarkar N (2020) Unveiling the behavior of curcumin in biocompatible microemulsion and its differential interaction with gold and silver nanoparticles. J Phys Chem C 124:3905–3914. https://doi.org/10.1021/acs.jpcc.9b11553
Ghouas H, Haddou B, Kameche M et al (2016) Removal of tannic acid from aqueous solution by cloud point extraction and investigation of surfactant regeneration by microemulsion extraction. J Surfactant Deterg 19:57–66. https://doi.org/10.1007/s11743-015-1764-9

Gibaud S, Attivi D (2012) Microemulsions for oral administration and their therapeutic applications. Expert Opin Drug Deliv 9:937–951. https://doi.org/10.1517/17425247.2012.694865

Guo Y, Mao X, Zhang J et al (2019) Oral delivery of lycopene-loaded microemulsion for brain-targeting: preparation, characterization, pharmacokinetic evaluation and tissue distribution. Drug Deliv 26:1191–1205. https://doi.org/10.1080/10717544.2019.1689312

Gursoy RN, Benita S (2004) Self-emulsifying drug delivery systems (SEDDS) for improved oral delivery of lipophilic drugs. Biomed Pharmacother 58:173–182. https://doi.org/10.1016/j.biopha.2004.02.001

He G, Pan Q, Rempel GL (2003) Synthesis of poly(methyl methacrylate) nanosize particles by differential microemulsion polymerization. Macromol Rapid Commun 24:585–588. https://doi.org/10.1002/marc.200390089

He CX, He ZG, Gao JQ (2010) Microemulsions as drug delivery systems to improve the solubility and the bioavailability of poorly water-soluble drugs. Expert Opin Drug Deliv 7:445–460. https://doi.org/10.1517/17425241003596337

Hoar TP, Schulman JH (1943) Transparent water-in-oil dispersions: the oleopathic hydro-micelle. Nature 152:102–103. https://doi.org/10.1038/152102a0

Housaindokht MR, Nakhaei Pour A (2012) Study the effect of HLB of surfactant on particle size distribution of hematite nanoparticles prepared via the reverse microemulsion. Solid State Sci 14:622–625. https://doi.org/10.1016/j.solidstatesciences.2012.01.016

Hu L, Jia Y, Niu F et al (2012) Preparation and enhancement of oral bioavailability of curcumin using microemulsions vehicle. J Agric Food Chem 60:7137–7141. https://doi.org/10.1021/jf204078t

Johnson EJ, Krinsky NI (2009) Carotenoids and coronary heart disease. In: Carotenoids. Birkhäuser, Basel, pp 287–300

Jomova K, Valko M (2013) Health protective effects of carotenoids and their interactions with other biological antioxidants. Eur J Med Chem 70:102–110. https://doi.org/10.1016/j.ejmech.2013.09.054

Kajbafvala A, Salabat A, Salimi A (2018) Formulation, characterization, and in vitro/ex vivo evaluation of quercetin-loaded microemulsion for topical application. Pharm Dev Technol 23:741–750. https://doi.org/10.1080/10837450.2016.1263995

Karppi J, Laukkanen JA, Mäkikallio TH et al (2012) Low β-carotene concentrations increase the risk of cardiovascular disease mortality among Finnish men with risk factors. Nutr Metab Cardiovasc Dis 22:921–928. https://doi.org/10.1016/j.numecd.2012.01.008

Kasperczyk S, Dobrakowski M, Kasperczyk J et al (2014) Beta-carotene reduces oxidative stress, improves glutathione metabolism and modifies antioxidant defense systems in lead-exposed workers. Toxicol Appl Pharmacol 280:36–41. https://doi.org/10.1016/j.taap.2014.07.006

Kelkel M, Schumacher M, Dicato M, Diederich M (2011) Antioxidant and anti-proliferative properties of lycopene. Free Radic Res 45(8):925–940

Kerdudo A, Dingas A, Fernandez X, Faure C (2014) Encapsulation of rutin and naringenin in multilamellar vesicles for optimum antioxidant activity. Food Chem 159:12–19. https://doi.org/10.1016/j.foodchem.2014.03.005

Kitagawa S, Tanaka Y, Tanaka M et al (2009) Enhanced skin delivery of quercetin by microemulsion. J Pharm Pharmacol 61:855–860. https://doi.org/10.1211/jpp.61.07.0003

Klampfl CW (2003) Solvent effects in microemulsion electrokinetic chromatography. Electrophoresis 24:1537–1543. https://doi.org/10.1002/elps.200305379

Kogan A, Garti N (2006) Microemulsions as transdermal drug delivery vehicles. Adv Colloid Interf Sci 123–126:369–385. https://doi.org/10.1016/j.cis.2006.05.014

Kohli K, Chopra S, Dhar D et al (2010) Self-emulsifying drug delivery systems: an approach to enhance oral bioavailability. Drug Discov Today 15:958–965. https://doi.org/10.1016/j.drudis.2010.08.007

Kong B, Yu J, Savino K et al (2012) Synthesis of α-calcium sulfate hemihydrate submicron-rods in water/n-hexanol/CTAB reverse microemulsion. Colloids Surf A Physicochem Eng Asp 409:88–93. https://doi.org/10.1016/j.colsurfa.2012.05.041

Kreilgaard M (2002) Influence of microemulsions on cutaneous drug delivery. Adv Drug Deliv Rev 54:S77–S98. https://doi.org/10.1016/S0169-409X(02)00116-3

Kuentz M (2011) Oral self-emulsifying drug delivery systems, from biopharmaceutical to technical formulation aspects. J Drug Deliv Sci Technol 21:17–26. https://doi.org/10.1016/S1773-2247(11)50002-4

Lawrence MJ, Rees GD (2012) Microemulsion-based media as novel drug delivery systems. Adv Drug Deliv Rev 64:175–193. https://doi.org/10.1016/j.addr.2012.09.018

Le Marchand L (2002) Cancer preventive effects of flavonoids - a review. Biomed Pharmacother 56:296–301. https://doi.org/10.1016/S0753-3322(02)00186-5

Lehmann L, Keipert S, Gloor M (2001) Effects of microemulsions on the stratum corneum and hydrocortisone penetration. Eur J Pharm Biopharm 52:129–136. https://doi.org/10.1016/S0939-6411(01)00160-6

Liu CH, Lee WS, Wu WC (2016) Photodynamic inactivation against Pseudomonas aeruginosa by curcumin microemulsions. RSC Adv 6:63013–63022. https://doi.org/10.1039/c6ra10193c

Lopes L (2014) Overcoming the cutaneous barrier with microemulsions. Pharmaceutics 6:52–77. https://doi.org/10.3390/pharmaceutics6010052

Lopes LB, Vandewall H, Li HT et al (2010) Topical delivery of lycopene using microemulsions: enhanced skin penetration and tissue antioxidant activity. J Pharm Sci 99:1346–1357. https://doi.org/10.1002/jps.21929

Lu CT, Zhao YZ, Wong HL et al (2014) Current approaches to enhance CNS delivery of drugs across the brain barriers. Int J Nanomedicine 9:2241–2257. https://doi.org/10.2147/IJN.S61288

Lv X, Liu T, Ma H et al (2017) Preparation of essential oil-based microemulsions for improving the solubility, pH stability, photostability, and skin permeation of quercetin. AAPS PharmSciTech 18:3097–3104. https://doi.org/10.1208/s12249-017-0798-x

Magno LM, Angelescu DG, Sigle W, Stubenrauch C (2011) Microemulsions as reaction media for the synthesis of Pt nanoparticles. Phys Chem Chem Phys 13:3048–3058. https://doi.org/10.1039/c0cp01085e

Malik MA, Wani MY, Hashim MA (2012) Microemulsion method: a novel route to synthesize organic and inorganic nanomaterials. 1st Nano update. Arab J Chem 5:397–417. https://doi.org/10.1016/j.arabjc.2010.09.027

Mandal SD, Mandal S, Patel J (2016) Brain targeting efficiency of Curcumin loaded mucoadhesive microemulsion through intranasal route. J Pharm Investig 46:179–188. https://doi.org/10.1007/s40005-016-0227-2

McAllister K, Sazani P, Adam M et al (2002) Polymeric nanogels produced via inverse microemulsion polymerization as potential gene and antisense delivery agents. J Am Chem Soc 124:15198–15207. https://doi.org/10.1021/ja027759q

McClements DJ (2012) Nanoemulsions versus microemulsions: terminology, differences, and similarities. Soft Matter 8:1719–1729. https://doi.org/10.1039/c2sm06903b

Mehnert W (2001) Solid lipid nanoparticles production, characterization and applications. Adv Drug Deliv Rev 47:165–196. https://doi.org/10.1016/S0169-409X(01)00105-3

Mobasheri A, Henrotin Y, Biesalski H-K, Shakibaei M (2012) Scientific evidence and rationale for the development of curcumin and resveratrol as nutraceutricals for joint health. Int J Mol Sci 13:4202–4232. https://doi.org/10.3390/ijms13044202

Mouri A, Diat O, El Ghzaoui A et al (2014) Phase behavior of reverse microemulsions based on Peceol®. J Colloid Interface Sci 416:139–146. https://doi.org/10.1016/j.jcis.2013.10.058

Müller RH, Mäder K, Gohla S (2000) Solid lipid nanoparticles (SLN) for controlled drug delivery – a review of the state of the art. Eur J Pharm Biopharm 50:161–177. https://doi.org/10.1016/S0939-6411(00)00087-4

Nagula RL, Wairkar S (2019) Recent advances in topical delivery of flavonoids: a review. J Control Release 296:190–201. https://doi.org/10.1016/j.jconrel.2019.01.029

Nastiti CMRR, Ponto T, Abd E et al (2017) Topical nano and microemulsions for skin delivery. Pharmaceutics 9:37. https://doi.org/10.3390/pharmaceutics9040037

Nazar MF, Shah SS, Khosa MA (2011) Microemulsions in enhanced oil recovery: a review. Pet Sci Technol 29:1353–1365. https://doi.org/10.1080/10916460903502514

Padmanaban G, Rangarajan PN (2016) Curcumin as an adjunct drug for infectious diseases. Trends Pharmacol Sci 37:1–3. https://doi.org/10.1016/j.tips.2015.09.007

Pangeni R, Sahni JK, Ali J et al (2014) Resveratrol: review on therapeutic potential and recent advances in drug delivery. Expert Opin Drug Deliv 11:1285–1298. https://doi.org/10.1517/17425247.2014.919253

Parvathy KS, Negi PS, Srinivas P (2009) Antioxidant, antimutagenic and antibacterial activities of curcumin-β-diglucoside. Food Chem 115:265–271. https://doi.org/10.1016/j.foodchem.2008.12.036

Pavel FM (2004) Microemulsion polymerization. J Dispers Sci Technol 25:1–16. https://doi.org/10.1081/DIS-120027662

Peinado C, Bosch P, Martín V, Corrales T (2006) Photoinitiated polymerization in bicontinuous microemulsions: fluorescence monitoring. J Polym Sci Part A Polym Chem 44:5291–5303. https://doi.org/10.1002/pola.21649

Pepe D, Phelps J, Lewis K et al (2012) Decylglucoside-based microemulsions for cutaneous localization of lycopene and ascorbic acid. Int J Pharm 434:420–428. https://doi.org/10.1016/j.ijpharm.2012.06.016

Perlman ME, Murdande SB, Gumkowski MJ et al (2008) Development of a self-emulsifying formulation that reduces the food effect for torcetrapib. Int J Pharm 351:15–22. https://doi.org/10.1016/j.ijpharm.2007.09.015

Pillai AB, Nair JV, Gupta NK, Gupta S (2015) Microemulsion-loaded hydrogel formulation of butenafine hydrochloride for improved topical delivery. Arch Dermatol Res 307:625–633. https://doi.org/10.1007/s00403-015-1573-z

Pouton CW (2000) Lipid formulations for oral administration of drugs: non-emulsifying, self-emulsifying and "self-microemulsifying" drug delivery systems. Eur J Pharm Sci 11:S93–S98. https://doi.org/10.1016/S0928-0987(00)00167-6

Praça FG, Viegas JSR, Peh HY et al (2020) Microemulsion co-delivering vitamin a and vitamin E as a new platform for topical treatment of acute skin inflammation. Mater Sci Eng C 110:110639. https://doi.org/10.1016/j.msec.2020.110639

Rao J, McClements DJ (2011) Food-grade microemulsions, nanoemulsions and emulsions: fabrication from sucrose monopalmitate & lemon oil. Food Hydrocoll 25:1413–1423. https://doi.org/10.1016/j.foodhyd.2011.02.004

Rao AV, Rao LG (2007) Carotenoids and human health. Pharmacol Res 55(3):207–216. https://doi.org/10.1016/j.phrs.2007.01.012

Riccioni G (2009) Carotenoids and cardiovascular disease. Curr Atheroscler Rep 11:434–439. https://doi.org/10.1007/s11883-009-0065-z

Rigotti A (2007) Absorption, transport, and tissue delivery of vitamin E. Mol Asp Med 28:423–436. https://doi.org/10.1016/j.mam.2007.01.002

Roohinejad S, Middendorf D, Burritt DJ et al (2014a) Capacity of natural β-carotene loaded microemulsion to protect Caco-2 cells from oxidative damage caused by exposure to H2O2. Food Res Int 66:469–477. https://doi.org/10.1016/j.foodres.2014.10.012

Roohinejad S, Oey I, Everett DW, Niven BE (2014b) Evaluating the effectiveness of β-carotene extraction from pulsed electric field-treated carrot pomace using oil-in-water microemulsion. Food Bioprocess Technol 7:3336–3348. https://doi.org/10.1007/s11947-014-1334-6

Roussaki M, Gaitanarou A, Diamanti PC et al (2014) Encapsulation of the natural antioxidant aureusidin in biodegradable PLA nanoparticles. Polym Degrad Stab 108:182–187. https://doi.org/10.1016/j.polymdegradstab.2014.08.004

Sanchez-Dominguez M, Pemartin K, Boutonnet M (2012) Preparation of inorganic nanoparticles in oil-in-water microemulsions: a soft and versatile approach. Curr Opin Colloid Interface Sci 17:297–305. https://doi.org/10.1016/j.cocis.2012.06.007

Santanna VC, Curbelo FDS, Castro Dantas TN et al (2009) Microemulsion flooding for enhanced oil recovery. J Pet Sci Eng 66:117–120. https://doi.org/10.1016/j.petrol.2009.01.009

Santos P, Watkinson AC, Hadgraft J, Lane ME (2008) Application of microemulsions in dermal and transdermal drug delivery. Skin Pharmacol Physiol 21:246–259. https://doi.org/10.1159/000140228

Scalbert A, Morand C, Manach C, Rémésy C (2002) Absorption and metabolism of polyphenols in the gut and impact on health. Biomed Pharmacother 56:276–282. https://doi.org/10.1016/S0753-3322(02)00205-6

Seljak KB, Berginc K, Trontelj J et al (2014) A self-microemulsifying drug delivery system to overcome intestinal resveratrol toxicity and presystemic metabolism. J Pharm Sci 103:3491–3500. https://doi.org/10.1002/jps.24114

Shah RM, Malherbe F, Eldridge D et al (2014) Physicochemical characterization of solid lipid nanoparticles (SLNs) prepared by a novel microemulsion technique. J Colloid Interface Sci 428:286–294. https://doi.org/10.1016/j.jcis.2014.04.057

Shah B, Khunt D, Misra M, Padh H (2016) Non-invasive intranasal delivery of quetiapine fumarate loaded microemulsion for brain targeting: formulation, physicochemical and pharmacokinetic consideration. Eur J Pharm Sci 91:196–207. https://doi.org/10.1016/j.ejps.2016.05.008

Sintov AC, Shapiro L (2004) New microemulsion vehicle facilitates percutaneous penetration in vitro and cutaneous drug bioavailability in vivo. J Control Release 95:173–183. https://doi.org/10.1016/j.jconrel.2003.11.004

Smoliga JM, Baur JA, Hausenblas HA (2011) Resveratrol and health – a comprehensive review of human clinical trials. Mol Nutr Food Res 55:1129–1141. https://doi.org/10.1002/mnfr.201100143

Solans C, Kunieda H (1997) Industrial applications of microemulsions. Marcel Dekker, New York

Spernath A, Yaghmur A, Aserin A et al (2002) Food-grade microemulsions based on nonionic emulsifiers: media to enhance lycopene solubilization. J Agric Food Chem 50:6917–6922. https://doi.org/10.1021/jf025762n

Talegaonkar S, Azeem A, Ahmad F et al (2008) Microemulsions: a novel approach to enhanced drug delivery. Recent Pat Drug Deliv Formul 2:238–257. https://doi.org/10.2174/187221108786241679

Tang H, Xiang S, Li X et al (2019) Preparation and in vitro performance evaluation of resveratrol for oral self-microemulsion. PLoS One 14:e0214544. https://doi.org/10.1371/journal.pone.0214544

Terao J (2009) Dietary flavonoids as antioxidants. In: Food factors for health promotion. KARGER, Basel, pp 87–94

Thilakarathna S, Rupasinghe H (2013) Flavonoid bioavailability and attempts for bioavailability enhancement. Nutrients 5:3367–3387. https://doi.org/10.3390/nu5093367

Tønnesen HH, Másson M, Loftsson T (2002) Studies of curcumin and curcuminoids. XXVII. Cyclodextrin complexation: solubility, chemical and photochemical stability. Int J Pharm 244:127–135. https://doi.org/10.1016/S0378-5173(02)00323-X

Traber MG, Atkinson J (2007) Vitamin E, antioxidant and nothing more. Free Radic Biol Med 43:4–15. https://doi.org/10.1016/j.freeradbiomed.2007.03.024

Tsai YH, Lee KF, Huang YB et al (2010) In vitro permeation and in vivo whitening effect of topical hesperetin microemulsion delivery system. Int J Pharm 388:257–262. https://doi.org/10.1016/j.ijpharm.2009.12.051

Vicentini FTMC, Vaz MMOLL, Fonseca YM et al (2011) Characterization and stability study of a water-in-oil microemulsion incorporating quercetin. Drug Dev Ind Pharm 37:47–55. https://doi.org/10.3109/03639045.2010.491078

Walle T, Hsieh F, DeLegge MH et al (2004) High absorption but very low bioavailability of oral resveratrol in humans. Drug Metab Dispos 32:1377–1382. https://doi.org/10.1124/dmd.104.000885

Wang C, Bu Y, Zhao L (2017) Synthesis of water and oil dual-absorption materials by bicontinuous microemulsion polymerization. Polym Sci – Ser B 59:292–299. https://doi.org/10.1134/S1560090417030162

Wilhelm EA, Vogt AG, Reis AS et al (2018) The efficacy of microemulsion-based delivery to improve vitamin E properties: evaluation of the antinociceptive, antioxidant, antidepressant- and anxiolytic-like activities in mice. J Pharm Pharmacol 70:1723–1732. https://doi.org/10.1111/jphp.13018

Winsor P (1954) Solvent properties of amphiphilic compounds. Butterworths, London

Wolf S, Feldmann C (2016) Microemulsions: options to expand the synthesis of inorganic nanoparticles. Angew Chemie – Int Ed 55:15728–15752. https://doi.org/10.1002/anie.201604263

Woodside JV, McGrath AJ, Lyner N, McKinley MC (2015) Carotenoids and health in older people. Maturitas 80:63–68. https://doi.org/10.1016/j.maturitas.2014.10.012

Xiao Y, Chen X, Yang L et al (2013) Preparation and oral bioavailability study of curcuminoid-loaded microemulsion. J Agric Food Chem 61:3654–3660. https://doi.org/10.1021/jf400002x

Xing Q, Song J, You X et al (2016) Microemulsions containing long-chain oil ethyl oleate improve the oral bioavailability of piroxicam by increasing drug solubility and lymphatic transportation simultaneously. Int J Pharm 511:709–718. https://doi.org/10.1016/j.ijpharm.2016.07.061

Yen FL, Wu TH, Tzeng CW et al (2010) Curcumin nanoparticles improve the physicochemical properties of curcumin and effectively enhance its antioxidant and antihepatoma activities. J Agric Food Chem 58:7376–7382. https://doi.org/10.1021/jf100135h

Yutani R, Komori Y, Takeuchi A et al (2016) Prominent efficiency in skin delivery of resveratrol by novel sucrose oleate microemulsion. J Pharm Pharmacol 68:46–55. https://doi.org/10.1111/jphp.12497

Zhao X, Wang Z (2019) A pH-sensitive microemulsion-filled gellan gum hydrogel encapsulated apigenin: characterization and in vitro release kinetics. Colloids Surf B Biointerfaces 178:245–252. https://doi.org/10.1016/j.colsurfb.2019.03.015

Zhao X, Wang Z, Li X (2016) Preparation, in-vitro release and antioxidant potential of formulation of apigenin with hydroxypropyl-β-cyclodextrin modified microemulsion. J Incl Phenom Macrocycl Chem 86:93–102. https://doi.org/10.1007/s10847-016-0644-x

Chapter 6
Membrane Emulsification for Encapsulation of Bioactives: Application to the Encapsulation of Antioxidants

Océane Alliod and Catherine Charcosset

6.1 Membrane Emulsification (ME)

Emulsions are usually prepared using high-pressure homogenizers, ultrasounds and rotor/−stator systems. In the dispersing zone of these machines, high shear stresses are applied to deform and disrupt large droplets into the smaller ones. These high shear stresses result in an increase in temperature. Therefore, temperature or shear-sensitive ingredients such as proteins or starches may lose their functional properties.

In addition, these processes usually suffer from a lack of precision in droplet size control and result in droplet polydispersity. The production of monodispersed emulsions has been investigated by several new techniques based on microfluidics. However, scaling up to industrial volumes is a major limitation of these processes. Unlike these methods, membrane emulsification (ME) has the potential for scaling up while producing droplets of well-defined size.

ME has received increasing attention over the last 30 years as an alternative to other methods of emulsification. In ME, small oil droplets are formed at the extremity of the membrane pores and are detached by the continuous phase flowing on top of the membrane. The droplets size can be controlled by membrane pore size and the flowrate can be increased by increasing the membrane surface. This process is called direct membrane emulsification. In another configuration (premix membrane emulsification (PME), a first emulsion with large droplets is prepared and passed through the membrane. Small and homogenous droplets are then obtained. Many membranes have been used for ME (polymeric and ceramic), with regular and irregular pores. In addition, several configurations are available that add vibration or rotation to the classical processes, and several scales are possible for some milliliters to several liters. These different types of emulsification, membranes and set-ups

O. Alliod · C. Charcosset (✉)
Université Claude Bernard Lyon 1, CNRS, Villeurbanne, France
e-mail: catherine.charcosset@univ-lyon1.fr

M. A. Aboudzadeh (ed.), *Emulsion-based Encapsulation of Antioxidants*, Food Bioactive Ingredients, https://doi.org/10.1007/978-3-030-62052-3_6

are detailed in the next sections (Nazir et al. 2010; Charcosset 2012; Vladisavljević 2016).

6.1.1 Direct Membrane Emulsification

Direct membrane emulsification (DME) is used to generate emulsions. In DME, a dispersed phase is injected through membrane pores into a continuous phase (Fig. 6.1). Both phases are immiscible, so droplets form at the interface of both liquids and the membrane. Droplets grow at pore openings until they detach when having reached a certain size. Emulsifier molecules in the continuous phase stabilize the newly formed interface immediately after formation in order to prevent droplet coalescence. The resulting droplet size is controlled primarily by the choice of the membrane, which allows a precise size control. Usually a shear stress is applied on the membrane surface in order to facilitate the detachment of the droplets.

This technique can be used to form oil-in-water (O/W), water-in-oil (W/O), bubbles, double emulsions or can be the first step to form solid particles (Fig. 6.2). It can also be coupled with spontaneous emulsification or microemulsification techniques; thus, this process is called membrane micromixing. In membrane micromixing, the organic phase is introduced into the aqueous phase through a microporous membrane under controlled injection rate and shear on the membrane/aqueous phase interface. Once they are in contact, a solvent displacement occurs that creates nanoemulsions, liposomes or microemulsions. Therefore, the membrane is not aimed at emulsifying but at controlling the addition of one phase into another.

As the interface between the two liquids and the membrane is of great importance in DME, the affinity of the membrane with both phases should be considered

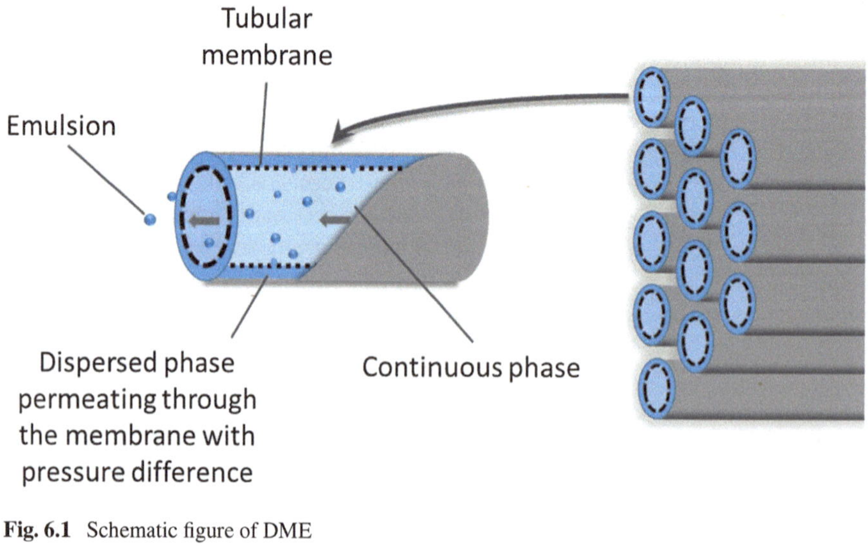

Fig. 6.1 Schematic figure of DME

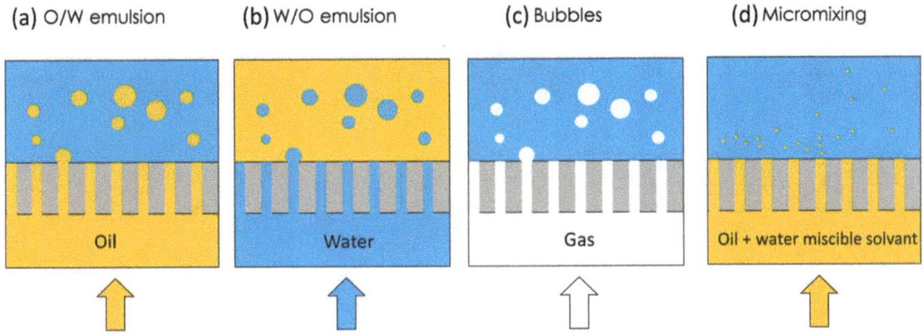

Fig. 6.2 Different formulations that can be produced with DME: (**a**) O/W emulsion, (**b**) W/O emulsion, (**c**) bubbles, (**d**) micromixing

according to the formulation desired. O/W emulsions require hydrophilic membranes, so the oil having no affinity with the membrane can be pushed out of the pores to create the emulsions. In this context, W/O emulsions require hydrophobic membranes for the same reason.

Advantages of DME are as follow: the emulsion is generated at low shear stress and low pressure without the requirement of another process, the droplet size is controlled by the pore size and monodispersed droplets are created, moreover, this process is easy to scale up by increasing the membrane surface. Drawbacks are that the flux through the pores is extremely low, especially in the case of very small droplets, which cause membrane fouling. Moreover, droplets size is 3–10 times larger than the pore size depending on composition and shear stress, which can be an issue to obtain very small droplets.

In Fig. 6.2d, the membrane generates micromixing between two phases, one coming through the membrane pores and the other one flowing on the membrane surface either by crossflow or by stirred flow. The technique is called membrane dispersion or membrane mixing. Thanks to the jets coming from the membrane pores, micromixing and the properties of the final products could be optimized. The technique is used to produce crystals, organic and inorganic nanoparticles or nanocapsules, liposomes, to perform parallel or consecutive reactions and to synthesize polymers. Similar to ME, several parameters influence the process results such as membrane pore size, distances between adjacent pores, flowrate through the membrane, stirring speed, etc. (Jia and Liu 2013).

6.1.2 Premix Membrane Emulsification

Membranes are also used to modify emulsions. In this process, a coarse emulsion is passed through the membrane pores in order to reduce the droplets sizes, change the emulsion sense or demulsify it (Fig. 6.3). The most studied technique is premix membrane emulsification (PME) which used to reduce the size and homogenize the

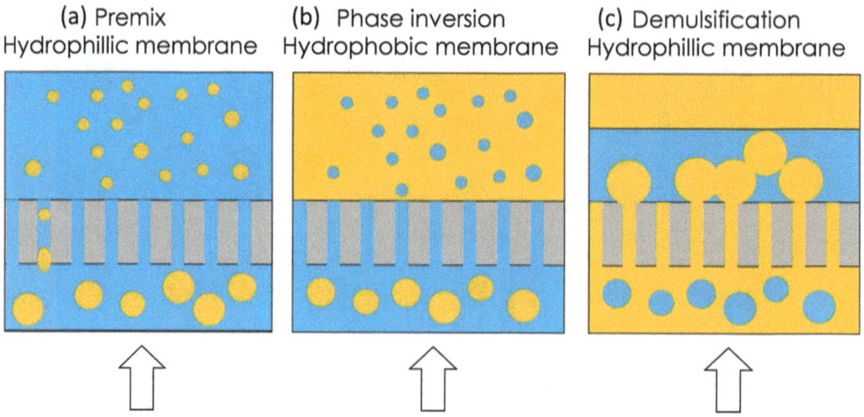

Fig. 6.3 Different types of utilization of PME: (**a**) premix with a hydrophilic membrane, (**b**) phase inversion with a hydrophobic membrane, (**c**) demulsification with a hydrophilic membrane

emulsion. In PME, the coarse emulsion is usually obtained by magnetic stirrer or rotor stator systems but it can be prepared with any process. The chemical composition of the membrane has an important impact in this process as well. In PME, a membrane with affinity for the continuous phase should be used, and thus, hydrophilic and hydrophobic membranes are needed to manufacture O/W and W/O emulsion, respectively. If the membrane has affinity for the dispersed phase, two phenomenon can be occurred depending on the concentration and the procedure used, there phenomenon are phase inversion or emulsification.

Advantages of PME are the same as other ME techniques: the emulsion is generated at low shear stress, the droplet size is controlled by the pore size, monodispersed droplets are produced, and the process is easy to scale up by increasing the membrane surface. Additional advantages are that flux through the pores can be much higher than in DME and if the flux is high enough, no fouling will occur. Drawbacks are that an additional step of premix emulsification is required and that a higher pressure than in DME is usually necessary depending on the pore size and composition of the emulsion.

6.2 Set-ups for Membrane Emulsification

6.2.1 Membrane Types

A membrane is a selective barrier commonly used for filtration e.g. microfiltration, ultrafiltration, nanofiltration, or reverse osmosis, requiring different types of membranes for each application. Only the most commonly used membranes for emulsification are presented here. Membranes used for emulsification should have certain properties that are uniform pores, different pore sizes available, high mechanical strength and good resistance to temperature or organic solvent (depending on the application), possibility to modify the surface to change hydrophilicity or charge for example.

6.2.1.1 Shirasu Porous Glass (SPG) Membranes

The SPG membrane is the most used type of membrane for emulsification. In Fig. 6.4, Scanning electron microscopy image of a SPG membrane is shown. It is a glass composed mainly of SiO_2 and Al_2O_2, which is obtained by reaction of Shirasu (volcanic ash), boric acid and limestone. After several heat treatments, additives and acid leaching, the final membrane is produced. Final pore size depends, for the same composition, on the heating time and temperature. A large range of mean pore size is achievable by this process: from 40 nm to 40 μm. SPG membranes are available in tubes or flat discs. They are resistant to pressure and chemicals. The pores are cylindrical, tortuous and interconnected.

SPG membranes are hydrophilic; however, they can be easily hydrophobized. Two main types of hydrophobization are usually used: a physical coating with silicone resin or a chemical reaction with organosilane (Vladisavljević et al. 2005; Kukizaki and Wada 2008).

A new type of asymmetric SPG membrane has been also investigated to increase flowrate through the membrane (Kukizaki and Goto 2007). The authors proved that it allows increasing the flowrate 20 times without any change in droplet size distribution. However, this membrane is more difficult to produce and is not commercially available. Advantages of SPG membranes are that a wide range of pore size is available as well as small modules for laboratory scale, they have a high porosity

Fig. 6.4 Scanning electron microscopy image of a SPG membrane with 200 nm pore size

and they are cheap. Disadvantages are mainly due to their tortuosity creating fouling and fragility but also decreasing the number of available pores for emulsification.

6.2.1.2 Polymeric Membranes

Polymeric membranes are traditionally used for liposomes extrusion or filtration. Many different kinds of polymers can be chosen for this purpose depending of the chemical affinities between the membrane and the liquids e.g. polycarbonate, polyethersulfone, polypropylene, nylon, polyester, cellulose acetate. Each polymeric membrane presents a specific geometry (porosity, tortuosity, pore spacing, etc.) (Gehrmann and Bunjes 2018). For example, the polyester membrane is a track-etched membrane with highly defined straight-through pores and on the contrary nylon and cellulose acetate membranes are branched membranes. Thickness is also very different from one polymeric membrane to another varying from 10 to 180 μm.

It is hard to conclude about general advantages or disadvantages as this family of membranes is very diverse. Nonetheless, one advantage is that they are usually cheap and many different chemical compositions are available for each application. The disadvantages are that they are quite fragile and not all of them are very effective for ME.

6.2.1.3 Metallic Membranes

Metallic membranes are usually micro-engineered membranes with controlled pore geometry and pore spacing. The aim is to avoid fouling and to achieve high transmembrane flux by lowering membrane hydraulic resistance. At the same time, they lower down drastically the pores surface area for the same membrane area, which can lead to lower flowrate than for SPG membranes (Vladisavljević 2016). Typical metallic membranes are made from nickel or stainless steel.

Advantages are that pores with controlled size produce more monodispersed droplets, they are chemically resistant, there is less fouling as the pores are not tortuous. Disadvantages are that they are produced by technologies that are still expensive. The number of pores per membrane area is low so lower flowrates for the same membrane surface are obtained. Only few pore sizes are available and all in the micron range.

6.2.1.4 Silicon Nitride Micro-Engineered Membranes

Another type of micro-engineered membrane which also used is silicon nitride micro-engineered membranes (Wagdare et al. 2010). They are usually made with photolithographic techniques and treated with air plasma to obtain a hydrophilic surface. Their advantages and drawbacks are the same as the ones presented for micro-engineered metallic membranes.

6.2.1.5 Ceramic Membranes

Tubular ceramic membranes are typically used for ultrafiltration and microfiltration. They are composed of two layers; a thin filtration layer with a thickness of 20–30 μm with a pore size from 0.1 to 10 μm for microfiltration and 2–50 nm for ultrafiltration, deposited on a macroporous layer with a thickness of some hundreds of microns. These traditional ceramic membranes are less used than SPG membranes for emulsification may be because they produce more polydispersed emulsions in the same conditions as SPG membranes (Vladisavljević et al. 2004a).

Their advantages and drawbacks are similar to SPG membranes, with the additional disadvantage that the interface between the two layers, the thin filtration layer and the macroporous layer, creates fragility and makes membranes less resistant to transmembrane pressure.

6.2.2 Membrane Cleaning

Membrane cleaning is a key point in membrane processes as the membrane undergoes internal or external fouling in almost every application. Several procedures are possible, depending on the membrane type and the fouling molecules. The main strategy is usually chemical cleaning with highly reactive compounds or physicochemical cleaning with surfactants. Regarding surfactant cleaning, Derquim+ is considered as the best cleaning agent for membrane emulsification (Trentin et al. 2012). Moreover, increasing temperature and number of cycles lead to a better cleaning.

6.2.3 Set-Ups with Tubular Membranes

Various set-ups are possible for DME and PME with tubular or flat disc membranes. The main configurations are summarized in Fig. 6.5.

6.2.3.1 Microkits

SPG Technology Co., Ltd. (Miyazaki, Japan) developed a special device to be used at lab scale with small SPG membranes of 20 mm length with an effective length of 10 mm. This module can be used in an external pressure configuration, i.e. the dispersed phase is pushed from the tube outside to the inside or in an internal pressure configuration, i.e. the dispersed phase is pushed from the tube inside to the outside. The external pressure type microkit is known to be more effective for droplet size reduction than the internal one. It is also more resistant to transmembrane pressure. The tank is filled with the dispersed phase of the emulsion or with a premix and

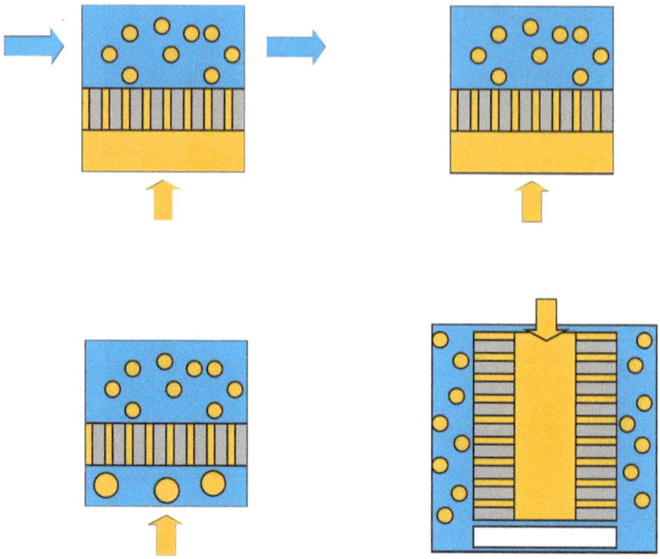

Fig. 6.5 Different configurations for DME and PME, (**a**) cross-flow ME, (**b**) dead-end ME, (**c**) PME in dead-end, (**d**) stirring, rotating or vibrating ME

pushed by N_2 flux. It allows passing only a few milliliters through the membrane as the tank capacity is typically 10 mL.

In DME, the device is immersed in the continuous phase within a beaker and the shear rate is generated by a magnetic stirrer. In order to facilitate the flow through the pores, the membrane should be wetted by the continuous phase. In PME, the homogenized emulsion just falls by gravity into a beaker. To our knowledge, most of the premix emulsions with SPG membranes were produced with this device (Joseph and Bunjes 2012; Vladisavljević et al. 2004) except for the first study in which they used a cross-flow module which is presented in the next section (Suzuki et al. 1996). Advantages of this device are that small volumes can be prepared with less than 1 mL of dead volume and both DME and PME can be performed without dilution. Disadvantages are that only small volumes batches can be prepared and that the set-up only resists up to 8 bars.

6.2.3.2 Cross-Flow

The cross-flow set-up is based on the ones used for microfiltration. A typical ME set-up consists in a pressurized (N_2) vessel filled with the dispersed phase that is pushed through a tubular membrane. Inside the membrane tube, the continuous phase circulates thanks to a pump. This configuration is a batch mode with recirculation of the emulsion that is created and concentrates. The emulsion obtained is kept under stirring in order to stay homogeneous. The flow of the dispersed phase is very low and usually the continuous phase flowrate is high to generate shear stress.

As seen previously, this set up can be used for premix emulsification as well but has the disadvantage of premix dilution (Suzuki et al. 1996). Advantages of this device are that large volumes can be produced and the set-up is very simple. Disadvantages are that the minimum continuous phase volume is usually several hundreds of milliliters and the process cannot be proceeded in a continuous way unless a very small amount of dispersed phase is required or a very long membrane is used.

6.2.4 Set-ups with Flat Disc Membranes

6.2.4.1 Dead-End

Typically, flat disc membranes are assembled in a set-up with a dead-end configuration. This set-up is used for micro-engineered membranes that are flat discs. The membrane is fixed in the device and the dispersed phase is pushed from the bottom of the set-up to the glass container above the membrane containing the continuous phase. Inside the glass container, a mechanical stirrer generates the shear stress at the membrane/continuous phase interface. This shear stress detaches the droplets as a result of both centrifugal force and viscous shear force. Advantages of this set-up are that it produces monodispersed droplets, small volumes can be produced at lab scale and no N_2 is required. Disadvantages are than only small batches can be produced, the membrane surface is smaller than with tubular membranes, and the flowrate is low.

6.2.5 Dynamic DME

In addition to static DME presented in the above section, dynamic DME has also been developed. The aim of dynamic DME is to improve detachment of the droplets and decrease membrane fouling. Indeed, fouling is a common and well-known problem in membrane processes. Typically, particles and/or molecules can be retained by the membrane by steric or adsorption effect. These phenomena decrease the permeate flux through the membrane and the membrane selectivity. In case of ME, fouling is mainly due to macromolecules (such as lipid and surfactant) adsorption on the membrane surface and/or inside the membrane pores.

6.2.5.1 Rotational Membranes

A dynamic membrane can be a rotating membrane; the membrane rotates on its axis in order to increase shear stress at the membrane continuous phase interface (Hancocks et al. 2016; Aryanti et al. 2009). This is typically done with tubular SPG membranes. The parameters of importance are the membrane diameter because the centrifuge force increases with the diameter of the tubes and the speed of rotation.

6.2.5.2 Vibrating or Oscillating Membranes

Dynamic DME can also be performed by addition of vibrations to the membrane. The membrane vibrates to help droplet detachment from the pore openings. It can be done either with tubular or flat disc membranes. This idea was first developed for filtration and then applied to emulsification (Zhu and Barrow 2005). A piezoactuator system is assembled to the membrane in order to produce vibrations. It was found that smaller droplets could be produced by introducing low frequency (0–100 Hz) membrane vibrations without widening the droplet size distribution.

6.2.6 *Commercial Membranes and Industrial Set-ups*

Typical commercially available membranes used for emulsification are presented in Table 6.1. Several industrial set-ups are available for example from SPG Technology Inc. (Japan), Micropore Technology (UK), Emulsar (France) and Kinematics (Swistzerland). They are based on rotation or pulsations (Micropore Technology), vibrations (Emulsar) and rotation (Kinematics).

Table 6.1 Summary of different types of commercial available membranes, set-ups and pore sizes available

Type	Geometry	Set-up	Pore size	Material
SPG	Tubular: 20–500 mm Flat disc	Cross-flow Rotational Dead-end Vibrating	0.1–20 μm	Shirasu porous glass
Ceramic	Tubular: 250–1178 mm	Cross-flow	0.002–10 μm	Ceramic
Micropore	Flat disc	Dead-end Vibrating	5–20 μm	Metal
Aquamarjin	Flat disc	Dead-end Vibrating	0.1–100 μm	Silicon nitride
Polymeric filter	Flat disc	Dead-end Vibrating	0.1–100 μm	Different polymers

6.3 Forces Involved and Parameters of Influence

6.3.1 Droplets Generation in DME

In DME, droplets are generated at the oil/water/membrane interface. In this section, forces that are generated during the process and dimensionless numbers to characterize the relative effect of these forces are presented.

In DME, the droplets formed at the membrane surface, grow until a critical size and then are carried away with the continuous phase, flowing parallel to the membrane surface for cross-flow or not for other set-ups. The final droplet size of the emulsion is determined at the membrane, dispersed phase and continuous phase interface. This droplet size is the result of different forces (Fig. 6.6).

Negligible forces:

- F_B, the buoyancy force due to the density difference between the dispersed and continuous phases
- F_G, the gravitational force

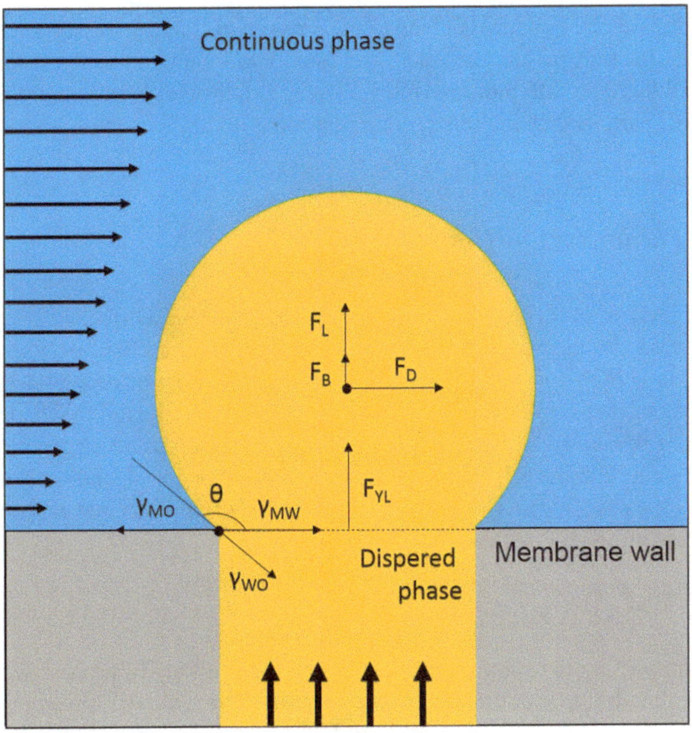

Fig. 6.6 Forces exerted at membrane, oil and water interface for an O/W emulsion and a hydrophilic membrane

Force due to transmembrane pressure:

- F_{SP}, due to the static pressure difference between the dispersed and continuous phases

Fluid forces due to shear stress:

- F_L, the dynamic lift force due to the asymmetric shear stress profile at the membrane surface
- F_D, Drag force, due to the continuous phase flow, parallel to the membrane surface

Interfacial forces:

- γ_{WO}, Interfacial tension between water and oil
- γ_{MO}, Wetting of the oil phase on the membrane
- γ_{MW}, Wetting of the water phase on the membrane

The resulting force of all interfacial contributions is called interfacial force F_γ or capillary force and is opposed to the static pressure difference.

6.3.1.1 Instabilities

Instabilities are phenomena that lead to the breakup of liquid jets. However, in classical DME, the transmembrane velocity is usually low enough not to create this kind of instabilities. The different instabilities leading to droplet break-up are presented in the PME part, as they are much more important in this configuration.

6.3.2 Parameters of Influence in DME

Vladisavljević et al. differentiates two main categories of DME: shear-controlled emusification, for circular rectilinear pores (Vladisavljević 2016) or when interfacial tension is high and the shear stress is the main source of droplet detachment, and interfacial- tension-driven DME in tortuous pores such as SPG where droplets detachment is mainly a result of low interfacial tension. Some parameters do not have the same influence if DME is shear-controlled or interfacial-tension-driven, when it is the case, a difference is made. Effects of each parameter on droplet size in DME are presented in Table 6.2.

(a)*Membrane parameters*

Wetting
Membrane wetting is essential for effective emulsification. To prepare O/W emulsions, the membrane should by hydrophilic in order to avoid oil spreading onto the membrane surface and creating large polydispered droplets. On the contrary, in order to produce water-in-oil emulsions, membranes should be hydrophobic to avoid water spreading on the membrane. Hydrophilicity can also be modified with

Table 6.2 Effect of each parameter on droplet size in DME, IFT: interfacial tension

Parameter	Effect on droplet size
Membrane pores	Proportional to pore size (factor 3–10)
Membrane wettability	Good conditions = Same wettability as dispersed phase
Transmembrane pressure	IFT driven: plateau until a critical value then fast growing Crossflow controlled: increasing with pressure then plateau phase
Continuous phase flowrate	Cross-flow controlled: higher flowrate given smaller droplets IFT driven: no effect
Surfactant	Lower IFT: smaller droplets Higher [surfactant]: smaller droplet Faster adsorption at the interface: smaller droplets
Viscosity of the continuous phase	Cross-flow controlled: Higher viscosity: smaller droplets IFT driven: Lower viscosity: larger droplets

time due to adsorption of molecules on the membrane. It can happen when cationic surfactants are used, because SPG membranes are negatively charged. Cetrimonium bromide, for example can be adsorbed onto the surface, which becomes more hydrophobic. This adsorption led to polydispersed emulsions (Kukizaki 2009).

Pore Size

Pore size is the main parameter of influence regarding the final droplet size in ME. Variation of droplet size with pore size is different if DME is shear stress or interfacial tension controlled (Sugiura et al. 2002).

For interfacial-tension-driven DME, droplet size varies linearly with pore size. For SPG membranes and microengineered membranes, the ratio between final droplet size and pore size is not less than 3, if the other parameters such as transmembrane pressure and surfactant are optimized (Kukizaki and Goto 2007; Nakashima et al. 2000). In shear-controlled droplet DME, the mean droplet size is determined by a balance between the shear force exerted at the liquid–liquid interface by the continuous phase and the interfacial force $F\gamma$. Increasing shear stress at the membrane interface decreases droplet size and typical droplet size to pore ratios are between 3 and 10 for SPG membranes and from 7 to 36 for silicon nitride membranes (Abrahamse et al. 2002) depending on the composition of the emulsion.

Activated Pores

Due to their process of fabrication, SPG membranes present a high tortuosity, thus only a low percentage of pores are actually activated and available for emulsification. The percentage of activated pores does not change from one membrane to another, however it was found to vary with pressure, linearly (Abrahamse et al. 2002) or exponentially (Vladisavljević and Schubert 2003).

Pore Spacing

If pore spacing is not sufficient, coalescence of droplets from adjacent pores can occur before detachment. It is more eager to happen if the interfacial force is too low, meaning high wetting of the dispersed phase onto the membrane or high interfacial tension or at low shear stress. However, in some cases, a push-off mechanism between droplets can occur leading to smaller droplet size with thinner pore spaces.

(b) *Process and formulation parameters*

Transmembrane Flux
In DME, emulsification occurs only if the capillary pressure or emulsification pressure is reached. For shear-controlled DME, the detachment is not spontaneous and requires a certain time. This time is constant so higher transmembrane flux produces larger droplets. At high fluxes, the push-of force as a result of droplet–droplet interaction on the membrane surface, facilitates the droplet detachment process, the droplet size becomes constant as the transmembrane flux increases (Egidi et al. 2008). For interfacial-tension-driven DME and low transmembrane pressure, droplets spontaneously detach from the membrane and their sizes do not change significantly with the flux. However, quite quickly, the regime changes from dripping to continuous outflow regime where droplets become a continuous jet. In this regime, the shear stress detaches the droplets. At these conditions, droplets are drastically larger at higher transmembrane flux and usually much more polydispersed due to random detachment from the membrane.

Continuous Phase Flowrate
In shear-controlled detachment, the flow of continuous phase creates the shear and a higher shear leads to smaller droplets. For interfacial-tension-driven DME, the flowrate of the continuous phase and shear have only a low impact as detachment is a spontaneous phenomenon. When the transition from dripping to outflow regime occurs, the continuous phase flowrate starts to have an effect. Droplets are getting smaller at higher flowrate but in a less controlled manner, meaning higher polydispersity.

Surfactants
Surfactants have two important characteristics: their ability to low down the interfacial tension between oil and water and also their kinetics of adsorption to the newly created interface. Locally, the interfacial tension depends on the nature of surfactant but also on its concentration. Low interfacial tension is essential for an effective detachment of droplets but also to avoid coalescence and stabilize droplets.

The influence of surfactant adsorption kinetics on droplet stabilization has been investigated in several studies (van der Graaf et al. 2004; Rayner et al. 2005). The faster the surfactant adsorbs at the interface, the smaller the droplets are. Fast adsorption allows faster detachment and avoids coalescence of newly created droplets from two close-by pores. However, surfactants must be chosen carefully. Cationic surfactants can adsorb on the SPG membrane surface and change its hydrophilicity. Thus, oil spreads over the membrane and creates large and polydispersed droplets. Zwitterionic surfactants can also generate this kind of problem as they present a positive charge.

Viscosity of the Continuous Phase

In few studies, the impact of phase viscosity in DME has been investigated. The shear stress exerted on the droplets depends on viscosity, higher viscosity leading to higher shear stress. In shear-control emulsification, higher viscosity led to smaller droplets (Hancocks et al. 2013). However, another effect competes with this one, higher viscosity results in slower adsorption kinetics of the surfactant, and the droplets have more time to grow resulting in larger droplets (Kukizaki 2009). Without shear stress, it was shown that an increase in dispersed phase viscosity leads to a decrease in droplet size (Kukizaki 2009).

6.3.3 Break-up Mechanisms in PME

In PME, membranes are used as homogenizers; mechanisms involved are different than the ones in DME. First, droplets break-up in PME occurs within the pores whereas in DME, break-up occurs at the membrane, continuous phase and dispersed phase interface. Moreover, in DME, the model geometry is a T-junction whereas in PME it is more complicated as disruption occurs within the pores of complex geometries (tortuosity, branching, dead-end, etc.). However, one major phenomenon, wall shear stress and four other phenomena have been identified as break-up mechanisms in PME (Nazir et al. 2010).

Wall Shear Stress

Droplet size reduction in PME is mainly due to wall shear stress. The shear stress depends on the velocity of the premix emulsion within the pores as well as pore geometry (size, porosity and tortuosity). A formulation parameter, viscosity, has also an impact on shear stress. At pressures below a critical pressure, emulsions do not permeate trough the membrane. At moderate shear stress, droplets are homogenized at the diameter of the pores or slightly higher due to droplets deformation. At high shear stress, an oil jet phenomenon generates impact of the droplets on the pore walls and leads to droplet size smaller than the pore size (Vladisavljević et al. 2004).

Localized Shear Forces

Shear stress is also generated at the pores intersection due to localized shear forces (van der Zwan et al. 2006). The difference in flux in the branches generates a shear stress that helps reducing the droplet size of the emulsion. However this effect is difficult to measure or predict. In a membrane composed of several complex branched pores, the effect of localized shear force on droplet disruption becomes impossible to predict. According to van der Zwan et al., this localized shear stress is more important at high flowrate because at low flowrate no branched pores are activated (van der Zwan et al. 2006).

Interfacial Tension Instabilities

Two types of instabilities can occur due to interfacial tension according to flow conditions. Laplace instabilities occur when a droplet is elongated at low flowrate, a difference in Laplace pressure is generated, creating dumbbell-shape and snap-off effects that disrupt the droplet. At high flowrate of the continuous phase, another phenomenon takes place inside the channel, Rayleigh instabilities. The droplets remain elongated within the pores, which then may lead to break-up into polydispersed droplets forces (van der Zwan et al. 2006).

Steric Hindrance Between Droplets
Nearby droplets may have an impact on each other too, if the interface is well stabilized, coalescence does not occur between droplets but steric hindrance can induce break-up. The more droplets accumulate within the pores the more likely steric hindrance occurs.

6.3.4 Parameters of Influence in PME

(a) Membrane parameters

Membrane properties have a great influence on droplet disruption. Porosity, tortuosity and pore size are major properties regarding final droplet size. Moreover, like in DME, hydrophilic properties of the membrane have great influence on the PME process. The membrane should be hydrophilic if the continuous phase is water and hydrophobic if the continuous phase is oil. If the dispersed phase has a high affinity with the membrane, demulsification occurs instead of PME.

Pore Size
Pore size has a direct influence on wall shear stress. Depending of the transmembrane pressure and pore size, the droplet to pore size ratio typically varies from 1 to 1.5 for SPG membranes (Vladisavljević et al. 2006).

Pore Geometry
Higher tortuosity increases the shear stress. Moreover, membranes like SPG membranes present a lot of branching, which can create higher shear stress and more effective size reduction. On the contrary, in this type of geometry, many pores are not active, typically only 1% (Vladisavljević et al. 2007) leading to higher pressure requirement for the same surface of membrane and same pore size.

(b) Process and formulation parameters

Transmembrane Flux
As the premix emulsion is a mixture of hydrophobic and hydrophilic liquids, membrane pore walls have affinity with one of the two phases. Usually as mentioned previously for effective PME, the affinity is stronger with the continuous phase, so a minimum pressure is required to ensure that the premix emulsion goes through the membrane pores. Moreover, the minimum pressure for total premix emulsion to go through the pores is higher than in DME where only capillary pressure is present. In order to avoid filtration, with water permeating trough the membrane pores, as the chemical affinity is stronger, and oil fouling the membrane, a higher pressure is

required to disrupt the droplets and having a homogeneous emulsion flowing through the pores.

Vladisavljević et al. defined the transmembrane pressure, ΔP_{tm}, in ME as the addition of two different pressures, in their studies, the flow pressure required to overcome membrane resistance to the flow due to its pore sizes and geometry, ΔP_{flow}, and also the disruption pressure, ΔP_{disrup}, required to overcome interfacial tension between oil and water in order to reduce the emulsion size and go through the membrane pores. In PME, the optimal transmembrane pressure is much higher than the minimum pressure required, 10–50 times. Higher transmembrane pressure leads to higher flux within the membrane pores and thus higher shear stress and size reduction of the emulsion (Vladisavljević et al. 2004).

Number of Cycles

In a high-pressure homogenizer, several cycles are often required to reach the targeted droplet size or to reduce dispersity. In PME, the necessity of several cycles depends on the membrane type used. For tortuous and branched membranes like SPG membranes, shear stress is more important. Moreover they are thick membranes which explains why 1 extrusion cycle is required through SPG membranes (Vladisavljević et al. 2006; Joseph and Bunjes 2014), whereas polymeric membranes which are less tortuous and thinner, required up to 21 extrusion cycles (Joseph and Bunjes 2014) to get droplets of a diameter close to pore size and monodispersed.

Dispersed Phase Content and Premix Size Distribution

In PME, the final droplet size does not depend on the premix properties such as size distribution or oil content (if interfacial tension is constant). This is not the case for typical emulsification processes such as high-pressure homogenization where the final droplet size highly depends on dispersed phase content. However, the transmembrane pressure depends highly on dispersed phase content due to increase in droplets number to be reduced in size and viscosity (Vladisavljević et al. 2004), which can be a feasibility issue.

Viscosities

Viscosity of the premix emulsion has an important effect on transmembrane pressure and flux, the permeate flux being inversely proportional to viscosity (Vladisavljević et al. 2004). At the same time, the continuous phase viscosity increases the wall shear stress leading to smaller droplets at higher viscosity (Vladisavljević et al. 2006).

Surfactants

During PME, new oil/water interfaces are created. Surfactants should be in excess in the premix in order to stabilize the smaller droplets of the final emulsion. Moreover, like in DME, their ability to low down the interfacial tension and their kinetics of adsorption is important. Locally, interfacial tension depends on the nature of surfactant but also on its concentration. Low interfacial tension is essential for an effective stabilization of droplets and to avoid any coalescence. Adsorption of the surfactant at the newly created interface, like in all other homogenization

Table 6.3 Effect of each factor on droplet size in PME

Parameter	Effect on droplet size
Pore size	SPG: Proportional to pore size (factor 1–1.5) Others: less clear
Pore geometry	Tortuosity and branching = improve size reduction
Transmembrane pressure	Higher pressure or flux: More effective shear stress and size reduction
Number of cycle	SPG: Slight decrease after 1 cycle Others: 10 or 20 cycles required for monodispersity
Dispersed phase content	No important effect on size At same pressure decrease of transmembrane flux
Viscosity	Increase in continuous phase viscosity = size decrease Increase in premix viscosity = increase in transmembrane pressure
Surfactant	Higher concentration = smaller droplets Faster adsorption = smaller droplets

processes, such as high-pressure homogenization, has to be fast enough. In high-pressure homogenization, repeated cycles are used to counterbalance the fact that the new surface area cannot be covered fast enough by surfactants. In PME, the processing time is longer allowing more time for coverage of the new interface by the surfactant. However, this phenomenon is not negligible in PME.

Effects of each parameter on droplet size in PME are presented in Table 6.3.

6.4 Antioxidant Encapsulation

Antioxidants are natural or synthesize compounds present in fruits and vegetables (Bazinet and Doyen 2017; Jia et al. 2016). They have the property to prevent or delay the oxidation of other substances. Phenolic antioxidants act mainly by retarding the formation of free radicals. They can synthetic antioxidants (like butylated hydroxyanisol, butylated hydroxytoluene and tertiary butyl hydroquinone or natural antioxidants like vegetables extracts like rosemary, sage and tea extracts. Natural phenolic antioxidants include tocopherols, polyphenols and protein and peptides. Tocopherols are present mainly in vegetable oils. The four types of tocopherols are α-, β-, γ- and δ- tocopherols; their difference is due to the position of the substituted methyl group (CH_3). The most popular is vitamin E, which is the α-form. Polyphenols are present in plants; they include more than 10,000 different compounds. Flavonoids are a major class of polyphenols, and among them antocyanins are the most popular. Antioxidant activities have also been reported for protein and peptides, from milk, soy, egg and fish. Some other antioxidants show a lower or even no direct activities, like chelating agents, ascorbic and isoascorbic acid, sulfites and proteins and peptides.

To improve the stability of antioxidants that are sensible to temperature, light, oxygen and pH, encapsulation is particularly attractive. It has also the advantage to limit their possible unpleasant taste such as bitter or astringent and increase their bioavailability after administration (i.e. insufficient gastric residence time, low permeability, and/or solubility within the gut). Compounds soluble in water like polyphenols are usually encapsulated in W/O emulsions and compounds soluble in oil, such as tocopherol, are loaded in O/W emulsions. Water-in-oil-in-water (W/O/W) emulsions are also an alternative to encapsulate both molecules soluble in water and molecules soluble in oil.

6.4.1 Membrane Emulsification for Antioxidant Encapsulation

ME is an attractive emulsification technique for antioxidant encapsulation because of the very low pressure and shear stress applied. Some articles have reported the encapsulation of antioxidants by PME or DME, such as W/O emulsions and W/O/W double emulsions (Fig. 6.7).

O/W Emulsions
ME has been reported for the preparation of O/W emulsions with antioxidant properties. The antioxidant compound can be either a surfactant (like lecithin) that is used to stabilize the oil/water interface or a natural compound (like beta-carotene) that is encapsulated in the emulsion. PME was used by Surh et al. to prepare lecithin-stabilized O/W emulsions using a SPG membrane (Surh et al. 2008). Lecithin has recognized antioxidant activities and is extracted from vegetables and fruits like soybeans and rapeseed and also eggs. It is a negatively charged food grade emulsifier widely used in the food industry which can produce small oil droplets during conventional emulsification techniques. However, these authors concluded that lecithin tended to foul SPG membranes by blocking the membrane pores, which was attributed to the possible interaction between positive groups on the lecithin molecules with anionic silanol groups on the membrane surface.

Several antioxidants compounds have been encapsulated in emulsions like beta-carotene, catechol and hydroxytyrosol and astaxanthin. Repeated PME was used to produce emulsions with controlled droplet size or ME with pulsed back-and-forward

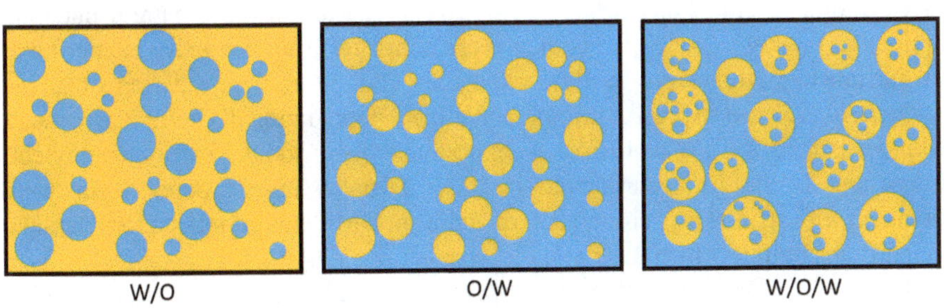

Fig. 6.7 Different formulations for encapsulation of antioxidants that can be produced by ME method

to reduce membrane fouling. Beta-carotene was encapsulated in O/W emulsions using PME and polymeric microfiltration membranes (Trentin et al. 2011). Several protein surfactants were tested such as bovine serum albumin and whey protein concentrate, as well as a nonionic surfactant Tween 20 as a co-surfactant. Like lecithin, these protein surfactants generated membrane fouling which was reduced by the addition of Tween 20. In addition, it was shown that whey protein concentrate and bovine serum albumin showed a higher protective effect against degradation of beta-carotene than Tween 20. Catechol was encapsulated in O/W emulsions using pulsed back-and-forward ME (Piacentini et al. 2016). Catechol was used a model of biophenols found in olive mill wastewaters produced from biologic olive oil (obtained without toxic pesticides). Stable emulsions with high dispersion phase contents were obtained (30% v/v) and high encapsulation efficiencies (respectively 98% for catechol and 92% for biophenol fractions purified and concentrated from olive mill wastewaters). Moreover, hydroxytyrosol, a compound of the olive biophenol family, was encapsulated in solid lipid particles using cold pulsed back-and-forward ME (Bazzarelli et al. 2017). The encapsulation rate obtained by ME was around 90% while the one obtained by homogenization was around 40%. O/W emulsions containing astaxanthin were also produced by repeated PME (Ribeiro et al. 2005). Astaxanthin is a carotenoid found mainly in seaweeds belonging to the species *Hematococcus pluvialis*. Three passes were necessary to decrease the droplet size and increase emulsion dispersity. However, encapsulation was not sufficient to limit oxidative degradation of astaxanthin which was approximatively 30% after 3 weeks. Limits of the ME technique were pointed out to be reduced flux and membrane fouling.

W/O/W Double Emulsions

W/O/W double emulsions have numerous applications in food, cosmetic and pharmaceutics for encapsulation and delivery of active ingredients. They can be produced by a two-step process in which a premix emulsion is first formed using surfactant(s) and then dispersed in an aqueous phase by adding other surfactant(s). ME can be used for both steps or only for the second one to control precisely the emulsion droplet size and dispersity. Double emulsions are attractive antioxidant encapsulation systems.

DME was used by Matos et al. for the encapsulation of *trans*-resveratrol in W/O/W double emulsions (Matos et al. 2015). Resveratrol is a natural phenolic antioxidant found in many plants; however, it changes from the active isomer *trans* to the inactive isomer *cis* by light exposure. Therefore, encapsulation of *trans*-resveratrol is aimed to protect this molecule from degradation due to light exposure. Monodispersed W/O/W emulsions were obtained using sodium carboxymethylcellulose and Tween 20 as surfactants in DME. Droplets sizes around 60 μm were obtained with span values around 0.9. In addition, DME was compared to a classical agitation technique which gave polydispersed emulsions.

In addition, PME was tested for the preparation of double emulsions containing procyanidin rich extracts (Berendsen et al. 2015). Procyanidins are found for example in grape seed and have several health benefits. The first emulsion was obtained by rotor stator homogenization (Ultra Turrax) and the double emulsion was produced using a Microkit with a 10 μm pore size SPG membrane. Finally, spray drying enabled to produce microcapsules containing procyanidins in powder form. Different surfactants were tested such as whey protein, whey protein carboxyl-methyl cellulose, to stabilize the interface of the emulsion droplets before spray drying.

Nanocapsules

Membrane mixing can be used to produce micro or nanocapsules, or nano or microparticles, usually obtained by nanoprecipitation between an organic phase containing solvent and polymer and an aqueous phase. Nanoprecipitation involves dispersion of preformed polymers, based on interfacial deposition of polymers following displacement of semi-polar solvent miscible with water. Molecules like antioxidant compounds can be added to the organic phase to produce loaded capsules (if oil is added) or loaded particles (without oil).

Very few studies have reported the preparation of nanocapsules using a membrane technique. Only vitamin E (α-tocopherol) and rosemary essential oils with reported antioxidant properties have been encapsulated using SPG membranes (Khayata et al. 2012; Ephrem et al. 2014). In these studies, membrane mixing with hydrophilic 1 μm pore size SPG membranes was shown to produce nanocapsules with very small sizes (around 200 nm), polydispersity indexes below 0.25, indicating an adequate homogeneity of the system and high encapsulation efficiencies (98–99%).

6.5 Conclusion

ME is a suitable technique for encapsulation of sensitive compounds in simple or double emulsions by applying moderate pressures (below 50 bars). Commercial ME systems include vibrations, rotation or pulsations that detache the droplets at the outlets of the membrane pores. In these set-ups, higher shear rates are created at the membrane surface and ME may lose some of its advantages such as low energy and suitability for shear fragile ingredients. Repeated PME set-ups are currently investigated for large scale applications. For example, Alliod et al. proposed a set-up including a high pressure syringe pump that controls the transmembrane pressure up to 60 bars and the flowrate up to 200 mL/min (Alliod et al. 2018). Two high pressure syringe pumps in parallel can be used to produce continuously nanoemulsions with controlled droplet sizes.

Several antioxidants were encapsulated using ME in W/O emulsions, W/O/W double emulsions and nanocapsules. To our knowledge, there is no comparison between ME and other techniques like high pressure homogenization to show the effect of the process on the stability of the emulsion and active compound during production and storage (at different temperature, oxygen, and light conditions). Such studies would be highly needed to choose the best process for antioxidant encapsulation.

References

Abrahamse A, Van Lierop R, Van der Sman R, Van der Padt A, Boom R (2002) Analysis of droplet formation and interactions during cross-flow membrane emulsification. J Membr Sci 204:125–137

Alliod O, Valour JP, Urbaniak S, Fessi H, Dupin D, Charcosset C (2018) Preparation of oil-in-water nanoemulsions at large-scale using premix membrane emulsification and Shirasu Porous Glass (SPG) membranes. Colloids Surf A Physicochem Eng Asp 557:76–84

Aryanti N, Hou R, Williams RA (2009) Performance of a rotating membrane emulsifier for production of coarse droplets. J Membr Sci 326:9–18

Bazinet L, Doyen A (2017) Antioxidants, mechanisms, and recovery by membrane processes. Crit Rev Food Sci Nutr 57:677–700

Bazzarelli F, Piacentini E, Giorno L (2017) Biophenols-loaded solid lipid particles (SLPs) development by membrane emulsification. J Membr Sci 541:587–594

Berendsen R, Güell C, Ferrando M (2015) Spray dried double emulsions containing procyanidin-rich extracts produced by premix membrane emulsification: effect of interfacial composition. Food Chem 178:251–258

Charcosset C (2012) Membrane processes in biotechnology and pharmaceutics. In: Charcosset C (ed) Membranes for the preparation of emulsions and particles. Elsevier, Oxford, Great Britain pp 213–251

Egidi E, Gasparini G, Holdich RG, Vladisavljević GT, Kosvintsev SR (2008) Membrane emulsification using membranes of regular pore spacing: droplet size and uniformity in the presence of surface shear. J Membr Sci 323:414–420

Ephrem E, Greige-Gerges H, Fessi H, Charcosset C (2014) Optimisation of rosemary oil encapsulation in polycaprolactone and scale-up of the process. J Microencapsul 31:746–753

Gehrmann S, Bunjes H (2018) Influence of membrane material on the production of colloidal emulsions by premix membrane emulsification. Eur J Pharm Biopharm 126:140–148

Hancocks R, Spyropoulos F, Norton I (2016) The effects of membrane composition and morphology on the rotating membrane emulsification technique for food grade emulsions. J Membr Sci 497:29–35

Hancocks R, Spyropoulos F, Norton I (2013) Comparisons between membranes for use in cross-flow membrane emulsification. J Food Eng 116:382–389

Jia Z, Dumont MJ, Orsat V (2016) Encapsulation of phenolic compounds present in plants using protein matrices. Food Biosci 15:87–104

Jia Z, Liu Z (2013) Membrane-dispersion reactor in homogeneous liquid process. J Chem Tech Biotech 88:163–168

Joseph S, Bunjes H (2012) Preparation of nanoemulsions and solid lipid nanoparticles by premix membrane emulsification. J Pharm Sci 101:2479–2489

Joseph S, Bunjes H (2014) Evaluation of Shirasu Porous Glass (SPG) membrane emulsification for the preparation of colloidal lipid drug carrier dispersions. Eur J Pharm Biopharm 87:178–186

Khayata N, Abdelwahed W, Chehna MF, Charcosset C, Fessi H (2012) Preparation of vitamin E loaded nanocapsules by the nanoprecipitation method: from laboratory scale to large scale using a membrane contactor. Int J Pharm 423:419–427

Kukizaki M (2009) Shirasu porous glass (spg) membrane emulsification in the absence of shear flow at the membrane surface: Influence of surfactant type and concentration, viscosities of dispersed and continuous phases, and transmembrane pressure. J Membr Sci 327:234–243

Kukizaki M, Goto M (2007) Preparation and characterization of a new asymmetric type of shirasu. J Membr Sci 299:190–199

Kukizaki M, Wada T (2008) Effect of the membrane wettability on the size and size distribution of microbubbles formed from shirasu-porous-glass (spg) membranes. Coll Surf A 317:146–154

Matos M, Gutiérrez G, Iglesias O, Coca J, Pazos C (2015) Enhancing encapsulation efficiency of food-grade double emulsions containing resveratrol or vitamin B12 by membrane emulsification. J Food Eng 166:212–220

Nakashima T, Shimizu M, Kukizaki M (2000) Particle control of emulsion by membrane emulsification and its applications. Adv Drug Deliv Rev 45:47–56
Nazir A, Schroën K, Boom R (2010) Premix emulsification: a review. J Membr Sci 362:1–11
Piacentini E, Poerio T, Bazzarelli F, Giorno L (2016) Microencapsulation by membrane emulsification of biophenols recovered from olive mill wastewaters. Membranes 6:1–11
Rayner M, Trägårdh G, Trägårdh C (2005) The impact of mass transfer and interfacial expansion rate on droplet size in membrane emulsification processes. Colloids Surf A Physicochem Eng Asp 266:1–17
Ribeiro HS, Rico LG, Badolato GG, Schubert H (2005) Production of O/W emulsions containing astaxanthin by repeated premix membrane emulsification. J Food Sci 70:E117–E123
Sugiura S, Nakajima M, Kumazawa N, Iwamoto S, Seki M (2002) Characterization of spontaneous transformation-based droplet formation during microchannel emulsification. J Phys Chem B 106:9405–9409
Surh J, Jeong YG, Vladisavljević GT (2008) On the preparation of lecithin-stabilized oil-in-water emulsions by multi-stage premix membrane emulsification. J Food Eng 89:164–170
Suzuki K, Shuto I, Hagura Y (1996) Characteristics of the membrane emulsification method combined with preliminary emulsification for preparing corn oil-in-water emulsions. Food Sci Tech Int 2:43–47
Trentin A, De Lamo S, Güell C, López F, Ferrando M (2011) Protein-stabilized emulsions containing beta-carotene produced by premix membrane emulsification. J Food Eng 106:267–274
Trentin A, Güell C, Gelaw T, de Lamo S, Ferrando M (2012) Cleaning protocols for organic microfiltration membranes used in premix membrane emulsification. Sep Purif Technol 88:70–78
van der Graaf S, Schroën C, van der Sman R, Boom R (2004) Influence of dynamic interfacial tension on droplet formation during membrane emulsification. J Colloid Interface Sci 277:456–463
van der Zwan E, Schroën K, van Dijke K, Boom R (2006) Visualization of droplet break-up in premix membrane emulsification using microfluidic devices. Colloids Surf A Physicochem Eng Asp 277:223–229
Vladisavljević G, Kobayashi I, Nakajima M, Williams R, Shimizu M, Nakashima T (2007) Shirasu Porous Glass membrane emulsification: characterization of membrane structure by high resolution X-ray microtomography and microscopic observation of droplet formation in real time. J Membr Sci 302:243–253
Vladisavljević G, Shimizu M, Nakashima T (2004) Preparation of monodisperse multiple emulsions at high production rates by multi-stage premix membrane emulsification. J Membr Sci 244:97–106
Vladisavljević G, Shimizu M, Nakashima T (2005) Permeability of hydrophilic and hydrophobic Shirasu porous-glass (SPG) membranes to pure liquids and its microstructure. J Membr Sci 250:69–77
Vladisavljević GT (2016) Integrated membrane processes for the preparation of emulsions particles and bubbles. In: Basile A, Charcosset C (eds) Integrated membrane systems and processes. Wiley-Blackwell, Chichester, pp 79–140
Vladisavljević GT, Lambrich U, Nakajima M, Schubert H (2004a) Production of O/W emulsions using SPG membranes, ceramic α-aluminium oxide membranes, microfluidizer and a silicon microchannel plate a comparative study. Colloids Surf A Physicochem Eng Asp 232:199–207
Vladisavljević GT, Schubert H (2003) Influence of process parameters on droplet size distribution in spg membrane emulsification and stability of prepared emulsion droplets. J Membr Sci 225:15–23
Vladisavljević GT, Shimizu M, Nakashima T (2006) Production of multiple emulsions for drug delivery systems by repeated SPG membrane homogenization: influence of mean pore size, interfacial tension and continuous phase viscosity. J Membr Sci 284:373–383
Wagdare NA, Marcelis AT, Ho OB, Boom RM, van Rijn CJ (2010) High throughput vegetable oil-in-water emulsification with a high porosity micro-engineered membrane. J Membr Sci 347:1–7
Zhu J, Barrow D (2005) Analysis of droplet size during crossflow membrane emulsification using stationary and vibrating micromachined silicon nitride membranes. J Membr Sci 261:136–144

Chapter 7
Encapsulation of Antioxidants Using Double Emulsions

María Matos, Rocío Díaz-Ruiz, Ali Marefati, Marilyn Rayner, and Gemma Gutiérrez

7.1 Introduction

Interfacial phenomena have become increasingly relevant, mainly because of their influence on the hydrodynamics and mass transfer (e.g. Marangoni effect), on mineral separation (e.g. flotation), on solid-gas or solid-liquid interactions (e.g. fluid phase catalysis) and encapsulation process (e.g. biomedicine, food, and cosmetic industries). A mixture composed of immiscible liquid phases is highly unstable. To avoid the coalescence of the drops, stabilizing agent (such as surfactant) is added to the mixture that locates at the interface. The surfactant reduces the interfacial tension and provides repulsion at the fluid interfaces to enhance the stability of the mixture.

An emulsion is a macroscopically homogeneous mixture of two (or more) immiscible liquids. Two main types of emulsions can be found, oil-in-water (O/W) and water in oil (W/O), which possess the capacity to transport or solubilize antioxidant compounds of other bioactive compounds in their inner phase. Multiple emulsions were first reported in 1925 by Seifriz (1925). The simplest multiple emulsions are double emulsions; which are ternary systems where the dispersed droplets contain smaller droplets of a different phase. They have either a water-in-oil-in-water ($W_1/O/W_2$) or an oil-in-water-in-oil ($O_1/W/O_2$) structure (Aserin 2008), as shown in Fig. 7.1. Double emulsions are complex systems and further instabilities than that of simple emulsions can be observed.

M. Matos · R. Díaz-Ruiz · G. Gutiérrez (✉)
Department of Chemical and Environmental Engineering, University of Oviedo, Oviedo, Spain
e-mail: gutierrezgemma@uniovi.es

A. Marefati · M. Rayner
Department of Food Technology, Engineering, and Nutrition, Lund University, Lund, Sweden

M. A. Aboudzadeh (ed.), *Emulsion-based Encapsulation of Antioxidants*, Food Bioactive Ingredients, https://doi.org/10.1007/978-3-030-62052-3_7

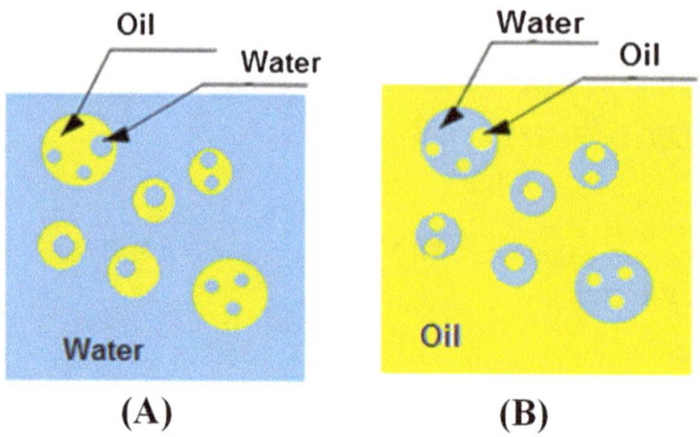

Fig. 7.1 Structure of double emulsions: (**a**) water-in-oil-in-water emulsion ($W_1/O/W_2$) and (**b**) oil-in-water-in-oil emulsion $O_1/W/O_2$

These emulsions have found applications in a variety of areas: separation technology (Rajasimman and Sangeetha 2009; Jiao et al. 2013; Anarakdim et al. 2020), encapsulation of sensitive molecules (Hemar et al. 2010; Matos et al. 2018) cosmetic, pharmaceutical, functional food products (Jiménez-Colmenero 2013), etc.

Double emulsions are highly desirable as drug delivery systems, when the therapy requires repeated administration, via ingestion or injection and for delivery of compounds with very short half-life. The single administration of the active component through a double emulsion has shown prolonged-release properties, avoiding alterations caused by the environment (e.g. oxidation, light, enzymatic degradation) or during food digestion (Rosen 2004; Aserin 2008). Because of their complex structure, multiple emulsions can be viewed as systems that control the transport of molecules from an external to an internal phase or vice versa.

Antioxidants can be considered as special compounds capable of inhibiting or delaying the oxidation process. These bioactive compounds have, among others, the function to prevent a variety of diseases. Due to their radical scavenging property, antioxidants have attracted considerable attention in treating diseases related to oxidative stress including atherosclerosis, cancer, cataracts, prophylaxis and therapy, longevity, Alzheimer's and Parkinson (Butterfield et al. 2002; Moharram and Youssef 2014). However, due to the natural limitations of antioxidants for use in food systems (Muschiolik and Dickinson 2017), such as low solubility or losses due to ambient light, they have to be protected through encapsulation technology.

Encapsulation technology was defined as a promising approach that has been employed for the protection and controlled release of different bioactive compounds including natural antioxidants. However, due to some intrinsic characteristics e.g. low solubility, short shelf life, difficultly in their packaging and handling, losses due to environmental stresses and food processes, undesirable flavors and odors, untargeted release and instability in various conditions during digestion in the gastrointestinal tract, application of these bioactive compounds in real food products, pharmaceuticals, and cosmetics has been limited. The use of double emulsions can overcome these restrictions (Maqsoudlou et al. 2020).

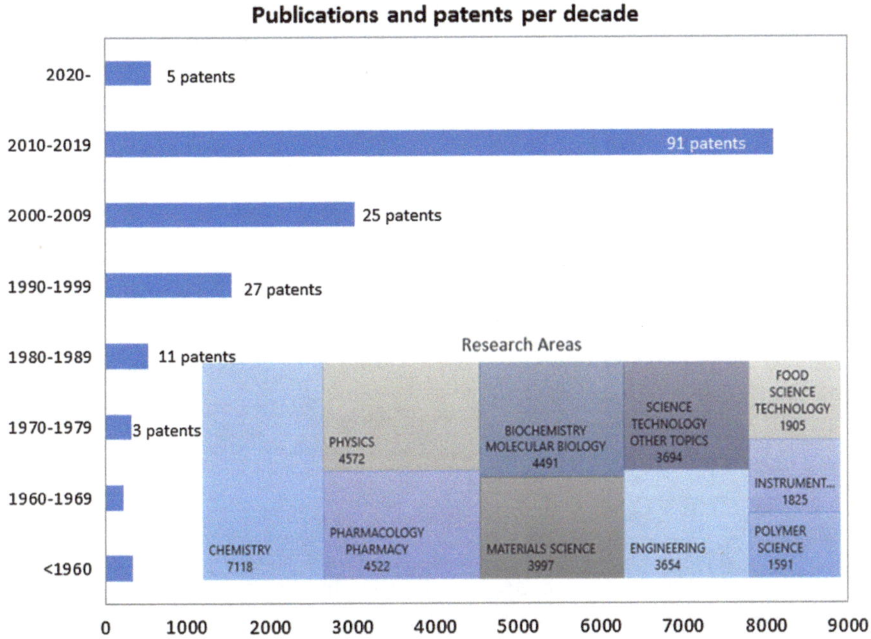

Fig. 7.2 Publications per decade found in Web of Science® [v.5.35] using search terms (TS = "ouble AND emulsion*" OR TS = "multiple AND emulsion*") as of August 4th, 2020. Refined by document type = patents. Inset: the fraction of the total number of publications per topic

To get an overview of the current status in the area of double and multiple emulsions, a basic bibliometric analysis of the publication frequency in this research area was carried out (Fig. 7.2). Despite the first rise of multiple emulsions in 1925 (Seifriz 1925), the majority of the works in this field had been carried out in the last 20 years. The total number of publications in the area currently exceeds 14,000 and includes a large number of patents (162), the majority of which have been filed since 2010. Many publications are in the area of fundamental research (chemistry and physics), however looking into applied research, publications on double or multiple emulsions have been mainly in the area of pharmacology where the delivery of actives and therapeutics can be achieved or improved by formulation into the inner phase(s) or makes use of double emulsions in diagnostics.

However, the use of double emulsions has been restricted by the fact that they are unstable thermodynamic systems due to an excess of free energy associated with the emulsion droplets surface (Garti 1997a; Aserin 2008). To formulate a $W_1/O/W_2$ double emulsion, at least two stabilizers are introduced into the system: a lipophilic one, to form the primary W_1/O emulsion, and another hydrophobic stabilizer, as a secondary emulsifier to form the final multiple emulsion.

The main objective of this chapter is to describe the most relevant works carried out in recent years dealing with double emulsions with special attention on the final applications in encapsulating compounds with antioxidant properties.

7.2 Multiple Emulsions Formulation

Numerous studies have been conducted to investigate the feasibility of encapsulation of different hydrophobic and hydrophilic bioactive compounds and nutrients using double emulsions in the form of $O_1/W/O_2$ (Benichou et al. 2007) or $W_1/O/W_2$ (Lamprecht et al. 2000; Wroński et al. 2012; Giroux et al. 2013; Tamnak et al. 2016; Ilyasoglu Buyukkestelli and El 2019).

7.2.1 *$W_1/O/W_2$ Double Emulsions*

$W_1/O/W_2$ double emulsions consist of small water droplets trapped within larger oil droplets that are themselves dispersed in a continuous water phase. The structural properties of this kind of multiple emulsions permit the controlled release of a component from the inner to the outer phase. Table 7.1 summarize some case studies of $W_1/O/W_2$ double emulsions in which the encapsulated compounds, stabilizer (lipophilic and hydrophilic) used and the weigth/volume percetnatge of each phase is reported.

Encapsulated substance

A wide range of hydrophilic compounds are encapsulated within W_1 phase of $W_1/O/W_2$ emulsions. These compounds include: iron (Fe^{+3}) (Choi et al. 2009; Ilyasoglu Buyukkestelli and El 2019), epigallocatechin-3-gallate (EGCG) (Evageliou et al. 2019), quercetin (Chouaibi et al. 2019), oleuropein (Gharehbeglou et al. 2019), vitamin C (Kheynoor et al. 2018), *Zygosaccharomyces rouxii* (Devanthi et al. 2018), lamivudine (Jena et al. 2018), magnesium chloride (Bonnet et al. 2009, 2010; Zhu et al. 2018), insulin (Mutaliyeva et al. 2017), crocin (Mehrnia et al. 2017), beetroot juice (Eisinaite et al. 2016, 2017), garlic extract (Ilić et al. 2017), azocasein hydrolysates (Giroux et al. 2016), hydroxytyrosol (Flaiz et al. 2016), apigenin (Kim et al. 2016), tartrazine dye (Tamnak et al. 2016), carmine (Matos et al. 2013b; Marefati et al. 2015), trans-resveratrol (Matos et al. 2014, 2015, 2018), vitamin B_{12} (O'Regan and Mulvihill 2010; Giroux et al. 2013; Matos et al. 2015), saffron extract (picrocrocin, saffranal and crocin) (Esfanjani et al. 2015), procyanidin (Berendsen et al. 2015a, b), fluorescein isothiocyanate-conjugated bovine serum albumin (FITC-BSA) (Jaimes-Lizcano et al. 2013), HP-b-CyD-ibuprofen (Hattrem et al. 2015), sodium chloride (Lutz et al. 2009; Sapei et al. 2012), xylitol (Santos et al. 2014), asprin (Tang et al. 2013), copper sulfate (Dragosavac et al. 2012), cisplatin (Ashjari et al. 2012), cadmium (Palencia and Rivas 2011), copper (Rivas and Palencia 2011),

Table 7.1 An overview of the formulations of $W_1/O/W_2$ double emulsions used in the most recent studies for encapsulation

W_1 phase		*O phase*		*W_2 phase*		
Encapsulated	W_1/ emulsion	Stabilizer	O/ emulsion	Stabilizer	W_2/ emulsion	References
EGCG	8 wt%	PGPR	32 wt%	Bacterial cellulose and WPI	60 wt%	Evageliou et al. (2019)
Lamivudine	25% v/v	Span 80	25% v/v	Tween 80	50% v/v	Jena et al. (2018)
Vitamin C	4 wt%	PGPR	16 wt%	Tween 80	80 wt%	Kheynoor et al. (2018)
Crocin	1 wt%	PGPR	9 wt%	WPC, GA and Angum Gum	90 wt%	Mehrnia et al. (2017)
Freeze-dried beetroot juice	8 wt%	PGPR	32 wt%	WPI	60 wt%	Eisinaite et al. (2017)
Hydroxytyrosol	8 wt%	PGPR	32 wt%	NaCAS	60 wt%	Flaiz et al. (2016)
Azocasein hydrolysates	7 wt%	PGPR	28 wt%	NaCAS	65 wt%	Giroux et al. (2016)
Vitaflavan	6 wt%	PGPR	14 wt%	WPI and Chitosan	80 wt%	Berendsen et al. (2015b)
Saffron extract	2.5 wt%	Span 80	22.5 wt%	Maltodextrin, WPC and citrus pectin	75 wt%	Esfanjani et al. (2015)
Resveratrol	4% v/v	PGPR	16% v/v	Tween 20 and CMC	80% v/v	Matos et al. (2014)
Riboflavin	8 wt%	PGPR	32 wt%	NaCAS	60 wt%	Cofrades et al. (2014)
Aspirin	10 wt%	Span 80	15 wt%	Cremophore EL®	75 wt%	Tang et al. (2013)
Vitamin B12	7 wt%	PGPR	28 wt%	Milk protein	65 wt%	Giroux et al. (2013)
$CuSO_4$	1.5% v/v	PGPR	3.5% v/v	Tween 20	95% v/v	Dragosavac et al. (2012)
Cisplatin	9% v/v	Span 60 and PLGA (poly- lactic-co-glycolic acid)	26% v/v	Tween 60 and poly vinyl alcohol	65% v/v	Ashjari et al. (2012)
Copper	2.4% v/v	Span 80	2.4% v/v	Tween 80	95.2% v/v	Rivas and Palencia (2011)
PTSA	4 wt%	PGPR	16 wt%	Tween 20	80 wt%	Mun et al. (2011)
KCl	15 wt%	Saturated monoglyceride and tripalmitin	35 wt%	NaCAS	50 wt%	Frasch-Melnik et al. (2010)

(continued)

Table 7.1 (continued)

W_1 phase		*O phase*		*W_2 phase*		
Encapsulated	W_1/ emulsion	Stabilizer	O/ emulsion	Stabilizer	W_2/ emulsion	References
NaCl	60% v/v	Span 80 and Tween 80	20% v/v	Tween 80 and XG	20% v/v	ElShafei et al. (2010)
Lactobacillus rhamnosus	9% v/v	PGPR	21% v/v	Sweet whey	70% v/v	Pimentel-González et al. (2009)
Polymeric dye R478	3% v/v	PGPR	12% v/v	WPI	85% v/v	Scherze et al. (2005)
PTSA	15% v/v	Hexaglyceryl condensed ricinoleate	35% v/v	Decaglycerol monolaurate	50% v/v	Shima et al. (2004)
$MgCl_2$	4 wt%	PGPR	6 wt%	NaCAS	90 wt%	Bonnet et al. (2009)
Vitamin B12	8 wt%	PGPR	32 wt%	NaCAS - Maltodextrin	60 wt%	O'Regan and Mulvihill (2010)
FITC-BSA	33.3% v/v	Abil® EM90 and Abil® Care 85	33.3% v/v	Tween 80 and Carbomer	33.3% v/v	Jaimes-Lizcano et al. (2013)

water soluble fluorescent dye p-toluenesulfonic acid (PTSA), potassium chloride (Frasch-Melnik et al. 2010), sodium ascorbate (Lutz et al. 2009), riboflavin (Bou et al. 2014), lysozyme (Meng et al. 2003) and even probiotic bacteria such as *Lactobacillus rhamnosus* and *Lactobacillus acidophilus* (Shima et al. 2006).

Internal aqueous phase

Studies revealed that several types of internal phases have been used as the internal aqueous phase, depending on the final application of the double emulsion and the encapsulated compound. These include using purified water (Hernández-Marín et al. 2013; Neumann et al. 2017), ethanol (Kim et al. 2016), mixture of water and alcohol (20%) (Matos et al. 2014, 2015; Chouaibi et al. 2019), sodium chloride solution (0.1–8%) (Wen and Papadopoulos 2001; Serdaroğlu et al. 2016), water and propyl paraben (0.02 wt%) (Gharehbeglou et al. 2019), potassium chloride solution (0.1–1 M) (Frasch-Melnik et al. 2010; Balcaen et al. 2016; Nelis et al. 2019), phosphate buffer (1–20 mM) (Tabatabaee Amid and Mirhosseini 2014; Marefati et al. 2015; Giroux et al. 2016), glycerine solution (5% w/w) (Tamnak et al. 2016; Nabavi et al. 2017; Zhao-Miao et al. 2018), hydrated gelatin (O' Regan and Mulvihill 2009; O'Regan and Mulvihill 2010), magnesium chloride (Zhu et al. 2018), citrate buffer (Panagopoulou et al. 2017), acetic acid (0.01 M) (Berendsen et al. 2015a, b), 4-(2-hydroxyethyl) piperazine-1-ethanesulfonic acid (HEPES) buffer (16.66 mM) (Jaimes-Lizcano et al. 2013).

To increase the stability, sometimes thickening agents such as *k*-carrageenan (Poyato et al. 2013), gelatin (5% w/w) (Scherze et al. 2005; Sapei et al. 2012; Schuster et al. 2012), gellan gum (Carrillo-Navas et al. 2012) have been used. In order to avoid bacterial growth, 0.01–0.05 wt% sodium azide has been added (Berendsen et al. 2015a, b; Flaiz et al. 2016).

The volume fraction of water phase in primary emulsion has been reported to be up to 30% (Kim et al. 2016). However, the volume fraction of 1–20% are more frequently used (Panagopoulou et al. 2017; Kheynoor et al. 2018; Evageliou et al. 2019).

Inner emulsifier

The most commonly used lipophilic stabilizer was polyglycerol polyricinoleate (PGPR) which has been predominantly applied to stabilize primary W_1/O emulsions (Cofrades et al. 2014; Díaz-Ruiz et al. 2020). The concentration of PGPR varied between 0.8–30 wt% (in most cases under 10 wt%). Span 80 has also been used on several occasions at a concentration of 1–31 wt% (Zou et al. 2013; Carrillo et al. 2015; Jena et al. 2018). Other stabilizers that were used to stabilize primary emulsions of W_1/O type include Dow Corning 749 Fluid (Nabavi et al. 2017), Abil® EM90 or Abil care 85 (Jaimes-Lizcano et al. 2013) and Span 60 (Ashjari et al. 2012).

Oil phase

Different oil types have been used as the oil phase of double emulsions. For example, medium-chain triglyceride (Lutz et al. 2009; Matos et al. 2014, 2015) is a very commonly used one. Vegetable oils such as olive (Evageliou et al. 2019; Ilyasoglu Buyukkestelli and El 2019), sunflower, partially hydrogenated sunflower oil (PHSO)

(Eisinaite et al. 2017), canola (Schuch et al. 2014a; Tamnak et al. 2016), rapseed oil (Neumann et al. 2017) and soybean oil (Iqbal et al. 2013; Tabatabaee Amid and Mirhosseini 2014; Carrillo et al. 2015) have frequently been used in the formulation of double emulsions. Other types of vegetable oils have been used but less frequently than the previously mentioned ones. For instance, red pepper seed oil (Chouaibi et al. 2019), butter oil (Giroux et al. 2013), soft palm mid faction oil (Nelis et al. 2019), sunflower and pumpkin seed oil mixture (Ilić et al. 2017), mixture of butter, linseed and mineral oil (Giroux et al. 2016), corn oil (Choi et al. 2009; Santos et al. 2014; Hattrem et al. 2015), castor oil (Tang et al. 2013), perilla oil (Flaiz et al. 2016), orange oil (Kim et al. 2016), solid shea nut oil (Marefati et al. 2015), Chia essential oil (Carrillo-Navas et al. 2012; Cofrades et al. 2014), eucalyptus oil (ElShafei et al. 2010), mixture of marjoram oil and linalool oil (ElShafei et al. 2010) and olein (Bonnet et al., 2009).

In some studies, mineral-based or synthetic substances have been employed as the oil phase such as silicone oils (Jaimes-Lizcano et al. 2013), mixture of silicone and polydimethyl siloxane (Zhao-Miao et al. 2018), styrene (Jaimes-Lizcano et al. 2013; Zou et al. 2013; Lei et al. 2016), styrene mixed with other lipophilic compounds e.g. lignin (Pan et al. 2015), dichloromethane (Ashjari et al. 2012; Zafar et al. 2016), paraffin oil (Gaitzsch et al. 2011), decane (Palencia and Rivas 2011; Rivas and Palencia 2011), hexane (Virtudazo et al. 2011), toluene and tetrahydrofurane (Foster et al. 2010), octanoic acid triacylglycerol (Shima et al. 2004, 2006), ethyl acetate (Meng et al. 2003) and methylene chloride (Lamprecht et al. 2000).

The volume of oil phase in relation to other phases in formulation of double emulsion ranges from 2.4% to 60% of the total double emulsion.

Outer emulsifier

The more frequently hydrophilic emulsifier used to stabilize $W_1/O/W_2$ emulsions were milk-based proteins in the form of sodium caseinate (NaCAS) (Serdaroğlu et al. 2016; Nelis et al. 2019) and whey protein isolate (WPI) (Berendsen et al. 2015a; Ilyasoglu Buyukkestelli and El 2019) and whey protein concentrate (WPC) (Esfanjani et al. 2015).

In addition to milk proteins, non-ionic surfactants such as Tween 20 and Tween 80 (Pawlik et al. 2010; Matos et al. 2014, 2015) have been frequently used as outer/hydrophilic emulsifier while ionic surfactants like sodium dodecyl sulfate (Pays et al. 2002) has been used rarely.

Thickening agents such as sodium carboxymethylcellulose (CMC) (Murillo-Martínez et al. 2011; Matos et al. 2014, 2015), gum arabic (GA) (Su et al. 2008; Leal-Calderon et al. 2012; Mehrnia et al. 2017), xanthan gum (XG) (ElShafei et al. 2010; Nelis et al. 2019) and pectin (Gharehbeglou et al. 2019) have been used in combination with other emulsifiers to improve the stability of double emulsions. Moreover, natural polysaccharides such as chitosan (Berendsen et al. 2015a, b) have also been used together with other emulsifiers and stabilizers to enhance emulsion stability.

Particles such as modified quinoa starch granules had been used to stabilize O/W_2 interface (Mun et al. 2011; Matos et al. 2013b; Marefati et al. 2015). Moreover, other types of particles such as cellulose nanofibrils (Carrillo et al. 2015), modified

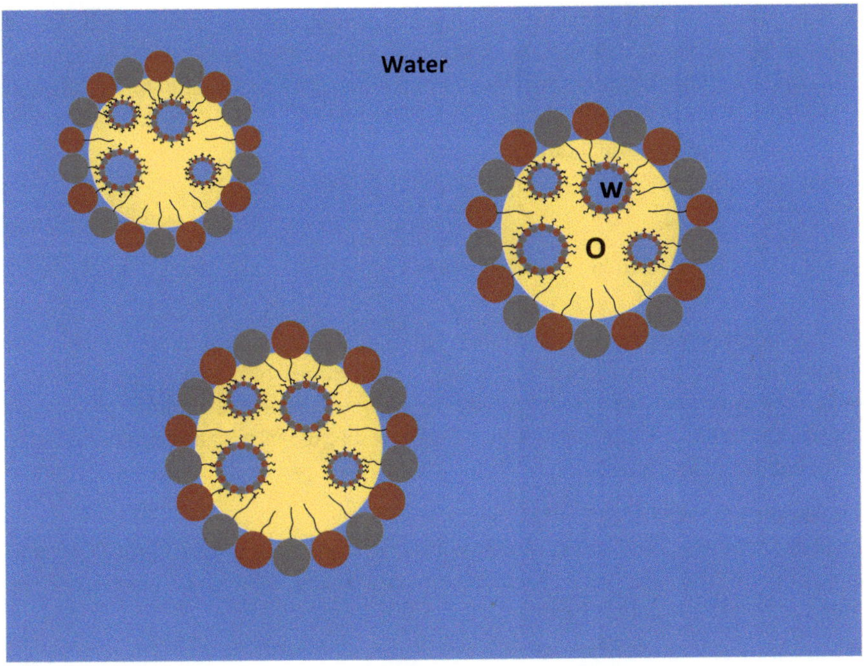

Fig. 7.3 Aqueous droplets in oil drops of $W_1/O/W_2$ double emulsion, both interfaces stabilize by hydrophobic and hydrophilic surfactant molecules

lignin (Pan et al. 2015) and iron oxide nano-particles (Zou et al. 2013) were used to fabricate hollow microspheres based on template Pickering double emulsions.

Figure 7.3 presents the orientation of the surfactant in the water and oil droplets of the $W_1/O/W_2$ double emulsion.

External water phase

As external aqueous phase purified water (Kheynoor et al. 2018; Evageliou et al. 2019; Gharehbeglou et al. 2019) has been frequently used in of $W_1/O/W_2$ double emulsions formulation.

In some cases, lactose (Bonnet et al. 2009; Chouaibi et al. 2019) or glucose (Shima et al. 2004; Frasch-Melnik et al. 2010; Pawlik et al. 2010) were added to the external aqueous phase in order to regulate osmotic pressure. Similar to internal aqueous phase, the use of buffers as the external aqueous phase is very common as well, such as phosphate (Matos et al. 2013b; Esfanjani et al. 2015; Marefati et al. 2015), citrate (Panagopoulou et al. 2017) sodium chloride (Flaiz et al. 2016; Serdaroğlu et al. 2016; Neumann et al. 2017) and potassium chloride (Nelis et al. 2019). However, it is necessary to balance ionic strength between both aqueous phases in order to avoid undesired swelling or deswelling of the inner aqueous drops (Iqbal et al. 2013; Schuch et al. 2014a; Khadem et al. 2020). In some cases, up to 0.05 wt% sodium azide is added to prevent bacterial growth (Mehrnia et al. 2017; Panagopoulou et al. 2017; Chouaibi et al. 2019).

The pH of the external aqueous phase is occasionally adjusted to the desired value using HCl or NaOH (Shima et al. 2004; Esfanjani et al. 2015; Mehrnia et al. 2017). The volume of the external water phase has been reported to be in the range of 40–90% (Pays et al. 2002; Bonnet et al. 2009; Schuch et al. 2014a) of total weight or volume of double emulsions.

7.2.2 $O_1/W/O_2$ Double Emulsions

$O_1/W/O_2$ double emulsions consist of small oil droplets trapped within larger water droplets that are themselves dispersed in a continuous oil phase. Like emulsions $W_1/O/W_2$, it is possible to encapsulate biocompounds in the internal phases and control the release. Table 7.2 summarize some studies, in which this type of emulsions had been developed.

Encapsulated substance

As stated before $O_1/W/O_2$ emulsions are less commonly used than $W_1/O/W_2$ double emulsions. Hence, just a few hydrophobic compounds, such as Flumethrin® are found to be encapsulated (Benichou et al. 2007).

Inner oil phase

As internal oil phase, vegetable oils (Bernewitz et al. 2014; Estrada-Fernández et al. 2018), medium-chain triglyceride oil (Benichou et al. 2007; Zhu et al. 2016) and mineral oil (Pal 2007) have been used as the inter oil phase of $O_1/W/O_2$ double emulsions. The volume of the internal oil phase was reported to be anywhere between 6% and 37%.

Table 7.2 An overview of the formulations of $O_1/W/O_2$ double emulsions used in the most recent studies for encapsulation

O_1 phase		*W phase*		*O_2 phase*		
Encapsulated	O_1/ emulsion	Stabilizer	W/ emulsion	Stabilizer	O_2/ emulsion	References
–	9 wt%	WPC-GA	21 wt%	WPC-GA solid particles	70 wt%	Estrada-Fernández et al. (2018)
–	25% v/v	Silane modified Fe_3O_4 nanoparticles and Pluronic F68	8% v/v	PGPR	67% v/v	Zhu et al. (2016)
–	25 wt%	Tween 20, dry glucose syrup and pectin	25 wt%	PGPR	50 wt%	Bernewitz et al. (2014)
Flumethrin®	6 wt%	Fenugreek gum and XG	14 wt%	3225C	80 wt%	Benichou et al. (2007)
–	6.44–37.12% v/v	Triton X-100	16.56–20.88% v/v	Emsorb 2500	77–42% v/v	Pal (2007)

Inner emulsifier
As a hydrophilic emulsifier to stabilize the O_1/W interface, whey protein (Benichou et al. 2007; Estrada-Fernández et al. 2018), iron oxide (Fe_3O_4) nanoparticles (Zhu et al. 2016), Tween 20 and polyethylene glycol p-(1,1,3,3-tetramethylbutyl)-phenyl ether (also known as Triton X-100) have been used (Pal 2007). In addition, thickeners such as GA (Estrada-Fernández et al. 2018) or pectin (Bernewitz et al. 2014) have been added to the aqueous phase to optimize the stability of the emulsions.

Figure 7.4 presents the orientation of the surfactant in the droplets trapped in the water droplets.

Aqueous phase
Purified water has been used as the aqueous phase in the volume of 8–25% as the intermediate aqueous phase of double O_1/W/O_2 emulsions.

Outer emulsifier
Several emulsifiers were used as to stabilize the outer W/O_2 interface such as insoluble WPC-GA complex (Estrada-Fernández et al. 2018), PGPR (Bernewitz et al. 2014; Zhu et al. 2016), silicone surfactant 3225C (Benichou et al. 2007) and sorbitan ooleate Emsorb 2500 (Pal 2007).

External oil phase
As external oil phase vegetable oils (Bernewitz et al. 2014; Estrada-Fernández et al. 2018) and medium-chain triglyceride oil (Benichou et al. 2007; Zhu et al. 2016) were used as the external oil phase of O_1/W/O_2 double emulsions. The volume of the external oil phase was in the range of 42–80%.

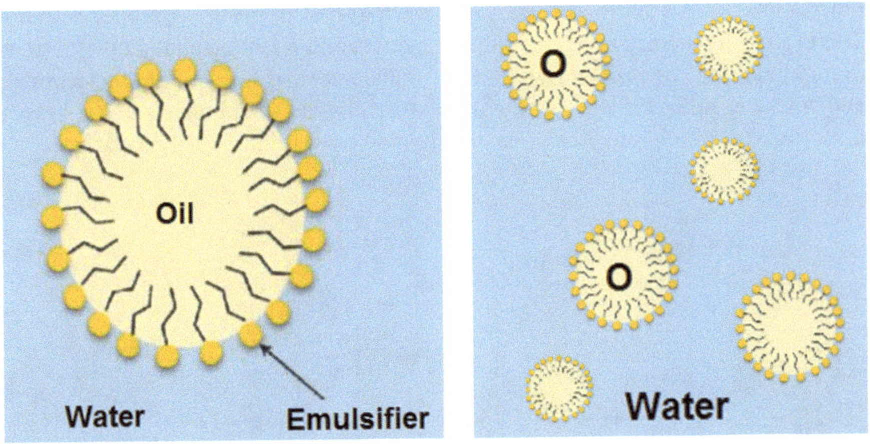

Fig. 7.4 Schematic diagram of oil droplets with hydrophobic emulsifier dispersed in an aqueous phase

7.3 Preparation Methods of Double Emulsions

Double emulsions are usually formed by a two-stage emulsification process using a variety of techniques. Two emulsifiers are needed, one to stabilize the primary emulsions and the other to stabilize the secondary emulsions. Inner droplet sizes need to be considerably smaller than the external ones. Schematic preparation of $W_1/o/W_2$ double emulsions by two steps process, the most common case, is described in Fig. 7.5.

The internal emulsion is commonly prepared by mechanical agitation methods such as magnetic stirrer, helix mixer, rotor-stator high-speed mixer (HSM) and high-pressure homogenizer (HPH) followed by, in some cases, a sonication step which allows reduction of the final droplet size and separate possible flocs.

The final double emulsion is produced by several techniques depending on their final application. The most commonly used are the mechanical agitation methods: magnetic stirrers, Couette cells, rotor-stator HSM, HPH and colloids mill (CM). The secondary emulsification step is carried out with lower shear to avoid rupture of the interface between the innermost and outermost phase. However, this could often result in polydisperse drops. Therefore, other techniques that allow more control over the final droplet size are used, such as sonication, membrane emulsification (ME) and microchannel emulsification (MCE) (Okochi and Nakano 1997; Rayner and Trägårdh 2002; Sugiura et al. 2002; Rayner et al. 2004). Table 7.3 summarizes the techniques used in the most recent research works for preparation of $W_1/O/W_2$ double emulsions.

Emulsions must fulfill certain conditions of stability to keep their functional properties. Droplet size and its distribution may determine the final properties of an emulsion. In some applications, such as drug delivery systems, good control of droplet size, and a narrow distribution are required to ensure a proper activity. Therefore, it is important to study the properties and phenomena that may lead to destabilization. The problem with most emulsions used in food, pharmaceutical, and cosmetic industries, is that the droplets of the dispersed phase do not have a uniform size, which may lead to instability and loss of functional properties.

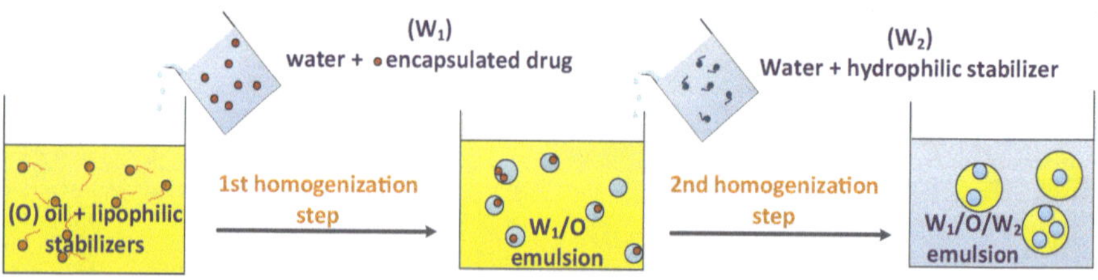

Fig. 7.5 Two steps $W_1/O/W_2$ double emulsions preparation

Table 7.3 An overview of the techniques used in preparation of $W_1/O/W_2$ double emulsions employed in the most recent studies

Type	Preparation method: W_1/O	Preparation method: $W_1/O/W_2$	References
W/O/W	Rotor-stator HSM	Rotor-stator HSM	Pimentel-González et al. (2009), Hattrem et al. (2015) Ilyasoglu Buyukkestelli and El (2019)
O/W/O	Rotor-stator HSM	HPH	Benichou et al. (2007))
W/O/W	Magnetic stirrer	Rotor-stator HSM	Lamprecht et al. (2000), Esfanjani et al. (2015)
W/O/W	HPH	Rotor-stator HSM	Tabatabaee Amid and Mirhosseini (2014)
W/O/W	Rotor-stator HSM and Sonication	Rotor-stator HSM and Sonication	Gharehbeglou et al. (2019)
W/O/W	Rotor-stator HSM	Rotor-stator HSM and microfluidics device	Balcaen et al. (2016), Nelis et al. (2019)
W/O/W	Magnetic stirrer	Rotor-stator HSM and HPH	Mehrnia et al. (2017)
W/O/W	Rotor-stator HSM	ME	Berendsen et al. (2015a, b), Eisinaite et al. (2017)
W/O/W	Rotor-stator HSM	HPH	O'Regan and Mulvihill (2010), Schuch et al. (2014b), Tamnak et al. (2016)
W/O/W	Rotor-stator HSM and HPH	Rotor-stator HSM and HPH	Giroux et al. (2013)
W/O/W	Sonication	Rotor-stator HSM and further vacuum evaporation	Zafar et al. (2016)
W/O/W	Rotor-stator HSM	ME	Scherze et al. (2005), Dragosavac et al. (2012), Matos et al. (2015), Ilić et al. (2017)
W/O/W	HPH	HPH	Su et al. (2008), Cofrades et al. (2013, 2014), Bou et al. (2014), Flaiz et al. (2016)
W/O/W	Sonication	Sonication	Mun et al. (2011), Ashjari et al. (2012), Tang et al. (2013)
W/O/W	Rotor-stator HSM	CM	Schuster et al. (2012)
W/O/W	Stirrer	CM	Schuch et al. (2014a)
W/O/W	CM	CM	Schuch et al. (2013)
W/O/W	Helix mixer and couette cell	Couette cell	Chouaibi et al. (2019)
W/O/W	Couette cell	Couette cell	Bonnet et al. (2009)
W/O/W	Sonication	Magnetic stirrer	Mutaliyeva et al. (2017)
W/O/W	Rotor-stator HSM	Magnetic stirrer	ElShafei et al. (2010), Jaimes-Lizcano et al. (2013)
W/O/W	Magnetic stirrer and sonication	Magnetic stirrer and sonication	Jena et al. (2018)

(continued)

Table 7.3 (continued)

Type	Preparation method: W_1/O	Preparation method: $W_1/O/W_2$	References
O/W/O	CM	Magnetic stirrer	Bernewitz et al. (2014)
W/O/W	Handshaking	Handshaking	Carrillo et al. (2015)
W/O/W	Sonication	Handshaking	Pan et al. (2015)
W/O/W	Magnetic stirrer	Handshaking	Zou et al. (2013)
W/O/W	Dropwise addition	ME	Palencia and Rivas (2011), Rivas and Palencia (2011)
W/O/W	Microfluidic device	ME	Choi et al. (2009)
W/O/W	Microfluidic device	Microfluidic device	Foster et al. (2010), Zhao-Miao et al. (2018)
W/O/W	Catastrophic inversion process	Catastrophic inversion process	Zhang et al. (2014)
W/O/W	Dropwise addition	Dropwise addition	Leal-Calderon et al. (2012)

ME and MCE require less shear than conventional emulsification processes and a high monodispersity and encapsulation efficiency can be achieved (van der Graaf et al. 2005). However, the main drawback of these techniques is the low throughput as it takes a long time to obtain concentrated emulsions due to the low dispersed phase flux. Thus, improvements in micro-engineered membrane techniques for premix emulsification are being studied (Sahin et al. 2014).

The most common techniques used for both internal and external emulsion preparation are described in following subsections.

7.3.1 Rotor-Stator High-Speed Mixer (HSM)

Rotor-stator HSM is the most commonly used technique; it is easy to use and allows well-controlling the operating parameters such as volume of sample, time, and rotational velocity.

This technique apply shear rate at the sample, which leads to rupturing the interface and producing drops. The conditions of the second emulsification step, to prepare double emulsions, should be softer than the first emulsification step to produce the inner emulsion. Therefore, in order to avoid that the inner drops reach the external phase, normally less velocity is applied in the second step than in the first step (Shima et al. 2004; Matos et al. 2015).

The main disadvantage of this technique is the wide size distribution obtained. Depending on the formulation and the operating conditions, the final droplet size can be varied from 0.5 μm (Shima et al. 2004) to 60 μm (Tabatabaee Amid and Mirhosseini 2014).

7.3.2 High-Pressure Homogenizer (HPH)

Before emulsifying with HPH (Fig. 7.6), a coarse emulsion needs to prepare via another method (at least gentle agitation). The coarse emulsion is fed into the instrument, and then it passed through a narrow conduct at high pressure which will reduce the droplets sizes of the emulsions significantly.

Different pressure values can be applied and samples can be passed through the system several times reducing the emulsion droplet size each time. Droplet size between 0.8 μm (O'Regan and Mulvihill 2010) and 12 μm (Giroux et al. 2013) have been registered. One of the limitations of this equipment is the large amount of sample that is required. Generally, 200 mL of sample is prepared (Flaiz et al. 2016), which can be a limitation for studies where expensive components are involved.

Similar to rotor-stator HSM, the preparing conditions for the double emulsions should be less vigorous than the ones for the inner emulsion. Since the disruption of the inner drops is undesired, less pressure is commonly used for the preparation of the double emulsions than for the inner one (Cofrades et al. 2013, 2014; Bou et al. 2014).

7.3.3 Colloid Mill (CM)

This technique is used to prepare fine emulsions from a coarse emulsion which passes through the narrow channel that exists between the rotor part and the static part of the system. The rotor moves at high velocity applying high centrifugal force on the disperse phase, which leads to disruption of the droplets and producing the fine emulsion.

The particle size obtained ranged from 10 μm (Schuster et al. 2012) to 150 μm (Schuch et al. 2013), depending on the selected formulation and the rotational velocity of the rotor. The necessary volume to produce fine emulsion is quite high for laboratory scale (around 300 mL) (Schuch et al. 2013, 2014a).

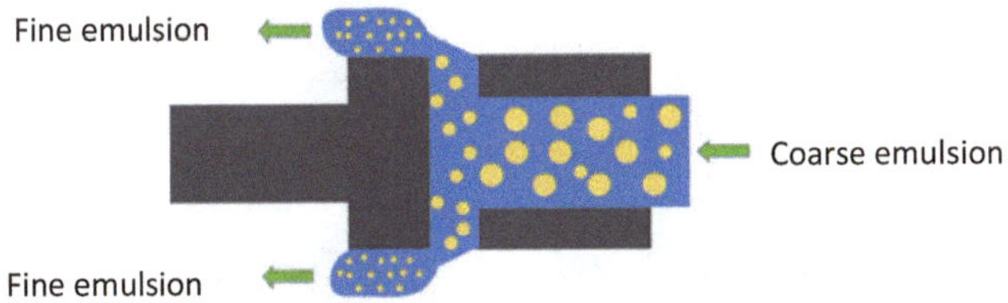

Fig. 7.6 Schematic diagram of emulsification process via a HPH device

7.3.4 Sonication

Sonication (Fig. 7.7) is frequently used to reduce the droplet size. The sonication probe is submerged into the samples and vibration is created, which produce droplet size reduction. Application of this technique produces a wider size distribution, which could be undesired for come applications such as drug delivery systems.

The time and selected frequency are the main operating conditions that need to be determined to obtain the desired final droplet size. As a general trend, droplet size is smaller than the one obtained by other techniques and can be varied between 0.2 and 3 μm (Zafar et al. 2016; Gharehbeglou et al. 2019; Nelis et al. 2019).

This technique normally is used for small volumes of samples, which are frequently no more than 30 mL (Ashjari et al. 2012; Pan et al. 2015; Mutaliyeva et al. 2017).

7.3.5 Catastrophic Inversion

Catastrophic inversion is a low-energy method that is less-frequently used for the preparation of double emulsions but is widely used for the preparation of nanoemulsions (Solans and García-Celma 1997; Antonio and Preziosi 2018) and allows obtaining double emulsions in a single step.

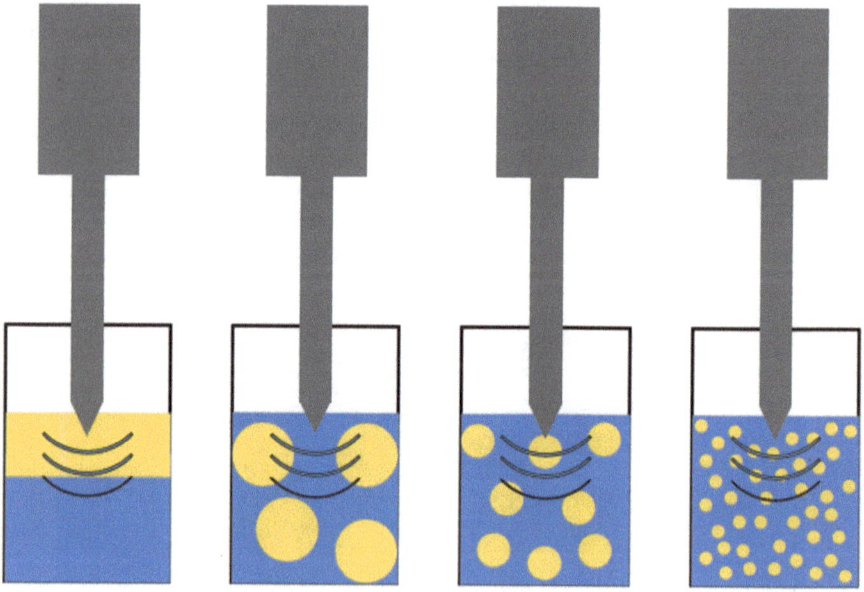

Fig. 7.7 Schematic diagram of sonication process in emulsion systems

The ratio of oil and water, salinity and the type of stabilizer used are the key factor on the type of emulsions produced (single or double emulsion). For example, in a study first a single O/W emulsion is produced and when the water portion increased a double W/O/W emulsion tended to appear spontaneously. The aqueous phase contained salt, which increased the salinity. On the other hand, the use of a surfactant with high dependency on the salinity as a stabilizer is crucial for double emulsions production through catastrophic method. High salinity concentration turns a high hydrophilic surfactant in a surfactant with medium hydrophilicity character and hence it is able to stabilize O/W and the W/O interfaces, producing the double emulsion via just one-step procedure (Fig. 7.8). External droplets of around 10 μm were obtained (Zhang et al. 2014).

7.3.6 Membrane Emulsification (ME)

ME was developed for the first time in Japan at the beginning of the 1990s and it has been widely studied for controlled size simple emulsions during the last two decades. In this process, the dispersed phase passes through the membrane pores under pressure, while the continuous phase moves over the membrane surface enhancing droplet detachment when the droplets reach a certain size. Droplet size is highly under control in ME and the droplet diameters are 2–10 times bigger than the membrane pore diameter. Moreover, in this method the strain on liquid phases is

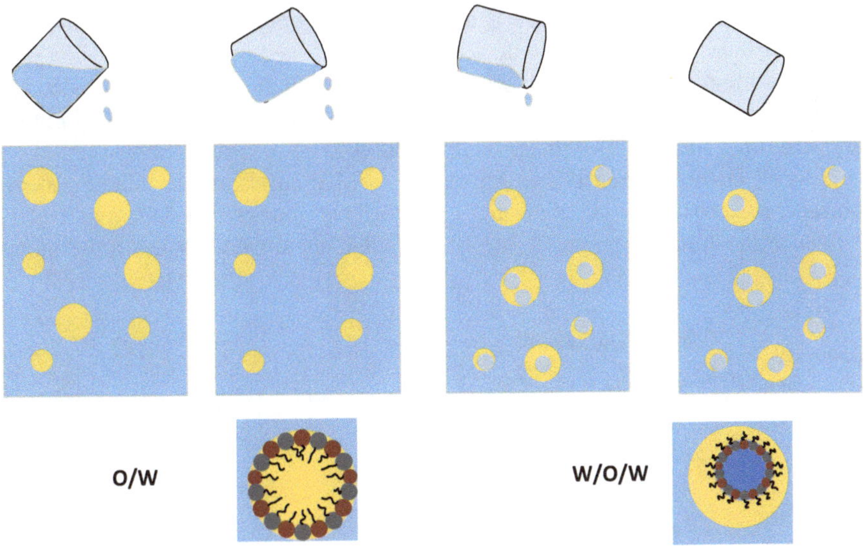

Fig. 7.8 Schematic diagram of double emulsion formation through catastrophic phase inversion technique

reduced, and droplet size can be narrowly distributed with less shear stress and energy consumption. The preparation of uniform and tailored droplets has enormous commercial significance in many industries for special applications (Charcosset 2009).

There are two types of ME (SUZUKI et al. 1996; Kawakatsu et al. 1997): (1) direct membrane emulsification (DME), when a dispersed phase passes through the membrane into the continuous phase; (2) premix membrane emulsification (PME), when a coarse emulsion is forced by pressure through the membrane to reduce the droplet size.

Although ME has been frequently used for simple emulsions preparation, it has also been widely tested in $W_1/O/W_2$ double emulsions production (Muschiolik 2007).

DME

DME is frequently used for the preparation of the external double emulsion. Therefore, a first step is required through which the primary (W_1/O) emulsion is prepared. The controlled size of the oil drops of the external emulsion is essential from the point of view of the final production, while it is not the case of the inner water emulsion (Matos et al. 2015).

Two types of membranes are used: (i) cross-flow ME with tubular membranes in which the W_1/O emulsion is injected from the outer part of the membrane, while the continuous phase (W_2) flows through the membrane lumen, and (ii) flat membranes where the W_1/O emulsion is injected from the bottom side of the membrane and the continuous phase is placed on the upper part. Figure 7.9 shows a schematic diagram of both types of dispositions.

In ME some problems may be happened e.g. droplet coalescence, droplets clogging over the membrane surface and difficulty in circulation of the product. To overcome these challenges, new devices have been tested on simple emulsion production such as rotating and vibrating membranes which enhance droplet detachment (Yuan et al. 2009; Pawlik and Norton 2012). However, still these techniques are scare used for multiple emulsions production.

Several factors influence the droplet size in ME: emulsion formulation, membrane characteristics, equipment and operating parameters. All these parameters influence the forces responsible for droplet attachment/detachment on the mem-

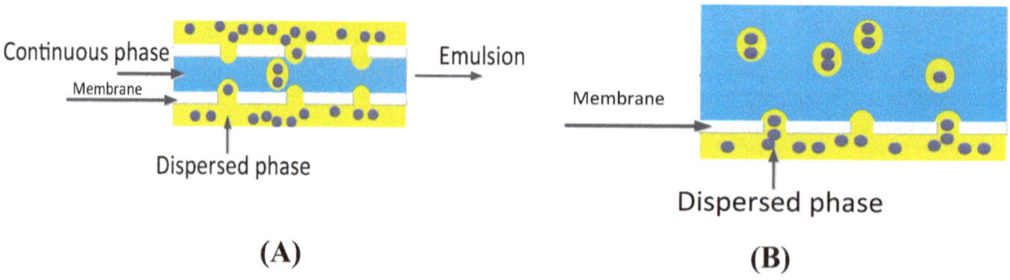

Fig. 7.9 Schematic diagram of direct ME with (**a**) cross-flow unit and (**b**) flat membranes

brane surface, resulting in different droplet sizes. Change the process conditions and the membrane pore size yield in emulsions with a certain mean droplet size. High entrapment efficiency and monodispersity can be achieved in the production of multiple emulsions. The particle sizes registered on double emulsions produced by ME can range from 40 to 60 µm (Matos et al. 2015; Ilić et al. 2017). Membranes of different materials have been tested such as metallic, ceramic or *Shirasu Porous Glass* (SPG) (Vladisavljevic et al. 2006; Matos et al. 2015; Ilić et al. 2017).

PME

PME has been described in several studies for simple and multiple emulsions production (Ramakrishnan et al. 2012; Berendsen et al. 2014) to encapsulate bio compounds, and its performance differs slightly from conventional direct ME.

A coarse double emulsion is previously prepared by conventional methods (e.g. rotor-stator HSM or CM) and then it is forced through the pores of a membrane, as shown in Fig. 7.10. This process can be repeated several times but after several cycles, the droplet size cannot be further decreased (Vladisavljevic et al. 2004; Eisinaite et al. 2016; Na et al. 2019; Kenji et al. 2020).

In this case, the membrane determines droplet size, but it is not as critical as in direct emulsification and the final droplet size does not depend so much on the operating parameters. The main disadvantage is that the first conventional emulsification stage can alter organic shear-sensitive molecules. However, high emulsion production rates can be obtained (Shima et al. 2004; Vladisavljevic et al. 2004; Sahin et al. 2014).

Ceramic and metallic membranes are the most commonly used materials in this method. In this context, membrane material and stabilizer are the key factors on the final droplet size and distribution, while the composition of the primary emulsion (W_1/O) has a scare effect (Gehrmann and Bunjes 2018; Na et al. 2019).

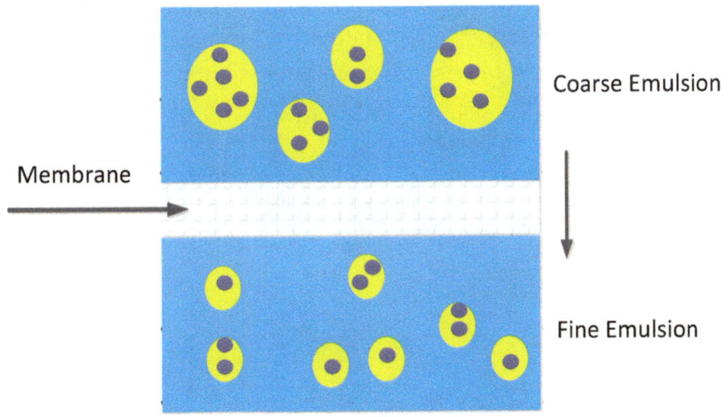

Fig. 7.10 Schematic diagram of PME

7.3.7 *Microchannel Emulsification (MCE)*

Microchannels have a wide variety of shapes and some of them (straight-through microchannels) have some similarities with membranes (Kobayashi et al. 2003; Sugiura et al. 2004). Two main types can be distinguished (i) T-shape microchannels and (ii) flow focusing systems (Fig. 7.11). In MCE, double emulsions can be produced in just one step yielding high monodispersity of the final droplet size distribution (Chong et al. 2015; Nabavi et al. 2017).

Using T-shape microchannels, (Fig. 7.11a), each involved phase (inner, middle, and external) is injected at a different point (T-junction) of the channel (Okushima et al. 2004).

Other authors (Utada et al. 2005) have described other type of microchannel based on the flow focusing. The inner and the medium phases are injected into the microchannel through two concentric tubes while the external phase is injected on the opposite direction (Fig. 7.11b). Flow focusing allows a good control on the emulsions production and also this technique has been widely used for the production of other types of colloids such as particles or vesicles.

Wetting the microchannel surface by the continuous phase is crucial in this method. To this purpose, usually two different materials as the continuous phase are used, a hydrophobic one in the section where the W_1/O emulsion is produced and another hydrophilic material where $W_1/O/W_2$ double emulsion is produced (Ding et al. 2019).

MCE technique do not only have a good control on the droplet size formation (internal and external) but allows good controlling how many droplets of the internal phase enter each external drop.

The only limitation of the MCE technique is the low rate production. In some studies in order to increase this rate, the parallelization of several individual microfluidic system was used (Mulligan and Rothstein 2012).However, the system was far to get large production of double emulsion with controlled size as can be necessary in food industry, but offered a good alternative for specific cases of pharmaceutical or medical applications, in which the required volume is low.

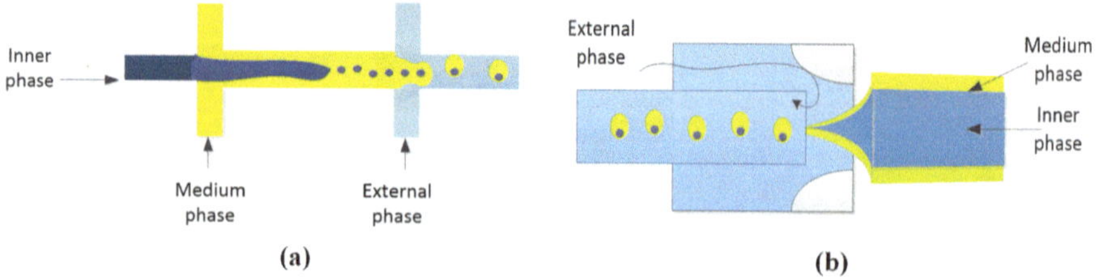

Fig. 7.11 Schematic diagram of MCE (**a**) T-shape and (**b**) flow focusing

7.4 Characterization of Double Emulsions

7.4.1 Stability and Droplet Size Distribution

To obtain a stable double emulsion, the stability of the primary emulsion must be ensured. This stability depends on droplet size (normally it must be lower than 1 μm), amount of the dispersed and continuous phases (in the case of $W_1/O/W_2$ double emulsions water is usually in the range 20–30% (v/v) both for W_1 and W_2 and the hydrophilic-lipophilic balance (HLB) value of the emulsifier for both phases (Reynolds et al. 2009; Campbell et al. 2012). The size of W_1 droplets ranged from 10 to 18 nm (Mehrnia et al. 2017; Gharehbeglou et al. 2019) to 2.6–250 μm (Dragosavac et al. 2012; Jaimes-Lizcano et al. 2013; Berendsen et al. 2015b; Pan et al. 2015). Stability of double emulsions is normally analyzed through checking the droplet size distribution to monitor if any destabilization phenomenon is taking place (e.g., coalescence, Ostwald ripening or even phase separation). For this purpose, droplet size distribution is monitored with time using laser light scattering techniques taking into account the refractive index (RI) of the dispersed phase, e.g., water in the case of primary W_1/O emulsions and oil for the $W_1/O/W_2$ double emulsions (Leister and Karbstein 2020).

Moreover, the stability can be monitored with time by visual inspection and more accurately by Turbiscan instrument which is based on static multiple light scattering (SMLS) (Márquez et al. 2010; Matos et al. 2014). Turbiscan instrument monitors transmitting (TS) and backscattering (BS) lights by providing TS and BS data at every 40 μm in percentage relative to standards (suspension of monodisperse spheres and silicone oil) as a function of the sample height (in mm). These profiles build up a macroscopic fingerprint of the emulsion at a given time, providing useful information about changes in droplet size distribution and the appearance of a creaming/clarification front with time (Matos et al. 2012, 2013a).

7.4.2 Morphological and Rheological Characterizations

Micrographs are commonly used (Chu et al. 2003; Matos et al. 2015) to confirm the formation of the double emulsion droplets by visual inspection and to compare them with the mean sizes obtained by laser light scattering.

Control the rheological properties of emulsions is critical since it determines the manufacturing condition (e.g., pumping, mixing), the final properties (e.g., the organoleptic and textural characteristics) and the applications of the product. For example, emulsions used in personal care application could vary in their consistency from fluid-like consistency (e.g., body lotions) to semi-solid consistency (e.g., hand creams), depending on their rheological properties (Tadros 2004; Lobato-Calleros et al. 2008; Gabriele et al. 2009; Gutiérrez et al. 2014). In addition, rheological measurements provide useful information on the physical stability of the double emulsions. Different type of measurements can be done: steady state shear stress vs. shear rate, constant controlled stress, dynamic oscillatory analysis and also performing these measurements as a function of temperature Accelerated storage testing is needed for prediction of the long-term physical stability of the formulations as well as the change of consistency with time (Tadros 2004).

7.4.3 Encapsulation Efficiency (EE) and Encapsulation Stability (ES)

The EE is related to the loss of encapsulated compound from the W_1 phase into the W_2 phase during the second emulsification step, which creates the final double emulsions. On the other hand, the ES is a measure of how much of the entrapped compound leaks from the internal W_1 phase into the outer continuous W_2 phase during storage time. It may be considered that a double emulsion has good stability if the initial EE is around 95% and after a few weeks of storage, it is still around 70–80% (O' Regan and Mulvihill 2009; Dickinson 2011).

Some authors stated that in order to properly calculate the EE or ES of a system, the recovery yield (R_y) (Eq. 7.2) of the encapsulated compound should be determined first (Marefati et al. 2015; Ilić et al. 2017; Matos et al. 2018). In this context, the encapsulation (or entrapment) efficiency is given by Eq. 7.1:

$$EE(\%) = 100 - \frac{C_{recovered} x100}{R_y C_0} \tag{7.1}$$

where C_0 is the expected concentration of the encapsulated compound based on the amount of marker added. The R_y may be defined as the concentration of marker found in the aqueous phase recovered from an emulsion that has been separated into a cream phase and an aqueous phase by centrifugation relative to the concentration of marker present in (or added to) the external aqueous phase after emulsion preparation. For this purpose, a standard emulsion is required, in which 100% of the W_1 is present in W_2. Therefore, an oil-in-water emulsion (O/W_2) is prepared using the same formulation as in the experiments. Next, this O/W_2 emulsion is diluted at the same ratio with W_1, which contained the appropriate amount of encapsulated compound. Then, the concentration of the compound in the recovered aqueous phase ($C_{recovered}$) is determined. For this analysis, a blank reference is also needed consisting of an O/W_2 emulsion diluted with W_1, in which the compound of interest was not present. Finally, the R_y is calculated by Eq. 7.2:

$$R_y(\%) = \frac{C_{recovered} \times 100}{C_0} \tag{7.2}$$

The ES can be defined as the level of the marker added or compound of interest encapsulated in the inner aqueous phase (W_1) which remains entrapped in it after storage time or after exposure of the double emulsion to environmental stresses (O' Regan and Mulvihill 2009). The ES is given by Eq. 7.3 as follows:

$$ES(\%) = 100 - \frac{C_t x100}{R_y C_0} \tag{7.3}$$

where C_t is the concentration of marker found in the external aqueous phase recovered by centrifugation after passing storage time (t).

However, a lot of authors estimate the EE or ES without considering Ry of the encapsulated compound. In those cases, the obtained values could be overestimated if part of the compound is not recovered after performing the separation process to recover W_2 from W_1/O due to some losses during the process or interactions with filter materials.

In order to determine the concentration of the encapsulated compounds, different analytical methods are used. A fast and easy way is to determine the electrical conductivity (ElShafei et al. 2010; Pawlik et al. 2010; Ilyasoglu Buyukkestelli and El 2019). Another commonly used method is direct determination of the concentration by Ultraviolet-Visible (UV-VIS) absorption (Mutaliyeva et al. 2017; Jena et al. 2018; Evageliou et al. 2019) or fluorescence spectroscopy (Shima et al. 2004; Mun et al. 2011), or by chromatographic techniques such as HPLC (high-performance liquid chromatography) separation (Pando et al. 2015; Chouaibi et al. 2019; Gharehbeglou et al. 2019) and gas chromatography equipped with a mass spectrometer (GC–MS) or with a flame ionization detector (Santos et al. 2014; Hattrem et al. 2015). Finally, in a lower extension, other techniques that have also been used are differential scanning calorimetry (DSC) (Schuch et al. 2013, 2014b), atomic absorption spectroscopy (AAS) (Palencia and Rivas 2011; Rivas and Palencia 2011; Dragosavac et al. 2012), Lowry-Peterson- assay for quantification of protein content (Lamprecht et al. 2000; Meng et al. 2003) or counting methods in case of encapsulation of bacteria (Shima et al. 2006).

7.4.4 *Antioxidant Activity*

Some authors have studied the antioxidant activity using the 1,1-diphenyl-2-picrylhydrazyl (DPPH) test according to the method of Blois (Blois 1958). Briefly in this method, a volume of the sample is dissolved in MeOH (concentrations ranging from 0.05 to 1.0 mg mL^{-1}). Then, DPPH (2.5 mL of 75 μmol L^{-1} in MeOH) is added. The solution is kept at room temperature for 30 min and the absorbance was measured at 517 nm. The DPPH scavenging effect (%) is defined according to Eq. 7.4 as follows:

$$DPPH_{scavenging\ effect}(\%) = 100 \cdot \frac{A_0 - (A - A_b)}{A_0} \tag{7.4}$$

where A_0 is the absorbance at 517 nm of DPPH without sample, A is the absorbance at 517 nm of sample and DPPH and A_b is the absorbance at 517 nm of sample without DPPH (Pinheiro et al. 2015). Based on the results of the antioxidant activity, the authors concluded that layer-by-layer deposition of chitosan/fucoidan on a sacrificial template could achieve biodegradable hollow nanocapsules with potential antioxidant and antimicrobial activity.

Other scientists (Leiva-Vega et al. 2020) assessed the influence of light on the antioxidant activity of emulsified curcumin. The samples were maintained at 25 °C for 6 days and were subjected to conventional consumer lighting for different periods of time. The curcumin was then extracted and its antioxidant potential was evaluated. Particularly, an aliquot was mixed with hexane and DPPH solution and was subsequently sonicated to promote emulsion leakage and the release of curcumin. The final solution was kept for 30 min at room temperature in the dark conditions to complete the reaction. The absorbance was measured at 515 nm using a spectrophotometer. These authors reported that free curcumin (dissolved in hexane) exhibited a considerable decrease ($p < 0.05$; ≈85%) in its antioxidant potential at day 6. Light is a recognized to help curcumin deterioration via auto-oxidative transformation into different chemical species, so that reducing its biological activity. The emulsified curcumin showed an increase in its antioxidant potential in primary, secondary and tertiary systems during exposing it to the light, reaching the antioxidant activity of free curcumin only after 6 days. Probably, curcumin was well encapsulated in the emulsion droplets, and not all of it could be extracted by hexane. Nevertheless, by passing time, due to particle instability, more curcumin became available and was extracted by hexane, so expressing its antioxidant activity. Therefore, the curcumin encapsulation and the floc formation during storage could have influenced the resulting antioxidant power (Leiva-Vega et al. 2020).

In another study, the effect of resveratrol-loaded emulsion on DPPH radical was evaluated through the same method as described by Pinheiro et al. (2015). The authors showed that storage only caused small variations in the antioxidant activity of the resveratrol-loaded emulsion (Acevedo-Fani et al. 2017).

Similarly, the antioxidant activity of β-carotene-encapsulated double emulsions was assayed against DPPH radical. In the study of this activity, after 2 years of storage, the antioxidant activity of the capsules made from PLGA was remained almost unchanged at 2.3 mg/ml (Gimenez-Rota et al. 2019).

7.5 Applications of Multiple Emulsions

Depending on the final application of double emulsions, their formulation and preparation methods are determined (Fig. 7.12). The main applications are presented in the following subsections.

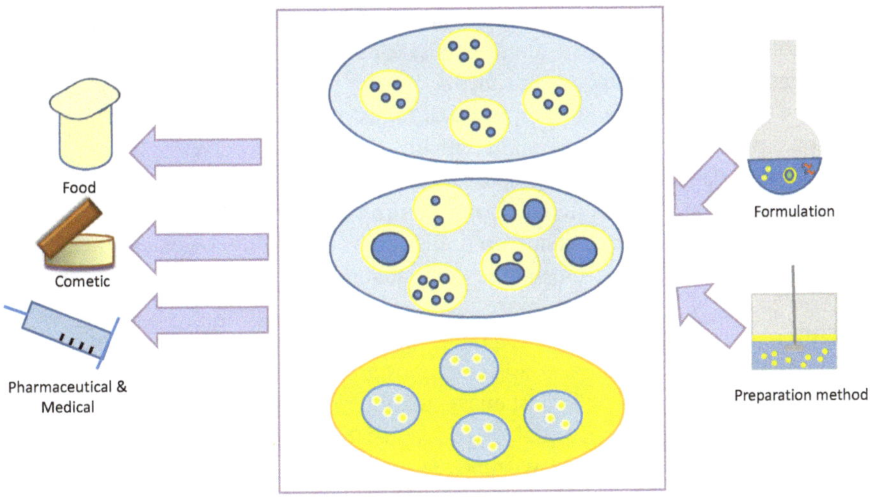

Fig. 7.12 Applications of double emulsions

7.5.1 Food Emulsions

Double emulsions find applications in the food industry. Studies have been performed to entrap flavor components in a release system. It was determined that sensitive food materials and flavors can be encapsulated successfully in W/O/W emulsions. Moreover, it was also stated that the organoleptic properties can be altered by the presence of active compounds and that the encapsulation of a functional ingredient in a W/O/W emulsion gives the product a different taste and delayed release of flavor in comparison to when it is formulated through O/W emulsions (Garti 1997a, b).

7.5.2 Delivery of Antioxidants

Bioactive compounds are defined as essential and nonessential compounds (e.g., vitamins or polyphenols) that exist in nature. They are part of the food chain and can affect the human health (Biesalski et al. 2009). Bioactive substances present in food provide health benefits beyond the basic nutritional value of the product. Until recently, vitamins and other micronutrients have been recommended to avoid deficiency symptoms. Nowadays, the most extensively studied compounds are antioxidants, which prevent the risk of chronic diseases including cancer and cardiovascular disorders (Biesalski et al. 2009).

Antioxidant compounds, such as carotenoids, polyphenols, phytosterols, omega-3 fatty acids, etc., protect the cells from the oxidative stress caused by the action of free radicals, and they are used as active ingredients in several food products. Carotenoids are compounds soluble in lipids (lipophilic antioxidants) and are responsible for the color of fruits and vegetables. Among the most important carotenoids for the organism, β-carotenes, α-carotenes, lycopene, cryptoxanthin, lutein and zeaxanthin can be mentioned.

For treatments that require repeated administration(via ingestion or injection) and for antioxidants with a very short half-life, the possibility of a single administration of a W/O/W emulsion followed by a slow and controlled release is an improvement compared to conventional forms of drug delivery. The W/O/W emulsion protects the bioactive antioxidant from alterations caused by the external environment (oxidation, light, enzymatic degradation) or during food digestion (Aserin 2008).

Different technologies have been developed to encapsulate antioxidants, including spray drying, cocertation, emulsions, liposomes, micelle, nanoparticles, freeze-drying, cocrystallization and yeast encapsulation (Fang and Bhandari 2010; Munin and Edwards-Levy 2011). Lu et al. 2016 specify that each of these technologies has its own specific strengths and weaknesses in terms of encapsulation, protection, delivery, cost, regulatory status, ease of use, biodegradability and biocompatibility. Among these methods, emulsions are widely considered as one of the most popular encapsulation and delivery systems for a wide range of lipophilic, hydrophilic and amphiphilic bioactive molecules (McClements and Li 2010), due to their high-efficiency encapsulation, maintenance of chemical stability of encapsulated molecules (Klinkesorn et al. 2005) and controlled release (Mao et al. 2013). Furthermore, some emulsion-encapsulated antioxidant presented even higher biological activities compared with pure free molecules (Wang et al. 2008). Several bioactive and functional antioxidants have been encapsulated using double emulsions both with hydrophilic (Fechner et al. 2007; Hemar et al. 2010; Matos et al. 2014) or lipophilic character (McClements et al. 2007; Berendsen et al. 2014, 2015b).

7.5.3 *Microspheres and Microcapsules*

Most of chemotherapy drugs are administrated as emulsions because they are water-soluble. W/O/W emulsion systems are suitable carriers because the drug can be encapsulated in the inner phase and the viscosity of the external water phase is low. The controlled droplet size of the emulsion allows a targeted delivery to a specific site in the body providing sustained release of the drug. Although there are some examples of drugs that can be dissolved directly in the dispersed phase of a simple emulsion or double emulsions (in the primary emulsion part) (Higashi et al. 1999), typically more complex carriers are required.

Carriers can be classified as microcapsules and microspheres, according to their structure, although in many cases both terms are used indistinctly. Microcapsules and microspheres are generally round and deliver the active compound to the site where their function is needed.

Microcapsules (Fig. 7.13) are carriers with a lipid core where the molecule is encapsulated and a solid polymeric shell is protecting this core. The shell can be produced through a simple emulsion polymerization or by interfacial polymerization

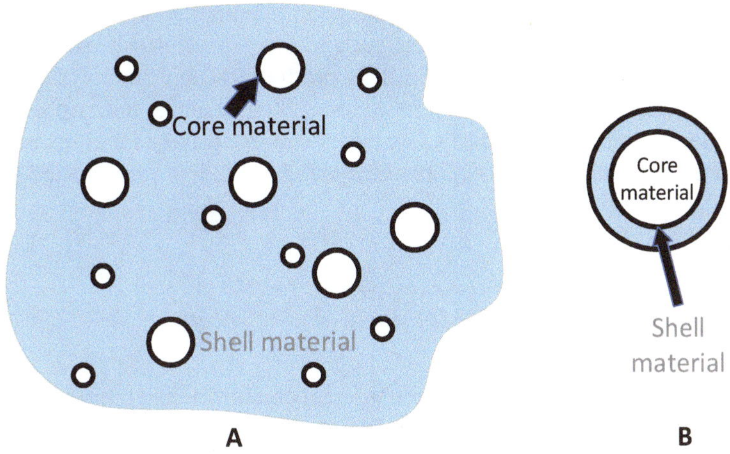

Fig. 7.13 Schematic diagrams of two types of microcapsules: (**a**) multinuclear capsules and (**b**) continuous core/shell microcapsule (Gharsallaoui 2009)

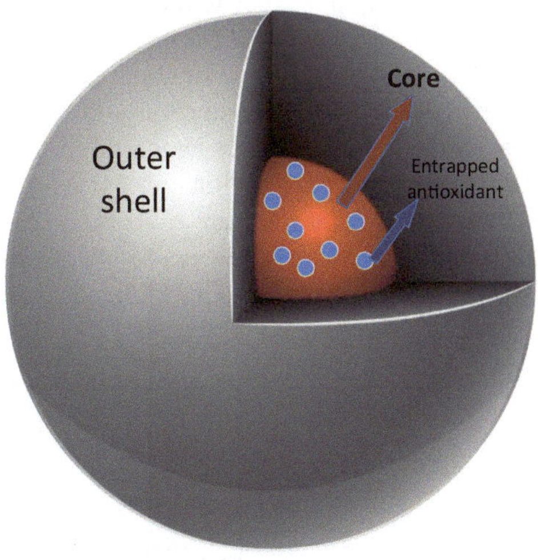

Fig. 7.14 Schematic diagram of a microsphere

of multiple emulsions. In the first case, the structure can be controlled by the composition of dispersed phase, so that phase dispersion takes place inside the droplet and either a microcapsule or a microsphere can be produced (Ma et al. 1999). Freeze-drying of simple (Quispe-Condori et al. 2011; Marefati et al. 2013) and double emulsions (Marefati et al. 2015) are other options for the shell production. Moreover, preparation of microspheres and microcapsules by removal of the hydrophobic polymer-solvent using evaporation has also been reported (Donnell and Mcginity 1997; Freitas et al. 2005).

Microspheres (Fig. 7.14) are micron-sized particles that entrap active compounds inside a homogeneous matrix. These matrices usually have a lipid nature to protect the drug or in some cases to facilitate skin absorption. In this context, microspheres are mostly based on natural products such as alginate, chitosan and albumin or formed by polymeric materials such as n-isopropyl acrylamide or poly(acrylamide-co-acrylic acid). For example, the properties of biodegradable microspheres based on poly(lactic acid) (PLA) and PLGA have been extensively investigated for the encapsulation of highly water-soluble compounds including proteins and peptides (Donnell and Mcginity 1997).

7.5.4 *Liquid Extraction Membranes*

Double emulsion are commonly named as emulsion liquid membrane (ELM) when they are used in extraction processes (Rajasimman et al. 2009; Anarakdim et al. 2020). In this regard, $W_1/O/W_2$ double emulsions have found applications in separation of metal and other contaminants from dilute solutions, which are always hard to recover due to their low concentration. By this technique, the waste material present in water, which behaves as the external aqueous phase (W_2), is transferred through the oily phase (O) to the inner aqueous phase (W_1), which is in less proportion than the external W_2,. Normally a reactant is used on the oily phase in order to enhance the transport of the contaminant, which in this case the technique is named reactive extraction. This technique offers certain advantages over conventional solvent extraction that requires larger volumes of solvent and the corresponding equipment size when the metal ion concentration in the effluent stream is low. This type of double emulsions require a specific conditions of stability, because once the extraction process took place, the emulsions needs to be destabilized in order to recover the W_1 which contains higher concentration of contaminant than the initial waste water solution (W_2).

7.6 Conclusions

Over the last decades, double emulsions have been used as promising platforms for encapsulation of compounds with antioxidant properties. Antioxidants with hydrophilic character are encapsulated more commonly in this type of colloidal system, since the W/O/W type of double emulsions is the most extensively used. The type of phases and stabilizers used are the key factors determining the emulsion stability, EE and release behaviour. A correct selection of stabilizer(s) is important to ensure the stability of double emulsion during preparation and storage. In double emulsions, the two interfaces are very close to each other and stabilizers movements such as the one produced by Marangoni effect could enhance drop coalesce producing emulsion instability and undesired release of the encapsulated compounds. In this regard, the use of particles, polymers, proteins or even combination of them offers a clear advantage since their stability is found to be higher, due to the limited movement of these types of stabilizers from the interface once they located there adequately.

Characterization of double emulsions in terms of morphology, stability, rheological behaviour, EE and antioxidant activity is very important for the final application as antioxidant carriers. Their characterizations present clear difficulties in comparison to conventional single emulsions, since destabilization can easily occur during the characterization, optical and light scattering techniques offer a clear advantage in this aspect. Simultaneous use of several techniques is required for a correct double emulsion physical characteristics determination.

Preparation method of double emulsions is crucial in determining the size and uniformity of the formed droplets. In this context, to avoid the destabilization of the inner emulsions, low/soft energy needs to be applied during the second emulsification step; hence, obtaining small droplet size is not always possible by conventional mechanical methods. To solve this issue, employing low-energy consuming techniques such as ME and MCE offers higher control on the droplet size and could prevent destabilization of the inner emulsion. Also, the use of one step and low-energy techniques, such as catastrophic inversion presents a promising alternative for double emulsion production, however, stabilizers with medium lipophilic-hydrophilic character are required, which is not always possible for all formulations. Notwithstanding, further research is necessary to carry out to ensure the production of double emulsions with desired droplet size and distribution through one step low-energy methods. Moreover, new formulations need to be developed by including the use of natural stabilizers without unpleasant taste for consumers but with enough efficiency to ensure emulsion stability and encapsulation efficiency.

Acknowledgments This study was supported by the Consejería de Economía y Empleo del Principado de Asturias (Plan de Ciencia, Tecnología e Innovación 2013-2017), under the Grant IDI/2018/000185.

References

Acevedo-Fani A, Silva HD, Soliva-Fortuny R et al (2017) Formation, stability and antioxidant activity of food-grade multilayer emulsions containing resveratrol. Food Hydrocoll 71:207–215. https://doi.org/10.1016/j.foodhyd.2017.05.007

Anarakdim K, Gutiérrez G, Cambiella Á et al (2020) The effect of emulsifiers on the emulsion stability and extraction efficiency of Cr(VI) using emulsion liquid membranes (ELMs) formulated with a green solvent. Membranes (Basel) 10:76. https://doi.org/10.3390/membranes10040076

Antonio P, Preziosi V (2018) Chapter 3 – Catastrophic phase inversion techniques for nanoemulsification. In: Nanoemulsions – formulation, applications, and characterization. pp 53–76. https://doi.org/10.1016/B978-0-12-811838-2.00003-5

Aserin A (2008) Multiple emulsion: technology and applications. Wiley Interscience, Hoboken

Ashjari M, Khoee S, Mahdavian AR (2012) Controlling the morphology and surface property of magnetic/cisplatin-loaded nanocapsules via W/O/W double emulsion method. Colloids Surf A Physicochem Eng Asp 408:87–96. https://doi.org/10.1016/j.colsurfa.2012.05.035

Balcaen M, Vermeir L, Declerck A, Van Der Meeren P (2016) Influence of internal water phase gelation on the shear- and osmotic sensitivity of W/O/W-type double emulsions. Food Hydrocoll 58:356–363. https://doi.org/10.1016/j.foodhyd.2016.03.011

Benichou A, Aserin A, Garti N (2007) O/W/O double emulsions stabilized with WPI-polysaccharide conjugates. Colloids Surf A Physicochem Eng Asp 297:211–220. https://doi.org/10.1016/j.colsurfa.2006.10.048

Berendsen R, Güell C, Henry O, Ferrando M (2014) Premix membrane emulsification to produce oil-in-water emulsions stabilized with various interfacial structures of whey protein and carboxymethyl cellulose. Food Hydrocoll 38:1–10. https://doi.org/10.1016/j.foodhyd.2013.11.005

Berendsen R, Güell C, Ferrando M (2015a) Spray dried double emulsions containing procyanidin-rich extracts produced by premix membrane emulsification: effect of interfacial composition. Food Chem 178:251–258. https://doi.org/10.1016/j.foodchem.2015.01.093

Berendsen R, Güell C, Ferrando M (2015b) A procyanidin-rich extract encapsulated in water-in-oil-in-water emulsions produced by premix membrane emulsification. Food Hydrocoll 43:636–648. https://doi.org/10.1016/j.foodhyd.2014.07.023

Bernewitz R, Schmidt US, Schuchmann HP, Guthausen G (2014) Structure of and diffusion in O/W/O double emulsions by CLSM and NMR-comparison with W/O/W. Colloids Surf A Physicochem Eng Asp 458:10–18. https://doi.org/10.1016/j.colsurfa.2014.01.002

Biesalski H-K, Dragsted LO, Elmadfa I et al (2009) Bioactive compounds: definition and assessment of activity. Nutrition 25:1202–1205. https://doi.org/10.1016/j.nut.2009.04.023

Blois MS (1958) Antioxidant determinations by the use of a stable free radical. Nature 181:1199–1200. https://doi.org/10.1038/1811199a0

Bonnet M, Cansell M, Berkaoui A et al (2009) Release rate profiles of magnesium from multiple W/O/W emulsions. Food Hydrocoll 23:92–101. https://doi.org/10.1016/j.foodhyd.2007.11.016

Bonnet M, Cansell M, Placin F et al (2010) Influence of the oil globule fraction on the release rate profiles from multiple W/O/W emulsions. Colloids Surf B Biointerfaces 78:44–52. https://doi.org/10.1016/j.colsurfb.2010.02.010

Bou R, Cofrades S, Jiménez-Colmenero F (2014) Physicochemical properties and riboflavin encapsulation in double emulsions with different lipid sources. LWT – Food Sci Technol 59:621–628. https://doi.org/10.1016/j.lwt.2014.06.044

Butterfield DA, Castegna A, Pocernich CB et al (2002) Nutritional approaches to combat oxidative stress in Alzheimer' s disease. J Nutr Biochem 13:444–461. https://doi.org/10.1016/S0955-2863(02)00205-X

Campbell SB, Larson T, Smeets NMB et al (2012) Miniemulsification by catastrophic phase inversion. Chem Eng J 183:534–541. https://doi.org/10.1016/j.cej.2011.12.092

Carrillo CA, Nypelö TE, Rojas OJ (2015) Cellulose nanofibrils for one-step stabilization of multiple emulsions (W/O/W) based on soybean oil. J Colloid Interface Sci 445:166–173. https://doi.org/10.1016/j.jcis.2014.12.028

Carrillo-Navas H, Cruz-Olivares J, Varela-Guerrero V et al (2012) Rheological properties of a double emulsion nutraceutical system incorporating chia essential oil and ascorbic acid stabilized by carbohydrate polymer-protein blends. Carbohydr Polym 87:1231–1235. https://doi.org/10.1016/j.carbpol.2011.09.005

Charcosset C (2009) Preparation of emulsions and particles by membrane emulsification for the food processing industry. J Food Eng 92:241–249. https://doi.org/10.1016/j.jfoodeng.2008.11.017

Choi SJ, Decker EA, McClements DJ (2009) Impact of iron encapsulation within the interior aqueous phase of water-in-oil-in-water emulsions on lipid oxidation. Food Chem 116:271–276. https://doi.org/10.1016/j.foodchem.2009.02.045

Chong D, Liu X, Ma H et al (2015) Advances in fabricating double-emulsion droplets and their biomedical applications. Microfluid Nanofluidics 19:1071–1090

Chouaibi M, Mejri J, Rezig L et al (2019) Experimental study of quercetin microencapsulation using water-in-oil-in-water (W1/O/W2) double emulsion. J Mol Liq 273:183–191. https://doi.org/10.1016/j.molliq.2018.10.030

Chu L, Xie R, Zhu J et al (2003) Study of SPG membrane emulsification processes for the preparation of monodisperse core – shell microcapsules. J Colloid Interface Sci 265:187–196. https://doi.org/10.1016/S0021-9797(03)00350-3

Cofrades S, Antoniou I, Solas MT et al (2013) Preparation and impact of multiple (water-in-oil-in-water) emulsions in meat systems. Food Chem 141:338–346. https://doi.org/10.1016/j.foodchem.2013.02.097

Cofrades S, Santos-López JA, Freire M et al (2014) Oxidative stability of meat systems made with W1/O/W2 emulsions prepared with hydroxytyrosol and chia oil as lipid phase. LWT – Food Sci Technol 59:941–947. https://doi.org/10.1016/j.lwt.2014.06.051

Devanthi PVP, Linforth R, El Kadri H, Gkatzionis K (2018) Water-in-oil-in-water double emulsion for the delivery of starter cultures in reduced-salt moromi fermentation of soy sauce. Food Chem 257:243–251. https://doi.org/10.1016/j.foodchem.2018.03.022

Díaz-Ruiz R, Martínez-Rey L, Laca A, et al (2020) Enhancing trans-resveratrol loading capacity by forcing W1/O/W2 emulsions up to its colloidal stability limit. Colloids Surf B Biointerfaces. In press. https://doi.org/10.1016/j.colsurfb.2020.111130

Dickinson E (2011) Double emulsions stabilized by food biopolymers. Food Biophys 6:1–11. https://doi.org/10.1007/s11483-010-9188-6

Ding S, Serra CA, Vandamme TF et al (2019) Double emulsions prepared by two – step emulsification: history, state-of- the-art and perspective. J Control Release 295:31–49. https://doi.org/10.1016/j.jconrel.2018.12.037

Donnell PBO, Mcginity JW (1997) Preparation of microspheres by the solvent evaporation technique. Adv Drug Deliv Rev 28:25–42. https://doi.org/10.1016/S0169-409X(97)00049-5

Dragosavac MM, Holdich RG, Vladisavljević GT, Sovilj MN (2012) Stirred cell membrane emulsification for multiple emulsions containing unrefined pumpkin seed oil with uniform droplet size. J Membr Sci 392–393:122–129. https://doi.org/10.1016/j.memsci.2011.12.009

Eisinaite V, Juraite D, Schroën K, Leskauskaite D (2016) Preparation of stable food-grade double emulsions with a hybrid premix membrane emulsification system. Food Chem 206:59–66. https://doi.org/10.1016/j.foodchem.2016.03.046

Eisinaite V, Juraite D, Schroën K, Leskauskaite D (2017) Food-grade double emulsions as effective fat replacers in meat systems. J Food Eng 213:54–59. https://doi.org/10.1016/j.jfoodeng.2017.05.022

ElShafei GMS, El-Said MM, Attia HAE, Mohammed TGM (2010) Environmentally friendly pesticides: essential oil-based w/o/w multiple emulsions for anti-fungal formulations. Ind Crop Prod 31:99–106. https://doi.org/10.1016/j.indcrop.2009.09.010

Esfanjani AF, Jafari SM, Assadpoor E, Mohammadi A (2015) Nano-encapsulation of saffron extract through double-layered multiple emulsions of pectin and whey protein concentrate. J Food Eng 165:149–155. https://doi.org/10.1016/j.jfoodeng.2015.06.022

Estrada-Fernández AG, Román-Guerrero A, Jiménez-Alvarado R et al (2018) Stabilization of oil-in-water-in-oil (O1/W/O2) Pickering double emulsions by soluble and insoluble whey protein concentrate-gum Arabic complexes used as inner and outer interfaces. J Food Eng 221:35–44. https://doi.org/10.1016/j.jfoodeng.2017.10.006

Evageliou V, Panagopoulou E, Mandala I (2019) Encapsulation of EGCG and esterified EGCG derivatives in double emulsions containing whey protein isolate, bacterial cellulose and salt. Food Chem 281:171–177. https://doi.org/10.1016/j.foodchem.2018.12.105

Fang Z, Bhandari B (2010) Encapsulation of polyphenols – a review. Trends Food Sci Technol 21:510–523. https://doi.org/10.1016/j.tifs.2010.08.003

Fechner A, Knoth A, Scherze I, Muschiolik G (2007) Stability and release properties of double-emulsions stabilised by caseinate – dextran conjugates. Food Hydrocoll 21:943–952. https://doi.org/10.1016/j.foodhyd.2006.10.021

Flaiz L, Freire M, Cofrades S et al (2016) Comparison of simple, double and gelled double emulsions as hydroxytyrosol and n-3 fatty acid delivery systems. Food Chem 213:49–57. https://doi.org/10.1016/j.foodchem.2016.06.005

Foster T, Dorfman KD, Ted Davis H (2010) Giant biocompatible and biodegradable PEG-PMCL vesicles and microcapsules by solvent evaporation from double emulsion droplets. J Colloid Interface Sci 351:140–150. https://doi.org/10.1016/j.jcis.2010.05.020

Frasch-Melnik S, Spyropoulos F, Norton IT (2010) W1/O/W2 double emulsions stabilised by fat crystals – formulation, stability and salt release. J Colloid Interface Sci 350:178–185. https://doi.org/10.1016/j.jcis.2010.06.039

Freitas S, Merkle HP, Gander B (2005) Microencapsulation by solvent extraction / evaporation: reviewing the state of the art of microsphere preparation process technology. J Control Release 102:313–332. https://doi.org/10.1016/j.jconrel.2004.10.015

Gabriele D, Migliori M, Di R et al (2009) Characterisation of dairy emulsions by NMR and rheological techniques. Food Hydrocoll 23:619–628. https://doi.org/10.1016/j.foodhyd.2008.05.002

Gaitzsch F, Gäbler A, Kraume M (2011) Analysis of droplet expulsion in stagnant single water-in-oil-in-water double emulsion globules. Chem Eng Sci 66:4663–4669. https://doi.org/10.1016/j.ces.2011.06.020

Garti N (1997a) Double emulsions — scope, limitations and new achievements. Colloids Surf A Physicochem Eng Asp 123–124:233–246. https://doi.org/10.1016/S0927-7757(96)03809-5

Garti N (1997b) Progress in stabilization and transport phenomena of double emulsions in food applications. LWT – Food Sci Technol 30:222–235. https://doi.org/10.1006/fstl.1996.0176

Gehrmann S, Bunjes H (2018) Influence of membrane material on the production of colloidal emulsions by premix membrane emulsification. Eur J Pharm Biopharm 126:140–148. https://doi.org/10.1016/j.ejpb.2016.11.006

Gharehbeglou P, Jafari SM, Homayouni A et al (2019) Fabrication of double W1/O/W2 nano-emulsions loaded with oleuropein in the internal phase (W1) and evaluation of their release rate. Food Hydrocoll 89:44–55. https://doi.org/10.1016/j.foodhyd.2018.10.020

Gharsallaoui A (2009) Ingredients: microencapsulation of food ingredients: still a field to be explored? New Food Mag 3:241

Gimenez-Rota C, Palazzo I, Scognamiglio MR et al (2019) β-Carotene, α-tocoferol and rosmarinic acid encapsulated within PLA/PLGA microcarriers by supercritical emulsion extraction: encapsulation efficiency, drugs shelf-life and antioxidant activity. J Supercrit Fluids 146:199–207. https://doi.org/10.1016/j.supflu.2019.01.019

Giroux HJ, Constantineau S, Fustier P et al (2013) Cheese fortification using water-in-oil-in-water double emulsions as carrier for water soluble nutrients. Int Dairy J 29:107–114. https://doi.org/10.1016/j.idairyj.2012.10.009

Giroux HJ, Robitaille G, Britten M (2016) Controlled release of casein-derived peptides in the gastrointestinal environment by encapsulation in water-in-oil-in-water double emulsions. LWT – Food Sci Technol 69:225–232. https://doi.org/10.1016/j.lwt.2016.01.050

Gutiérrez G, Matos M, Benito JM et al (2014) Preparation of HIPEs with controlled droplet size containing lutein. Colloids Surf A Physicochem Eng Asp 442:111–122. https://doi.org/10.1016/j.colsurfa.2013.05.077

Hattrem MN, Kristiansen KA, Aachmann FL et al (2015) Ibuprofen-in-cyclodextrin-in-W/O/W emulsion - improving the initial and long-term encapsulation efficiency of a model active ingredient. Int J Pharm 487:1–7. https://doi.org/10.1016/j.ijpharm.2015.03.059

Hemar Y, Cheng LJ, Oliver CM et al (2010) Encapsulation of resveratrol using water-in-oil-in-water double emulsions. Food Biophys 5:120–127. https://doi.org/10.1007/s11483-010-9152-5

Hernández-Marín NY, Lobato-Calleros C, Vernon-Carter EJ (2013) Stability and rheology of water-in-oil-in-water multiple emulsions made with protein-polysaccharide soluble complexes. J Food Eng 119:181–187. https://doi.org/10.1016/j.jfoodeng.2013.05.039

Higashi S, Tabata N, Kondo K et al (1999) Size of lipid microdroplets effects results of hepatic arterial chemotherapy with an anticancer agent in water-in-oil-in-water emulsion to hepatocellular carcinoma. J Pharmacol Exp Ther 289:816–819

Ilić JD, Nikolovski BG, Petrović LB et al (2017) The garlic (A. sativum L.) extracts food grade W1/O/W2 emulsions prepared by homogenization and stirred cell membrane emulsification. J Food Eng 205:1–11. https://doi.org/10.1016/j.jfoodeng.2017.02.006

Ilyasoglu Buyukkestelli H, El SN (2019) Development and characterization of double emulsion to encapsulate iron. J Food Eng 263:446–453. https://doi.org/10.1016/j.jfoodeng.2019.07.026

Iqbal S, Baloch MK, Hameed G, McClements DJ (2013) Controlling W/O/W multiple emulsion microstructure by osmotic swelling and internal protein gelation. Food Res Int 54:1613–1620. https://doi.org/10.1016/j.foodres.2013.09.035

Jaimes-Lizcano YA, Wang Q, Rojas EC, Papadopoulos KD (2013) Evaporative destabilization of double emulsions for effective triggering of release. Colloids Surf A Physicochem Eng Asp 423:81–88. https://doi.org/10.1016/j.colsurfa.2013.01.054

Jena AK, Nayak AK, De A et al (2018) Development of lamivudine containing multiple emulsions stabilized by gum odina. Future J Pharm Sci 4:71–79. https://doi.org/10.1016/j.fjps.2017.10.002

Jiao H, Peng W, Zhao J, Xu C (2013) Extraction performance of bisphenol A from aqueous solutions by emulsion liquid membrane using response surface methodology. Desalination 313:36–43. https://doi.org/10.1016/j.desal.2012.12.002

Jiménez-Colmenero F (2013) Potential applications of multiple emulsions in the development of healthy and functional foods. Food Res Int 52:64–74. https://doi.org/10.1016/j.foodres.2013.02.040

Kawakatsu T, Kikuchi Y, Nakajima M (1997) Regular-sized cell creation in microchannel emulsification by visual microprocessing method. J Am Oil Chem Soc 74:317–321. https://doi.org/10.1007/s11746-997-0143-8

Kenji R, Luhede L, Fritsching U et al (2020) Premix membrane emulsification using flat microfiltration inorganic membranes with tailored structure and composition. J Membr Sci 608:118124. https://doi.org/10.1016/j.memsci.2020.118124

Khadem B, Khellaf M, Sheibat-othman N (2020) Investigating swelling-breakdown in double emulsions. Colloids Surf A 585:124181. https://doi.org/10.1016/j.colsurfa.2019.124181

Kheynoor N, Hosseini SMH, Yousefi GH et al (2018) Encapsulation of vitamin C in a rebaudioside-sweetened model beverage using water in oil in water double emulsions. Food Sci Technol 96:419–425. https://doi.org/10.1016/j.lwt.2018.05.066

Kim BK, Cho AR, Park DJ (2016) Enhancing oral bioavailability using preparations of apigenin-loaded W/O/W emulsions: in vitro and in vivo evaluations. Food Chem 206:85–91. https://doi.org/10.1016/j.foodchem.2016.03.052

Klinkesorn U, Sophanodora P, Chinachoti P et al (2005) Increasing the oxidative stability of liquid and dried tuna oil-in-water emulsions with electrostatic layer-by-layer deposition technology. J Agric Food Chem 53:4561–4566. https://doi.org/10.1021/jf0479158

Kobayashi I, Nakajima M, Mukataka S (2003) Preparation characteristics of oil-in-water emulsions using differently charged surfactants in straight-through microchannel emulsification. Colloids Surf A Physicochem Eng Asp 229:33–41. https://doi.org/10.1016/j.colsurfa.2003.08.005

Lamprecht A, Ubrich N, Hombreiro-Perez M et al (2000) Influences of process parameters on nanoparticle preparation performed by a double emulsion ultrasonication technique. Int J Pharm 196:177–182. https://doi.org/10.1016/S0378-5173(99)00422-6

Leal-Calderon F, Homer S, Goh A, Lundin L (2012) W/O/W emulsions with high internal droplet volume fraction. Food Hydrocoll 27:30–41. https://doi.org/10.1016/j.foodhyd.2011.09.004

Lei L, Zhang Q, Shi S, Zhu S (2016) High internal phase emulsion with double emulsion morphology and their templated porous polymer systems. J Colloid Interface Sci 483:232–240. https://doi.org/10.1016/j.jcis.2016.08.034

Leister N, Karbstein HP (2020) Evaluating the stability of double emulsions — a review of the measurement techniques for the systematic investigation of instability mechanisms. Colloid Interface 4:1–18. https://doi.org/10.3390/colloids4010008

Leiva-Vega J, Villalobos-Carvajal R, Ferrari G et al (2020) Influence of interfacial structure on physical stability and antioxidant activity of curcumin multilayer emulsions. Food Bioprod Process 121:65–75. https://doi.org/10.1016/j.fbp.2020.01.010

Lobato-Calleros C, Sosa-Pérez A, Rodríguez-Tafoya J et al (2008) Structural and textural characteristics of reduced-fat cheese-like products made from W1/O/W2 emulsions and skim milk. LWT – Food Sci Technol 41:1847–1856. https://doi.org/10.1016/j.lwt.2008.01.006

Lu W, Kelly AL, Miao S (2016) Emulsion-based encapsulation and delivery systems for polyphenols. Trends Food Sci Technol 47:1–9. https://doi.org/10.1016/j.tifs.2015.10.015
Lutz R, Aserin A, Wicker L, Garti N (2009) Release of electrolytes from W/O/W double emulsions stabilized by a soluble complex of modified pectin and whey protein isolate. Colloids Surf B Biointerfaces 74:178–185. https://doi.org/10.1016/j.colsurfb.2009.07.014
Ma G, Nagai M, Omi S (1999) Effect of lauryl alcohol on morphology of uniform polystyrene-poly(methyl methacrylate) composite microspheres prepared by porous glass membrane emulsification technique. J Colloid Interface Sci 19:110–128. https://doi.org/10.1006/jcis.1999.6467
Mao L, Roos YH, O'Callaghan DJ, Miao S (2013) Volatile release from whey protein isolate-pectin multilayer stabilized emulsions: effect of pH, salt, and artificial salivas. J Agric Food Chem 61:6231–6239. https://doi.org/10.1021/jf4011615
Maqsoudlou A, Assadpour E, Mohebodini H, Jafari SM (2020) Improving the efficiency of natural antioxidant compounds via different nanocarriers. Adv Colloid Interf Sci 278:102122. https://doi.org/10.1016/j.cis.2020.102122
Marefati A, Rayner M, Timgren A et al (2013) Freezing and freeze-drying of Pickering emulsions stabilized by starch granules. Colloids Surf A Physicochem Eng Asp 436:512–520. https://doi.org/10.1016/j.colsurfa.2013.07.015
Marefati A, Sjöö M, Timgren A et al (2015) Fabrication of encapsulated oil powders from starch granule stabilized W/O/W Pickering emulsions by freeze-drying. Food Hydrocoll 51:261–271. https://doi.org/10.1016/j.foodhyd.2015.04.022
Márquez AL, Medrano A, Panizzolo LA, Wagner JR (2010) Effect of calcium salts and surfactant concentration on the stability of water-in-oil (w / o) emulsions prepared with polyglycerol polyricinoleate. J Colloid Interface Sci 341:101–108. https://doi.org/10.1016/j.jcis.2009.09.020
Matos M, Lobo A, Benito JM et al (2012) Extending the useful life of metalworking fluids in a copper wire drawing industry by monitoring their functional properties. Tribol Trans 55:685–692. https://doi.org/10.1080/10402004.2012.694580
Matos M, Suárez MÁ, Gutiérrez G et al (2013a) Emulsification with microfiltration ceramic membranes: a different approach to droplet formation mechanism. J Membr Sci 444:345–358. https://doi.org/10.1016/j.memsci.2013.05.033
Matos M, Timgren A, Sjöö M et al (2013b) Preparation and encapsulation properties of double Pickering emulsions stabilized by quinoa starch granules. Colloids Surf A Physicochem Eng Asp 423:147–153. https://doi.org/10.1016/j.colsurfa.2013.01.060
Matos M, Gutiérrez G, Coca J, Pazos C (2014) Preparation of water-in-oil-in-water (W1/O/W2) double emulsions containing trans-resveratrol. Colloids Surf A Physicochem Eng Asp 442:69–79. https://doi.org/10.1016/j.colsurfa.2013.05.065
Matos M, Gutiérrez G, Iglesias O et al (2015) Enhancing encapsulation efficiency of food-grade double emulsions containing resveratrol or vitamin B12 by membrane emulsification. J Food Eng 166:212–220. https://doi.org/10.1016/j.jfoodeng.2015.06.002
Matos M, Gutiérrez G, Martínez-Rey L et al (2018) Encapsulation of resveratrol using food-grade concentrated double emulsions: emulsion characterization and rheological behaviour. J Food Eng 226:73–81. https://doi.org/10.1016/j.jfoodeng.2018.01.007
McClements DJ, Li Y (2010) Structured emulsion-based delivery systems: controlling the digestion and release of lipophilic food components. Adv Colloid Interf Sci 159:213–228. https://doi.org/10.1016/j.cis.2010.06.010
McClements DJ, Decker EA, Weiss J (2007) Emulsion-based delivery systems for lipophilic bioactive components. J Food Sci 72:109–124. https://doi.org/10.1111/j.1750-3841.2007.00507.x
Mehrnia MA, Jafari SM, Makhmal-Zadeh BS, Maghsoudlou Y (2017) Rheological and release properties of double nano-emulsions containing crocin prepared with Angum gum, Arabic gum and whey protein. Food Hydrocoll 66:259–267. https://doi.org/10.1016/j.foodhyd.2016.11.033
Meng FT, Ma GH, Qiu W, Su ZG (2003) W/O/W double emulsion technique using ethyl acetate as organic solvent: effects of its diffusion rate on the characteristics of microparticles. J Control Release 91:407–416. https://doi.org/10.1016/S0168-3659(03)00273-6
Moharram HA, Youssef MM (2014) Methods for determining the antioxidant activity: a review. Alex J Food Sci Technol 11:31–42. https://doi.org/10.12816/0025348

Mulligan MK, Rothstein JP (2012) Scale-up and control of droplet production in coupled microfluidic flow-focusing geometries. Microfluid Nanofluidics 13. https://doi.org/10.1007/S10404-012-0941-7

Mun S, Choi Y, Shim JY et al (2011) Effects of enzymatically modified starch on the encapsulation efficiency and stability of water-in-oil-in-water emulsions. Food Chem 128:266–275. https://doi.org/10.1016/j.foodchem.2011.03.014

Munin A, Edwards-Levy F (2011) Encapsulation of natural polyphenolic compounds; a review. Pharmaceutics 3:793–829. https://doi.org/10.3390/pharmaceutics3040793

Murillo-Martínez MM, Pedroza-Islas R, Lobato-Calleros C et al (2011) Designing W1/O/W2 double emulsions stabilized by protein-polysaccharide complexes for producing edible films: rheological, mechanical and water vapour properties. Food Hydrocoll 25:577–585. https://doi.org/10.1016/j.foodhyd.2010.06.015

Muschiolik G (2007) Multiple emulsions for food use. Curr Opin Colloid Interface Sci 12:213–220. https://doi.org/10.1016/j.cocis.2007.07.006

Muschiolik G, Dickinson E (2017) Double emulsions relevant to food systems: preparation, stability, and applications. Compr Rev Food Sci Food Saf 16:532–555. https://doi.org/10.1111/1541-4337.12261

Mutaliyeva B, Grigoriev D, Madybekova G et al (2017) Microencapsulation of insulin and its release using w/o/w double emulsion method. Colloids Surf A Physicochem Eng Asp 521:147–152. https://doi.org/10.1016/j.colsurfa.2016.10.041

Na X, Zhou W, Li T et al (2019) Preparation of double-emulsion-templated microspheres with controllable porous structures by premix membrane emulsification. Particuology 44:22–27. https://doi.org/10.1016/j.partic.2018.07.007

Nabavi SA, Vladisavljević GT, Manović V (2017) Mechanisms and control of single-step microfluidic generation of multi-core double emulsion droplets. Chem Eng J 322:140–148. https://doi.org/10.1016/j.cej.2017.04.008

Nelis V, Declerck A, De Neve L et al (2019) Fat crystallization and melting in W/O/W double emulsions: comparison between bulk and emulsified state. Colloids Surf A Physicochem Eng Asp 566:196–206. https://doi.org/10.1016/j.colsurfa.2019.01.019

Neumann SM, van der Schaaf US, Schuchmann HP (2017) The Diffusion and Coalescence Time Analyzer (DCTA): a novel experimental setup for investigating instability phenomena in double emulsions. Food Struct 12:103–112. https://doi.org/10.1016/j.foostr.2017.02.002

O' Regan J, Mulvihill DM (2009) Water soluble inner aqueous phase markers as indicators of the encapsulation properties of water-in-oil-in-water emulsions stabilized with sodium caseinate. Food Hydrocoll 23:2339–2345. https://doi.org/10.1016/j.foodhyd.2009.06.009

O'Regan J, Mulvihill DM (2010) Sodium caseinate-maltodextrin conjugate stabilized double emulsions: encapsulation and stability. Food Res Int 43:224–231. https://doi.org/10.1016/j.foodres.2009.09.031

Okochi H, Nakano M (1997) Comparative study of two preparation methods of W/O/W emulsions: stirring and membrane emulsification. Chem Pharm Bull 45:1323–1326. https://doi.org/10.1248/cpb.45.1323

Okushima S, Nisisako T, Torii T, Higuchi T (2004) Controlled production of monodisperse double emulsions by two-step droplet breakup in microfluidic devices. Langmuir 20:9905–9908. https://doi.org/10.1021/la0480336

Pal R (2007) Rheology of double emulsions. J Colloid Interface Sci 307:509–515. https://doi.org/10.1016/j.jcis.2006.12.024

Palencia M, Rivas BL (2011) Metal-ion retention properties of water-soluble amphiphilic block copolymer in double emulsion systems (w/o/w) stabilized by non-ionic surfactants. J Colloid Interface Sci 363:682–689. https://doi.org/10.1016/j.jcis.2011.05.089

Pan J, Yin Y, Gan M et al (2015) Fabrication and evaluation of molecularly imprinted multi-hollow microspheres adsorbents with tunable inner pore structures derived from templating Pickering double emulsions. Chem Eng J 266:299–308. https://doi.org/10.1016/j.cej.2014.11.126

Panagopoulou E, Evageliou V, Kopsahelis N et al (2017) Stability of double emulsions with PGPR, bacterial cellulose and whey protein isolate. Colloids Surf A Physicochem Eng Asp 522:445–452. https://doi.org/10.1016/j.colsurfa.2017.03.020

Pando D, Beltrán M, Gerone I et al (2015) Resveratrol entrapped niosomes as yoghurt additive. Food Chem 170:281–287. https://doi.org/10.1016/j.foodchem.2014.08.082

Pawlik AK, Norton IT (2012) Encapsulation stability of duplex emulsions prepared with SPG cross-flow membrane, SPG rotating membrane and rotor-stator techniques — a comparison. J Membr Sci 415–416:459–468. https://doi.org/10.1016/j.memsci.2012.05.032

Pawlik A, Cox PW, Norton IT (2010) Food grade duplex emulsions designed and stabilised with different osmotic pressures. J Colloid Interface Sci 352:59–67. https://doi.org/10.1016/j.jcis.2010.08.049

Pays K, Giermanska-Kahn J, Pouligny B et al (2002) Double emulsions: how does release occur? J Control Release 79:193–205. https://doi.org/10.1016/S0168-3659(01)00535-1

Pimentel-González DJ, Campos-Montiel RG, Lobato-Calleros C et al (2009) Encapsulation of Lactobacillus rhamnosus in double emulsions formulated with sweet whey as emulsifier and survival in simulated gastrointestinal conditions. Food Res Int 42:292–297. https://doi.org/10.1016/j.foodres.2008.12.002

Pinheiro AC, Bourbon AI, Cerqueira MA et al (2015) Chitosan/fucoidan multilayer nanocapsules as a vehicle for controlled release of bioactive compounds. Carbohydr Polym 115:1–9. https://doi.org/10.1016/j.carbpol.2014.07.016

Poyato C, Navarro-Blasco I, Calvo MI et al (2013) Oxidative stability of O/W and W/O/W emulsions: effect of lipid composition and antioxidant polarity. Food Res Int 51:132–140. https://doi.org/10.1016/j.foodres.2012.11.032

Quispe-Condori S, Saldaña MDA, Temelli F (2011) Microencapsulation of fl ax oil with zein using spray and freeze drying. LWT – Food Sci Technol 44:1880–1887. https://doi.org/10.1016/j.lwt.2011.01.005

Rajasimman M, Sangeetha R (2009) Optimization of process parameters for the extraction of chromium (VI) by emulsion liquid membrane using response surface methodology. J Hazard Mater 168:291–297. https://doi.org/10.1016/j.jhazmat.2009.02.044

Rajasimman M, Sangeetha R, Karthik P (2009) Statistical optimization of process parameters for the extraction of chromium (VI) from pharmaceutical wastewater by emulsion liquid membrane. Chem Eng J 150:275–279. https://doi.org/10.1016/j.cej.2008.12.026

Ramakrishnan S, Ferrando M, Güell C (2012) Food grade microcapsules produced by membrane emulsification. Procedia Eng 44:1530–1533. https://doi.org/10.1016/j.proeng.2012.08.855

Rayner M, Trägårdh G (2002) Membrane emulsification modelling: how can we get from characterisation to design? Desalination 145:165–172. https://doi.org/10.1016/S0011-9164(02)00403-4

Rayner B, Bergenstahl L, Massarelli G, Trägårdh G (2004) Double emulsions prepared by membrane emulsification: stability and entrapment degree in a flavour release system. In: Abstract food colloids conference. Harrogate

Reynolds PA, Gilbert EP, Henderson MJ, White JW (2009) Structure of high internal phase aqueous-in-oil emulsions and related inverse micelle solutions. 4. Surfactant mixtures. J Phys Chem 113:12243–12256. https://doi.org/10.1021/jp903475z

Rivas BL, Palencia M (2011) Removal-concentration of pollutant metal-ions by water-soluble polymers in conjunction with double emulsion systems: a new hybrid method of membrane-based separation. Sep Purif Technol 81:435–443. https://doi.org/10.1016/j.seppur.2011.08.021

Rosen MJ (2004) Surfactants and interfacial phenomena, 3rd edn. Wiley Interscience, Hoboken

Sahin S, Sawalha H, Schroën K (2014) High throughput production of double emulsions using packed bed premix emulsification. Food Res Int 66:78–85. https://doi.org/10.1016/j.foodres.2014.08.025

Santos MG, Carpinteiro DA, Thomazini M et al (2014) Coencapsulation of xylitol and menthol by double emulsion followed by complex coacervation and microcapsule application in chewing gum. Food Res Int 66:454–462. https://doi.org/10.1016/j.foodres.2014.10.010

Sapei L, Naqvi MA, Rousseau D (2012) Stability and release properties of double emulsions for food applications. Food Hydrocoll 27:316–323. https://doi.org/10.1016/j.foodhyd.2011.10.008

Scherze I, Knöfel R, Muschiolik G (2005) Automated image analysis as a control tool for multiple emulsions. Food Hydrocoll 19:617–624. https://doi.org/10.1016/j.foodhyd.2004.10.029

Schuch A, Deiters P, Henne J et al (2013) Production of W/O/W (water-in-oil-in-water) multiple emulsions: droplet breakup and release of water. J Colloid Interface Sci 402:157–164. https://doi.org/10.1016/j.jcis.2013.03.066

Schuch A, Leal LG, Schuchmann HP (2014a) Production of W/O/W double emulsions. Part I: visual observation of deformation and breakup of double emulsion drops and coalescence of the inner droplets. Colloids Surf A Physicochem Eng Asp 461:336–343. https://doi.org/10.1016/j.colsurfa.2013.11.043

Schuch A, Wrenger J, Schuchmann HP (2014b) Production of W/O/W double emulsions. Part II: influence of emulsification device on release of water by coalescence. Colloids Surf A Physicochem Eng Asp 461:344–351. https://doi.org/10.1016/j.colsurfa.2013.11.044

Schuster S, Bernewitz R, Guthausen G et al (2012) Analysis of W 1/O/W 2 double emulsions with CLSM: statistical image processing for droplet size distribution. Chem Eng Sci 81:84–90. https://doi.org/10.1016/j.ces.2012.06.059

Seifriz W (1925) Studies in emulsions. III-V. J Phys Chem 29:738–749. https://doi.org/10.1021/j150252a009

Serdaroğlu M, Öztürk B, Urgu M (2016) Emulsion characteristics, chemical and textural properties of meat systems produced with double emulsions as beef fat replacers. Meat Sci 117:187–195. https://doi.org/10.1016/j.meatsci.2016.03.012

Shima M, Kobayashi Y, Fujii T et al (2004) Preparation of fine W/O/W emulsion through membrane filtration of coarse W/O/W emulsion and disappearance of the inclusion of outer phase solution. Food Hydrocoll 18:61–70. https://doi.org/10.1016/S0268-005X(03)00042-0

Shima M, Morita Y, Yamashita M, Adachi S (2006) Protection of Lactobacillus acidophilus from the low pH of a model gastric juice by incorporation in a W/O/W emulsion. Food Hydrocoll 20:1164–1169. https://doi.org/10.1016/j.foodhyd.2006.01.001

Solans C, García-Celma M (1997) Surfactants for microemulsions. Opin Colloid Interface 2:464–471. https://doi.org/10.1016/S1359-0294(97)80093-3

Su J, Flanagan J, Singh H (2008) Improving encapsulation efficiency and stability of water-in-oil-in-water emulsions using a modified gum arabic (Acacia (sen) SUPER GUM™). Food Hydrocoll 22:112–120. https://doi.org/10.1016/j.foodhyd.2007.03.005

Sugiura S, Nakajima M, Kumazawa N et al (2002) Characterization of spontaneous transformation-based droplet formation during microchannel emulsification. J Phys Chem B 106:9405–9409. https://doi.org/10.1021/jp0259871

Sugiura S, Nakajima M, Yamamoto K et al (2004) Preparation characteristics of water-in-oil-in-water multiple emulsions using microchannel emulsification. J Colloid Interface Sci 270:221–228. https://doi.org/10.1016/j.jcis.2003.08.021

Suzuki K, Shuto I, Hagura Y (1996) Characteristics of the membrane emulsification method combined with preliminary emulsification for preparing corn oil-in-water emulsions. Food Sci Technol 2:43–47. https://doi.org/10.3136/FSTI9596T9798.2.43

Tabatabaee Amid B, Mirhosseini H (2014) Stabilization of water in oil in water (W/O/W) emulsion using whey protein isolate-conjugated durian seed gum: enhancement of interfacial activity through conjugation process. Colloids Surf B Biointerfaces 113:107–114. https://doi.org/10.1016/j.colsurfb.2013.08.042

Tadros T (2004) Application of rheology for assessment and prediction of the long-term physical stability of emulsions. Adv Colloid Interf Sci 108–109:227–258. https://doi.org/10.1016/j.cis.2003.10.025

Tamnak S, Mirhosseini H, Tan CP et al (2016) Encapsulation properties, release behavior and physicochemical characteristics of water-in-oil-in-water (W/O/W) emulsion stabilized with pectin-pea protein isolate conjugate and Tween 80. Food Hydrocoll 61:599–608. https://doi.org/10.1016/j.foodhyd.2016.06.023

Tang SY, Sivakumar M, Nashiru B (2013) Impact of osmotic pressure and gelling in the generation of highly stable single core water-in-oil-in-water (W/O/W) nano multiple emulsions of aspirin assisted by two-stage ultrasonic cavitational emulsification. Colloids Surf B Biointerfaces 102:653–658. https://doi.org/10.1016/j.colsurfb.2012.08.036

Utada AS, Lorenceau E, Link DR et al (2005) Monodisperse double emulsions generated from a microcapillary device. Science 308:537–541. https://doi.org/10.1126/science.1109164

van der Graaf S, Schroën C, Boom RM (2005) Preparation of double emulsions by membrane emulsification — a review. J Membr Sci 251:7–15. https://doi.org/10.1016/j.memsci.2004.12.013

Virtudazo RVR, Fuji M, Shirai T (2011) Fabrication of calcined hierarchical porous hollow silicate micro-size spheres via double emulsion process. Mater Lett 65:3112–3115. https://doi.org/10.1016/j.matlet.2011.06.085

Vladisavljevic GT, Shimizu M, Nakashima T (2004) Preparation of monodisperse multiple emulsions at high production rates by multi-stage premix membrane emulsification. J Membr Sci 244:97–106. https://doi.org/10.1016/j.memsci.2004.07.008

Vladisavljevic G, Surh J, McClements J (2006) Effect of emulsifier type on droplet disruption in repeated Shirasu porous glass membrane homogenization. Langmuir 22:4526–4533. https://doi.org/10.1021/la053410f

Wang X, Jiang Y, Wang YW et al (2008) Enhancing anti-inflammation activity of curcumin through O/W nanoemulsions. Food Chem 108:419–424. https://doi.org/10.1016/j.foodchem.2007.10.086

Wen L, Papadopoulos KD (2001) Effects of osmotic pressure on water transport in W1/O/W2 emulsions. J Colloid Interface Sci 235:398–404. https://doi.org/10.1006/jcis.2000.7384

Wroński S, Vladimirov V, Adach A (2012) Modelling of mass transfer from multiple emulsions. Int J Heat Mass Transf 55:4241–4245. https://doi.org/10.1016/j.ijheatmasstransfer.2012.03.065

Yuan Q, Aryanti N, Gutiérrez G, Williams RA (2009) Enhancing the throughput of membrane emulsification techniques to manufacture functional particles. Ind Eng Chem Res 48:8872–8880. https://doi.org/10.1021/ie801929s

Zafar N, Bitar A, Valour JP et al (2016) Elaboration of ammonio methacrylate copolymer based spongy cationic particles via double emulsion solvent evaporation process. Mater Sci Eng C 61:85–96. https://doi.org/10.1016/j.msec.2015.12.017

Zhang Y, Gou J, Sun F et al (2014) Impact of electrolytes on double emulsion systems (W/O/W) stabilized by an amphiphilic block copolymer. Colloids Surf B Biointerfaces 122:368–374. https://doi.org/10.1016/j.colsurfb.2014.07.008

Zhao-Miao LIU, Yu DU, Yan PANG (2018) Generation of water-in-oil-in-water (W/O/W) double emulsions by microfluidics. Chinese J Anal Chem 46:324–331. https://doi.org/10.1016/S1872-2040(17)61072-7

Zhu Y, Zheng Y, Wang F, Wang A (2016) Fabrication of magnetic porous microspheres via (O1/W)/O2 double emulsion for fast removal of Cu2+ and Pb2+. J Taiwan Inst Chem Eng 67:505–510. https://doi.org/10.1016/j.jtice.2016.08.006

Zhu Q, Qiu S, Zhang H et al (2018) Physical stability, microstructure and micro-rheological properties of water-in-oil-in-water (W/O/W) emulsions stabilized by porcine gelatin. Food Chem 253:63–70. https://doi.org/10.1016/j.foodchem.2018.01.119

Zou S, Liu H, Yang Y et al (2013) Multihollow nanocomposite microspheres with tunable pore structures by templating Pickering double emulsions. React Funct Polym 73:1231–1241. https://doi.org/10.1016/j.reactfunctpolym.2013.06.006

Chapter 8
Stability and Release Behavior of Bioactive Compounds (with Antioxidant Activity) Encapsulated by Pickering Emulsion

Bakht Ramin Shah

8.1 Introduction

Recently, emulsions have attracted tremendous scientific attentions owing to their wide range of useful applications in the field of cosmetics, food, pharmaceuticals and paint, etc. The word "emulsion" is derived from the Latin word mulgeo, mulgere, which means "milk", as milk is a typical example of emulsion containing fat and water, along with other constituents. Emulsions are thermodynamically unstable systems comprising of droplets of a liquid dispersed in another immiscible or partially miscible liquid (Chen et al. 2011). The phase that is present in the form of droplets is known as the dispersed phase, and the phase in which the droplets are suspended is called the continuous phase, whereas the boundary between them is called the "interface". During their passage through the emulsions, these interfaces emit lights which give cloudy appearances to the emulsions (Loi et al. 2019). Common emulsions formed spontaneously are not stable and tend to destabilization because of droplets coalescence. However, the potential applications of emulsions are strongly dependent on their stability, which is to maintain their characteristics as long as possible. Therefore, in order to stabilize these emulsions, the oil and water mixtures require (i) the addition of emulsifiers in the form of amphiphiles which adsorb to the bare oil-water interface, thus preventing droplets coalescence, and (ii) energy input– through exposure to prolong periods of mechanical agitation, stirring, homogenizing or power ultrasound (Kentish et al. 2008). In simple words, it can be stated that an emulsion consists of oil, water and stabilizer (amphiphiles). These emulsions may be of the oil-in-water (O/W) or water-in-oil (W/O) types depending

B. R. Shah (✉)
University of South Bohemia in Ceske Budejovice, Faculty of Fisheries and Protection of Waters, South Bohemian Research Centre of Aquaculture and Biodiversity of Hydrocenoses, Institute of Aquaculture and Protection of Waters, České Budějovice, Czech Republic
e-mail: bshah@frov.jcu.cz

M. A. Aboudzadeh (ed.), *Emulsion-based Encapsulation of Antioxidants*, Food Bioactive Ingredients, https://doi.org/10.1007/978-3-030-62052-3_8

on whether the oil is dispersed as droplets in water, or vice versa. If a droplet of the emulsion is dispersed in pure water it is of O/W type, conversely if a droplet is dispersed in pure oil it is regarded as W/O (McClements 2012). Nowadays the most commonly known emulsions used as delivery systems are nano, micro or solid particles stabilized Pickering emulsions. It is important to specify precisely the kind of emulsion used in a particular study, because this affects the most appropriate method used to synthesize them, the foremost factors affecting their stability (such as pH, temperature, presence of salts, etc.) and their physicochemical and functional properties.

Before going to detail about Pickering emulsion, it is important to clarify briefly the confusion between nano and micro emulsions, which are often, miscomprehend due to the prefixes nano and micro relating their droplets size. A nanoemulsion is a conventional surfactant stabilized emulsion with very small particles ($r < 100$ nm) (Tadros et al. 2004). Actually, these kinds of emulsions can be fabricated without surfactants as stabilizer, but practically they will be highly unstable to droplet coalescence and hence surfactants are needed to impart them kinetic stability during storage (McClements 2015). On the other hand, microemulsions are conventional surfactant stabilized emulsions which may also have very small particles ($r < 100$ nm) but are thermodynamically stable contrary to the nanoemulsions which are kinetically stable.

Another important class of emulsions, which are stabilized by solid particles instead of surfactants, is called Pickering emulsions and is discussed in the next section in detail.

8.2 Pickering Emulsions (PEs)

The phenomenon of PE was first introduced one century ago by Ramsden (Ramsden 1904) and later by Pickering (Pickering 1907) Substituting solid particles for traditional surfactants in these emulsions not only make them more stable against coalescence but also impart them many useful properties. For example, some food grade particles as PE stabilizers have lower toxicity, and thus are safe for usage in vivo. In addition, the solid particles confer useful characteristics of enhanced conductivity, responsiveness and porosity. The significant stability of PEs against coalescence can be contributed to the irreversible adsorption of the solid particles onto the interfaces of the dispersed and continuous phases (Low et al. 2017).

The high resistance against coalescence and Oswald ripening makes it possible to preserve the droplets under high concentration of dispersed phase and even they are allowed to dry and re-disperse (Akartuna et al. 2008; Frelichowska et al. 2009). Furthermore, they also have shown enhanced stability against the influence of environmental factors such as pH, temperature, oil composition, ionic strengths and so forth (Shah et al. 2016a). In short, these promising properties make PEs as useful candidates in various disciplines especially in food and nutrition, pharmaceutics and cosmetics where the use of toxic surfactants are undesirable. Therefore, in recent years scientific community has been paid tremendous attention in the development of cost effective, facile and novel PEs. A simple sketch of the particles stabilized O/W Pickering emulsion has been given in Fig. 8.1.

A wide range of materials has been employed as particulate emulsifier for the fabrication of the PEs including organic particles (e.g. polymer latex and polymer micelle) and inorganic particles (such as silica, hydroxides, and clay particles). To be adsorbed at the interfaces, the particles used should be partially wetted by both the oil and water phases. Depending on the degree of relative wettability of the particles, the emulsions can be classified as either O/W or W/O. The relative wettability of the liquid phases for the solid particles is determined by the three-phase contact angle θ and should be greater than 0° and less than 180°. If the contact angle measured through the aqueous phase is greater than 90°, then the solid particles are relatively more wetted by the oil phase, and in such situations, W/O emulsions are generally formed. When the contact angle θ is less than 90°, the particles are preferentially wetted by the aqueous phase and in such situations, O/W emulsions are formed. In case of θ = 90°, both liquid phases equally wet the particles equally, and in such situations, there is no preferred emulsion O/W or W/O (Fig. 8.2).

Generally, the contact angle is measured through sessile drop method. Briefly, the particles with a specified volume fraction (φ_p) are dispersed in water and are spin-coated onto glass cover slips. The samples are then air-dried overnight before use. Both advancing and receding contact angle measurements are made, and the average used is regarded as the equilibrium contact angle (French et al. 2015).

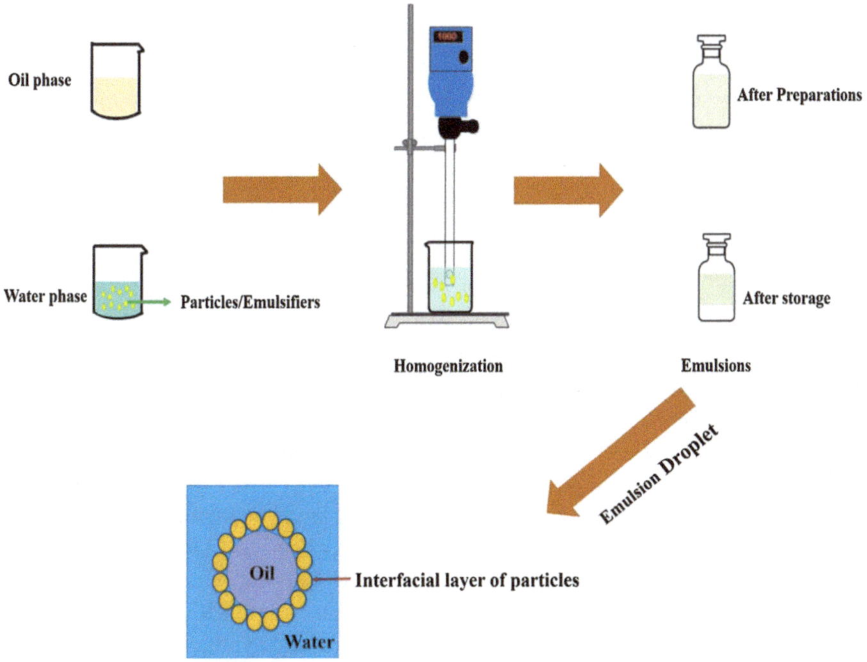

Fig. 8.1 Schematic diagram of the formation of an O/W Pickering

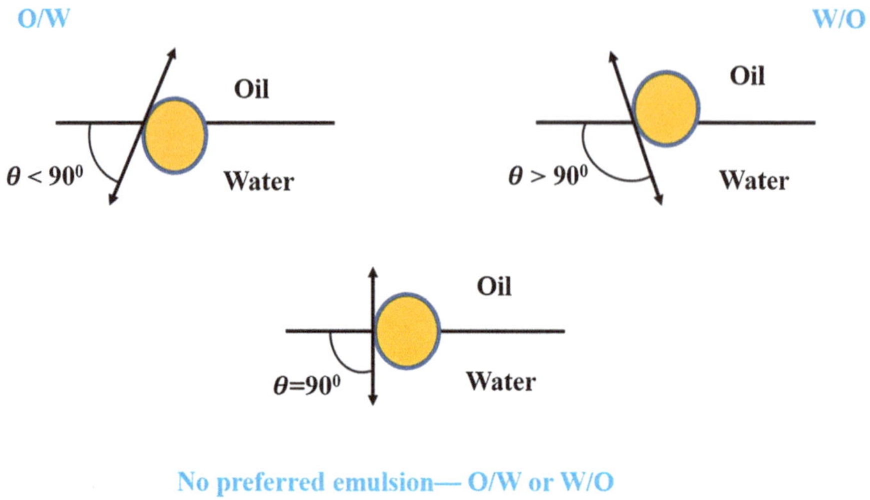

Fig. 8.2 Emulsion types based on wetting conditions of the particles described in terms of contact angle θ between the phases

8.3 Different Types of Particles Used as Stabilizers/ Emulsifiers for PEs

As stated earlier, PEs are the emulsions that are stabilized by solid particles instead of surfactants. These particles are partly wetted by both the phases of PEs i.e. oil and water. Available literature shows that so far various types of particles have been used as emulsifiers to kinetically stabilize the PEs. The particles could be either from inorganic or organic sources and have been discussed below in details.

8.3.1 Inorganic Particles

Inorganic particles such as silica particles have been used extensively as Pickering emulsifiers due to their simple preparation and modification (Jiang et al. 2020). Besides, these particles are of interest as they are commercially available with desirable characteristics such as varying well-defined sizes, surface areas as well as hydrophobicity. Although, these particles were first used as model emulsifiers in the preparation of non-food grade emulsions (Gautier et al. 2007; Horozov and Binks 2006), later on some studies, they were used as food grade emulsifiers as well (Pichot et al. 2010; Skelhon et al. 2012). Both United States Food and Drug Administration (FDA) and the European Food Safety Authority (EFSA) have approved silica and other insoluble forms of silicates as safe food additives (up to 1500 mg SiO_2 per day) (EFSA. 2009; FDA. 1979). However, their inorganic and synthetic origin still left a huge gap for critics on their suitability for food applications. This provided a turning point to the scientific community for paying substantial attention towards biocompatible and biodegradable organic nutrient-based particles as particulate emulsifiers for stabilizing PEs.

8.3.2 Protein-Based Particles

Various protein-based particles have been successfully applied as emulsifiers for the preparation of O/W PEs (McClements 2004). This includes both plant proteins such as zein or soy protein and animal proteins such as whey protein. Zein is a major food grade protein found in corn and is capable of self-assembling to form nano or micro particles. Zein is insoluble in water but soluble in aqueous alcoholic solutions. Consequently, it was supposed that zein-based colloidal particles hold great potential as stabilizers for PEs without surface modification (de Folter et al. 2012). Unfortunately, PEs stabilized solely by zein were unstable and separation was observed only after 3 days of storage. The reason was attributed to the overly hydrophobic nature of the particles obtained through antisolvent approach that facilitated the formation of agglomerates at the aqueous medium. This behavior hinders the particles to be adsorbed at oil droplet surface. Favorably, the problem was solved by combing zein with other hydrophobic compounds such as sodium stearate, chitosan, sodium caseinates (NaCas) etc. that impart significant stability to the synthesized PEs (Gao et al. 2014), (Feng and Lee 2016; Wang et al. 2016). Similarly, PEs stabilized by soy protein for instance soy protein isolate (SPI) or its principal component glycinin (Liu and Tang 2016; Luo et al. 2013) and whey protein isolate (WPI) based PEs have been reported previously (Wu et al. 2015) .

8.3.3 Lipid-Based Particles

In fact, biomaterials with lower environmental influence are considered better candidates in the synthesis of stable PEs. In this context, lipid nanoparticles (NPs) have found a privileged place in the field. Different lipid-based particles have been reported to be efficient PEs stabilizers (Pawlik et al. 2016). The lipid or lipophilic molecules that used for the formation of particulate emulsifiers generally have one or several polar groups (e.g, phytosterols, flavonoids, glyceryl stearyl citrate, or lactylate). The particles from these molecules in aqueous media can be synthesized by mechanical treatments in same way as emulsification process and sometimes may require high temperatures in the case of lipids with high melting points (Gupta and Rousseau 2012).

Lipid based particles were found to stabilize the emulsions of either type i.e., W/O or O/W. For example, fat crystals stabilized W/O PEs like edible spreads were prepared by various research groups (Dickinson 2012; Rousseau 2013). Most recently W/O as well as O/W PEs as a delivery systems stabilized by solid lipid NPs were reported (Dieng et al. 2019; Pawlik et al. 2016).

8.3.4 Carbohydrate-Based Particles

Modified starch-based particles comprise a large group of food grade emulsifiers for PEs. Particular attention has been given to these materials owing to their striking advantages of being natural ingredient, ample in nature, renewable, biocompatible, biodegradable, sustainable and comparatively cheaper (Dufresne 2014). In order to use them for stabilizing oil-water interfaces, it is necessary to modify them chemically because starches are originally highly hydrophilic. The chemical modification is aimed to create hydrophobicity in these starches and is mostly achieved by partial hydrophobization with octenyl succinic anhydride (OSA) (Miao et al. 2014; Timgren et al. 2013). Consequently, the wetting properties and affinity of the particles for both phases are improved. Most commonly used carbohydrates-based particles include chitin, cocoa powder and cocoa fibers, cellulose and chitin nanocrystals, chitosan and so forth. Among them, in recent years chitosan (CS) has attracted much attention due to its wide range of useful applications in many fields such as biomedical, pharmaceuticals, metal chelation, food additives, and other industrial applications (Guibal et al. 2001; Kumar et al. 2004; Rabea et al. 2003).

CS is a renewable linear cationic polysaccharide composed of randomly distributed β-(1–4)-linked D-glucosamine and N-acetyl-D- glucosamine units (Fig. 8.3a), produced by the deacetylation of chitin. The degree of deacetylation (DD) and molecular weight of CS are the two main characteristics which have a significant impact on its physical and chemical characteristics such as emulsification capacity, aggregation activity, rheological and solution properties (Bodnar et al. 2005). Furthermore, CS has been known as a pH responsive polymer and its amine groups are protonated and positively charged at low pH, giving CS a water-soluble cationic polyelectrolyte character. On contrary, CS amines are deprotonated at high pH and it loses its charge thereby making it insoluble in aqueous medium. Considering this information, a number of attempts were made to synthesize pH tunable CS-NPs as stabilizers for PEs (Asfour et al. 2017; Liu et al. 2012; Wang and Heuzey 2016). In this regard, another important approach for the synthesis of CS-NPs is ionic gelation that involves the ionic interaction between the positively charged primary amino groups of CS and negatively charged groups of poly anions (Konecsni et al. 2012). As an example, using ionic gelation, tripolyphosphate (TPP) crosslinked CS-NPs were synthesized. These NPs were then used as stabilizers for PEs, the preparation of which was optimized first by preparing the PEs at fixed concentration (5 wt%) of medium chain triglyceride (MCT) and subsequently by increasing MCT content to 10, 20, 30 and 50 wt%. As a result, PEs prepared with CS:TPP ratio of 5:5 (w/w) and 50 wt% MCT showed best qualities in term of stability against the tested parameters such as storage time, salts, pH, etc. (Shah et al. 2016a).

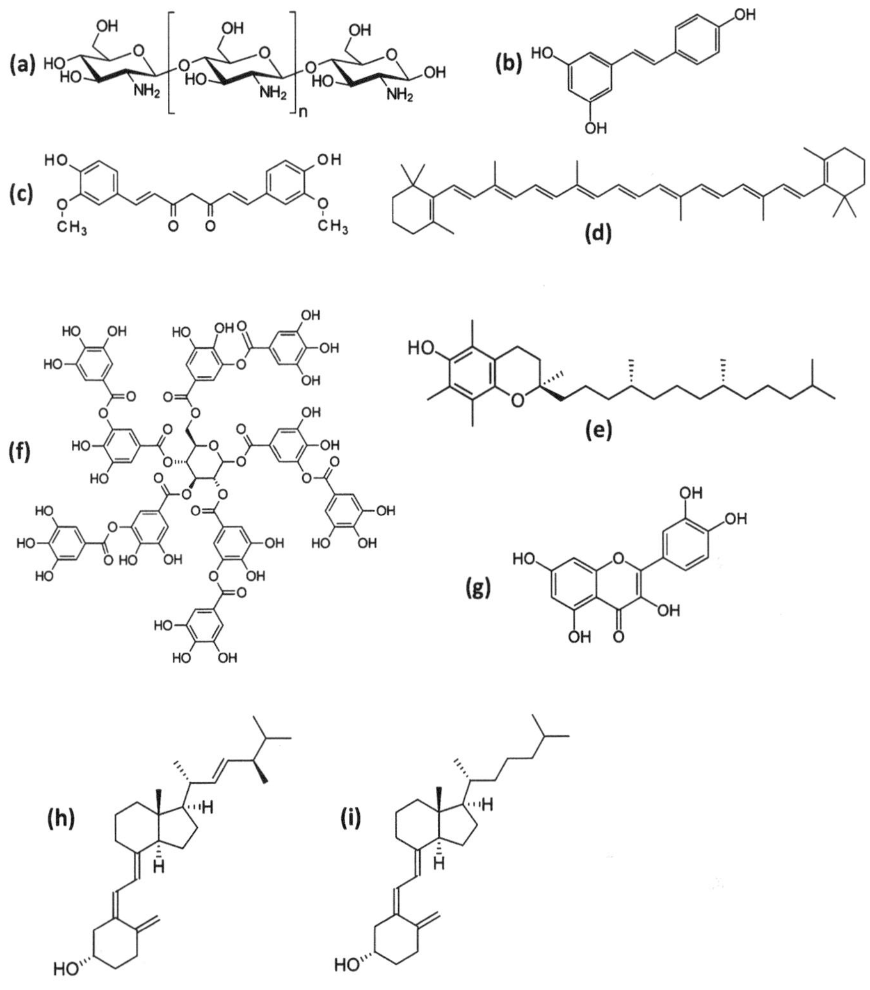

Fig. 8.3 Chemical structures of (a) Chitosan, (b) Resveratrol, (c) Curcumin in Keto form, (d) β-carotene, (e) α-tocopherol, (f) Tannic acid, (g) Quercetin, (h) vitamina D_2 and (i) vitamin D_3

8.4 Factors Influencing Formation and Stability of Pickering Emulsions

In a broader way, emulsion stability denotes the capability of an emulsion to retain its characteristics unchanged over a period of time and against different influencing factors, such as pH, ionic strength, temperature, oil-water ratios, nature of biopolymers and medium, presence of other agents (e.g. surfactants) in the system and charge of biopolymers, etc. Stability of an emulsion is prerequisite for its applications in different industries. Therefore, numerous experimental studies have been conducted on evaluating the stability of PEs to understand the possible physical mechanisms that prevent droplet coalescence (Frith et al. 2008). This implies that controlling and modification of these procedures will help in enhancing stability and hence significant usage of the PEs. There are various underlying mechanisms corresponding to the destabilization of PEs, which may be influenced by the above-mentioned factors. These mechanisms are as follow:

Gravitational Separation The first common and visually observable mechanism is gravitation separation (GS) of the emulsion into upper cream layer and lower serum layer. The separation occurs because of the difference in density of the continuous and dispersed phases and is explained by Stokes law as below (Eq. 8.1) (Pal 2019):

$$v = \frac{2gr^2\left(\rho_2 - \rho_1\right)}{9\eta_1}, \tag{8.1}$$

where v is the GS rate (m/s); g is the gravitational acceleration (m/s^2); r is the droplet radius (m); ρ_1 and ρ_2 are the densities of the dispersed and continuous phases (kg/m^3), respectively and η_1 is the continuous phase viscosity (kg/m·s).

From the Stokes equation, certain important presumptions can be withdrawn as follow; (i) depending on the density differences, emulsion droplets will either sediment or cream and (ii) GS rate is directly proportional to the droplet size but inversely proportional to the viscosity of the continuous phase. This means that GS can be slowed down by reducing the droplet size and/or increasing the viscosity of the continuous phase and can be used as a strategy to delay creaming.

Flocculation: is the process of flocs formation that happens due to the aggregation of two or more emulsion droplets, though the droplets maintain their individual identity. It can occur if the attractive forces between dispersed phase droplets overcome the repulsive forces. The droplets stay in close proximity to each other but they do not reach close enough to flocculate by merging into each other. The phenomenon can be explained by Van der Waals interactions/forces, which are always attractive in these dispersions and needs to be counterbalanced by either electrostatic or steric repulsions. The electrostatic interactions can be greatly influenced by the solvent conditions, particularly ionic strength and pH (Mcclements 2007).

Coarsening or Ostwald ripening: is defined as the growth of larger droplets at the expense of smaller ones, either due to the diffusion of disperse phase molecules through the continuous phase or because the solubility of the material within a spherical droplet in the surrounding continuous phase increases as the radius of the droplet decreases. Simply stated, Ostwald ripening is the process of disappearance of small particles or droplets by dissolution and deposition on the larger particles or droplets (Bommana et al. 2019).

Corresponding to the preceding statements, a research group attempted to evaluate the effect of pH, NaCl and oil contents on the PE gels stabilized by wheat protein NPs. In their study, 70% oil content, pH 5.5 and 6.0 in the absence of NaCl were the favorable conditions to obtain a stable PE (Zhu et al. 2018). In another study, it was found that PEs stabilized by TPP cross linked CS-NPs showed enhanced stability to all the tested parameters including storage time, oil-water ratio, pH and salts (Shah et al. 2016a). Similarly, particle concentration also plays an important role in emulsion stability by preventing Ostwald ripening. One research group defined a minimum particle concentration required to prevent Ostwald ripening, if water-soluble substances are partially in the oil phase. In this study, toluene-in-water emulsions formed at very low concentrations of silanised fumed silica NPs. It was observed that droplets size increased and flocculated together even after gentle rotation of the emulsions. However, increasing the particle concentration to 1 wt%, significantly reduced the rate of droplets coarsening and prevented droplets flocculation. These findings suggested the possible approach for controlling the stability of emulsions formulated with polar, slightly hydrophilic oils at low silica particle concentrations (Juárez and Whitby 2012). Another research group demonstrated similar correlation by describing the influence of particle concentration on the average drop diameter in the emulsion. The particles used in this study were based on partially hydrophobic silica at primary diameter and concentration of 25 nm and 0.7 wt%, respectively. In order to evaluate the effect of particle concentration, all emulsions were prepared by homogenization of the water and oil phases (at an oil volume fraction of 0.33) together for 3 min. The results showed that with increasing particle concentration, emulsion droplet size decreased until a minimum size ($\approx$ 5–20 μm) is reached as the extent of coalescence during drop formation is reduced (Binks and Whitby 2004).

8.5 PEs as Delivery Systems for Bioactive Compounds with Antioxidant Properties

Antioxidants also called free-radical scavengers (FRS) are substances/compounds that can prevent or slow down oxidation. Oxidation is a chemical process damaging the cells caused by free radicals, unstable molecules that the body produces as a reaction to environmental and other pressures (Lobo et al. 2010).

There are two types of antioxidants i.e. endogenous antioxidants that are produced by the body and exogenous ones, which come from outside the body. In exogenous, particular attention is being paid to natural antioxidants including polyphenols, carotenoids, glucosinolates and different kinds of vitamins such as vitamin E, C, etc.

It is a well-known fact that antioxidants can aid in preventing life threatening pathologies including heart disease, liver disease and some cancers (such as oral, oesophageal, stomach and bowel cancers). Many people have common practice of taking antioxidants in the form of supplements as a defense against these diseases. However, antioxidants available in the form of commercial food additives are prepared synthetically and may contain high contents of preservatives. Several reports claim that synthetic antioxidants, such as butylated hydroxyanisole (BHA), butylated hydroxytoluene (BHT), and tertiary butyl hydroquinone (TBHQ), can produce toxins or act as carcinogens resulting in the development and progression of cancer

(Nieva-Echevarría et al. 2015). At this point, it is inevitable to discover and rely on vital natural antioxidant sources as an alternative for the synthetic ones to ensure better health. Food is the source of essential nutrients for growth and maintenance; however, other phytochemicals promote health by combating the aging process and preventing disease. Consequently, this provoked the attention of scientific community, food manufacturers, cultivators, and consumers towards the antioxidant constituents from plant materials and their magnificent roles in maintaining human health. Nevertheless, most of these materials based on dietary sources or other phytochemicals have problem concerning their bioavailability implying that they may not be adequately absorbed to stimulate any biological effects (Aqil et al. 2013). This low absorption could most probably because of their hydrophobic nature and therefore, an utmost need was left to develop effective strategies to encapsulate and deliver them to enhance their stability, solubility and bioactivity.

In this regard, biocompatible, biodegradable and food grade materials-based emulsion technology (e.g. PEs) to encapsulate, protect, and release these compounds provided a well-suited platform to accomplish the task. Some examples of the natural lipophilic compounds, which have been known for their potential antioxidants activities encapsulated in PEs, are discussed in the following subsections.

8.5.1 Resveratrol

Resveratrol is a naturally occurring polyphenolic compound (*trans*-3,5,4′-trihydroxystilbene) (Fig. 8.3b) with efficient antioxidant, cardioprotective, anti-inflammatory and anticancer potentials. It is mainly found in red grapes, red wine, white wine, peanuts, blueberries and pistachios.

According to some reports, comparatively peanuts have more resveratrol contents than red grapes which could maybe due to more water contents in red grapes than peanuts (Burns et al. 2002; Sanders et al. 2000). It is believed that low solubility of resveratrol in water (<0.05 mg/mL), caused by its chemical structure affects its absorption in aqueous medium. Moreover, several other problems hindering its biological activities are its photosensitive nature, short biological half-life, and rapid metabolism and elimination. To overcome these issues, numerous resveratrol nano-formulations have been synthesized and evaluated, including liposomes, solid lipid NPs, polymeric NPs and cyclodextrins which have been studied in great detail (Summerlin et al. 2015). However, PEs based formulations to encapsulate and deliver resveratrol demonstrated better outcome among them as hallmarked by several studies.

Similarly, the same research group formulated quinoa starch particles stabilized PEs with a mixture of miglyol and orange oil (in ratio 1:9) as dispersed phase. In comparison to the surfactant (Tween 20) stabilized O/W emulsions, the Pickering emulsions showed higher stability against creaming phenomena and superior encapsulation efficiency (up to 98%), although both types of emulsions had similar droplet sizes. Their formulated system confirmed to be an appropriate resveratrol carrier system for further use in functional food formulations (Matos et al. 2018).

8.5.2 *β-Carotene*

The term carotenoids is collectively applied to a class of natural coloring agents generally found in fruits and vegetables, and have been used in various food, cosmetic, and pharmaceutical products so far. Besides their role as colorants, carotenoids are also well-appreciated for their health benefits as pro-vitamin A, as well as antioxidants to prevent many chronic diseases, such as cancer, cardiovascular disease, macular degeneration (Rodriguez-Amaya 2015).

Among different carotenoids, β-carotene (Fig. 8.3d) has attracted special attention, due to its abundance and potential bioactive nature. However, the applicability of naturally occurring β-carotene is detained by its highly hydrophobic characteristics as consequence of its unsaturated structure, which makes it insoluble in water and liable to degradation. Furthermore, β-carotene commonly participates in making protein complexes, which hinder its adsorption by human body, thereby significantly lowering its bioavailability. The discussion implies that incorporation of β-carotene into food systems is no doubt challenging but a mandatory job (Rodriguez-Amaya 2015). Therefore, intensive research work has thus been performed in recent years to incorporate β-carotene into functional foods by encapsulating it in PEs in particular, to fulfil its health potentials. In this context, two kinds of PEs stabilized by different concentrations of either whey protein isolate (WPI) or sodium caseinate (NaCas) (0.1 to 2.0 wt%) in 30 wt% sucrose aqueous solution were synthesized with the aim of seeing their comparative protective effects on β-carotene. The outcome of the study showed that the system formulated with 0.8 wt% concentrations of protein had high β-carotene stability. Nevertheless, NaCas provided a better barrier than WPI, probably due to the different amino acid composition and interface structure which significantly reduced β-carotene degradation rate (Cornacchia and Roos 2011). In line with this study, other researchers synthesized O/W PEs based on pea protein isolate (PPI) at pH 3.0 to be used as a delivery system for β-carotene. The required emulsions were produced by microfluidization at a specified protein concentration of 6.0% (w/v) and varying oil fractions (ϕ) between 0.2–0.6. The same procedure was used to prepare β-carotene-loaded PPI emulsions, but the oil phase used here contained β-carotene (30 wt%) which was directly mixed with preheated (~ 45 °C) soy oil to a final concentration of 0.2 wt%. The results demonstrated that increasing ϕ favored the gel-like network strengthening of these emulsions. Most importantly, by conducting the in vitro simulated digestion analysis, it was found that the release of β-carotene during the intestinal digestion of these emulsions was controllable by changing ϕ. The gel-like emulsion at higher oil fractions ($\phi = 0.6$) showed much lower release of β-carotene, but higher stability towards degradation during the digestion, than that at $\phi = 0.3$. The authors concluded that the reported formulation could be an important tool for the design of novel delivery systems for lipophilic bioactive components in general and for the development of plant protein-based formulations in particular (Shao and Tang 2016).

The same research group in another study prepared an O/W emulsion stabilized by soy glycinin particles as a delivery system for β-carotene. They evaluated the release behavior of β-carotene under simulated intestinal conditions. Their findings suggested that β-carotene in the PE was released at a much lower rate than that in a conventional emulsion, and β-carotene was rather stable during the digestion process (Liu and Tang 2016). Most recently, β-carotene was encapsulated in wheat gluten nanoparticles (WGNP) or wheat gluten nanoparticle-xanthan gum (WGNP-XG) complexes. Comparing the two formulations, it was found that the WGNP-XG emulsions had larger initial mean particle diameters (23.9 μm) than the WGNP ones (9.4 μm), but they were still stable against aggregation in wide range of pH values (4–8) and ionic strengths (0–1000 mM NaCl). Furthermore, these PEs demonstrated enhanced protection of the encapsulated β-carotene from chemical degradation during storage, with around 94.3% and 70.1% of the carotenoids being retained after one-month storage at 25 and 37 °C, respectively. Conducting the in vitro digestion experiment, it was found that that β-carotene had a higher bioaccessibility in the WGNP-XG emulsions than in the WGNP ones (Fu et al. 2019).

8.5.3 Curcumin

Curcumin, is a natural and typical flavonoid compound extracted from turmeric *Curcuma longa*, predominantly exists in keto-enol form (1,7-bis(4-hydroxy-3-methoxyphenyl)1,6-heptadiene-3,5-dione) (Fig. 8.3c).

Curcumin is widely used as a spice and food coloring agent in different cuisines and food products as well as in traditional medicine for many centuries in countries such as India and China. Over the past decades, numerous research studies have signified the importance of curcumin as antioxidant, anti-inflammatory, antiarthritic, anti-amyloid, hepatoprotective, thrombo-suppressive, anti-HIV, antimicrobial and antitumor agent (Patra and Sleem 2013).

Although the exact mechanism(s) through which curcumin performs these activities is still unknown, the antioxidant ability of this yellow pigment appears to be an indispensable constituent underlying its pleiotropic biological activities. In fact, curcumin has been reported to hinder lipid peroxidation and to efficiently scavenge superoxide anion and hydroxyl radicals (Ruby et al. 1995). Besides its inherent ability to attenuate the reactivity of oxygen free radical species, curcumin has been shown in vivo to enhance the activities of detoxifying enzymes such as glutathione-S-transferase (Piper et al. 1998). Furthermore, curcumin has established potent inducing effect on heme oxygenase-1 (HO-1), which is one of the genes encoding for proteins having antioxidant characteristics. This pathway supported the enhanced heme oxygenase activity to be an important pillar in curcumin-mediated cytoprotection against oxidative stress (Motterlini et al. 2000).

However, the clinical advancement of curcumin is hindered by its low water solubility (i.e. 0.0004 mg mL^{-1} at pH 7.3) and degradation under physiological conditions. To improve its solubility and hence bioavailability, encapsulation and delivery approaches based on nanotechnology have been remained the fundamental interest of scientific community from time to time. A wide range of particles have been used for encapsulation of curcumin including liposome, silk fibroin and chitosan, chitosan, β-cyclodextrin inclusion complex, PLGA NPs, nanospheres, phospholipids, cyclodextrin, silica particles and polymeric NPs, etc. (Patra and Sleem 2013; Zhao et al. 2012). Among different formulations, PEs have been shown to efficiently fulfill the duties of encapsulation and delivery. A research group evaluated stability and release behavior of curcumin encapsulated in silica NPs stabilized PEs during storage and simulated gastric and intestinal digestion. Stability and release kinetics of curcumin were characterized describing encapsulated curcumin with stability approximately 100 fold higher than the stability of curcumin suspended in distilled water. Furthermore, the steady release profile confirmed sustained release of over 80% of the encapsulated curcumin in 36 h. During simulated gastric digestion model experiment (2 h), above 80% of the encapsulated curcumin was retained. In addition, incubation in simulated intestinal environment resulted in destabilization of the emulsion and approximately 60% of the encapsulated curcumin was released within 2 h of incubation. Overall, these results demonstrated that PEs has a potential for effective delivery of bioactive compounds (Tikekar et al. 2013). In another report, preparation of TPP crosslinked CS-NPs stabilized PEs for the encapsulation and delivery of curcumin was reported. The preparation of the PEs was optimized, and the emulsions showed enhanced stability during storage and against different pH ranges and salts concentration. The in vitro release profile of the encapsulated curcumin form the PEs confirmed its sustained release over extended period, thereby supporting these PEs as efficient delivery vehicles for curcumin and other bioactive compounds (Shah et al. 2016a). Similarly, in another study, curcumin encapsulated PEs or nanoemulsions were prepared by dissolving curcumin at a concentration of 0.1 wt% either in MCT or LCT (long chain triglyceride) such as corn oil. PE was prepared by homogenizing the aqueous phase containing TPP crosslinked CS-NPs and the oil phase at a speed of 10,000 rpm for 3 min. On the other hand, curcumin encapsulated nanoemulsions were prepared using nonionic surfactants (Span80 and Tween80 at ratio of 1.5:8.5) and the oil phase (5 wt% MCT or LCT) at surfactant to oil ratio (SOR) of 2:1. Results demonstrated slower rate of digestion and consequently lower bioaccessibility values of curcumin in PEs than for nanoemulsions. The authors also indicated that in comparison to the free curcumin, curcumin encapsulated in PEs had higher radical scavenging potentials, which acknowledged the protective effect of the emulsion systems on antioxidant activity of curcumin (Shah et al. 2016b). Another research group synthesized gel-like PE (50%, v/v, oil) stabilized by zein/Tannic acid (TA) complex colloidal particles as a new encapsulation system for lipophilic ingredients such as curcumin. Compared with NaCas-stabilized emulsions and bulk oil, the emulsions stabilized by zein/TA exhibited enhanced shielding effects on the chemical stability of the encapsulated curcumin after exposure to UV light, where the lipid oxidation rate also decreased significantly in these emulsions. It was postulated that the zein particle layers loaded with TA around the oil droplets could protect them versus severe gastric environment, decelerating the release of free fatty acids (FFA) and curcumin at the time of in vitro simulated digestion. The authors concluded that zein/TA stabilized PEs are promising encapsulating agent to protect bioactive compounds from degradation and control their release during digestion, which can further enhance the bioavailability of these ingredients (Zou et al. 2017). Most recently, PEs stabilized by milled starch particles were fabricated to enhance the bioaccessibility of curcumin via controlling the digestion of lipids in the human gastrointestinal (GI) tract. Through obtaining data from two different evaluating techniques, it was found that the bioaccessibility of encapsulated curcumin in PEs was 27.6% and 50.7%, respectively, in comparison to free curcumin suspended in bulk oil phase, which was 22.1% and 7.8%, respectively. Based on this huge difference in bioaccessibilities of encapsulated and free curcumin, the synthesized PEs were regarded as potential delivery systems for lipophilic bioactive compounds such as curcumin (Lu et al. 2019).

8.5.4 α-Tocopherol

Vitamin E is a collective name used for a group of fat-soluble vitamins that have been known for their distinctive antioxidant activities. Various sources of vitamin E include dietary sources (nuts, such as almonds, peanuts and hazelnuts and vegetable oils such as sunflower, wheat germ, safflower, corn and soybean oils). Some leafy vegetables such as spinach and broccoli also contain vitamin E. Naturally vitamin E occurs in 8 different forms, with 4 tocopherols (alpha, beta, gamma and delta) and 4 tocotrienols. Among these, alpha- (or α-) tocopherol (Fig. 8.3e) is the most common and most potent form that is recognized to meet human requirements (Alqahtani et al. 2015). Substantial evidences acknowledge that daily intake of α-tocopherol may benefit human health due to its antioxidant potency and ability to inhibit various diseases. However, being a strongly hydrophobic molecule, making it difficult to disperse directly into foods and beverages that have an aqueous continuous phase. Additionally, exposure to light, heat, and oxygen promotes the chemical degradation of α-tocopherol during storage, leading to a reduction in its biological activity and nutritional benefits. To overcome these challenges, α-tocopherol can be encapsulated and protected using colloidal delivery systems such as PEs. A research group fabricated WPI stabilized O/W emulsion for the encapsulation and delivery of both α-tocopherol and β-carotene. Summarizing their results, they concluded that highly concentrated O/W emulsions were synthesized with oil fractions of up to 60%. Encapsulation of α-tocopherol and β-carotene did not affect stability of the emulsions rather interestingly confirmed enhanced protection of the encapsulants i.e. α-tocopherol and β-carotene against degradation most likely by the protein layer surrounding the oil droplets (Gaspar et al. 2017).

8.5.5 Vitamin D

Vitamin D is a lipophilic compound, which not only contributes in maintaining normal calcium metabolism, but also plays a vital role in a wide range of non-classic actions. Previous studies have shown that vitamin D has both anti-inflammatory and antiperoxidative activity (Ke et al. 2016). Normally, vitamin D accumulates in adipose tissues and it is believed that a typical adult adipose contain sufficient amounts equivalent to its several months of daily reference intake (DRI) (Hengist et al. 2019). There are two main forms of this vitamin, namely vitamin D_2 (ergocalciferol) (Fig. 8.3h) and vitamin D_3 (cholecalciferol) (Fig. 8.3i). Although foods such as beef liver, dairy products, egg yolk and fish contain small amounts of vitamin D, the main source is sunlight which is needed to modify its precursor 7-dehydrocholesterol into a bio-functional form known as vitamin D_3 (Borel et al. 2015). This implies that people who have little exposure to sunlight or have underlying pathological conditions such as obesity, hyperparathyroidism or gastrointestinal diseases are at high risk of vitamin D_3 deficiency (Cashman 2019). Therefore, there is an intense need to formulate vitamin D_3 enriched functional foods with particular focus on emulsions based encapsulated systems to overcome its oxidative instability as well as to enhance its bioavailability in aqueous environments (Winuprasith et al. 2018). In this regard, in a study O/W PE stabilized by nanofibrillated cellulose (NFC) extracted from mangosteen was developed to encapsulate vitamin D_3. The formulated emulsions contained 10 wt% oil (0.01 wt% vitamin D_3 and 9.99 wt% soybean oil) and 1 wt% NFC as emulsifier. In order to evaluate the impact of NFC on lipid digestion and vitamin bioaccessibility, the in vitro gastrointestinal (GIT) environment consisted of all the three phases i.e. mouth, stomach and small intestine was simulated. The authors observed that an increase in NFC concentration led to a decrease in lipid digestion and vitamin bioaccessiblity. In addition, the results indicated that mangosteen fiber can be used as potential stabilizer for an O/W PE, that exhibited minor effect on lipid digestion and encapsulated vitamin D_3 bioaccessibility when used at relatively low levels (Winuprasith et al. 2018). One year later, a similar study was conducted by the same authors, in which vitamin D_3 was encapsulated in 10% wt soybean O/W PEs stabilized by either NFC or WPI at 0.3 wt%, 0.5 wt% and 0.7 wt%. Stability of the prepared vitamin D_3-loaded emulsions were tested against temperature (30 °C to 90 °C), pH (2 to 8), and ionic strength (0 to 500 mM NaCl). Based on the obtained results, it was concluded that NFC can be used as an efficient emulsifier for producing vitamin enriched emulsions with good long-term stability (Mitbumrung et al. 2019). In a very recent study, vitamin D_3 fortified PEs stabilized by nanochitin (NCh) were prepared. The authors evaluated the effect of emulsifier format (molecular vs particles) on the GIT fate of the emulsions by examining their physicochemical properties, microstructure, digestibility, and bioaccessibility using an in vitro human GIT model. PEs were prepared by homogenization of a 90 wt% aqueous NCh (0.11 wt%) suspension with a 10 wt% oil phase (2 wt% vitamin D_3 in corn oil). The final stock PE contained 0.1 wt% NCh and 10.0 wt% oil. The behavior of these emulsions was compared to those of a

Tween 80-stabilized emulsion, as well as to emulsions containing a combination of nanochitin and Tween 80. After analyzing the results, it was found that the NCh-emulsions experienced much more droplet aggregation within the simulated GIT as compared to the Tween 80-stabilized ones. Vitamin D_3 bioaccessibility was 45% less and lipid digestion was 30% less and for the NCh-emulsions than for the Tween 80-stabilized ones. In conclusion, their findings proposed that NCh decelerate lipid digestion, which may be useful for developing high-satiety foods, however, on the other hand as it also decrease vitamin D_3 bioaccessibility, it could be a bottleneck of the current system from nutritional point of view (Zhou et al. 2020).

8.5.6 Tannic Acid

Tannic acid (TA), is a specific form of tannin and is a naturally occurring plant polyphenol, composed of a central glucose molecule derivatized at its hydroxyl groups with one or more galloyl residues (Gülçin et al. 2010) (Fig. 8.3f). It can be found in practically all aerial plant tissues but mostly in tea, nettle, wood, berries, Chinese galls and oak is believed to be its richest source (Robles 2014). Early studies have shown that TA inhibited skin, lung and forestomach tumors induced by polycyclic aromatic hydrocarbon carcinogens and N-methyl-N-nitrosourea in experimental mice models (Bance and Teel 1989). Furthermore, a line of reports have described that TA has antimutagenic and anticarcinogenic, antihypertensive activities which could be related in part to its antioxidant potential, being a polyphenol (Andrade Jr. et al. 2005). To confirm this claim, the antioxidant and radical scavenging properties of TA with different analytical methodology such as DPPH (2,2-diphenyl-1-picryl-hydrazyl-hydrate) and ABTS (2,2′-azino-bis(3-ethylbenzothiazoline-6-sulfonic acid)) assays were studied. It was concluded that TA has significant antioxidant potential compared with standard antioxidant compounds e.g. BHA, BHT, tocopherol and trolox (Gülçin et al. 2010). The addition of TA to conventional emulsions can also improve the chemical stability of lipophilic bioactives, which was again attributed to the antioxidant properties of TA (Li et al. 2019; Liu et al. 2020). In this context, some researchers incorporated TA in the formulation of PEs, for example in one of pioneering studies, zein/TA colloidal particles were synthesized based on the hydrogen-bonding interaction between zein and TA in aqueous ethanol solution. Then using these particles at different concentrations (0.25–1.5 wt %), stable PE gels with high corn oil volume fraction (> 50%) were prepared (Zou et al. 2015). Rheological behavior of these PE gels, formulated over a wide range of zein/TA particle concentration (1–5%, w/v) and oil fractions (5–60%, v/v), was investigated as well. Based on the obtained results, it was concluded that the microstructure and rheological properties of the synthesized PE can be switched by altering both the particle concentration as well as the oil content (Zou et al. 2018). In another study, PEs stabilized by zein/TA NPs with weight ratios of zein to TA of 4:1, 2:1 and 1:1

with varying zein concentrations (0.1%, 0.2% or 0.3%, w/v) were prepared. Where the optimum ratio was found to be 0.3% concentration of zein and zein to TA of 1:1. And the emulsion formed at this ratio showed dramatically improved the oxidative stability as compared to others (Zhou et al. 2019).

8.5.7 *Quercetin*

Quercetin, is an another important representative of polyphenols (Fig. 8.3g), which is mainly found in a variety of human foods including red onions, grapes, apples, berries, cherries, broccoli, citrus fruits, tea (*Camellia sinensis*) and at considerable high concentrations (180 mg per 100 g) in capers and lovage. Quercetin has shown a wide range of biological benefits such as antioxidant activity in radical scavenging, lowering of blood pressure and ameliorating hyperglycemia-related diseases (Bischoff 2008). In aim to enhance the efficacy of the quercetin by enhancing its stability and accessibility, quercetin was encapsulated in W/O PEs stabilized by an interfacial complex of water-insoluble polyphenol crystals and protein. The outcomes found that polyphenol crystals of either curcumin or quercetin absorb at the interface and stabilized water droplets for several days when used alone; however, when WPI was added to the polyphenols, the water-oil interface exhibited a significant improvement in the stabilization of the system (Zembyla et al. 2019). Quercetin was also encapsulated in olive oil in SPI/Pectin-stabilized O/W emulsion. To fabricate the desired emulsions, first SPI/pectin complex particle dispersions were prepared by blending SPI aqueous dispersions (5.0% protein w/w) with specified concentrations of pectin samples (1.0% w/v) in SPI/pectin ratio of 1/1(v/v). Thereafter, this SPI/pectin complex particle suspension was homogenized with oil phase (olive oil 50%) by ultrasonic for 6 min to obtain the required O/W emulsions. The same procedure was applied to formulated quercetin loaded emulsions where 2.0 mL quercetin (0.1 mg/mL) was first dissolved in the oil phase and added to the preceding SPI/pectin complex and was homogenized as mentioned above to prepare the ultimate SPI/pectin complex particle stabilized emulsions with encapsulated quercetin. Stability of the prepared emulsions against pH was further evaluated in different conditions (pH 3.0 to pH 9.0). The results showed that emulsion at pH 3.0 exhibited enhanced stability stable after storage for 30 days' at 4 °C and also showed best freeze-thaw stability after 3 cycles. Moreover, rheological measurements of these emulsions revealed a broad viscoelasticity zone and had the best viscoelasticity stability. In vitro intestinal digestion experiment was performed too and quercetin availability reached 15.94% at pH 7.0 and 7.8% at pH 3.0. Regarding these values for availability, the authors concluded that quercetin can be consumed in a green and healthful way, being encapsulated in SPI/Pectin-stabilized emulsions (Wang et al. 2020).

8.6 Conclusion

Since long time ago, plant-based compounds have been a great source of materials used in beneficial medical treatments. Many plant extracts have been shown antioxidant potentials thereby aiding in treatment and prevention of pathologies like cardiovascular diseases, liver diseases, cancers and other related conditions. However, issues of poor oral bioavailability of these compounds hinder their clinical advancements, and hence need delivery systems to ensure their efficient delivery.

Synthesis of these delivery vehicles are principally aimed to protect the encapsulants from harsh environments e.g. human gut and also to ensure their targeted delivery. Furthermore, most importantly, by maintaining their sustained and controlled release, these systems are supposed to enhance bioavailability of the encapsulated hydrophobic bioactive compounds which have low solubility in aqueous medium that hinder their potential applications in food, pharmaceutical and cosmetic industries. These advantages can be significantly achieved if the synthesized system is of better quality in terms of long-term stability during storage and against different influencing factors such as pH, salts, heat and so on, as well as have reduced toxicity towards the organisms.

In this perspective, among different formulations, PEs have shown to perform the duties proficiently due to their better qualities in terms of long-term stabile and non-toxic nature as compared to the others. PEs are actually-stabilized by biodegradable components derived from different nutrients (e.g. polymers, proteins, fats etc.), differing them from conventional emulsions which are stabilized by surfactants where toxicity is an issue. On these grounds, therefore, in the current chapter we particularly focused on the PEs, firstly on their synthesis, characterization and factors influencing their stability, and thereafter their vital role for encapsulating and delivery of bioactive compounds with well-known antioxidant potentials.

References

Akartuna I, Studart AR, Tervoort E, Gonzenbach UT, Gauckler LJ (2008) Stabilization of oil-in-water emulsions by colloidal particles modified with short amphiphiles. Langmuir 24:7161–7168. https://doi.org/10.1021/la800478g

Alqahtani S et al (2015) Cellular uptake, antioxidant and antiproliferative activity of entrapped α-tocopherol and γ-tocotrienol in poly (lactic-co-glycolic) acid (PLGA) and chitosan covered PLGA nanoparticles (PLGA-Chi). J Colloid Interface Sci 445:243–251. https://doi.org/10.1016/j.jcis.2014.12.083

Andrade RG Jr, Dalvi LT, Silva JMC Jr, Lopes GK, Alonso A, Hermes-Lima M (2005) The antioxidant effect of tannic acid on the in vitro copper-mediated formation of free radicals. Arch Biochem Biophys 437:1–9. https://doi.org/10.1016/j.abb.2005.02.016

Aqil F, Munagala R, Jeyabalan J, Vadhanam MV (2013) Bioavailability of phytochemicals and its enhancement by drug delivery systems. Cancer Lett 334:133–141. https://doi.org/10.1016/j.canlet.2013.02.032

Asfour MH, Elmotasem H, Mostafa DM, Salama AAA (2017) Chitosan based Pickering emulsion as a promising approach for topical application of rutin in a solubilized form intended for wound healing: in vitro and in vivo study. Int J Pharm 534:325–338. https://doi.org/10.1016/j.ijpharm.2017.10.044

Bance R, Teel R (1989) Effect of tannic acid on rat liver S9 mediated mutagenesis, metabolism and DNA binding of benzo [a] pyrene. Cancer Lett 47:37–44. https://doi.org/10.1016/0304-3835(89)90174-2

Binks BP, Whitby CP (2004) Silica particle-stabilized emulsions of silicone oil and water: aspects of emulsification. Langmuir 20:1130–1137. https://doi.org/10.1021/la0303557

Bischoff SC (2008) Quercetin: potentials in the prevention and therapy of disease. Curr Opin Clin Nutr Metab Care 11:733–740. https://doi.org/10.1097/MCO.0b013e32831394b8

Bodnar M, Hartmann JF, Borbely J (2005) Preparation and characterization of chitosan-based nanoparticles. Biomacromolecules 6:2521–2527. https://doi.org/10.1021/bm0502258

Bommana MM, Kirthivasan B, Gupta SS, Shikhar A, Shah A (2019) Nanostructures in pharma: elixir to oral medicine. In: Nanoparticles in pharmacotherapy. Elsevier, Cambridge, MA, pp 23–44. https://doi.org/10.1016/B978-0-12-816504-1.00001-6

Borel P, Caillaud D, Cano N (2015) Vitamin D bioavailability: state of the art. Crit Rev Food Sci Nutr 55:1193–1205. https://doi.org/10.1080/10408398.2012.688897

Burns J, Yokota T, Ashihara H, Lean ME, Crozier A (2002) Plant foods and herbal sources of resveratrol. J Agric Food Chem 50:3337–3340. https://doi.org/10.1021/jf0112973

Cashman KD (2019) Vitamin D deficiency: defining, prevalence, causes, and strategies of addressing. Calcif Tissue Int:1–16. https://doi.org/10.1007/s00223-019-00559-4

Chen J, Vogel R, Werner S, Heinrich G, Clausse D, Dutschk V (2011) Influence of the particle type on the rheological behavior of Pickering emulsions. Colloids Surf A Physicochem Eng Asp 382:238–245. https://doi.org/10.1016/j.colsurfa.2011.02.003

Cornacchia L, Roos YH (2011) Stability of β-carotene in protein-stabilized oil-in-water delivery systems. J Agric Food Chem 59:7013–7020. https://doi.org/10.1021/jf200841k

de Folter JW, van Ruijven MW, Velikov KP (2012) Oil-in-water Pickering emulsions stabilized by colloidal particles from the water-insoluble protein zein. Soft Matter 8:6807–6815. https://doi.org/10.1039/C2SM07417F

Dickinson E (2012) Use of nanoparticles and microparticles in the formation and stabilization of food emulsions. Trends Food Sci Technol 24:4–12. https://doi.org/10.1016/j.tifs.2011.09.006

Dieng SM et al (2019) Pickering nano-emulsions stabilized by solid lipid nanoparticles as a temperature sensitive drug delivery system. Soft Matter. https://doi.org/10.1039/C9SM01283D

Dufresne A (2014) Crystalline starch based nanoparticles. Curr Opin Colloid Interface Sci 19:397–408. https://doi.org/10.1016/j.cocis.2014.06.001

EFSA (2009) Scientific opinion of the panel on food additives and nutrient sources added to food on calcium silicate, silicon dioxide and silicic acid gel added for nutritional purposes to food supplements following a request from the European Commission. EFSA J 1132:1–24. https://doi.org/10.2903/j.efsa.2009.1132

FDA (1979) Select Committee on GRAS Substances (SCOGS) Opinion: silicon dioxides. SCOGS Rep 61, US Food Drug Admin

Feng Y, Lee Y (2016) Surface modification of zein colloidal particles with sodium caseinate to stabilize oil-in-water pickering emulsion. Food Hydrocoll 56:292–302. https://doi.org/10.1016/j.foodhyd.2015.12.030

Frelichowska J, Bolzinger M-A, Valour J-P, Mouaziz H, Pelletier J, Chevalier Y (2009) Pickering w/o emulsions: drug release and topical delivery. Int J Pharm 368:7–15. https://doi.org/10.1016/j.ijpharm.2008.09.057

French DJ, Taylor P, Fowler J, Clegg PS (2015) Making and breaking bridges in a Pickering emulsion. J Colloid Interface Sci 441:30–38. https://doi.org/10.1016/j.jcis.2014.11.032

Frith W, Pichot R, Kirkland M, Wolf B (2008) Formation, stability, and rheology of particle stabilized emulsions: influence of multivalent cations. Ind Eng Chem Res 47:6434–6444. https://doi.org/10.1021/ie071629e

Fu D et al (2019) Encapsulation of β-carotene in wheat gluten nanoparticle-xanthan gum-stabilized Pickering emulsions: enhancement of carotenoid stability and bioaccessibility. Food Hydrocoll 89:80–89. https://doi.org/10.1016/j.foodhyd.2018.10.032
Gao Z-M, Yang X-Q, Wu N-N, Wang L-J, Wang J-M, Guo J, Yin S-W (2014) Protein-based pickering emulsion and oil gel prepared by complexes of zein colloidal particles and stearate. J Agric Food Chem 62:2672–2678. https://doi.org/10.1021/jf500005y
Gaspar RB, Nele M, Ferraz HC (2017) Encapsulation of α-tocopherol and β-carotene in concentrated oil-in-water beverage emulsions stabilized with whey protein isolate. J Dispers Sci Technol 38:89–95. https://doi.org/10.1080/01932691.2016.1144196
Gautier F et al (2007) Pickering emulsions with stimulable particles: from highly-to weakly-covered interfaces. Phys Chem Chem Phys 9:6455–6462. https://doi.org/10.1039/B710226G
Guibal E, Sweeney NVO, Zikan M, Vincent T, Tobin J (2001) Competitive sorption of platinum and palladium on chitosan derivatives. Int J Biol Macromol 28:401–408. https://doi.org/10.1016/S0141-8130(01)00130-1
Gülçin İ, Huyut Z, Elmastaş M, Aboul-Enein HY (2010) Radical scavenging and antioxidant activity of tannic acid. Arab J Chem 3:43–53. https://doi.org/10.1016/j.arabjc.2009.12.008
Gupta R, Rousseau D (2012) Surface-active solid lipid nanoparticles as Pickering stabilizers for oil-in-water emulsions. Food Funct 3:302–311. https://doi.org/10.1039/C2FO10203J
Hengist A et al (2019) Mobilising vitamin D from adipose tissue: the potential impact of exercise. Nutr Bull 44:25–35. https://doi.org/10.1111/nbu.12369
Horozov TS, Binks BP (2006) Particle-stabilized emulsions: a bilayer or a bridging monolayer? Angew Chem Int Ed 45:773–776. https://doi.org/10.1002/ange.200503131
Jiang H, Sheng Y, Ngai T (2020) Pickering emulsions: versatility of colloidal particles and recent applications. Curr Opin Colloid Interface Sci 49:1–15. https://doi.org/10.1016/j.cocis.2020.04.010
Juárez JA, Whitby CP (2012) Oil-in-water Pickering emulsion destabilisation at low particle concentrations. J Colloid Interface Sci 368:319–325. https://doi.org/10.1016/j.jcis.2011.11.029
Ke C-Y et al (2016) Vitamin D3 reduces tissue damage and oxidative stress caused by exhaustive exercise. Int J Med Sci 13:147. https://doi.org/10.7150/ijms.13746
Kentish S, Wooster TJ, Ashokkumar M, Balachandran S, Mawson R, Simons L (2008) The use of ultrasonics for nanoemulsion preparation. Innov Food Sci Emerg Technol 9:170–175. https://doi.org/10.1016/j.ifset.2007.07.005
Konecsni K, Low N, Nickerson M (2012) Chitosan–tripolyphosphate submicron particles as the carrier of entrapped rutin. Food Chem 134:1775–1779. https://doi.org/10.1016/j.foodchem.2012.03.070
Kumar MR, Muzzarelli RA, Muzzarelli C, Sashiwa H, Domb A (2004) Chitosan chemistry and pharmaceutical perspectives. Chem Rev 104:6017–6084. https://doi.org/10.1021/cr030441b
Li R et al (2019) Bioaccessibility and stability of β-carotene encapsulated in plant-based emulsions: impact of emulsifier type and tannic acid. Food Funct 10:7239–7252. https://doi.org/10.1039/C9FO01370A
Liu F, Tang C-H (2016) Soy glycinin as food-grade Pickering stabilizers: part. I. Structural characteristics, emulsifying properties and adsorption/arrangement at interface. Food Hydrocoll 60:606–619. https://doi.org/10.1016/j.foodhyd.2015.04.025
Liu H, Wang C, Zou S, Wei Z, Tong Z (2012) Simple, reversible emulsion system switched by pH on the basis of Chitosan without any hydrophobic modification. Langmuir 28:11017–11024. https://doi.org/10.1021/la3021113
Liu Y et al (2020) Enhancement of beta-carotene stability by encapsulation in high internal phase emulsions stabilized by modified starch and tannic acid. Food Hydrocoll 109:106083. https://doi.org/10.1016/j.foodhyd.2020.106083
Lobo V, Patil A, Phatak A, Chandra N (2010) Free radicals, antioxidants and functional foods: impact on human health. Pharmacogn Rev 4:118. https://doi.org/10.4103/0973-7847.70902
Loi CC, Eyres GT, Birch EJ (2019) Protein-stabilised emulsions. In: Melton L, Shahidi F, Varelis P (eds) Encyclopedia of food chemistry. Academic, Oxford, pp 404–409. https://doi.org/10.1016/B978-0-08-100596-5.22490-6

Low LE, Tey BT, Ong BH, Chan ES, Tang SY (2017) Palm olein-in-water Pickering emulsion stabilized by Fe3O4-cellulose nanocrystal nanocomposites and their responses to pH. Carbohydr Polym 155:391–399. https://doi.org/10.1016/j.carbpol.2016.08.091

Lu X, Zhu J, Pan Y, Huang Q (2019) Assessment of dynamic bioaccessibility of curcumin encapsulated in milled starch particle stabilized Pickering emulsions using TNO's gastrointestinal model. Food Funct 10:2583–2594. https://doi.org/10.1039/C8FO02495B

Luo L-J, Liu F, Tang C-H (2013) The role of glycinin in the formation of gel-like soy protein-stabilized emulsions. Food Hydrocoll 32:97–105. https://doi.org/10.1016/j.foodhyd.2012.11.031

Matos M, Laca A, Rea F, Iglesias O, Rayner M, Gutiérrez G (2018) O/W emulsions stabilized by OSA-modified starch granules versus non-ionic surfactant: stability, rheological behaviour and resveratrol encapsulation. J Food Eng 222:207–217. https://doi.org/10.1016/j.jfoodeng.2017.11.009

McClements DJ (2004) Protein-stabilized emulsions. Curr Opin Colloid Interface Sci 9:305–313. https://doi.org/10.1016/j.cocis.2004.09.003

Mcclements DJ (2007) Critical review of techniques and methodologies for characterization of emulsion stability. Crit Rev Food Sci Nutr 47:611–649. https://doi.org/10.1080/10408390701289292

McClements DJ (2012) Nanoemulsions versus microemulsions: terminology, differences, and similarities. Soft Matter 8:1719–1729. https://doi.org/10.1039/C2SM06903B

McClements DJ (2015) Food emulsions: principles, practices, and techniques. CRC Press, Boca Raton

Miao M, Li R, Jiang B, Cui SW, Zhang T, Jin Z (2014) Structure and physicochemical properties of octenyl succinic esters of sugary maize soluble starch and waxy maize starch. Food Chem 151:154–160. https://doi.org/10.1016/j.foodchem.2013.11.043

Mitbumrung W, Suphantharika M, McClements DJ, Winuprasith T (2019) Encapsulation of vitamin D3 in Pickering emulsion stabilized by Nanofibrillated Mangosteen cellulose: effect of environmental stresses. J Food Sci 84:3213–3221. https://doi.org/10.1111/1750-3841.14835

Motterlini R, Foresti R, Bassi R, Green CJ (2000) Curcumin, an antioxidant and anti-inflammatory agent, induces heme oxygenase-1 and protects endothelial cells against oxidative stress. Free Radic Biol Med 28:1303–1312. https://doi.org/10.1016/S0891-5849(00)00294-X

Nieva-Echevarría B, Manzanos MJ, Goicoechea E, Guillén MD (2015) 2, 6-Di-tert-butylhydroxytoluene and its metabolites in foods. Compr Rev Food Sci Food Saf 14:67–80. https://doi.org/10.1111/1541-4337.12121

Pal R (2019) Modeling of sedimentation and creaming in suspensions and pickering emulsions. Fluids 4:186. https://doi.org/10.3390/fluids4040186

Patra D, Sleem F (2013) A new method for pH triggered curcumin release by applying poly (l-lysine) mediated nanoparticle-congregation. Anal Chim Acta 795:60–68. https://doi.org/10.1016/j.aca.2013.07.063

Pawlik A, Kurukji D, Norton I, Spyropoulos F (2016) Food-grade Pickering emulsions stabilised with solid lipid particles. Food Funct 7:2712–2721. https://doi.org/10.1039/C6FO00238B

Pichot R, Spyropoulos F, Norton I (2010) O/W emulsions stabilised by both low molecular weight surfactants and colloidal particles: The effect of surfactant type and concentration. J Colloid Interface Sci 352:128–135. https://doi.org/10.1016/j.jcis.2010.08.021

Pickering SU (1907) Cxcvi.—emulsions. J Chem Soc Transactions 91:2001–2021

Piper JT, Singhal SS, Salameh MS, Torman RT, Awasthi YC, Awasthi S (1998) Mechanisms of anticarcinogenic properties of curcumin: the effect of curcumin on glutathione linked detoxification enzymes in rat liver. Int J Biochem Cell Biol 30:445–456. https://doi.org/10.1016/S1357-2725(98)00015-6

Rabea EI, Badawy ME-T, Stevens CV, Smagghe G, Steurbaut W (2003) Chitosan as antimicrobial agent: applications and mode of action. Biomacromolecules 4:1457–1465. https://doi.org/10.1021/bm034130m

Ramsden W (1904) Separation of solids in the surface-layers of solutions and 'suspensions'(observations on surface-membranes, bubbles, emulsions, and mechanical coagulation).—preliminary account. Proc R Soc London 72:156–164. https://doi.org/10.1098/rspl.1903.0034

Robles H (2014) Tannic acid. In: Wexler P (ed) Encyclopedia of toxicology, 3rd edn. Academic, Oxford, pp 474–475. https://doi.org/10.1016/B978-0-12-386454-3.00542-X

Rodriguez-Amaya DB (2015) Nomenclature, structures, and physical and chemical properties. Food Carotenoids: Chemistry, Biology, and Technology. Chichester: John Wiley & Sons, Ltd. 1–23

Rousseau D (2013) Trends in structuring edible emulsions with Pickering fat crystals. Curr Opin Colloid Interface Sci 18:283–291. https://doi.org/10.1016/j.cocis.2013.04.009

Ruby AJ, Kuttan G, Babu KD, Rajasekharan K, Kuttan R (1995) Anti-tumour and antioxidant activity of natural curcuminoids. Cancer Lett 94:79–83. https://doi.org/10.1016/0304-3835(95)03827-j

Sanders TH, McMichael RW, Hendrix KW (2000) Occurrence of resveratrol in edible peanuts. J Agric Food Chem 48:1243–1246. https://doi.org/10.1021/jf990737b

Shah BR et al (2016a) Preparation and optimization of Pickering emulsion stabilized by chitosan-tripolyphosphate nanoparticles for curcumin encapsulation. Food Hydrocoll 52:369–377. https://doi.org/10.1016/j.foodhyd.2015.07.015

Shah BR, Zhang C, Li Y, Li B (2016b) Bioaccessibility and antioxidant activity of curcumin after encapsulated by nano and Pickering emulsion based on chitosan-tripolyphosphate nanoparticles. Food Res Int 89:399–407. https://doi.org/10.1016/j.foodres.2016.08.022

Shao Y, Tang C-H (2016) Gel-like pea protein Pickering emulsions at pH 3.0 as a potential intestine-targeted and sustained-release delivery system for β-carotene. Food Res Int 79:64–72. https://doi.org/10.1016/j.foodres.2015.11.025

Skelhon TS, Grossiord N, Morgan AR, Bon SA (2012) Quiescent water-in-oil Pickering emulsions as a route toward healthier fruit juice infused chocolate confectionary. J Mater Chem 22:19289–19295. https://doi.org/10.1039/C2JM34233B

Summerlin N, Soo E, Thakur S, Qu Z, Jambhrunkar S, Popat A (2015) Resveratrol nanoformulations: challenges and opportunities. Int J Pharm 479:282–290. https://doi.org/10.1016/j.ijpharm.2015.01.003

Tadros T, Izquierdo P, Esquena J, Solans C (2004) Formation and stability of nano-emulsions. Adv Colloid Interface Sci 108:303–318. https://doi.org/10.1016/j.cis.2003.10.023

Tikekar RV, Pan Y, Nitin N (2013) Fate of curcumin encapsulated in silica nanoparticle stabilized Pickering emulsion during storage and simulated digestion. Food Res Int 51:370–377. https://doi.org/10.1016/j.foodres.2012.12.027

Timgren A, Rayner M, Dejmek P, Marku D, Sjöö M (2013) Emulsion stabilizing capacity of intact starch granules modified by heat treatment or octenyl succinic anhydride. Food Sci Nutr 1:157–171. https://doi.org/10.1002/fsn3.17

Wang L-J, Yin S-W, Wu L-Y, Qi J-R, Guo J, Yang X-Q (2016) Fabrication and characterization of Pickering emulsions and oil gels stabilized by highly charged zein/chitosan complex particles (ZCCPs). Food Chem 213:462–469. https://doi.org/10.1016/j.foodchem.2016.06.119

Wang Q, Wei H, Deng C, Xie C, Huang M, Zheng F (2020) Improving stability and accessibility of Quercetin in olive oil-in-soy protein isolate/pectin stabilized O/W emulsion. Foods 9:123. https://doi.org/10.3390/foods9020123

Wang XY, Heuzey MC (2016) Chitosan-based conventional and Pickering emulsions with long-term stability. Langmuir 32:929–936. https://doi.org/10.1021/acs.langmuir.5b03556

Winuprasith T, Khomein P, Mitbumrung W, Suphantharika M, Nitithamyong A, McClements DJ (2018) Encapsulation of vitamin D3 in Pickering emulsions stabilized by nanofibrillated mangosteen cellulose: impact on in vitro digestion and bioaccessibility. Food Hydrocoll 83:153–164. https://doi.org/10.1016/j.foodhyd.2018.04.047

Wu J et al (2015) Pickering emulsions stabilized by whey protein nanoparticles prepared by thermal cross-linking. Colloids Surf B Biointerfaces 127:96–104. https://doi.org/10.1016/j.colsurfb.2015.01.029

Zembyla M, Murray BS, Radford SJ, Sarkar A (2019) Water-in-oil Pickering emulsions stabilized by an interfacial complex of water-insoluble polyphenol crystals and protein. J Colloid Interface Sci 548:88–99. https://doi.org/10.1016/j.jcis.2019.04.010

Zhao L, Du J, Duan Y, Zhang H, Yang C, Cao F, Zhai G (2012) Curcumin loaded mixed micelles composed of Pluronic P123 and F68: preparation, optimization and in vitro characterization. Colloids Surf B Biointerfaces 97:101–108. https://doi.org/10.1016/j.colsurfb.2012.04.017

Zhou B, Gao S, Li X, Liang H, Li S (2019) Antioxidant Pickering emulsions stabilised by zein/tannic acid colloidal particles with low concentration. Int J Food Sci Technol 55:1924–1934. https://doi.org/10.1111/ijfs.14419

Zhou H et al (2020) Nanochitin-stabilized Pickering emulsions: influence of nanochitin on lipid digestibility and vitamin bioaccessibility. Food Hydrocoll 106:105878. https://doi.org/10.1016/j.foodhyd.2020.105878

Zhu YQ, Chen X, McClements DJ, Zou L, Liu W (2018) Pickering-stabilized emulsion gels fabricated from wheat protein nanoparticles: effect of pH NaCl and oil content. J Dispers Sci Technol 39:826–835. https://doi.org/10.1080/01932691.2017.1398660

Zou Y, Guo J, Yin S-W, Wang J-M, Yang X-Q (2015) Pickering emulsion gels prepared by hydrogen-bonded zein/tannic acid complex colloidal particles. J Agric Food Chem 63:7405–7414. https://doi.org/10.1021/acs.jafc.5b03113

Zou Y, Zhong J, Pan R, Wan Z, Guo J, Wang J, Yin S, Yang X (2017) Zein/tannic acid complex nanoparticles-stabilised emulsion as a novel delivery system for controlled release of curcumin. International Journal of Food Science & Technology 52:1221–1228

Zou Y, Yang X, Scholten E (2018) Rheological behavior of emulsion gels stabilized by zein/tannic acid complex particles. Food Hydrocoll 77:363–371. https://doi.org/10.1016/j.foodhyd.2017.10.013

Chapter 9
Advances in the Oxidative Stability Mechanisms of Emulsions

Parth Malik, Man Singh, and Rakesh Kumar Ameta

9.1 Introduction

The diversified abilities of natural resources have proved exceptional dedications to humankind and life on earth. Amongst the several natural events at macro or micro scale, emulsions are recognised as mixtures of two or more immiscible substances. The rationale of emulsion preparation pertains to restricting the access of constitutional phases, allowing null structural attenuations enabling homogenized distribution of electrostatic potential energies of static structures (Schick 1983). Nearly all medicines and several potential food ingredients (having a therapeutic value) are either entirely hydrophobic or at most semi-hydrophilic. 70% of our body content is water; this fact necessitates the aqueous dissolution of hydrophobic drugs to bring in their maximum benefits, from healthcare concern. Understanding emulsion science is all about being logical and cautiously predictive about several fundamental aspects at the same instant. First and foremost, it is essential to know how dispersion is different from solubilization and why we use the term dispersion specifically for emulsions, although its significance and intended purpose is similar to that of solubilisation (Walstra 1983; Shinoda and Friberg 1986). The hydrophobic and hydrophilic parts (most commonly oil and water for emulsions), can never be mixed completely with one another on its own. Therefore, it is a pre-requisite to have a common linker having both hydrophilic and hydrophobic proximities, which is

P. Malik
School of Nano Sciences, Central University of Gujarat, Gandhinagar, India

M. Singh
School of Chemical Sciences, Central University of Gujarat, Gandhinagar, India

R. K. Ameta (✉)
School of Chemical Sciences, Central University of Gujarat, Gandhinagar, India

SMMPISR, Kadi Sarva Vishwavidhyalaya, Gandhinagar, India

M. A. Aboudzadeh (ed.), *Emulsion-based Encapsulation of Antioxidants*, Food Bioactive Ingredients, https://doi.org/10.1007/978-3-030-62052-3_9

enabled by surfactants (Attwood and Florence 1983). Unlike solubilisation, dispersion is a gradual interaction facilitated by characteristic placements of hydrophilic head groups and hydrophobic tails in surfactants (Becher 1981). Since these features are present in a same compound, so there is a tendency that the interacting phases remain very close to each other, with the highest surfactant concentration along the separating region or periphery (Swarbrick 1967). Therefore, there remains a faint surface area domain, where the interactions occur during the formation of emulsions. Along this layer, there is a constant tendency of constituent hydrophilic and hydrophobic phases, to move towards the opposite side. A surfactant tries to minimize this interfacial tension by bringing the hydrophilic and hydrophobic phases closer but yet again, the interactions remains highly restricted and that's why the phases of emulsions can be separated using different mechanisms, that are presently not the focus of interest. In solutions, the solute and solvent phases interact on a greater extent, owing to which the interactions lead to a gradual loss of individual chemical identity of each of the constituents. Therefore, emulsions encompass interactions of hydrophilic and hydrophobic phases through the surfactant activities.

Next aspect regarding the emulsion terminology is the classification, which could be either composition driven or based on dispersed phase in dispersion medium droplet sizes (Gupta et al. 2016). Compositionally, the oil (hydrophobic) and water (hydrophilic) phases of an emulsion can give rise to oil-in-water (O/W) or water-in-oil (W/O) systems, based on their relative extents (McClements 2012). Broadly, this distinction is based on the choice of dispersed phase and dispersion media, with O/W regime for oil as dispersed phase while W/O system carries water as dispersed phase (Mason et al. 2006). This classification is better understood in terms of application prospects, with the pharmaceutical domain utilizing O/W systems to a higher extent (in compliance with physiological environment) while the industrial sector making use of emulsions for solubility enhancement or waste removal purposes, requires a more hydrophobic setup and thus employs W/O systems.

Naturally, the interactions of constituent phases generate surfactant stabilized oil dispersed in water or water dispersed in oil (droplets), where the size regime holds an immense significance with respect to thermodynamic and kinetic stabilities. Conventionally, emulsions can be formed spontaneously at room temperature and pressure conditions but the droplet sizes are quite large that lead to gradual sedimentation because of aggravated aggregation. This aggregation happens since there is no homogeneous distribution of the dispersed phase without providing energy from outside. To overcome this issue, external energy input is needed, enabling a uniform distribution of dispersed phase across the periphery of dispersion medium. This would ensure that across its entire composition, the same stoichiometries of dispersed phase and dispersion medium remain interactive. Fundamentally, there is no clear-cut distinction between micro and nanoemulsions based on particle size. As both comprise of same phases (oil, water and surfactant) and physicochemical properties (such as surface tension, viscosity, relative viscosity, interfacial tension, wetting coefficient), there is a faint distinction between them. Essentially, nanoemulsions have a high kinetic stability due to their comparatively small particle sizes, meaning thereby that each droplet in a nanoemulsion carries low amounts of oil, surfactant and dispersed drug (if they are being used as drug delivery vehicles). Contrary to this, microemulsions are thermodynamically stable systems owing to which there is no risk of phase separation at room temperature unlike nanoemulsions. This distinction can never be without differences in the particle sizes as it is highly difficult to keep the oil and water phases together in small sized droplets (due to the increasing internal Laplace pressure). The nanoemulsions are vulnerable to coalescence and phase separation only because it is very difficult to keep the small oil-water droplets as separated from each other under normal temperature and pressure conditions. Therefore, the stability of nanoemulsions always necessitates their storage at low temperatures, which keeps the kinetic energy of constituent phase molecules under control and avoids their separation. Readers are hereby referred to note that the differences on the particle sizes are not unanimous, vary vaguely in different literature sources and therefore, cannot be relied upon to commence the optimization attempts. However, those of thermodynamic and kinetic stability are more logical, unanimous and uniformly reported but they cannot be so without the difference in particle sizes. Conclusively, it can be said that the transient stability of low particle sizes prevents their long duration existence, gradually resulting in enhanced sizes that are much less kinetically stable.

Thermodynamic stability is decided by the comparison of free energies of the formed nano or micro emulsions to the free energy of separated constituent phases. The requirement of external energy in making nanoemulsions confers a high energy to the dispersed state owing to which these systems are thermodynamically unstable. Such constrictions are much less prevalent in microemulsions, where the dispersed state has a lower energy (and hence a high thermodynamic stability) than the free energies of separated constituent phases. Another aspect of interest in nanoemulsions formation is that it is not possible to reach a limit below threshold even if applied external energy is increased. This is so as the externally applied shear faces a continuous opposition from the within droplet Laplace pressure, making the droplet constriction non-feasible beyond a limit. The mention of droplet sizes is often inconsistent across literature and it is often used interchangeably for droplet diameters but it never means the droplet radii, as confused in some reports. The clarifications regarding the common aspects of these emulsified systems are addressed with reasonable clarity in at least two rigorous literature sources (Anton and Vandamme 2011; McClements 2012). Through these pioneering reports are in the domain of food emulsions, it has been illustrated that there is no peculiar change in physicochemical or thermodynamic properties on reducing the particles size of emulsion from the micrometer to nanometer range. Similarly, the suitability of materials that can be used as surfactants in these systems is also described.

With recent replacements of synthetic materials by the plant and polysaccharide derivatives, the use of these systems have been enhanced in pharmaceutical applications. These substitutes are often recognized as small molecule surfactants and are being increasingly preferred due to their natural origin and easier availability which together lower the toxicity and reduce the cost of developing intact emulsions as drug delivery vehicles. The microemulsions are formed only by the inclusion of these small molecules, as the conventional surfactants are unable to create ultra-low level of interfacial tensions at particular interaction region curvature. Apart from this, these species are peculiarly useful in ensuring the thermodynamic stability of interacting phases since their interactions are mediated via weaker binding forces contrary to the conventional and synthetic surfactants. So, it becomes inevitable to call the emulsions stabilized with these small molecules as microemulsions. One can call an emulsion as microemulsion if during the storage, no phase separation is observed at the room temperature and normal atmosphere pressure conditions (off course, not by adding something from outside randomly). If phase separation occurs during the room temperature storage, the formulated emulsion is a nanoemulsion (having low thermodynamic stability) and if it does not happen, the formulation is a microemulsion (with high thermodynamic stability) (McClements 2012). Thus, despite both micro and nanoemulsions essentially comprise of oil, surfactant and water as constituent ingredients, the distinct mechanisms are not easy to be noted physically or through some peripheral characterization. In general, a higher surfactant to oil ratio (SOR) is required to prepare a microemulsion, contrary to a nanoemulsion. Some very basic and experimental steps also distinguish these two systems, as highlighted in a highly informative contribution by Anton and colleagues. The scientists have noted the pattern of surfactant, oil and water intermixing as decisive factor in calling an emulsion as nano or micro. The distinction recognizes an emulsion as nanoemulsion only if surfactants (or their amicable substitutes as described above) are first mixed with the hydrophobic (i.e. oil) phase. A deviation from this pattern, meaning in case we add surfactant in water instead of oil, will always lead to a microemulsion. This order is peculiarly noted for nanoemulsion formation, irrespective of which in all possibilities, microemulsions are formed (Anton and Vandamme 2011). Please note that the described order of mixing phases pertains to O/W systems.

Therefore, both O/W and W/O systems have their specialized positives and negatives, which become the guidelines for their peculiar applications. For encapsulating drugs, without any doubt, a small particle would be an advantage since it would carry only a little drug quantity that will help in minimizing the onset of abrupt toxic response (Acosta 2009). At the same time, creation of a too low particle diameter is energetically challenging and it leads to the particles with high kinetic energy. Such eventualities leave very little recourses to control their random interaction once these are inside the body, leaving a substantially risky scenario that could mature into alarming toxicities because of elevated oxidative stress (McClements 2005). In this reference, microemulsions are slightly better to handle and control if their droplet size is restricted to <500 nm. While small particle sizes indeed exhibit significant gravitational stability but a too small size typically below 50 nm is disadvantageous for use in physiological conditions. This is because a too small particle size makes it difficult to track and monitor the interactions with cell surface receptors and blood proteins (Sukhanova et al. 2018). A very small particle size confers a high kinetic stability, the instant consequences of which could be highly deleterious. It could complicate the toxic manifestations via wrongful and undesired interactions with circulating proteins or accidental generation of free radicals in the native biochemical pathways. This instability is due to high external pressure applied to keep the oil and water phases together in such a small volume. Though high-energy approaches require specific infrastructure to prepare nanoemulsions, still these exhibit significant physicochemical and storage stability if made using biocompatible emulsifiers (such as edible plant metabolite or proteins). Indeed, it is necessary to screen their optimum performance at a particular temperature and pH, failing which these could induce toxic responses (Xu et al. 2018; Khalid et al. 2017). Thus, it is very important to understand the commercial feasibility of micro and nano emulsions as drug delivery vehicles with reference to their formation approaches, discussed in detail in the subsequent section.

There has been an extensive use of emulsions as delivery vehicles for low- bioavailability natural polyphenols. In such systems, oxidative damage is of considerable importance as too high oxygen may disturb the shelf life of encapsulated compounds and tamper their native biological activities. As a result, it becomes imperative to know about the mechanisms of oxidative balance of emulsions, and optimize them through proper selection of surfactants and co-surfactants. Perhaps that seems to be the evident reason for choosing edible proteins (egg albumins and many others) and plant derivatives (such as gelatine nanoparticles, carotenoids and several others) as the emulsion stabilizers. These entities serve the dual purpose of stabilizing with respect to aggregation and synergize the encapsulated drugs in their antioxidant response. With such insights, this chapter focuses on understanding the relevance of oxidative stability of emulsions and also discusses the recent studies reporting oil-surfactant selection to guard against the time-spanned free radical damage.

9.2 Formation Methods: Energy Considerations for Thermodynamic Stability

Two broad approaches are generally optimized for making nanoemulsions, namely high and low-energy methods. The high-energy approaches make use of specific sophisticated designs, provisioned to supply shear from external means. On the other hand, low-energy approaches rely more on homogenizing the dispersed phase in dispersion medium through variations in physicochemical properties with low external energy requirements. Two most commonly used instruments used to make nanoemulsions using high energy methods are high pressure homogenizer (HPH) and microfluidizer (MF) (Forgiarini et al. 2000; Kabalnov 1998; Zhang et al. 2015). Apart from these devices, ultrasonic homogenization (USH) is also an effective approach to prepare nanoscale dispersed droplets via cavitations (Nejadmansouri et al. 2016; Sivakumar et al. 2014). The detailed working of these instruments could be traced in several literature sources, the common aim of which is to maximize the external shear to transform the intact hydrophilic and hydrophobic cohesive forces into distributed hydrophobic and hydrophilic force expressions. Since these approaches confer high energy to distribute the dispersed phase, the formed emulsions are often vulnerable to revert to separate phases on the removal of this energy. This is the reason for low thermodynamic stability of nanoemulsions, and an important factor limiting the commercial scale utility of nanoemulsions as drug delivery vehicles.

Scheme 9.1 distinguishes the two methods with a little more clarity of their working mechanisms. An important factor affecting the practical application and performance of prepared emulsion system is that how much energy input has been used in making it. The more is the energy input, the greater is the thermodynamic vulnerability. Therefore, a reliable bet seems to focus on improving the kinetic stability of microemulsions rather than making hard-core nanoemulsions. To minimize the concomitant kinetic and thermodynamic stability barriers, studies on making homogeneously dispersed emulsions have substantially focused on counterbalancing of constitutive hydrophilicities and hydrophobicities, so that excess of neither creates an obstruction in attaining a uniform distribution. The terminology in this regard is hydrophilic-lipophilic balance, based on which surfactant has been selected. However, low energy approaches generally require a greater time to achieve a uniform distribution of dispersed phase across the dispersion medium but the net energy of the formed emulsion is lower, conferring a higher stability. Furthermore, high-energy approaches are vulnerable with respect to long-term stabilities and practical applications as the emulsions produced by this methodology could be much more susceptible to Ostwald ripening and random coalescence (Kim et al. 2016; Friberg et al. 2003). It is important to understand these implication

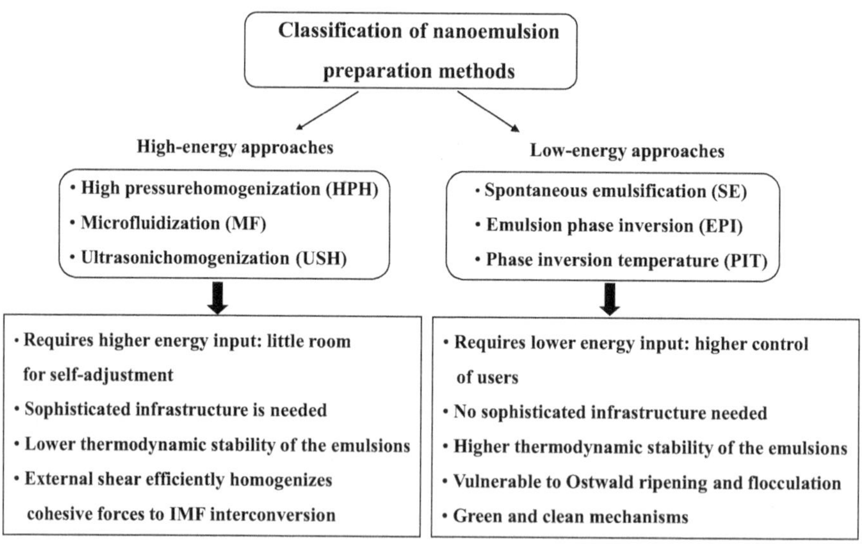

Scheme 9.1 A glimpse of high and low energy emulsification methods, comparing their infrastructural constraints and relative thermodynamic and kinetic stabilities

consequences as the low energy approaches are unable to produce similar reductions in particle sizes of oil dispersed water or water dispersed oil droplets. Therefore, nanoemulsions produced by low energy methods are very close to microemulsions in terms of their dispersion extents and readers are advised to go through the reference (Anton and Vandamme 2011) for this unexpected analogy.

9.3 Surfactants: The Backbone Elements of an Emulsion System

In the broadest sense, surfactants are the chemical substances (mostly organic compounds) having hydrophobic and hydrophilic regions accommodated together. Typical role of surfactants in an emulsion is to lower the surface tension, γ, failing which the constituent phases remain separate and non-interactive. Structurally, surfactants (more familiarly, surface active agents) comprise of a hydrocarbon chain (could be single or multiple) and a hydrophilic head group (Rosen 2004; Vold and Vold 1964; Mobius and Miller 2002). That's why surfactants are sometimes also referred to as amphiphiles. Functionally, the assortment tendency of hydrophobic chains is towards the centre or bulk contrary to the hydrophilic head groups that align towards the periphery and are engaged with water molecules. The separation of hydrophilic and hydrophobic regions is more vigorous along the surface of an emulsion, so surfactants are also frequently referred to as surface active agents.

Usually the arrangement of this head and tail assembly is spherical but spherical or cylindrical shapes are also reported with variation in hydrophilic or hydrophobic contributions. The hydrophobic region of a surfactant is known to induce the Brownian activities that are quantified in the dispersion analysis using dynamic light scattering (DLS) approach. Depending on head group ionic sensitivities, the surfactants could be cationic, anionic or non-ionic (Kosswig 2000; Rosen and Kunjappu 2012). Some other categories are also recognized based on single or multiple hydrophobic chains, which would be discussed in detail here. While reading about surfactants from relevant sources, readers would repeatedly get accustomed to the term "interface". In emulsions, interface is a region that is heterogeneous in its chemical composition; in fact, it is neither entirely hydrophobic nor hydrophilic. Due to this peculiarity, surfactants align along the interface with oppositely positioned hydrophilic and hydrophobic phases (Fig. 9.1).

Typically, surfactants modulate the interaction of dispersed phase and dispersion medium through adsorption at the philicphobic interface. The most common cationic surfactants are cetyltrimethyl ammonium bromide (CTAB) and dodecyltrimethyl ammonium bromide (DTAB), comprising of 16 and 12 carbon alkyl chains (C-AC) respectively. Both surfactants have quaternary ammonium ion as their hydrophilic moiety. The simplest and most widely used anionic surfactant is sodium dodecyl sulphate (SDS), containing a 12 C as hydrophobic domain and positively charged sodium as hydrophilic site.

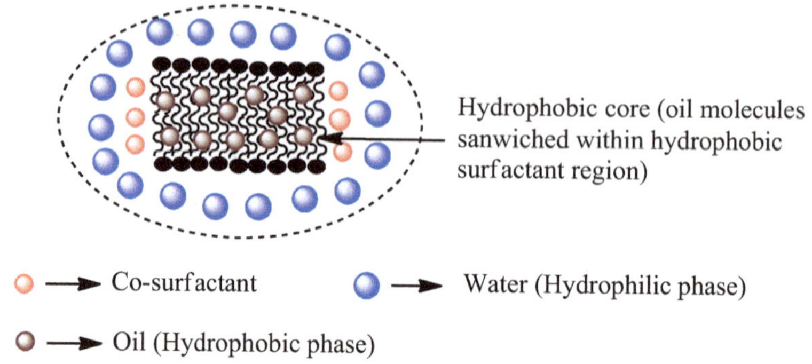

Fig. 9.1 Pictorial depiction of surfactant and co-surfactant/co-solvent placements in an O/W emulsion system

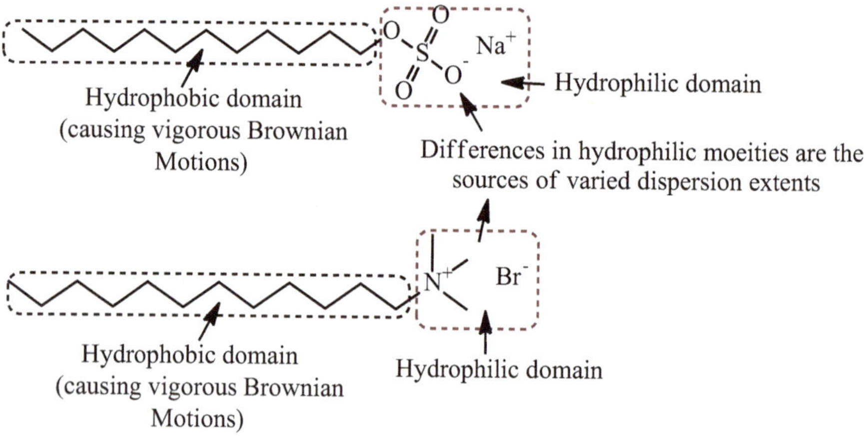

Fig. 9.2 Distinct aqueous interaction mechanisms of SDS and DTAB, with a greater hydrophilicity of SDS mediated by its SO_4^{-2} and Na^+ hydration sphere formation

Figure 9.2 illustrates a structural distinction of the SDS (anionic) and DTAB (cationic) surfactants, where it is shown that hydrophobic regions of the two molecules are identical with 12 C. So, if the two surfactants are dissolved in water at the same concentration and the two aqueous solutions are compared for the dispersion efficacies, the differences will be caused due to structurally distinct hydrophilic regions. The DTAB molecule having three electron releasing $–CH_3$ groups attached to its $\equiv \overset{+}{N}-$, weaken its hydrophilicity, making DTAB more hydrophobic than SDS, having no interfering species to inhibit its SO_4^{-2} and Na^+ structural activities. Unlike cationic and anionic surfactants, the nonionic surfactants do not have charge on their counterions, due to which their interactions with the dispersion medium are controlled by weaker non-Columbic forces. The most common examples of nonionic surfactants are Tweens, which are available in varying chemical functionalities characterized by their peculiar numeric values of 20, 40, 60 or 80. Due to their moderate binding forces and comparatively much lower aqueous structure breaking activities, the nonionic surfactants find an extensive usage in pharmaceutical and cosmetic domains, being the familiar constituents of skin care lotions, shampoos, creams and sometimes even in ointments. Other than Tweens, polaxamer-407 is also a widely used nonionic surfactant, frequently used in drug delivery purposes. The polymeric subunits (comprising of ethylene and propylene glycol species) in blocked assemblies confer a distributive philicphobic interaction ability to this surfactant (Azeem et al. 2009). The moderate binding forces in aqueous environments of nonionic surfactants enable a much better control of their dispersion activities than ionic surfactants, where it is energetically difficult to exert similar control on the interactions (Napper 1983). Readers are suggested to have a look at more specific literature sources regarding the understanding of aqueous structure breaking activities of ionic surfactants that are responsible for a weaker dispersion as compared to the nonionic surfactants.

Several studies have shown finer dispersion activities for additive surfactants than using them alone, with interesting differences being noted for the dual mixtures of cationic and nonionic or anionic and nonionic surfactants. The combinations of nonionic and ionic surfactants results in moderation of binding forces and enhances the dispersed phase distribution (Colombie et al. 2000). Similarly, if cationic and anionic surfactants having identically (opposite) charges on their counterions mix in the same concentrations, it is likely to nullify their net ionic contributions and enable a better dispersion than either of them alone (Vautier-Giongo et al. 2005; Kume et al. 2008). Due to such artefacts, several studies of recent origin include the edible grade (or in quantity) of neutral amphipathic molecules (such as short AC alcohols or polysaccharide derivatives such as sugar alcohols etc.) for making drug encapsulated nanoemulsions. Other than cationic and anionic variations, Gemini and zwitterionic surfactants (ZS) are also known, having variations that are intended to achieve the chemical functionalities of dual surfactant molecules (Kamenka et al. 1995). The Gemini surfactants (GS) are the surfactants having two hydrophilic head groups and two hydrophobic tails linked by an intervening spacer moiety placed close to the head groups. There is no particular restriction on the spacer for being hydrophobic or hydrophilic, however depends on the application intended for which one can have this interlinker as a fixed or rotating molecule. These features of GS make them superior dispersing agents than cationic or anionic surfactants, owing to which these molecules find overwhelming suitability in the drug delivery, solvent recovery, metallurgical operations, waste treatment, catalysis technology and several others (Brycki et al. 2017; Menger and Littau 1993). Similarly, ZS are the molecules having positive and negative charges on a single molecule (also known as amphoteric surfactants), either located adjacently or being farther apart. Typically, the positively-charged moiety can be a quaternary ammonium ion, a phosphonium ion, while the negatively-charged site can be anionic radicals such as sulfate, carboxylate, or sulfonate (Gavin 1978). The net charge of a ZS is zero at neutral pH condition, although some of these are susceptible to pH changes in solution state. Owing to these specialties, the ZS are preferential candidates while making emulsions intended to improve the bioavailability of poor aqueous solubility drugs

(Rosen and Zhu 1984; Kwaambwa and Nermark 2013). The investigations reporting improved structural expression of drugs via dispersion in nanoemulsions often characterize the drug loaded systems with the pH stability and ζ-potential analysis. These characterizations are based on the positive or negative charges orientation of drug and medium interacting regions owing to which there is a formation of electrostatic stern layer. So, if ZS are used in the dispersion medium to make the drug loaded nanoemulsions, it is much less likely to face the risk of charge driven immobilization, thereby resulting in higher kinetic stability and better bioactivity of dispersed phase through its uniform distribution (Kamenka et al. 1995). Before moving to stability, strengthening controls of nanoemulsions, it is imperative to understand the surfactant dynamics in varying hydrophilic and hydrophobic environments. Since the encapsulated compounds are mostly hydrophobic, a hydrophobic excess in the system often leads to coalescence (often recognized as Ostwald ripening), finally leading to settling. Furthermore, long-term storage of drug encapsulated nano and micro emulsions is a primitive factor to elongate a homogeneous distribution of constituent phases. In addition to this, the most important factor is that the encapsulated drugs should not undergo any structural degradation in the dispersed form. To ensure this, the interactions of drugs or biologically active compounds (BAC) must not be energetically strenuous or induce any harsh utilization of fed energy input. That is why emulsions for drug delivery purpose are often stabilized in a cautious manner that induces no harsh structural changes in the intact morphology. Such constraints could be matured only if the interactions are non-electrostatic and are controlled through van der Waals forces (vdW), London dispersionforces (LDF), steric stabilizations, dipole-dipole interactions (DDI) and hydrogen bonding (HB) (Malik et al. 2016; Malik and Singh 2017).

9.3.1 *Impact of Surfactant Choice on the Long Term Stability of Emulsions*

The surfactants function by their adsorption along the interface separating the hydrophilic and hydrophobic phases. In course of interface generation, the adequate adsorption of surfactant is hindered by its diffusion into either of the phases. Sometimes, repulsive interactions controlled by steric or electrostatic mechanisms create an energy barrier that opposes the surfactant adsorption, so the rheology of surfactants should be able to arrest its excessive interaction with hydrophilic or hydrophobic phase. For example, anionically sensitive SDS lowers the surface tension (γ) of water from 71.97 mN•m^{-1} (at 25 °C) to 51.98 mN•m^{-1} while non-ionic Tween 20 (at the same concentration) lowers the same to 38.23 mN•m^{-1} (Malik et al. 2014). Figure 9.3 distinguishes the aqueous interactions of SDS and Tween 20, where SDS is depicted to cause a vigorous structure breaking in the structured water-water IMFs. This is because of the stronger electrostatic interactions of water molecules with SO_4^{-2} and Na^+. Contrary to this, Tween 20 has no ionic centre and

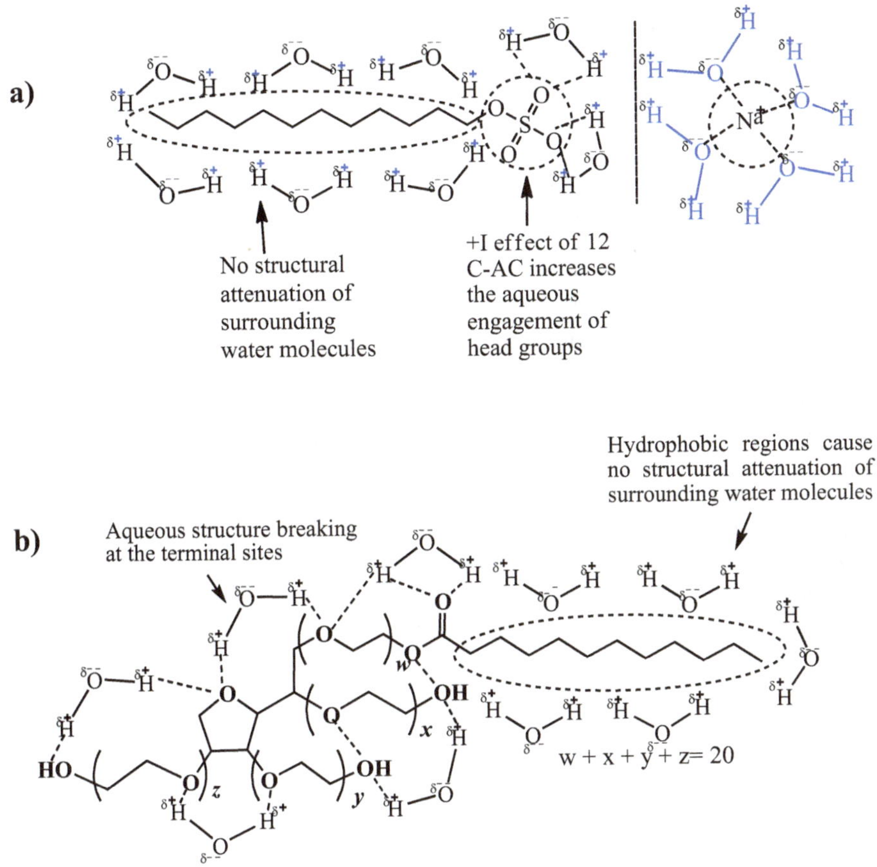

Fig. 9.3 Distinct aqueous interaction mechanisms for (**a**) SDS and (**b**) Tween 20. The sulphate (negatively charged) and sodium (positively charged) counterions in SDS contribute to aggressive coalescence via distinct alignment of water molecules. Contrary to this, Tween 20 interacts non-covalently with water molecules and there is no strong structure making or breaking, preventing any chance of immmobilization

merely reorients the water molecules via HB interactions involving its O and –OH groups. This is the reason why nanoemulsions stabilized with the blend of SDS and Tween 20 would provide a finer dispersion (lower particle size and polydispersity index and higher ζ-potential) than SDS alone. Also in this case, the surfactant activity of SDS is partially compensated by Tween 20 molecules unlike that of only SDS-stabilized systems. Now, since Tween 20 does not induce any hydration sphere formation and clustering due to its extensive hydrogen bonded interactions (the structural contributions of O and –OH groups), the emulsions stabilized with the blend of SDS and Tween 20 comprise of a smaller extent of non-covalent interactions compared to only SDS systems. Thus, dispersion is likely to be more uniform in (SDS + Tween 20) systems.

In order to understand the specific dispersion mechanisms of nonionic and ionic surfactants, knowledge regarding their aqueous solubility is a pre-requisite. Many studies have discussed this important fundamental aspect and unanimously concluded that aqueous solubility of a surfactant varies with increasing temperature. In this regard, readers are suggested to go through the extensive studies of Kahlweit et al. (1989). The compilation reports that while solubility of ionic surfactants increases with an increase in temperature, the reverse is true for nonionic surfactants. This is supposedly due to the fact that nonionic surfactants are extensively composed of propylene groups which get dehydrated at higher temperatures, consequently conferring a lipophilic character to these molecules. This information can be very vital in optimization of hydrophilic-lipophilic balance (HLB) value for nanoemulsions using ionic and nonionic surfactants and infers that although nonionic surfactants are more likely to exhibit an aqueous likeness at lower temperatures but at higher temperatures, this is not true. Therefore, when nanoemulsions are being used to deliver the drugs within the body, it is necessary that they should be designed with a more aqueous constitutional make up so that there is no undesired influence of physiological temperature range on the bioavailability of encapsulated drug molecule(s).

These solubility constraints must also be kept in mind while using ionic surfactants with the risk of greater aqueous structure breaking at physiological temperatures. It seems that the above-mentioned distinctions in aqueous interactions of ionic and nonionic surfactants are the main reasons why most studies report a greater homogeneity in the dispersed phase distribution with a mixture of the two moieties, which enables the attainment of adequate HLB values. In designing (W/O) nanoemulsions, generally, a hydrophobic tendency of the medium ingredients is preferred, owing to which nonionic surfactants would be preferable but again a too high hydrophobicity may induce coalescence, ultimately resulting in aggregation initiated sedimentation. So, in vitro or small-scale, optimizations are needed before implementing the same combinations on practical or pilot scale. On many occasions, it could be realised that the increasing hydrophobicity (in particular for O/W nanoemulsions) is a decisive factor in streamlining the existing hydrophilicity of the medium and only a particular limit of dispersed phase is able to induce a characteristic effect in a desired proportion. Take for example, the antioxidant activity determination of flavonoid dispersed formulation in an O/W nanoemulsion. Such activities are usually determined through standard in vitro radical scavenging assays, with the concurrent measurements of UV/Vis absorbance at characteristic wavelengths. It is exciting to note that the same compound when dispersed in the dispersion medium (comprising aqueous surfactant + co-surfactant or co-solvent) of an O/W nanoemulsion at lower concentration produces <50% radical scavenging, while the similar compound in enhanced concentration can even cause a scavenging of >80% in same dispersion medium (Malik et al. 2014; Malik and Singh 2017). As the dispersed compound is inherently hydrophobic, it implies that hydrophobicity plays a key role although not entirely as only the dispersed phase is pharmaceutically active ingredient of the formulation. Since the interactions of encapsulated bioactive compounds are highly random, their dispersion states can be variable with respect to time. Furthermore, it is highly essential that hydrophobic contributions balance with hydrophilic sensitivities. Therefore, encapsulated concentration of dispersed bioactive may remain unexpressed if the dispersion in the neat emulsion is not uniform. Furthermore, sensitivity towards instantaneous hydrophobicity may engage the encapsulated extents to localized interactions. Therefore, a given oil and surfactant composition is favourable only until a particular concentration of encapsulated bioactive compound encapsulation as excess of that may disturb the philic-phobic equilibration and hinder an adequate structural expression.

9.4 Why It Is Important to Have a Good Stability Against Oxidation in Emulsions

The hydrophobic phase of emulsions is normally rich of oils and fatty acids (FA), which are vulnerable to oxidation if bound by a combination of weak binding forces. The oxidation of these materials in the emulsions is even more risky since it could affect the chemical activities of encapsulated drugs. Thus, it is essential to ensure the oxidative stability of drug encapsulated emulsions (micro or nano), which could be accomplished most significantly by careful selection of surfactants or emulsifiers. This is because other than surfactants, co-surfactants and co-solvents, the aqueous phase is richly endowed with multifold materials that can promote the lipid oxidation, such as transition metals, enzymes and photosensitizers. These species play critical roles in affecting the redox balance of interfaces with a concurrent involvement of hydrophobic and hydrophilic phases. For instance, the transition metals in the aqueous phase are likely to promote the formation of lipid hydroxyl radicals from lipid peroxides located near the liquid droplet surfaces. Subsequently, these radicals may then diffuse into the inside of droplets and initiate the oxidation of other lipids.

Many interesting studies have reported the oil and FA as vulnerable constituents towards the oxidative degradation, likely to result in harmful flavours and rancidity. Since surfactants are the most intimate constituents in contact with oil phase in the emulsions, so to arrest the oxidative degradation of oils, the choice of surfactants should be seriously made and that is why these observations have given rise to the notion of antioxidant emulsifiers and co-surfactants. The main types of oils contributing to a greater risk of oxidative damage are the unsaturated FA, such as both linoleic and linolenic acid that are mostly found in edible oils. Although many edible oils also contain palmitic and stearic acids (both saturated) their proportion is lower, particularly in peanut and cottonseed oil. As toxicity concern mandates the selection of edible oils as the hydrophobic phase of drug delivering emulsions, so such risks caution against the selection of oils having only saturated FA as their constituent moieties (Miyashita et al. 1993, 1995). The following subsection discusses some of the studies focused at optimum selection of oil, surfactant and co-surfactant stoichiometries.

9.4.1 Discussion of the Oxidative Stability Strengthening Attempts in Emulsions

Early as in 1999, Mei et al. studied the effect of emulsion droplet and dispersed phenolic antioxidant charges along with the pH on the lipid oxidation rates. The negatively charged and neutral emulsions were prepared using SDS and Brij (polyethylene glycol dodecyl ether). Justification offered for gallamide, methyl gallate and gallic acid selection as antioxidants was the similar position of their functional groups which also conferred them a similarity with commonly used food antioxidants. The free radical scavenging activities of galloyl derivatives were assessed by their ability to arrest the peroxyl radical [originating from AAPH] initiated decay of porphyridiumcruentum β-phycoerythrin fluorescence. The inhibition in the lipid peroxidation with for Brij emulsions was noted to be pH dependent with significant activities for gallyol derivatives (at 5 and 500 μM) at neutral pH. The comparative efficacies followed the order methyl gallate > gallamide > gallic acid. At pH = 3, low concentrations of the galloyl derivatives were pro-oxidative while high concentrations were noted as antioxidant. Though both emulsions expressed significant radical scavenging, yet a lower oxidative stability was noted for SDS formulations. Both emulsions displayed pH dependent oxidative stability expressions, with SDS producing 4.8- and 6.0-fold higher TBARS (thiobarbituric acid reactive substances) concentrations at pH 3 and 7 respectively. Similarly, lipid peroxide concentrations were 13 and 19 fold higher at 3 and 7 pH, both counts were made after (48–54) hour oxidation. Notably, a greater lipid peroxide formation was observed for SDS formulations at acidic pH although TBARS formation remained same at acidic and neutral pH. Similarly, Brij emulsions also expressed a pH dependent oxidative stability, noted higher at acidic pH, screened via both TBARS and lipid peroxide generation. Upon increasing the galloyl derivative concentration to 500 μM, lipid oxidation rates sharply decreased with (88–97) wt% inhibition of TBARS formation and (75–98) wt% ranged lipid peroxide formation, after (48–54) hours of oxidation. Higher concentration distinguished the structural contributions of chosen antioxidants, with greater antioxidant potentials for gallic amide and methyl gallate than gallic acid. So, the methyl and amide group substitution facilitated an adequate H^+ population for a higher scavenging. Overall, Brij emulsions caused a greater antioxidant activity for high concentrations of galloyl derivatives (with lower concentrations inducing a pro-oxidant effect). On the contrary, in SDS formulations, the galloyl derivatives remained inactive for longer durations. The differences in antioxidant activities could be attributed to strongly ionic and non-electrostatic interactive forces with SDS and Brij, that affected greater and lower binding forces of the constituents. The greater activity with Brij infers a higher availability of galloyl derivatives to the free radical sources. Therefore, this study could be a platform to make an optimum choice of surfactant *vis-a-vis* pH stability and charge driven interactions of emulsion droplets with the dispersed phenolic antioxidants (Mei et al. 1999).

Rigorous studies by McClements and Decker group screened the oxidative stability of O/W emulsions through multiple likelihood factors, such as the effect of antioxidant species, variation in the hydrophilic-hydrophobic compositions of the interface, changes in the free FA expressions and activities, storage conditions, droplet charges and specific structural contributions of emulsifiers. Perhaps in earliest of these efforts, Hu et al. assessed the comparative oxidative and physical stability of positively charged O/W emulsions stabilized with casein, whey protein isolate (WPI), or soy protein isolate (SPI) at pH = 3. While a constant (5%) amount of corn oil was needed to make emulsions, the relative protein extents varied with 1.5% contribution of casein followed by SPI and the least with WPI (greater or equal to 0.5%). Significantly, though all emulsions developed positive charge in the order (WPI > casein > SPI), still the order of oxidative stability was different with maximum for casein and minimum for SPI (estimated using hydroperoxide and consequent derivative, hexanal formation). Thus, this investigation projected the role of other factors (besides oxidative stability and electrical charges) as being responsible for making oxidation stable emulsions. These factors include the interfacial film thickness, protein chelating properties and distinct free radical scavenging activities of the constituent amino acids. In summary, it was observed that casein aided in the formation of thick interfacial layers and immensely significant chelating attributes, contributing to enhanced formulation stability (Hu et al. 2003). From this research, the structural differences of casein, WPI and SPI with respect to hydrophilic and hydrophobic orientations of their constituent amino acids, emerges a key factor contributing to emulsion stability. Two subsequent and nearly similar investigations focused on free FA (specifically, oleic acid), studied oxidation of emulsion droplets in a pH dependent manner. Other than using oleic acid, these two investigations were also similar in the use of soybean oil. The first of these studies (in 2009) noted a pH dependent oleic acid mediated decrement of lipid oxidation, to characteristically low limits at low pH. This reduction in oleic acid lipid oxidation activity was ascertained through its inability to rise the negative charge of emulsion droplet. Sharp decrement in lipid oxidation extents with EDTA counter confirmed such observations, suggesting that hydrperoxides on free FA are not strong pro-oxidants in O/W emulsions and the pro-oxidant activity of free FA in O/W emulsions is due to their virtue to attract transition metals to the negatively charged emulsion droplet surface (Waraho et al. 2009). The second investigation presents a more focused study on studying the factors regulating the free FA pro-oxidant activity, making a critical case of unsaturation extent and the shape with respect to geometrical changes in the free FA chemical activities. This study assessed the pro-oxidant activity of linoleic, linolenic and oleic acids, wherein it was noted that the comparative pro-oxidant extent depended on the FA type with the order oxidation as, linolenic< linoleic < oleic. No significant distinctions in lipid oxidation rates were noted on comparison of *cis* and *trans* free FA isomers, ultimately concluding that the pro-oxidant action of free FA is due to their ability to attract pro-oxidant metals as well as co-oxidize the triglycerides in the oil phase. Interesting conclusions of this investigation were the suppressed oxidative actions with increased unsaturation extent of free FA, owing to which polyunsaturated FA caused a lower lipid oxidation than monounsaturated FA, pointing out at the likelihood more linear shape of di and tri unsaturated FA as the key structural distinction. Another fundamental link for pro-oxidant functioning of free FA is the charge on the emulsion droplet as a negatively charged droplet is expected to strongly interact with the positively charged transition metals. The study also left few unanswered trends, since a comparison of oleic, elaidic and linoelaidic acids revealed a much lower interaction of emulsion droplet negative charge with linoelaidic acid despite causing a highest lipid peroxidation. The scientists concerned cited the possibility of simultaneous oxidation (of their own) and co-oxidation of the oil triacylglycerols (Waraho et al. 2011).

Another study specifically describing the protection of olive oil oxidation in O/W emulsions was reported in 2010, via assessment of hydroperoxide concentration through a sensitive fluorimetric method. The lipophilic 2,2′-azobis(2,4-dimethylvaleronitrile) (AMVN) was used as a radical source while diphenyl-1-pyrenylphosphine (DPPP) was used as a probe owing to its stoichiometric reacting ability with hydroperoxides to form diphenyl-1-pyrenylphosphine oxide (DPPP-O), a fluorescent product. The interesting observations regarding oil oxidation revealed an enhanced oxidation rate in presence of a radical initiator and a large interface, predicting a higher activity of radical species across the interface than the oil or water phase. Apart from this, a role of surface area of dispersed droplets was found to be critical, since the addition of oil dispersed saturated ascorbic acid solutions caused a considerable reduction in the oxidation rate even in the presence of highest AMVN concentration. The radical source AMVN decomposed in a temperature dependent manner, with a dissociation constant of $2.5 \times 10–4\ min^{-1}$ at 40 °C and the rate of initiation of oxidation was computed using vitamin E as antioxidant. Although a different radical source would induce the oxidation in a strikingly different manner and may not exhibit the similar reaction kinetics with vitamin E (α-tocopherol). Nevertheless, a lower droplet size is likely to reduce the risk of oil oxidation and if there are any risks of such a possibility, these are more certain along the interface than in the oil or water alone (Mosca et al. 2010).

Another pertinent investigation reported the pro-oxidant activity of photosensitized riboflavin and was valuable in understanding light mediated enhancement in free radical generation in O/W emulsions (Lee and Decker 2011). Emulsions were prepared with and without riboflavin and visible light irradiation, where emulsions possessing them showing a greater extent of higher lipid hydroperoxides and volatiles compared to those made without them. The assessment of oxidation sources was undertaken through monitoring the metal chelator, sodium azide and superoxide dismutase activities as functions of transition metals, singlet oxygen and superoxide anion activities. The reduced lipid hydroxide formation by EDTA and sodium azide inclusion created a suspicion of lipid hydroperoxides and superoxide anion aggravated pro-oxidant response of photosensitized riboflavin. The results of this study could be extended to optimum storage conditions of O/W emulsions as well as structurally similar to riboflavin compounds for a better understanding of emulsion storage stability.

Two other studies in 2012 further confer importance to the oxidative stability of O/W emulsions (Charoen et al. 2012; Panya et al. 2012). The first one reported the storage stability (on the basis of pH and biopolymer emulsifier type) of oxidatively vulnerable rice bran O/W emulsions. In this study, WPI, modified starch (MS) and gum arabic (GA) were used as emulsifier and could be therefore specially suitable to decide the most stable of these natural emulsifiers, a relatively uncommon and unaddressed prospect. Significantly, all developed formulations had particle size <0.5 μm with high emulsifier concentrations (0.45 wt%, 1 wt% and 10 wt% for WPI, MS and GA respectively). The fact that all formulations remained (except little deviations for MS stabilized formulation) stable until 20 days, distinguishes a mechanism of dispersion and oxidative stability. To ascertain their oxidative stability, the formulations were stored at 37 °C, for ~3 weeks, in the presence of a pro-oxidant (iron/EDTA). Monitoring the formulation response at pH 3 and 7, it was noted that GA stabilized emulsions showed an accelerated oxidative tendency at 3 as well as 7 pH, whereas MS and WPI formulations accelerated the oxidation only at 7 pH. Furthermore, amongst all, the GA formulations had maximum percentage of biopolymer, inferring a highest extent of non-adsorbed biopolymer. The differences in the biopolymer activities could be most likely due to the relatively thicker and more porous interfacial layer formation by GA and MS, contrary to WPI, forming a thin and dense interfacial layer. Since, physicochemical mechanisms illustrating the differences in the oxidative stability of biopolymer coated lipid droplets remain poorly understood, so how the cationic state iron got more oxidatively activated by the biopolymer coated lipid droplets, remains an unanswered question. Nevertheless, the results definitely provided vital inputs for selection of structurally amicable bio-emulsifiers with respect to oil phase of emulsions (Charoen et al. 2012).

The second study noted an unfamiliar antioxidant response, with lower lipid oxidation activity for non-polar eicosylrosmarinate (esters carrying 20 C-AC) in O/W emulsions compared to 4, 8 and 12 C-chain esters. Notably, the antioxidant activity of 20 C-ester significantly increased within the surfactant micelles whereas decreased for 4 and 12 C-chain esters. Owing to repetitive mention of interfacial region as the site for lipid peroxidation, the surfactant micelles displayed similar mode of action. Such a similarity of performance index in surfactant micelles was ascertained through partitioning models, fluorescence behaviour and interaction of 20 carbon chain ester with 4-hexadecylbenzenediazonium (the standard interfacial probe). The reason for this unusual greater lipid peroxidation for higher AC esters in surfactant micelles has been postulated as micelle enabled localization of a certain region (in the 20 C-ester) inside the emulsion droplet core, enabling the surfactant micelles to increase its interfacial concentrations, owing to which a stronger free radical scavenging happened via decomposition of interfacial lipid hydroperoxides. The results of this study could be extended to understand the antioxidant actions of structurally similar (to eicosylrosmarinate) antioxidants such as tocopherols and many others (Panya et al. 2012).

In 1999, Freiría-Gándara *et al* reported a similar study, in which they evaluated the antioxidant efficacies and phase distribution of gallic acid and a series of alkyl gallates (as a function of progressive enhancement in hydrophobicity) in fish oil-in-water emulsions. These scientists noted a key role of phobic-phobic interactions towards inducing an antioxidant activity whereby the degradation of fish oil lipids was prevented. This was concluded by the highest antioxidant activity for the octyl derivative of alkyl gallates, although beyond this, the activity decreased for 12-carbon chain derivative. Most important observation of this study was the incremental interfacial concentration (rather than oil region) of the alkyl gallates, which were nearly (20–100) times more than stoichiometrically added antioxidant concentration. Contrary to this, the concentration in the oil region was either same or slightly lower (1–6 fold) and that in aqueous phase was much lower (0.8–10 fold). Apart from this, increasing oil to water stoichiometry stimulated the interfacial activity of hydrophilic antioxidant and decreased the contribution of hydrophobic compounds. Thus, by this study it can be concluded that oxidative stability of oil phase in the emulsions is prolonged by a stronger antioxidant expression in the interphase rather than in the hydrophobic core. This implies that philicphobic force gradients of surfactants and other moieties in the emulsions are pre-requisite to ensure their adequate structural expression. The results of this study could be valuable databases to include the optimum oil and surfactant concentration for designing stable emulsions so that the oil phase does not undergo any undesirable chemical deterioration (Freiría-Gándara et al. 2018).

Another significant effort by Pan *et al* in 2013 sheds light on the antioxidant activity of lecithin and Tween 20 to strengthen oxidative stability of curcumin in O/W emulsions. The authors have cited a critical role of the permeation of free radicals generated at the emulsion interface. While studying the interdependence of radical permeation rates (from oil to emulsion interface), the scientists noted that the rate with which peroxyl radical permeated from aqueous to oil phase in an emulsion was inversely proportional to the antioxidant expression of the emulsifiers. Native lecithin expressed a higher antioxidant activity compared to its oxidized state combination with Tween 20, owing to which higher stability of encapsulated curcumin was observed for emulsions stabilized with lecithin than Tween 20. This study therefore explains the selection of suitable emulsifiers pertaining to a higher oxidative stability of encapsulated bioactive compound as well as arresting the permeation of free radicals generated near the hydrophobic phase to the oil-water interface periphery (Pan et al. 2013).

In another study, Yamamoto et al. demonstrated the antioxidant activities of a series of lipophilic antioxidants, δ-tocopherol, epigallocatechin gallate, quercetin, green tea extract and rooibos tea extract using Tween 20, polyglycerol and sucrose esters of FA as surfactants. In course of respective antioxidant activity evaluation, tocopherol was found to exhibit a stronger effect than green tea extract by suppressing the increase in peroxide value of emulsified milk fat (the oil phase of studies emulsions). Interestingly, the antioxidant effect of green tea extract was enhanced by the inclusion of polyglycerol and sucrose esters of FA. At 40 °C in dark with AAPH, the antioxidant activity of tocopherol was noted as highest (greatest suppression of peroxide value). Significantly, some additional emulsifiers also exhibited suppressed incremental extents of peroxide expression even in the absence of antioxidants, whereas similar activity was enhanced in the presence of antioxidants.

Therefore, it could be concluded that appropriate blending of surfactant (type) and polyphenolic antioxidant is the decisive factor in modulating the stability of emulsion oil phase (Yamamoto et al. 2014). We also reported similar observations in our curcumin encapsulated nanoemulsions, which were made sequentially with SDS, DTAB, poloxamer-407 and Tween 20. The prevention of undue cottonseed oil (the selected oil phase) oxidation was noted for DTAB and Tween 20 compared to anionic SDS and nonionic poloxamer-407. Poloxamer 407 is a hydrophilic non-ionic surfactant of the more general class of copolymers known as poloxamers. Although both surfactants are non-ionic, higher antioxidant activities with Tween 20 than Poloxamer-407, illustrated a key role of specific Tween 20 structure, richly endowed with multiple propylene groups that impart a lipophilic character to the surfactant at higher temperatures. Similarly, more hydrophobic DTAB (as discussed earlier) caused a higher DPPH scavenging than SDS stabilized formulations (Malik et al. 2014).

Another valuable attempt by Liu *et al* talks about the elongation in β-carotene stability profile in its encapsulated form in gum arabic-stabilized O/W emulsions containing α-tocopherol, tertiary butyl hydroquinone (TBHQ) and ascorbyl palmitate as antioxidants. The effect of antioxidants on retaining the oxidative stability of emulsions was studied at 4, 25, 45 and 65 °C in the dark, respectively. The chosen antioxidants varied in maintaining the β-carotene thermal stability in the order, α-tocopherol > TBHQ > ascorbyl palmitate, signifying a most robust antioxidant effect of α-tocopherol (Liu et al. 2015). The fact that β-carotene did not serve as oil phase in this study may seem to be unfitted in the theme of present article but it is equally important to note here that compounds like β-carotene and other polyphenols are sometimes solubilized in oil phase before being dispersed in the complete emulsions. Therefore, here the oil and dispersed polyphenol or low bioavailability drugs are operating together in the hydrophobic phase. The retention of oil stability (after considerable time durations) in such cases is obviously concerned with the fact that encapsulated polyphenols or drugs too, are not subjected to undesirable oxidative environment that may result in a loss in their chemical activity. A slightly different study focused on determining the effect of salt (additives) on lipid peroxidation and observed the changes in lipid peroxidation tendencies by supplementing Tween 20 stabilized corn oil based O/W emulsions with increasing concentrations of sodium and potassium chlorides at neutral pH (Mei et al. 1998). The emulsions were prepared with and without metal chelators, and sodium chloride exhibited a dose-dependent lipid oxidation (determined through concentration of lipid peroxidation product, hexanal). Notably, similar patterns of lipid oxidation were noted for potassium chloride, making this effect cation independent. It was also noted that addition of EDTA significantly reduced the lipid oxidation in emulsions, suggesting a salt chemical activity facilitated out of the interfacial region movement/ transport of emulsified oil droplets. Thus, salts stimulated the hydrophilic activities of emulsified droplet populations since the lipid peroxidation is usually the highest along the interfacial regions. Furthermore, since EDTA reduced this salt induced lipid peroxidation, there could be a possibility that it might have catalyzed the micelle formation through which interfacial movement of emulsified oil droplets could be driven (Mei et al. 1998).

Another significant investigation compared the oxidative stability conferred by two natural (lecithin and quillaja) and two synthetic (Tween 80 and SDS) surfactants, in their omega-3 FA nanoemulsions. All formulations were prepared using high-pressure homogenization approach, developed <90 nm particle sizes and negative surface charges at pH of 7. Except lecithin, all formulations exhibited insignificant changes in particle sizes and remained resistant to creaming within (3–8) pH range. The impact of surfactant nature on the lipid oxidation was determined in the presence and absence of singlet oxygen photosensitizers, riboflavin and rose bengal, amongst which the last two surfactants enhanced the lipid oxidation compared to samples not containing photosensitizers. The relative lipid hydroperoxide formation was found the highest with Tween 80 and the least for quillaja saponin while propanol formation was noted as the highest with lecithin, followed by Tween 80, SDS and quillaja saponin at 37 °C. Similar trend of oxidative stability was noted for photosensitized oxidation mediated by riboflavin. The lowest extent of oxidation product formation with quillaja saponins, comfortably inferred it as most efficient antioxidant emulsifier amongst the chosen alternatives. The results offer an insight of low toxicity and surfactant like usage potential of natural compounds similar to quillaja saponins. This was concluded based on comparison to other surfactants, quillaja saponins that conferred the physical stability to the formulations over a wider pH range (3–8), up to500 mM NaCl concentrations and thermal processing conditions until 90 °C. Contrary to this, Tween 80 (nonionic) and lecithin (natural phospholipid) were less stable to NaCl concentrations and thermal processing. Thus, this study explains differences in the dispersion potentials of quillaja saponins from Tween 80 and lecithin, although all are nonionic. The differences in higher lipid oxidation stability with quillaja saponins could be attributed to their polyphenolic constituents that readily furnished H^+ owing to resonance stability of phenoxide ion (Uluata et al. 2015).

In 2017, a specifically focussed study on physical and oxidative stability of two sunflower lecithins, namely Sunlipon 50 and 90, having 59 and 90% phosphatidylcholine extents for the formulation of flaxseed oil carrying nanoemulsions, was reported. The emulsions were prepared through varying stoichiometries of these emulsifiers and analysed during their storage for droplet sizes, charges, appearance, microstructure and oxidation while being kept in dark at 55 °C. The stability analysis of the formulations showed an increased physical and chemical stability with decreasing phosphatidylcholine content of the used lecithin blends. Furthermore, the oxidative stability of formulations was found greater as compared to SDS or Tween 20 catalyzed emulsions that could be due to a physiological similarity of Sunlipon phospholipids. The differences in particle sizes accounted for a higher physical stability of formulations made with Sunlipon 50, could be due to hydrophobically driven aggregation in the larger phosphatidyl comprising Sunlipon 90 formulations. Therefore, this study explains that dispersion abilities of lecithins are due to its constituent phosphatidylcholine contents that vary for different functionalities (Liang et al. 2017).

Another versatile effort by Yamamoto et al. puts forward the antioxidant attributes of Tween 20 in imparting an extending stability to corn oil-emulsions. The scientists selected 0.5, 2.5, 5.0 and 10 wt% of Tween 20 (vs. oil) and conducted the oxidation in the dark at 34 °C, accelerating using radical generators, 2,2-azobis (2,4-dimethylvaleronitrile) (AMVN) and 2,2′-azobis (2-methylpropionamidine) dihydrochloride (AAPH). The extent of oxidation was monitored through the formation of principal oxidation compound as peroxide value in the mEq•kg^{-1}. It was noted that Tween 20 suppressed the peroxidation in a dose dependent manner for both AMVN and AAPH containing emulsions, although the AMVN system showed a higher and better suppression. On changing the corn oil to tocopherol-stripped oil, similar extent of PV inhibition was noted for both AMVN and AAPH, increase in Tween 20 concentration led to decrease in the tocopherol content along the surfactant periphery before oxidation. These findings suggest an effectiveness of higher Tween 20 concentrations in mitigating the oil molecule from the hydrophobic core to the oil-surfactant interface, protecting the emulsions against oxidative damage. A significant finding of this study was the optimum Tween 20 concentration that could facilitate the transport of needed tocopherol concentration from the sheer oil phase to the oil dispersed water droplets. Another important clue is obtained from the discussion stating that low particle sizes of emulsion droplets are likely to keep the system stable against oxidative damage due to their higher kinetic energies and stability, contrary to more vulnerable larger sized droplets (Yamamoto and Misawa 2018).

Another rigorous effort of recent origin by Gasa-Falcon *et al* describes the gastro intestinal (GIT) stability of β-carotene encapsulated nanoemulsions using Tween 20, lecithin, sodium caseinate and sucrose palmitate, emulsifiers. The emulsifiers were included at 2–8 wt% while the stability of nanoemulsions was monitored via particle size and ζ-potential measurements. In the stomach environment, the nanoemulsions stabilized with Tw-20, lecithin and sodium caseinate remained stable and did not show any significant variation in the particle sizes. However, after passing the GIT region, all formulations exhibited physical changes, since considerable variations were noted in the particle sizes (these variations varied with nature and concentration of used surfactant). Interestingly, lecithin stabilized formulations exhibited a highest secretion of free FA, with increasing concentration. The highest FFA expression for lecithin stabilized nanoemulsions inferred their greater monodispersion, enabling a highest β-carotene bioaccessibility (23.5 wt%). Therefore, it could be concluded that out of chosen surfactants, lecithin could enhance the β-carotene (functional oil phase) bioavailability in its native form with least chemical degradation (Gasa-Falcon et al. 2019).

9.5 Summary

Precisely, preserving the oxidative stability of food grade emulsions is an important shelf-life enhancing criterion regarding their long-term usage. Emulsion formation is not cumbersome but the challenge is maintenance of its manifested force gradients distribution. The mechanisms of minimizing the oxidative damage to the encapsulated materials (food grade nutraceuticals or low bioavailability drugs) are briefly depicted in Scheme 9.2.

An important criterion amongst these cautions is the selection of proper surfactant quality so that the binding forces are not stronger and the system is competent and capable enough to act like a carrier only. In this respect, nonionic, plant based or protein derivatives providing optimum hydrophilic-lipophilic balances are safer and comparatively more reliable than their ionic counterparts. Apart from facilitating a moderate binding, these surfactants or surfactant like designates also possess antioxidant properties which provide protection to the dispersed phase (oil encapsulated drugs or other BAC). Biomolecules already prevailing within the body like cholesterol, phosphatidylcholine derivatives and structural analogues would be a boost as this will not improve the digestion cum absorption profiles but will also help the consumers to discard their undesired physiological depositions. Caution must also be practiced regarding the optimum functional pH of the encapsulated materials since variation in the pH could trigger a misfired reorientation of dispersed phase and dispersion medium constituents, leading to an aggravated coalescence. In this respect, use of edible and mild acidic or basic stabilizers is generally beneficial (such as citric acid and acetic acid), so that chemical balance of interacting controls is not tilted towards a hydrophobic or hydrophilic excess. Similarly, selected surfactant must be actively functional at the interface and not segregate in only one kind of force zone since the interface is most probable site for the action of free radical neutralizing antioxidants. The importance of choosing edible oils with respect to their saturated and unsaturated FA activities, also comprise an important dimension, whereby the triglyceridic activities could be exercised.

9.6 Conclusions and Future Insights

In this chapter, criterions for strengthening the oxidative stability of emulsions have been discussed with an adequate discussion of the completed studies. Distinguishing features of ionic and nonionic surfactants with optimal stoichiometries of structurally compatible stabilizers in making kinetically and thermodynamically stable nanoemulsions are discussed. The discussions on the completed investigations have inferred a better antioxidant activity for natural surfactants or small molecules like surfactants compared to the synthetically prepared alternatives. For optimum dispersion controls, it is necessary that binding mechanisms are stabilized by non-covalent interactions (via VDW, steric stabilizations, HB or LDF). Prevention of coalescence is aptly facilitated by the controlled generation of oppositely charged species, giving rise to stronger electrostatic interactive forces. The inclusion of biocompatible antioxidant molecules is also a benign remedy for the vulnerable oxidative stability of oil phase. Similarly, in place of surfactants, FA of edible oils (with needful saturation or unsaturation degrees) could be included. In summary, maintaining the oxidative stability of oil phase in an emulsion could be mitigated via

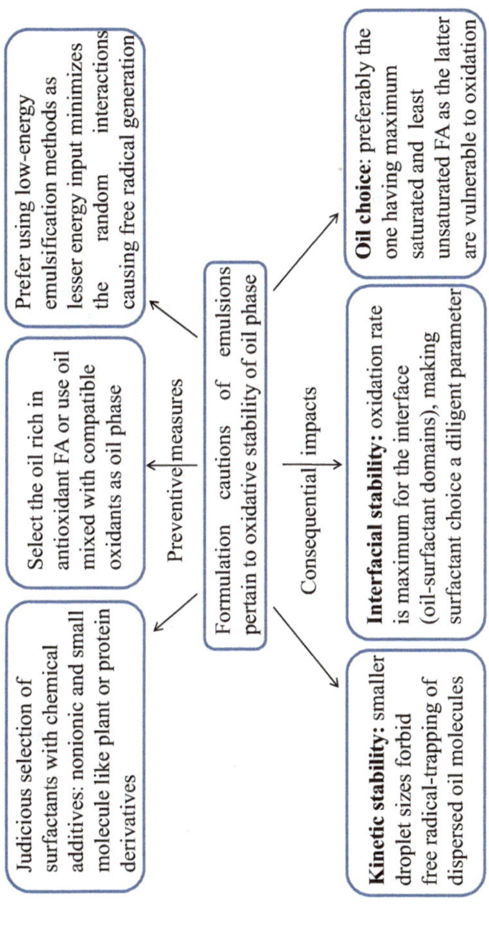

Scheme 9.2 Cautionary measures to arrest the oxidation of oil (phase) or FA in the emulsions

careful selection of surfactant-oil stoichiometry, structurally compatible co-surfactant and co-solvents and keeping the right pH controls. The combinative dispersion controls, *vis-à-vis* inclusion of GS, ZS or mixtures of nonionic and anionic/cationic stabilizers offer some novel insights to ensure long-term monodispersion controls.

References

Acosta E (2009) Bioavailability of nanoparticles in nutrient and nutraceutical delivery. Curr Opin Colloid Interface Sci 14(1):3–15. https://doi.org/10.1016/j.cocis.2008.01.002

Anton N, Vandamme TF (2011) Nano-emulsions and micro-emulsions: clarifications of the critical differences. Pharm Res 28(5):978–985. https://doi.org/10.1007/s11095-010-0309-1

Attwood D, Florence AT (1983) Surfactant systems: their chemistry, pharmacy and biology, 1st edn. Chapman and Hall, London. https://doi.org/10.1007/978-94-009-5775-6, Springer (Netherlands).

Azeem A, Rizwan M, Ahmad FJ et al (2009) Nanoemulsion components screening and selection: a technical note. AAPS Pharm Sci Tech 10(1):69–76. https://doi.org/10.1208/s12249-008-9178-x

Becher P (1981) In: Lucassen-Reynders EH (ed) Anionic surfactants: physical chemistry of surfactant action, Surfactant Science Series, Volume 11. Marcel Dekker, Inc., New York and Basel, pp 211–212. https://doi.org/10.1080/01932698208943635

Brycki BE, Kowalczyk IH, Szulc A et al (2017) Multifunctional Gemini surfactants: structure, synthesis, properties and applications. Chapter-4, 97–155. https://doi.org/10.5772/intechopen.68755

Charoen R, Jangchud A, Jangchud K et al (2012) Influence of interfacial composition on oxidative stability of oil-in-water emulsions stabilized by biopolymer emulsifiers. Food Chem 131(4):1340–1346. https://doi.org/10.1016/j.foodchem.2011.09.128

Colombie D, Sudol ED, El-Aasser MS (2000) Role of mixed anionic-nonionic systems of surfactants in the emulsion polymerisation of styrene: effect on particle nucleation. Macromolecules 33(20):7283–7291. https://doi.org/10.1021/ma9920588

Forgiarini A, Esquena J, Gonzalez J et al (2000) Studies of the relation between phase behavior and emulsification methods with nanoemulsion formation. In: Buckin V (ed) Trends in colloid and interface science XIV. Progress in colloid and polymer science, vol 115. Springer, Berlin, Heidelberg. https://doi.org/10.1007/3-540-46545-6_8.

Freiría-Gándara J, Losada-Barreiroa S, Paiva-Martins F et al (2018) Enhancement of the antioxidant efficiency of gallic acid derivatives in intact fish oil-in-water emulsions through optimization of their interfacial concentrations. Food Funct 9(8):4429–4442. https://doi.org/10.1039/C8FO00977E

Friberg S, Larsson K, Sjoblom J (2003) Food emulsions, 4th edn. Marcel Dekker, New York. CRC Press. doi: 0824746961 9780824746964

Gasa-Falcon A, Odriozola-Serrano I, Oms-Oliu G et al (2019) Impact of emulsifier nature and concentration on the stability of β-carotene enriched nanoemulsions during in vitro digestion. Food Funct 10(2):713–722. https://doi.org/10.1039/C8FO02069H

Gavin AM (1978) Edible oil deodorization. J Am Oil Chemists Soc 55:783–791. https://doi.org/10.1007/BF02682649

Gupta A, Eral HB, Hatton TA et al (2016) Nanoemulsions: formation, properties and applications. Soft Matter 12(11):2826–2841. https://doi.org/10.1039/C5SM02958A

Hu M, McClements DJ, Decker EA (2003) Lipid oxidation in corn oil-in-water emulsions stabilized by casein, whey protein isolate, and soy protein isolate. J Agric Food Chem 51(6):1696–1700. https://doi.org/10.1021/jf020952j

Kabalnov AS (1998) Coalescence in emulsions (Chapter 7). In: Binks BP (ed) Modern aspects of emulsion science. Royal Society of Chemistry, Cambridge, pp 205–260. https://doi.org/10.1039/9781847551474

Kahlweit M, Strey R, Schomacker R et al (1989) General patterns of the phase behavior of mixtures of water, nonpolar solvents, amphiphiles, and electrolytes. Langmuir 5(2):305–315. https://doi.org/10.1021/la00086a002.

Kamenka N, Chevallier Y, Zana RM (1995) Aqueous solutions of zwitterionic surfactants with varying carbon number of the intercharge group. 1. Micelle aggregation numbers. Langmuir 11(9):3351–3355. https://doi.org/10.1021/la00009a015

Khalid N, Shu G, Holland BJ et al (2017) Formulation and characterization of o/w nanoemulsions encapsulating high concentration of astaxanthin. Food Res Int 102:364–371. https://doi.org/10.1016/j.foodres.2017.06.019

Kim SH, Ji YS, Lee ES et al (2016) Ostwald ripening stability of curcumin-loaded MCT nanoemulsion: influence of various emulsifiers. Prev Nutr Food Sci 21(3):289–295. https://doi.org/10.3746/pnf.2016.21.3.289

Kosswig K (2000) Surfactants. In: Ullmann's encyclopedia of industrial chemistry. Weinheim, Wiley-VCH Verlag GmbH & Co. KGaA, pp 1–505. https://doi.org/10.1002/14356007.a25_747

Kume G, Gallotti M, Nunes G (2008) Review on anionic/cationic surfactant mixtures. J Surfact Deterg 11:1–11. https://doi.org/10.1007/s11743-007-1047-1

Kwaambwa HM, Nermark FM (2013) Interactions in aqueous solution of a zwitterionic surfactant with a water treatment protein from *Moringa oleifera* seeds studied by surface tension and ultrasonic velocity measurements. Green Sustain Chem 3(4):135–140. https://doi.org/10.4236/gsc.2013.34016

Lee JH, Decker EA (2011) Effects of metal chelator, sodium azide, and superoxide dismutase on the oxidative stability in riboflavin-photosensitized oil-in-water emulsion systems. J Agric Food Chem 59(11):6271–6276. https://doi.org/10.1021/jf2001537

Liang L, Chen F, Wang X et al (2017) Physical and oxidative stability of flaxseed oil-in-water emulsions fabricated from sunflower lecithins: impact of blending lecithins with different phospholipid profiles. J Agric Food Chem 65(23):4755–4765. https://doi.org/10.1021/acs.jafc.7b01469

Liu Y, Hou Z, Yang J et al (2015) Effects of antioxidants on the stability of β-carotene in O/W emulsion stabilized by Gum Arabic. J Food Sci Technol 52(6):3300–3311. https://doi.org/10.1007/s13197-014-1380-0

Malik P, Singh M (2017) Study of curcumin antioxidant activities in robust oil-water nanoemulsions. New J Chem 41(21):12506–12519. https://doi.org/10.1039/C7NJ02612A

Malik P, Ameta RK, Singh M (2014) Preparation and characterization of bionanoemulsions for improving and modulating the antioxidant efficacy of natural phenolic antioxidant curcumin. Chem Biol Interact 222:77–86. https://doi.org/10.1016/j.cbi.2014.07.013

Malik P, Ameta RK, Singh M (2016) Physicochemical study of curcumin in oil driven nanoemulsions with surfactants. J Mol Liq 220:604–622. https://doi.org/10.1016/J.MOLLIQ.2016.04.126

Mason TG, Wilking JN, Meleson K, Chang CB, Graves SM (2006) Nanoemulsions: formation, structure, and physical properties. J Phys Condens Matter 18:R635–R666. https://doi.org/10.1088/0953-8984/18/41/R01

McClements DJ (2005) Food emulsions: principles, practice and techniques, 2nd edn. CRC Academic Press, Boca Raton. https://doi.org/10.1201/9781420039436

McClements DJ (2012) Nanoemulsions versus microemulsions: terminology, differences, and similarities. Soft Matter 8(6):1719–1729. https://doi.org/10.1039/C2SM06903B

Mei L, Decker EA, McClements DJ (1998) Evidence of iron association with emulsion droplets and its impact on lipid oxidation. J Agric Food Chem 46(12):5072–5077. https://doi.org/10.1021/jf9806661

Mei L, McClements DJ, Decker EA (1999) Lipid oxidation in emulsions as affected by charge status of antioxidants and emulsion droplets. J Agric Food Chem 47(6):2267–2273. https://doi.org/10.1021/jf980955p

Menger FM, Littau CA (1993) Gemini surfactants: a new class of self-assembling molecules. J Am Chem Soc 115(22):10083–10090. https://doi.org/10.1021/ja00075a025
Miyashita K, Nara E, Ota T (1993) Oxidative stability of polyunsaturated fatty acids in an aqueous solution. Biosci Biotech Biochem 57(10):1638–1640. https://doi.org/10.1271/bbb.57.1638
Miyashita K, Azuma G, Ota T (1995) Oxidative stability of geometric and positional isomers of unsaturated fatty acids in an aqueous solution. J Jap Oil Chem Soc 44(6):425–430. https://doi.org/10.5650/jos1956.44.425
Mobius D, Miller R (2002) In: Mobius D, Miller R (eds) Organized monolayers and assemblies: structure, processes and function, vol 16, 1st edn. Elsevier Science, Amsterdam. ISBN (Hardcover): 9780444512567
Mosca M, Ceglie A, Ambrosone L (2010) Lipid oxidation in water-in-olive oil emulsions initiated by a lipophilic radical source. J Phys Chem B 114(10):3550–3558. https://doi.org/10.1021/jp911288e
Napper DH (1983) Polymeric stabilisation of colloidal dispersions. Academic, London
Nejadmansouri M, Hosseini SMH, Niakosari M et al (2016) Physicochemical properties and storage stability of ultrasound-mediated WPI-stabilized fish oil nanoemulsions. Food Hydrocoll 61:801–811. https://doi.org/10.1016/j.foodhyd.2016.07.011
Pan Y, Tikekar RV, Nitin N (2013) Effect of antioxidant properties of lecithin emulsifier on oxidative stability of encapsulated bioactive compounds. Int J Pharm 450(1–2):129–137. https://doi.org/10.1016/j.ijpharm.2013.04.038
Panya A, Laguerre M, Bayrasy C et al (2012) An investigation of the versatile antioxidant mechanisms of action of rosmarinate alkyl esters in oil-in-water emulsions. J Agric Food Chem 60(10):2692–2700. https://doi.org/10.1021/jf204848b
Rosen MJ (2004) Surfactants and interfacial phenomena, 3rd edn. Wiley-Interscience, New York. ISBN 0-471-47818-0
Rosen MJ, Kunjappu JT (2012) Surfactants and interfacial phenomena, 4th edn, Hoboken, Wiley. P.1. ISBN 978-1-118-22902-6
Rosen MJ, Zhu BY (1984) Structure/performance relationships in surfactants (Chapter-4) in JACS symposium series, vol 253. American Chemical Society, Washington, DC, pp 61–71. https://doi.org/10.1021/bk-1984-0253, ISBN13: 9780841208391
Schick MJ (1983) In: Becher P (ed) Encyclopedia of emulsion technology. New York, Marcel Dekker. https://doi.org/10.1002/pol.1983.130210820
Shinoda K, Friberg S (1986) Emulsions and solubilization. Wiley, Chichester. ISBN 0-471-03646-3. https://doi.org/10.1002/jctb.280420311.
Sivakumar M, Tang SY, Tan KW (2014) Cavitation technology - a greener processing technique for the generation of pharmaceutical nanoemulsions. Ultrason Sonochem 21(6):2069–2083. https://doi.org/10.1016/j.ultsonch.2014.03.025
Sukhanova A, Bozrova S, Sokolov P et al (2018) Dependence of nanoparticle toxicity on their physical and chemical properties. Nanoscale Res Lett 13(1):44. https://doi.org/10.1186/s11671-018-2457-x
Swarbrick J (1967) In: Schick MJ, Dekker M (eds) Nonionic Surfactants. Wiley-Interscience, New York. https://doi.org/10.1002/jps.2600560536
Uluata S, McClements DJ, Decker EA (2015) Physical stability, autoxidation, and photosensitized oxidation of ω-3 oils in nanoemulsions prepared with natural and synthetic surfactants. J Agric Food Chem 63(42):9333–9340. https://doi.org/10.1021/acs.jafc.5b03572
Vautier-Giongo C, Bakshi MS, Singh J et al (2005) Effects of interactions on the formation of mixed micelles of 1,2-diheptanoyl-sn-glycero-3-phosphocholine with sodium dodecyl sulfate and dodecyltrimethylammonium bromide. J Coll Interf Sci 282(1):149–155. https://doi.org/10.1016/j.jcis.2004.08.071
Vold MJ, Vold RD (1964) Colloid chemistry. The science of large molecules, small particles, and surfaces. Reinhold, New York, p 118
Walstra P (1983) Formation of emulsions. In: Becher P (ed) Encyclopedia of emulsion technology, Basic theory, vol 1. Marcel Dekker, New York, pp 57–128

Waraho T, Cardenia V, Rodriguez-Estrada MT et al (2009) Prooxidant mechanisms of free fatty acids in stripped soybean oil-in-water emulsions. J Agric Food Chem 57(15):7112–7117. https://doi.org/10.1021/jf901270m

Waraho T, Mcclements DJ, Decker EA (2011) Impact of free fatty acid concentration and structure on lipid oxidation in oil-in-water emulsions. Food Chem 129(3):854–859. https://doi.org/10.1016/j.foodchem.2011.05.034

Xu J, Mukherjee D, Chang SKC (2018) Physicochemical properties and storage stability of soybean protein nanoemulsions prepared by ultra-high pressure homogenization. Food Chem 240:1005–1013. https://doi.org/10.1016/j.foodchem.2017.07.077

Yamamoto Y, Misawa R (2018) Effect of emulsifier concentration on the oxidation of an o/w emulsion prepared from canola oil. Food Nutr Sci 9(6):683–692. https://doi.org/10.4236/fns.2018.96052

Yamamoto Y, Hiyama S, Takase Y et al (2014) Effects of antioxidants and additional emulsifiers on the stability of emulsified milk fat in the photo/radical oxidation system. J Oleo Sci 63(9):893–901. https://doi.org/10.5650/jos.ess14111

Zhang J, Peppard TL, Reineccius GA (2015) Preparation and characterization of nanoemulsions stabilized by food biopolymers using microfluidization. Flavour Fragr J 30(4):288–294. https://doi.org/10.1002/ffj.3244

Chapter 10
The Role of Antioxidants and Encapsulation Processes in Omega-3 Stabilization

Nor E. Rahmani-Manglano, Pedro J. García-Moreno, F. Javier Espejo-Carpio, A. Raúl Pérez-Gálvez, and Emilia M. Guadix-Escobar

10.1 Omega-3 Polyunsaturated Fatty Acids

Over the last decades the importance of ω-3 PUFAs, especially EPA and DHA, in human nutrition has been the focus of scientific research since these long chain fatty acids have been considered to improve human health as a result of their influence in cell and tissue function (Calder 2014). High ω-3 PUFAs intake has been associated with a reduced risk of suffering from major chronic diseases such as cardiovascular diseases (CVD) (e.g. coronary heart disease (CHD) and stroke), some cancers (e.g. colorectal, prostate and lung cancer), hypertension, inflammatory diseases (e.g. arthritis, psoriasis and asthma), neurodegenerative diseases (e.g. Alzheimer and Parkinson), type-2 diabetes and obesity (Calder 2014; Nguyen et al. 2019; Punia et al. 2019). In addition, it is well known that DHA plays an important structural role in the brain and eye since it is the main ω-3 PUFA in the brain (Layé et al. 2018) and it constitutes more than 50% of the fatty acids present in the outer segments of the retina (Calder 2014). Thus, DHA supply is essential, especially when these tissues are developing (from gestation to 18 months after birth) for optimal neurodevelopment (memory and learning ability) and visual development for fetus and infants (Calder 2014; Jacobsen 2016). Despite all the research carried out in this field, there are still some controversies regarding the role of ω-3 PUFAs and their beneficial effects in the prevention of some health conditions (e.g. mental disorders, diabetes or obesity) (Nguyen et al. 2019). However, there is a strong scientific evidence establishing a cause and effect relationship between ω-3 PUFAs consumption and some beneficial physiological effects (e.g. maintenance of normal brain function or cardiac function). Based on the later, the European Food Safety Authority

N. E. Rahmani-Manglano · P. J. García-Moreno (✉) · F. J. Espejo-Carpio
A. R. Pérez-Gálvez · E. M. Guadix-Escobar
Department of Chemical Engineering, University of Granada, Granada, Spain
e-mail: pjgarcia@ugr.es

M. A. Aboudzadeh (ed.), *Emulsion-based Encapsulation of Antioxidants*, Food Bioactive Ingredients, https://doi.org/10.1007/978-3-030-62052-3_10

(EFSA) has allowed the use of different health claims for EPA and/or DHA pursuant to Article 13(1) and Article 14 of Regulation (EC) No 1924/2006 (Table 10.1). In the USA, the Food and Drug Administration (FDA) has recently summited a letter of enforcement discretion, which allows the use of different qualified health claims regarding the relationship between ω-3 PUFAs intake and reduction of blood pressure in general population, provided that the dietary supplement and/or conventional food contains at least 0.8 g EPA and DHA combined per serving (FDA 2019). However, as the scientific evidence is considered insufficient and inconsistent by the agency, the qualified health claim must be accompanied by a disclaimer (FDA 2019).

Because of the recognized health benefits derived from ω-3 PUFAs consumption, several authoritative bodies and expert scientific organizations have set a series of recommendations to increase their intake. Internationally, the Food and Agriculture Organization of the United Nations (FAO) recommends a daily intake of 250 mg of EPA + DHA for adult men and non-pregnant and non-lactating adult women. The minimum intake recommended in case of adult pregnant or lactating women, in order to assure adult health and fetal and infant development, for EPA + DHA is 300 mg per day of which at least 200 mg per day should be DHA. The EPA + DHA intake recommendation for children is: (i) 100–150 mg/day for children from 2 to 4 years old, (ii) 150–200 mg/day for children from 4 to 6 years old

Table 10.1 Summary of EFSA's allowed health claims for ω-3 PUFAs

ω-3 PUFAs	Health Claim	Target Population	Dose	Reference
EPA + DHA	Maintenance of normal blood pressure	Adult men and women	3 g EPA and DHA per day	EFSA (2009)
	Maintenance of normal cardiac function	General population	250 mg EPA and DHA per day	EFSA (2010a)
	Maintenance of normal (fasting) blood concentration of triglycerides	Adult men and women	2 g EPA and DHA per day	EFSA (2010a)
DHA	Maintenance of normal (fasting) blood concentration of triglycerides	Adult men and women	2 g DHA per day in one or more servings	EFSA (2010b)
	Maintenance of normal brain function	General population	250 mg DHA in one or more servings	EFSA (2010b)
	Maintenance of normal vision	General population	250 mg DHA in one or more servings	EFSA (2010b)
	DHA contributes to normal brain development	Older infants and young children (>6–24 m.o.) Children (2–18 y.o.)	100 mg DHA in one or more servings 250 mg DHA in one or more servings	EFSA (2014)

Abbreviations: *m.o.* months old, *y.o.* years old

and (iii) 200–300 mg/day for children from 6 to 10 years old (FAO 2010). In Europe, the EFSA set an adequate intake (AI) of 250 mg per day for EPA + DHA in adults, while for pregnant or lactating women the recommended intake is that of the AI value plus 100–200 mg per day of DHA. In case of infants (from 6 months to 2 years), the AI is set to 100 mg per day of DHA. From 2 years onwards the AI is that for adults (250 mg per day of EPA + DHA) (EFSA 2010c). Moreover, the EFSA has set a recommended intake for EPA + DHA of 250–500 mg per day for European adults based on cardiovascular risk considerations (EFSA 2012).

EPA and DHA can be synthetized in the organism from α-linolenic acid (ALA; C18:3 n-3), which is the precursor of the ω-3 PUFAs family. This fatty acid is in turn synthesized from linoleic acid (LA; C18:2 n-6) in a desaturation reaction catalysed by Δ15-desaturase enzyme, which is only found in plants. Thus, ALA is considered as an essential fatty acid since it cannot be synthesized *de novo* by humans and it needs to be provided in the diet. ALA is naturally found in seeds, nuts and seed oils (e.g. flaxseeds, walnuts or soybean oil) (Calder 2013). When ALA enters the human body, it converts to EPA and DHA by the metabolic path shown in Scheme 10.1. This synthesis occurs mainly in the liver but in a very little extent since only about 8% of ALA convers to EPA and 1% to DHA (Layé et al. 2018). The synthesis reaction is also very slow. Moreover, the metabolic path of conversion from ALA to EPA is in direct competition with the metabolic path of conversion of LA (ω-6 PUFAs family precursor) to arachidonic acid (ARA; C20:4 n-6), since both reactions are catalysed by the same enzymes (Δ6-desaturase, Elongase and Δ5-desaturase) (Calder 2013).

Changes in food technology and eating habits in industrial societies have led to an unbalanced intake ratio omega-6/omega-3 of ~20/1 due to a high consumption of vegetables oils (e.g. corn oil) (Simopoulos 2011). This situation interferes with ω-3

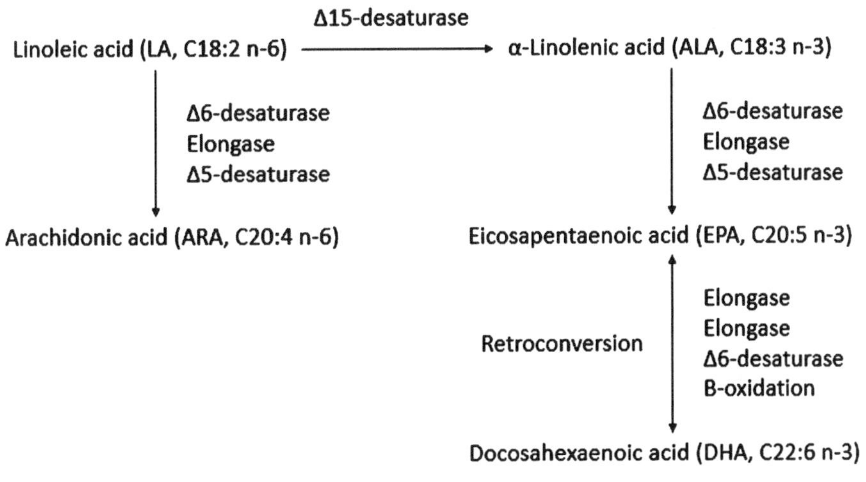

Scheme 10.1 Metabolic pathway of biosynthesis of α-linolenic acid (ALA), eicosapentaenoic acid (EPA) and docosahexaenoic acid (DHA)

PUFAs metabolism and, since ω-6 PUFAs are more present in current human diets, the synthesis of ω-6 PUFAs in the organism is quantitatively more important. The increase of ω-6 PUFAs in diet exert a proinflammatory and prothrombotic effect, which is a risk factor for major chronic diseases such as CHD, obesity or diabetes, among others (Simopoulos 2016). Thus, direct ω-3 PUFAs (e.g. EPA and DHA) intake is required for a balanced diet.

The main sources of dietary EPA and DHA are fish, krill, algae and other seafood (e.g. crab, prawns, lobsters or mussels) together with marine oils. However, these long chain polyunsaturated fatty acids (LC ω-3 PUFAs) are also modestly present in some other animal-derived foodstuff (e.g. meat, milk or eggs) (Calder 2014; Nguyen et al. 2019). Among fishes, two categories are distinguished: (i) lean fish, which stores lipid in the liver (e.g. cod, haddock and plaice) and (ii) fatty fish, which stores lipid in the flesh (e.g. salmon, mackerel, herring, sardines, trout and tuna) (Calder 2013). Due to differences in the diet and metabolism, the amount and ratio of EPA and DHA vary depending on fish type and species. Fatty fishes content per serving of dietary ω-3 PUFAs is up to 10-folds those of lean fishes (Calder 2013, 2014). On the other hand, marine oils are those obtained from the flesh or liver of the fish, crustaceans, cephalopods (e.g. squid) or marine mammals (Jacobsen 2016). Flesh fish oils are produced from the flesh of fatty fishes (e.g. sardine, anchovy, tuna, pollack, salmon and catfish) and are the most manufactured among the marine oils, while liver fish oils are obtained from the liver of lean fishes (e.g. cod, hake, halibut and shark) and constitutes less than 3% of the total marine oils production (Jacobsen 2016). In most fish oils (flesh fish oils and liver fish oils), EPA and DHA are present as triacylglycerides (TAG). However, the amount and ratio of these LC ω-3 PUFAs, once again, differs among oils. Other sources of marine oil are algae and krill. Krill oil contains ω-3 PUFAs in the form of phospholipids whilst algae oil contains such fatty acids in the form of TAG. Algae oils are of great interest, especially in infant formulas production, since their DHA content is up to 52% (Jacobsen 2016).

Nowadays, fish, krill, algae and marine oils consumption is insufficient to meet the requirements of a healthy diet. In this sense, food industry aims to produce added-value food products enriched mainly with fish oil containing high levels of EPA and DHA. However, due to the characteristic fishy flavour of marine oils and their prone to oxidation, efficient delivery systems capable of masking the taste and preventing lipid oxidation needs to be designed.

10.2 Lipid Oxidation

Due to the high degree of unsaturation, LC ω-3 PUFAs (e.g. EPA and DHA) are very prone to oxidation leading to the loss of their nutritional properties and to the formation of several off-odor and off-taste compounds (e.g. propanal) and other toxic compounds (e.g. malonaldehyde).

Lipid oxidation occurs through different mechanisms such as enzymatic oxidation, photo-oxidation and autoxidation, being the latter the most important one. While enzymatic oxidation takes place in presence of lipoxygenase enzymes and mainly affects vegetal oils, photo-oxidation occurs in presence of light and/or photosensitizers (e.g. pigments). Although photo-oxidation is not the main oxidation mechanism, it is of great importance since it influences autoxidation mechanism and the distribution of primary oxidation compounds (Frankel 2012a).

Autoxidation occurs as a free radical chain reaction when oxygen in its ground state reacts with unsaturated lipids under mild conditions (Frankel 2012b). The sequence of reactions taking place can be classified into three stages, namely: (I) initiation, (II) propagation and (III) termination (Scheme 10.2).

The initiation stage is characterized by the formation of very reactive lipid free radicals (or alkyl radicals) ($L^{\bullet}$) due to the loss of a hydrogen atom (H^{+}) by an unsaturated lipid (LH) in the presence of initiators (e.g. heat, light or redox metals) (reaction 10.1).

$$LH + I \rightarrow IH + L^{\bullet} \tag{10.1}$$

Depending on the initiator agent, lipid free radicals ($L^{\bullet}$) are produced by different mechanisms. The most likely and widely accepted initiation process is the transition metal-catalyzed decomposition of already present hydroperoxides (LOOH), either as impurities or as products of photosensitized oxidation. As a result of reactions 10.2 and 10.3, alkoxyl radicals ($LO^{\bullet}$) and peroxyl radicals ($LOO^{\bullet}$) are produced which, in turn, can further initiate lipid oxidation.

$$LOOH + M^{2+} \rightarrow LO^{\bullet} + OH^{-} + M^{3+} \tag{10.2}$$

$$LOOH + M^{3+} \rightarrow LOO^{\bullet} + H^{+} + M^{2+} \tag{10.3}$$

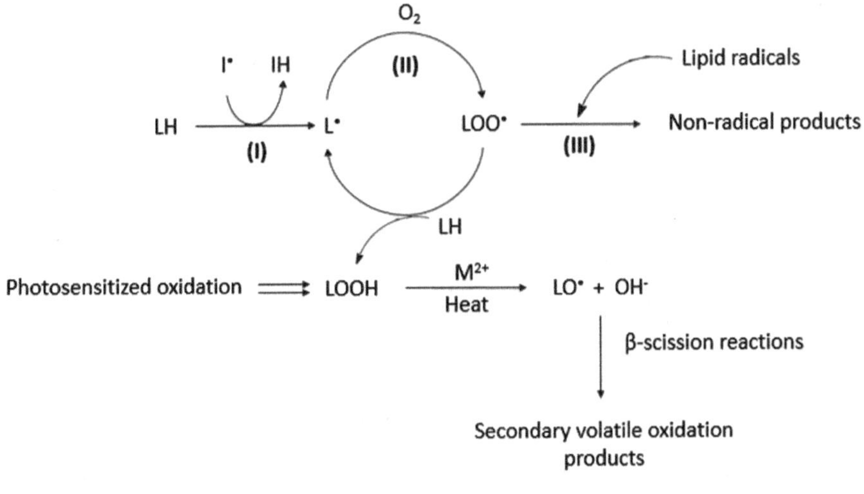

Scheme 10.2 Schematic representation of main lipid oxidation mechanisms

Subsequently, in the propagation stage, the alkyl radicals ($L^{\bullet}$) react with molecular oxygen to form peroxyl radicals ($LOO^{\bullet}$) (reaction 10.4). These products are also very unstable and reactive and can react with another unsaturated lipid (LH) in a hydrogen transfer reaction to form hydroperoxides (LOOH) and more alkyl radicals ($L^{\bullet}$) (reaction 10.5). The hydroperoxides (or peroxides) produced in this stage are known as primary oxidation products.

$$L^{\bullet} + O_2 \rightarrow LOO^{\bullet} \tag{10.4}$$

$$LOO^{\bullet} + LH \rightarrow LOOH + L^{\bullet} \tag{10.5}$$

Reaction 10.5 is rate-determining since it is slower than reaction 10.4 and will occur selectively for the weakest bounded hydrogen (allylic hydrogen) until the reaction is interrupted either by action of an external agent (e.g. antioxidant) or by unavailability of a hydrogen source. Thus, the probability of undergoing autoxidation in lipids will depend on the number of allylic methylene groups, i.e. the degree of unsaturations. In case of LC ω-3 PUFAs, the probability to undergo autoxidation is linearly related to the number of methylene-interrupted carbons (bis-allylic positions) present in the fatty esters and depending on their position, different hydroperoxides are obtained. EPA produces the eight 5-, 8-, 9-, 11-, 12-, 14-, 15-, and 18-hydroperoxides, while DHA produces the ten 4-, 7-, 8-, 10-, 11-, 13-, 14-, 16-, 17-, and 20-hydroperoxides (Frankel 2012b).

Finally, in the termination stage, the lipid radicals formed in the initiation and propagation stages accumulate and react with each other to produce non-radical stable compounds. Depending on the compounds present and the reaction conditions (e.g. pressure and temperature), different reactions will occur such as formation of alcohols or condensation reactions, thus, different compounds will be produced (e.g. alcohols or dimers) (Frankel 2012b). The main condensation reactions are showed below (reactions 10.6, 10.7, and 10.8).

$$LOO^{\bullet} + LOO^{\bullet} \rightarrow LOOL + O_2 \tag{10.6}$$

$$LO^{\bullet} + L^{\bullet} \rightarrow LOL \tag{10.7}$$

$$L^{\bullet} + L^{\bullet} \rightarrow L - L \tag{10.8}$$

During the initiation and propagation stages of autoxidation, lipid hydroperoxides accumulate; however, due to their high instability, they start to decompose leading to the formation of a large variety of organic compounds such as polymers, monomers and volatiles compounds. Among these, the focus has been traditionally placed on small molecular weight volatile compounds since these are responsible of the loss of the organoleptic properties of food lipids by the production of unpleasant odors and flavors. Hydroperoxides decompose through homolytic cleavage by action of heat or metal ions (e.g. haeme iron) producing alkoxyl radicals ($LO^{\bullet}$) and hydroxyl ions (OH^{-}). These radicals are very reactive and generally cause the cleavage of the aliphatic chain of fatty acids in what is known as β-scission reaction. As a result, low molecular weight compounds are produced such as alcohols,

Table 10.2 Fish oil volatile compounds as quality indicators (Frankel 2012c; Yeşiltaş 2019)

Compound	Fish oil quality indicator	Odour description	Thresholds, ppm
Substituted furans	2-Ethylfuran	Flower	2–27
Vinyl alcohols	1-Penten-3-ol	Sweet	0.5–3
Alkanals	Propanal	Sharp-irritating	0.04–1
2-Alkenals	2-Pentenal/2-Hexenal	Pungent, glue, green/sour, green	0.04–2.5
(E,E)-2,4-Alkadienals	(E,E)-2,4-Heptadienal	Green, rancid hazel nuts	0.04–0.3
Vinyl ketones	1-Penten-3-one	Pungent, rancid green, sharp fishy	0.00002–0.007

aldehydes, ketones, furans, alkanes and alkenes. The β-scission reactions products will vary depending on the original hydroperoxide structure. In case of fish oil, the most often identified volatile compounds are alcohols, aldehydes and ketones, and to a lesser extent, hydrocarbons, furans and aromatic compounds (Horiuchi et al. 1998; Frankel 2012c). Since these compounds are odor active, they are susceptible to be quality indicators of fish oil as well (Table 10.2).

Other LC ω-3 PUFAs secondary oxidation products that need attention are malonaldehyde (MDA) and 4-hydroxy-2-hexenal (HHE) due to their toxicity. Both aldehydes, bis-aldehyde and α,β-unsaturated aldehyde, respectively are very reactive and can interact with biomolecules such as DNA and proteins (structural and metabolic) leading to the formation of lipooxidation end products. Aldehyde modified molecules and their biological implications have been related to major chronic diseases such as cancer, neurodegenerative diseases, kidney disease or diabetes, among others (Guéraud et al. 2010; Vieira et al. 2017). MDA reacts with proteins and DNA by acting as a crosslinker and it is considered to be more mutagenic, whilst HHE can modify nucleophile species such as proteins, membranes or nucleic acids due to its electrophilic properties, thus exerting cytotoxic activity (Guéraud et al. 2010).

Alternatives to minimize lipid oxidation of LC ω-3 PUFAs during incorporation into food matrices include the addition of antioxidants and the development of delivery systems such as fish oil-in-water emulsions or fish oil nano-microencapsulates, which will be discussed in thoroughly in the next sections.

10.3 Antioxidants

Antioxidants have a key role in preventing lipid oxidation, since they can delay, control or inhibit oxidation reactions. Antioxidants can be classified by their mechanism of action as primary or secondary antioxidants. Primary antioxidants, known as chain-breaking antioxidants, are able to neutralize free radicals by either donating a hydrogen atom or by a single electron transfer mechanism. Therefore, these antioxidants play a crucial role in lipid oxidation since they can react with the formed lipid radicals ($L^•$) and convert them into more stable non-radical products avoiding further decomposition of the lipids. Meanwhile, secondary antioxidants prevent lipid oxidation by several mechanisms such as chelation of pro-oxidant metals ions, regeneration of primary antioxidants, decomposition of hydroperoxides (LOOH) and scavenging of oxygen, among others (Decker 2002).

According to food regulation, antioxidants must have a required daily intake value. This has limited their use in foods to a few compounds such as synthetic phenols and vitamins. The increasing interest towards natural ingredients, which has not the potential health hazards of synthetic compounds, has boosted the research on natural antioxidants. However, only few antioxidants from natural sources have been accepted by food regulation and some of them are not commercialized as antioxidants but as flavourants, binders, etc. Below we provide information about the most common used antioxidants in foods, which include those used in the stabilization of LC ω-3 PUFAs.

10.3.1 Carotenoids and Xanthophylls

Carotenoids are lipophilic pigments that contain and confer the yellow, orange or red color to a wide variety of foods. Furthermore, they comprise a group of important antioxidants in plant-derived food. Its antioxidant activity is based on their capacity to quench singlet oxygen and scavenge free radicals. However, some factors such as structure or the potential for interaction with other antioxidants could reduce their antioxidant effect or even induce a pro-oxidant effect in vitro and in vivo (Young and Lowe 2001). The main molecules used in foods are β-carotene (E160a), lycopene (E160d), astaxanthin (E160a), β-cryptoxanthin (E160a), capsanthin (E160c), lutein (E161) and zeaxanthin (E160a).

10.3.2 Synthetic Phenolic Antioxidants

Many phenolic compounds synthetically produced exhibit better antioxidant activity than natural antioxidants and are available at relatively cheaper prices. The inhibition of the lipid oxidation process is produced by trapping the peroxyl ($LOO^•$) or alkoxyl radicals ($LO^•$). Nevertheless, its toxicological effects have been the subject of controversy since some animal studies have linked synthetic phenolic antioxidants with an increased risk of cancer (Hocman 1988; Oikawa et al. 1998), being the cause for their replacement in food applications. The most common molecules are propyl gallate (PG; E310), octyl gallate (OG; E311), dodecyl gallate (DG; E312), butylated hydroxyltoluene (BHT; E321), butylated hydroxylanisole (BHA; E320) and tertiary-butylhydroquinone (TBHQ; E319). The use of these compounds in foods is strictly regulated and varies in different countries. In Europe, the directive

2006/52/EC allows their use in a limited number of food products. Particularly, BHT is only permitted in fats and oils.

10.3.3 Metal Chelators

Metal chelators are added into food owing to its ability to minimize the participation of metal ions in redox reactions. Food chelators usually have O-containing ligands, which would stabilize iron and copper in their oxidized form. In this way, chelators work as preventive antioxidants by: (i) obstructing the activity of catalytic metals and thus eliminating the initial oxidation step, and (ii) by avoiding the metal-catalyzed decomposition of peroxides (LOOH) to secondary oxidation products. Some of the most studied metal chelators are ethylenediaminetetraacetic acid (EDTA; E385), sodium tripolyphosphate (STPP; E451i), citric acid (E330) and its salts such as sodium citrate (E331), potassium citrate (E332) and calcium citrate (E333). These compounds are allowed as food additives in the European Union and United States. Particularly, sodium salts of EDTA are not allowed in Europe.

Furthermore, others compounds have been shown metal chelating capacity. Flavonoids are plant derived compounds with variable polyphenolic structure. Flavonoids such as kaempferol, quercetin, myricetin, luteolin, naringenin, and catechin are able of complexing Cu^{2+} and Fe^{3+} (Fernandez et al. 2002). Carnosine and other chelating peptides are able of chelating cupper ions and iron respectively (Velez et al. 2008; Guo et al. 2014). However, none of these compounds has been authorized as food additive by European regulations.

10.3.4 Protein Hydrolysates

Peptides are biopolymers formed by amino acids linked by peptide bonds. Hydrolysates are a complex mixture of peptides with different chain length and amino acids composition, which are preferably obtained by enzymatic hydrolysis. Numerous food proteins, such as milk, meat, fish, egg, seeds etc., have been described as source of antioxidant hydrolysates/peptides (Samaranayaka and Li-Chan 2011). Antioxidant hydrolysates/peptides have been reported to be able to inhibit lipid peroxidation in model food systems such as fish oil-in-water emulsions (García-Moreno et al. 2016) by exhibiting both radical scavenging and metal chelating activities. The structure and sequence of amino acids influence the antioxidant properties of peptides. Generally, short-chain peptides and the presence of amino acids such as tyrosine (Tyr), tryptophan (Trp), phenylalanine (Phe), lysine (Lys), methionine (Met) and histidine (His) are present in antioxidant peptides (Power et al. 2013). Food regulation do not consider protein hydrolysates as antioxidant additives; however, it could be used as a source of protein.

10.3.5 Tocopherols and Tocotrienols

Tocopherols and tocotrienols are one of the most well-known natural antioxidants. They are referred by the common term Vitamin E when they are coexisting with lipids in a biological setting. This group includes 11 different compounds, which differ in the number and position of methyl groups in the chromane ring. Moreover, for tocopherols the side chain in C-2 is saturated, while tocotrienols have an unsaturated side chain. All of them act as peroxyl (LOO•) or alkoxyl radical (LO•) scavengers. Normally, they are commercially extracted from vegetable oil sludge, especially soybean oil. The main food additives are α-tocopherol (E307), γ-tocopherol (E308), δ-tocopherol (E309), the rest of tocopherols and tocotrienols (Trolox, 2,2,5,7,8-Pentamethyl-6-chromanol, β-tocopherol, α-tocotrienol, β-tocotrienol, γ-tocotrienol, δ-tocotrienol and Plastochromanol) are regulated as tocopherol-rich extract (E-306). They can be employed in dietary foods and infant formula.

10.3.6 Ascorbic Acid and Its Derivatives

Ascorbic acid (vitamin C) and its derivatives are a group of antioxidant compounds widely used as a food additive. They have a resonance structure that provides a high reducing activity. Ascorbic acid (E300) and its sodium (E301) and calcium (E302) salts as well as Ascorbyl palmitate (E304) are considered as safe food additives. In contrast, in Europe, Erythorbic acid (E315) and Sodium erythorbate (E316) are only permitted in cured products and preserved products.

10.3.7 Spice and Plant Extracts

Recently, natural alternatives to synthetic antioxidants have been widely studied to protect the flavor/odor and nutritional value of foods. These extracts present radical scavenging activity, reducing power and metal chelating capacity (Embuscado 2015). Rosemary and sage extracts are interesting antioxidants because of its high content in antioxidant phenolic compounds such as carnosol, carnosic acid, rosmanol, episormanol, isorosmanol, rosmaridiphenol, rosmariquinone, rosmarinic acid or rosmadial (Berdahl and McKeague 2015). Tea extract is also an important source of polyphenols. The main compounds present in tea extract are epicatechin, epigallocatechin, epicatechin gallate, epigallocatechin gallate or theaflavin (Karaosmanoglu and Kilmartin 2015). Similarly, numerous antioxidants have been identified in herb and spice extracts, namely caffeic acid, capsaicin, carotol, curcumin, eriodictyol, eugenol, etc. However, some of these extracts such as rosemary and sage extracts have the drawback of having a detectable flavor. The Food and Drug Administration refers to natural ingredients as "ingredients extracted directly from plants or animal products as opposed to being produced synthetically." In Europe, the use of rosemary extract (E392) is regulated by directives 2010/67/EU and 2010/69/EU.

10.4 Delivery Systems

Besides the addition of antioxidants, several strategies have been investigated for the successful incorporation of LC ω-3 PUFAs into food matrices. Particularly, significant research has been carried out on the development of specific delivery LC ω-3 PUFAs systems such as fish oil-in-water emulsions and dried microencapsulates loaded with fish oil. The efforts have been focused on the enhancement of the oxidative stability of the delivery systems as well as on the chemical (e.g. oxidative stability) and physical stability (e.g. texture) of the enriched foods.

10.4.1 Oil-in-Water Emulsions

The formation and physical stabilization of oil-in-water emulsions requires the use of emulsifiers, which reduces interfacial intension favoring droplet disruption and provides steric and/or electrostatic repulsive forces (McClements and Jafari 2018). Physical and oxidative stability of oil-in-water emulsions are tightly related, and both are mainly affected by interfacial properties (e.g. thickness, porosity, charge, antioxidant activity) conferred by the emulsifier used (Berton-Carabin et al. 2018). Common emulsifiers for the stabilization of fish oil-in-water emulsions are biopolymers such as milk proteins (e.g. caseins and whey protein) (Horn et al. 2012b), and combinations of milk proteins and natural surfactants such as phospholipids (García-Moreno et al. 2014; Yesiltas et al. 2019a). Both milk proteins and phospholipids exhibit antioxidant activities such as radical scavenging and chelating activities (Díaz et al. 2003; García-Moreno et al. 2014), which are properties desired at the interface where lipid oxidation is initiated in oil-in-water emulsions. Lately, alternative protein emulsifiers showing antioxidant activities have been reported. For instance, García-Moreno et al. (2016) indicated the feasibility to use sardine protein hydrolysates with low degree of hydrolysis obtained by subtilisin for the physical and oxidative stabilization of fish oil-in-water emulsions. Recently, Yesiltas et al. (2018, 2019b) reported the advantages of using emulsifiers with covalently attached caffeic acid for the stabilization of fish oil-in-water emulsions. These studies confirmed the importance of having antioxidants at the interface for improving the oxidative stability of the emulsions.

Other factors affecting lipid oxidation in oil-in-water emulsions are droplet size, viscosity, pH and surface charge.

Droplet size determines the interfacial area in the emulsion, which is in fact the contact area between pro-oxidants and lipids. Theoretically, low droplet size is desired for a high physical stability of the emulsion and should imply low oxidative stability of the emulsion due to an increase in interfacial area. Nonetheless, contradictory results have been reported on the effect of droplet size on oxidative stability in emulsions (Jacobsen et al. 2000; Let et al. 2007b), which denoted that other factors, for instance interfacial properties have a greater influence on lipid oxidation in emulsion systems.

Viscosity of the emulsions is an important variable affecting both physical and oxidative stability of these heterogeneous systems. This is because viscosity controls both the mobility of oil droplets (e.g. by minimizing physical destabilization phenomena) (McClements 2005) and the mobility of pro-oxidants such as metal ions and lipid radicals decreasing rate of lipid oxidation (Shimada et al. 1996). Viscosity of oil-in-water emulsions is controlled by addition of stabilizers, commonly high molecular weight carbohydrates such as pectin, guar and xanthan gums.

Surface charge of oil droplets greatly affects lipid oxidation in oil-in-water emulsions. In the case of proteins used as emulsifiers, the pH of the emulsion will determine the charge of the interfacial protein layer. This might be of special importance in the case of metal-catalyzed lipid oxidation, since theoretically metal ions will be attracted to the interface when having negative surface charge, thus favoring lipid oxidation. On the contrary, trace of metals will be repelled from the interface in emulsions with positive surface charge, reducing lipid oxidation (Mei et al. 1998). However, at low pH, which is generally required to charge positively most of protein-based emulsifiers, the solubility of trace of metals increases. Thus, as consequence of an increased solubility, iron also becomes more active in acidic conditions to catalyze lipid oxidation via decomposition of hydroperoxides (Berton-Carabin et al. 2014).

In addition, oxidative stability of emulsions can be further improved by addition of antioxidants. For this purpose, phenolic compounds such as ferulic acid and caffeic acid have been widely used in oil-in-water emulsions to reduce lipid oxidation (Sørensen et al. 2008). These are hydrophilic compounds that will be located in the aqueous phase, thus mainly scavenging radicals in the continuous aqueous phase. In the last years, extensive research has been carried out on the lypophilization of phenolic compounds in order to modify their hydrophilic/hydrophobic ratio, which allows the location of these antioxidant compounds at the interface of oil-in-water emulsions. Interestingly, it has been reported that the optimum alkyl chain length for the phenolipids is determined by several factors such as the type of antioxidants to be lypophilized, type of emulsifiers used to stabilize the emulsion as well as the food system where the emulsion will be incorporated (Laguerre et al. 2010; Alemán et al. 2015).

A common strategy to protect the lipid ingredients of ω-3 PUFAs fortified foods against oxidation is their incorporation into food matrices as delivery systems, such as microcapsules or oil-in-water emulsions. Fish oil emulsions are often used as oil carrier in liquid or semisolid products such as milk, cheese cream, mayonnaise, salad dressing or yoghurt. The market of ω-3 PUFAs enriched dairy products has undergone a sharp increase since the 2000s, followed by bakery goods. Oil addition to dairy products and beverages is favored by their low storage temperature, commonly under refrigerated or frozen conditions, which restrain lipid oxidation.

Driven by the improvements in mixing and homogenization equipment, fish oil has been successfully incorporated into liquid and semiliquid products, ensuring acceptable levels of oxidative stability. The susceptibility of the fortified food to lipid oxidation depends to large extent on the nature of both the emulsifier protein and the food matrix. To this regard, Horn et al. (2012a) studied the effect of three emulsifiers including caseinates, whey proteins and a mixture of milk proteins and phospholipids on the oxidative stability of cream cheese fortified with 70% fish oil-in-water emulsion. The authors reported that both whey- and casein-based delivery emulsions provided poorer oxidative stability to the cheese cream, compared to direct addition of bulk oil. This fact was attributed to the lipid oxidation induced during emulsion preparation where temperature rose up to 65 °C. Under the same processing conditions, the emulsion employing a mixture of milk proteins and phospholipids as emulsifier presented better oxidative stability than direct addition of bulk oil. This result confirmed the role of milk phospholipids in reducing the exposure of oil droplets to oxidation agents.

Concerning the nature of the food matrix, Let et al. (2007a) studied the oxidative stability of milk, yoghurt and dressing fortified by 50% oil-in-water emulsions employing whey proteins as emulsifiers. Fortification of milk beverages with fish oil-in-water emulsions provided larger oxidative stability that direct incorporation of bulk oil. In contrast, the content of secondary volatile oxidation products (SVOP) and sensory data showed the opposite behavior for yoghurt and dressing, despite the fact that processing conditions and protein emulsifier were the same.

Early approaches to produce ω-3 PUFAs fortified mayonnaise, where vegetable oil was partially replaced by fish oil (Jacobsen et al. 2001), were hindered by the high content of protein-bound iron present in the egg yolk. Low pH levels activate iron as catalyzer of lipid oxidation reactions, limiting the stability of such emulsions. Further research on low pH systems proved that oxidative stability of ω-3 PUFAs fortified mayonnaises can be improved by adding antioxidants, either commercial synthetic chelators (Haahr and Jacobsen 2008; Alemán et al. 2015) or natural-based extracts (Hermund et al. 2015).

Margarines and spreadable fats are another range of foodstuffs where ω-3 PUFAs fortification leads to limited oxidative and sensory quality. To this regard, Kolanowski et al. (2004) incorporated up to 30 $g{\cdot}kg^{-1}$ of fish oil into a reduced-fat spread containing 450 $g{\cdot}kg^{-1}$ of rapeseed oil. The fortified product was stabilized by a combination of antioxidants (α-tocopherol, ascorbyl palmitate and rosemary extract) added at ratio 1:10 w/w with respect to the amount of fish oil. The enriched product could be stored for up to 3 months without significant decrease in both oxidative (e.g. peroxide and acid values) and sensory parameters (e.g. hardness, consistency, spreadability, fishy odor development).

Few studies deal with the ω-3 PUFAs enriched emulsions other than dairy products. For instance, Song et al. (2011) evaluated the oxidative stability of chocolate ice cream, which was fortified with omega-3 oil emulsion containing *Bificobacterium longum* as probiotics. Despite the good results on oxidative stability and probiotic bioavailability of the fortified product, the sensory analysis showed low acceptability scores due to the detection of fishy odors.

10.4.2 Microencapsulates

Encapsulation technology consist of coating substances (core material) within a homogeneous/heterogeneous matrix (encapsulating agent/s) at the micro- or nanoscale. Using this technology, a physical barrier is developed between the inner substance and the environment which prevents its degradation and facilitates its handling and transportation. It also can be used as a control time-release mechanism. Food industry has been using this technology over the last decades in the production of added-value food products fortified with different sensitive bioactive compounds such as vitamins, polyphenolic antioxidants, probiotics and ω-3 PUFAs. Encapsulation can be achieved by physical (e.g. spray-drying), physico-chemical (e.g. coacervation) or chemical (e.g. polymerization) means and several encapsulating agents are used for encapsulation of food bioactives, provided that they are biocompatible and food-grade. The most common wall materials in food applications are carbohydrates (e.g. maltodextrins, glucose syrup) and proteins (e.g. casein, whey protein), although gums (e.g. sodium alginate, guar gum), lipids (e.g. wax) and cellulose and its derivatives (e.g. carboxymethylcelullose) are also used (Desai and Park 2005). The choice of the encapsulation technique together with the encapsulating agent depends on different aspects such as the desired size of the particles, the food matrix in which the capsules will be incorporated and the controlled release of the substance, if required (Encina et al. 2016).

ω-3 PUFAs encapsulation and subsequent addition to complex food matrices still represents an important challenge to food industry due to their hydrophobic nature, which results in low solubility in most food systems, and utterly low oxidative stability. In this regard, fish oil encapsulation has gained great attention, not only to prevent its oxidation during processing, but to maintain the organoleptic properties of the fortified food by masking its flavor and oily texture. In the next sections, we provide the state-of-the art on the encapsulation of fish oil by spray-drying, as the most used encapsulation technique, and by electrospraying as an emerging encapsulation technology.

10.4.2.1 Spray-Dried Microencapsulates

Spray-drying is the preferred encapsulation technology in food industry due to its versatility, simplicity, low cost, ease to scale-up and good quality encapsulates. This technique consists of the atomization of a liquid feed (solution, dispersion or emulsion) into a hot gas stream (air or nitrogen), to obtain powder almost instantly in 5–30s (Desai and Park 2005). The size of the resulting capsules range in the microscale, ~ 10 μm – 3 mm, and depends on the liquid feed properties (e.g. viscosity) and the drying process conditions (e.g. atomization speed) (Gharsallaoui et al. 2007). Despite the use of high drying temperatures, this method is suitable to encapsulate heat-sensitive bioactive compounds with minimal thermal degradation due to the small droplets sizes generated during the atomization process (Espejo-Carpio et al. 2013). As a consequence: (i) the internal mass transfer rate, from the core of the droplet to the surface, is high enough to evaporate most of the solvent at the wet bulb temperature (T_{wb}) of the inlet air and, (ii) the external mass transfer, from the surface of the droplet to the air, is high enough to assure short contact time for drying, hence very short residence time of the capsules in the drying chamber. The resulting powder is collected in a cyclone where it is separated from the outer drying gas. Moreover, as it is a moisture removal technique, powders with low moisture content (MC, MC < 5%) and low water activity are produced, assuring their microbiological stability against chemical or biological degradations (Gharsallaoui et al. 2007).

Type of Encapsulating and Emulsifying Agents

Optimization of the feed formulation and the process variables is required to achieve an efficient encapsulation process to obtain good quality microcapsules in terms of high stability and high load capacity. The preparation of the liquid feed, regardless of the hydrophilic/hydrophobic nature of the core and wall materials, requires the dispersion of the bioactive compound in the encapsulating solution. Water is the preferred solvent in food applications, thus for encapsulating ω-3 PUFAs (e.g. fish oil), an emulsification process is commonly carried out to disperse the oil in the water-based biopolymer solutions. Encapsulating agents for fish oil microencapsulation purposes are desired to be water soluble biopolymers at high solids concentration leading to low viscosity solutions. Low viscosity is preferred since it allows to obtain smaller atomized droplets, hence smaller microcapsules. In addition, high coating and emulsifying properties are required to achieve high encapsulation efficiencies (EE) (Encina et al. 2016).

Proteins, carbohydrates and their combinations have been the most extensively used wall materials for fish oil microencapsulation, although the use of gums (e.g. gum arabic) and cellulose derivatives (e.g. sodium carboxymethyl cellulose) have also been reported (Table 10.3).

Proteins alone or mixtures have been used for the production of fish oil microcapsules due to their high emulsifying and film-forming properties. Among proteins, those of animal sources such as milk proteins (e.g. sodium caseinate or whey protein) or gelatin (e.g. fish gelatin) are the most commonly reported. Aghbashlo et al. (2013) investigated the influence of the wall material composition and the inlet drying temperature (140–180 °C) in microcapsules properties using milk proteins as the bulk materials namely, skim milk powder (SMP), whey protein concentrate (WPC), whey protein isolate (WPI) and WPI (80%) combined with milk protein concentrate (MPC) or sodium caseinate (NaCAS) (20%). The core to wall ratio was set to 1:2 leading to a theoretically oil load of 33.33%. The EE varied from

Table 10.3 Recent studies on ω-3 PUFAs encapsulation by spray-drying

Encapsulating Agent/s (EA*)	Emulsifier	FO:EA	Antioxidants	Gas: Inlet/Outlet (T°)	LC/EE (EO)	Capsules size	Oxidative Stability	References
SMP WPC WPI WPI:MPC WPI:NaCAS	–	1:2	–	Air: 140/NR °C Air: 160/NR °C Air: 180/NR °C	Theoretical LC: 33.33%; EE: 40.59–81.94%	1.37–4.59 μm	SMP capsules (inlet T: 180 °C) 25 °C/4 weeks PV	Aghbashlo et al. (2013)
Pure whey Pure whey: Whey permeate (WP)	Gum Acacia: Soybean lecithin	1:2.85	–	Air: 125/105 °C	LC: 13.7–18.4%; EE: 45.72–90.65%	10–150 μm	25 °C/30 days/ darkness PV	Lehn et al. (2018)
GE MD	–	1:2	–	Air: 160/80 °C	Theoretical LC: 33.33%; EE: 46.83–49.34%	–	Room temperature/6 months TBARS	Jeyakumari et al. (2016)
Hordein Glutelin Glutelin: Hordein = 1:2, 1:1, 2:1	–	1:1	–	Air: 120/60 ± 5 °C Air: 150/60 ± 5 °C Air: 180/60 ± 5 °C	LC: 46.5–50.1%; EE: 92.9–100.2%	1–5 μm	40 °C/8 weeks PV	Wang et al. (2011)
Soybean protein isolate (SPI)	–	1:1 1:2 1:3 1:4	–	Air: 180 ± 2/96 ± 8 °C	LC: 12.14–48.77%; EE: 57.73–88.74%	15–20 μm	No storage analysis PV TBARS RANCIMAT test	Di Giorgio et al. (2019)
SMP SMP:MD SMP:Lactose SMP:Sucrose	Tween 20	1:2	–	Air: 175/95–98 °C	Theoretical LC: 33.33%; EE: 76.22–85.12%	3.07–5.37 μm	No storage analysis PV	Aghbashlo et al. (2012)

GE:MD CS:MD GE:CS:MD GE:Microbial transgluminase:MD	–	1:4	–	Air: 180 ± 0.5/90 ± 5 °C	LC: 16.04–21.98%; EE: 67.35–88.01%	D[3,2]: 5.42–8.76 μm	–	Pourashouri et al. (2014)
GE:Xanthan gum (XG):Sucrose GE:XG:Sucrose: Trehalose Sucrose:Trehalose = 10:0–6:4	–	~ 1:2.6	–	Air: 110–120/NR °C	Theoretical LC: 28%; EE: 59.4–91.4%	–	No storage analysis PV	Huang et al. (2014)
GS MD Maltose	n-OSA starch	1:1.5	–	Air: 180/70 °C	Theoretical LC: 40%; EO: 3.42–7.00%	d90: 49.9–63.1 μm	20 °C/33% (RH)/8 weeks PV SVOP	Drusch et al. (2009)
GS	n-OSA starch	1:1.5	–	Air: 160/70 °C Air: 210/90 °C N2: 180/70 °C	Theoretical LC: 40%; EO: 3.60–7.50%	d90: 34.6–44.2 μm	20 °C/33%(RH)/56 days PV SVOP	Serfert et al. (2009a)
GS	n-OSA starch	1:1.5	α-Tocopherol δ-Tocopherol Ascorbyl palmitate Citrem Lecithin Rosemary extract	Air: 180/70 °C	Theoretical LC: 40%; EO: 2.63–3.83%	–	20 °C/33%(RH)/8 weeks PV SVOP	Serfert et al. (2009b)
GS	β-Lactoglobulin (β-LG) β-Lactoglobulin hydrolysate (β-LGH)	~ 1:2.2	–	Air: 180/70 °C	Theoretical LC: ~ 31%; EE: 98.7–99.5%	–	20 °C/33%(RH)/11 weeks PV	Tamm et al. (2015)

(continued)

Table 10.3 (continued)

Encapsulating Agent/s (EA*)	Emulsifier	FO:EA	Antioxidants	Gas: Inlet/Outlet (T°)	LC/EE (EO)	Capsules size	Oxidative Stability	References
GS	Fish protein hydrolysates (FPH)	1:6	–	Air: 180/70 °C	Theoretical LC: 14.33%; EE: 98.0%	–	20 °C/33%(RH)/12 weeks PV	Morales-Medina et al. (2016)
GS	Sugar beet pectin (SBP)	1:2 1:3	Extra virgin olive oil (EVOO)	Air: 180/80 °C	LC: 24.38–49.22%; EE: 90.42–97.87%	–	25 °C/3 months SVOP Induction period	Polavarapu et al. (2011)
MD	Soy protein isolate (SPI) Protein: Oil: 0.02:1–0.22:1	1:1.5	–	Air: 160/85 °C	LC: 37.34–40.71%; EE: 53.82–94.10%	d50,3: 74.1–168.9 μm	27 ± 2 °C/6 weeks PV	Linke et al. (2020)
NaCAS:GA	–	1:4	Sage polyphenols	Air: 160/80 °C	LC: 12.50–12.72%; EE: 68.99–73.21%	107–115 nm	60 °C/7 days PV TBARS	Binsi et al. (2017)
Hydroxypropyl methylcellulose (15cps + 5 cps)	–	1:1 1:0.75 1:0.5 1:0.25	–	Air: 180 ± 1/80 ± 1 °C	Theoretical LC: 50–80%; EE: 67.33–74.75%	D[4,3]: 18.32–54.67 μm	4 °C/28 days PV	Karim et al. (2016)
GA:Sodium Carboxymethyl cellulose (NaCMC) GA:NaCMC:Sodium polyphosphate	–	NR	–	Air: 180/90 °C	LC: 10.86–11.92%; EE: 75.20–82.81%	–	No storage analysis PV AV TOTOX TBARS after 28 days of storage	Patrick et al. (2013)

MD:BG CS:BG:MD	–	1:4	Oregano extract	Air: 180/90 °C	LC: 12.45–13.25%; EE: 68.94–81.88%	1.82–15 μm	60 °C/7 days; 28 ± 2 °C/4 weeks; 4 °C/4 weeks PV TBARS	Jeyakumari et al. (2018)
MD:GA: NaCAS M:GA:CAS:TPH MD:GA:TPH	–	1:5	Tuna protein hydrolysate	Air: 160/80 °C	Theoretical LC: ~ 17%; EE: 73.89–78.73%	1.03–15.3 μm	60 °C/1 week; 28 °C/4 weeks; 4 °C/4 weeks PV TBARS	Unnikrishnan et al. (2019)
Skipjack roe protein hydrolysate (SRPH)	–	1:4	Tannic acid (TA) Oxidized tannic acid (OTA)	Air: 200 ± 2/108 ± 2 °C	Theoretical LC: 20%; EE: 13–55%	6.16–17.07 μm	30 ± 1 °C/4 weeks PV TBARS SVOP	Intarasirisawat et al. (2015)
Vanilic acid grafted-CS	Tween 20	1:2.3	–	Air: 140/77 °C	Theoretical LC: 30%; EE: 84%	2.3 μm PDI: 0.345	Ambient temperature/4 weeks PV TBARS RANCIMAT test	Vishnu et al. (2017)
MD:CS	β-Lactoglobulin Thiol-modified β-lactoglobulin	1:1.65	–	Air: 160/NR °C Air: 170/NR °C Air: 180/NR °C	Theoretical LC: ~ 38%; EE: 71.63–94.90%	–	–	Chang et al. (2020)
Barley β-D-glucan:Waxy maize modified starch	Tween 80	1:3	–	Air: 154/70 °C	Theoretical LC: 25%; EE: 79.9%	PDI: 1.07	Storage temperature NR/5 days TBARS	Kurek et al. (2018)

*Abbreviations: *BG* bovine gelatin, *CS* chitosan, *EA* encapsulating agent, *EE* encapsulation efficiency, *EO* extractable oil, *FO* fish oil, *GA gum arabic*, *GE* fish gelatin, *GS glucose syrup*, *LC* load capacity, *MD* maltodextrin, *MPC* milk protein concentrate, *NaCAS* sodium caseinate, *NR* not reported, *PV* hydroperoxides content, *TBARS* thiobarbituric acid reactive substances, *TPH* tuna protein hydrolysate, *SVOP* secondary volatile oxidation products, *AV* anisidine value, *SMP* skim milk powder, *TOTOX* total oxidation, *WPC* whey protein concentrate, *WPI* whey protein isolate

40.59 ± 1.28% to 81.94 ± 0.24% and the particle size was in the range of 1.37–4.59 μm, being both significantly influenced by the drying temperature (the higher the drying temperature, the higher the EE and the larger the particle size). The highest EE (~ 82%) and lowest hydroperoxides content (PV) after drying at 180 °C corresponded to the SMP-based microcapsules, which led to increase their PV from 6.15 to 8.35 meq peroxide/kg oil over the storage time (25 °C during 4 weeks). The authors attributed this observation to the physical and chemical changes of the wall during storage time, which allowed oil diffusivity from the core to the surface (decreasing the EE to 59.35%) and oxygen diffusivity from the surface to the core thus favouring lipid oxidation. In a recent study, Lehn et al. (2018) investigated the suitability of pure whey alone or mixed with whey permeate (WP) as encapsulating agent(s) for ω-3 PUFAs microencapsulation (carp oil and chia oil) in presence of gum acacia and soybean lecithin as emulsion adjuvants. The oil load of the microcapsules was set to 26% and the drying conditions were fixed to 125/105 °C as the inlet and outlet temperature, respectively. The type of oil and the wall composition highly influenced the EE. In the case of fish oil, the EE varied from 45.72 ± 3.04% to 90.65 ± 0.96% when pure whey and WP were used as the bulk materials, respectively, and the particle size ranged from 10 to 150 μm. WP-based fish oil microcapsules were oxidatively stable during the storage time (30 days at 25 °C) being this fact attributed to its high encapsulation efficiency. Moreover, the use of fish gelatin (GE) alone as encapsulating agent for the production of fish oil microcapsules has also been reported (Jeyakumari et al. 2016). In this case the emulsion was produced using milk as the solvent and the core:wall ratio was set to 1:2, leading to theoretical oil load of 33.33% in the microcapsules. The EE after drying at 160/80 °C temperatures pair was of 46.83%. In this study, the oxidative stability of the encapsulates was not measured.

In the last years, plant proteins have gained great interest in order to substitute animal-derived proteins since they are sustainable, no subjected to any religious or diet restrictions (e.g. vegetarian or vegan diets), low cost and reduce the risk of spreading diseases such as mad cow disease (bovine spongiform encephalitis). Oilseed and cereals (e.g. barley) were used as proteins sources, although pulses-derived proteins (e.g. soy, peas, chickpeas and lentils) are the most commonly employed (Chang and Nickerson 2018). Wang et al. (2011) produced fish oil microcapsules (50% oil load) using barley proteins (glutelin, hordein and mixtures 1:2, 1:1, 2:1) from a water-based emulsion without the need of additional crosslinkers nor the use of organic solvents. The drying process was carried out at 150 °C (inlet temperature) and the EE ranged from 92.9% to 100%. During storage time (40 °C during 8 weeks), the PV of all encapsulates gradually increased reaching the maximum content after 3–4 weeks (45–76 meq peroxide/kg oil). The sample with the highest content of hordein (glutelin: hordein ratio of 1:2) showed the better protective effect against lipid oxidation. More recently, Di Giorgio et al. (2019) microencapsulated fish oil using soybean protein isolate (SPI) as the only wall constituent at different core:wall ratios (1:1–1:4). In this case, the concentration of the encapsulating agent in the emulsion and its production process played a major role on the microcapsules properties after drying (180/96 °C inlet/outlet temperature, respectively). The highest EE (88.74 ± 3.15%) was achieved at the 1:4 ratio when the emulsion was prepared by mechanical stirring (Ultra-turrax) followed by ultrasonic homogenization (USH). However, despite the higher EE, this sample had the highest PV and TBARS (thiobarbituric acid reactive substance) value after drying among those produced under the same conditions (Ultra-turrax and USH treatment of the emulsion prior drying) at lower core:wall ratios. This was attributed to: (i) the faster crust formation as consequence of an increase in the protein content, leading to an increased resistance to evaporation, thus an increase of particle temperature, and (ii) trace metals (iron and copper) present in the SPI. Nevertheless, these encapsulates were reported to maintain the oil oxidative stability over time, studied by RANCIMAT accelerated oxidation test method (90 °C and 20 L/h of air stream), since the induction period (IP, over 3 h) was similar to the non-encapsulated oil (IP = 3.9 h) and higher of those of lower core:wall ratios (except for the 1:1 sample).

Despite their high film-forming properties, proteins are frequently combined with low molecular weight carbohydrates since these act as a wall filling materials, which lead to less porous and more uniform matrices. Moreover, the presence of low molecular weight carbohydrates influences the drying behaviour of the droplets by modifying the wall material glass transition temperature (T_g), and thus influencing the crust formation. Moreover, the compounds derived from Maillard reaction occurring between proteins and reducing sugars at high temperatures (e.g. melanoidins) have been reported to change the physical and antioxidant properties of the wall leading to additional core stabilization. Aghbashlo et al. (2012) encapsulated fish oil using SPM alone or in combination with maltodextrin (MD), lactose or sucrose (70% SMP + 30% carbohydrate) with the inlet drying temperature and the core:wall ratio set to 175 °C and 1:2, respectively. The blends SMP-lactose and SMP-sucrose enhanced the EE of the microcapsules by decreasing the wall composites T_g, thus favouring a fast crust formation and preventing the oil diffusivity across the encapsulating wall (84.96 ± 0.03% and 84.92 ± 0.19%, respectively). The faster crust formation also led to larger microcapsules (5.21 and 5.37 μm, respectively) and lowers PV after drying (5.95 ± 0.05 and 5.95 ± 0.15 meq peroxide/kg oil, respectively) since less oil was exposed to the drying air at high temperature. However, the similarities in EE for both powders, despite the different nature of the sugar composites, led to conclude that Maillard reaction between the protein and carbohydrate constituents of the wall matrix did not occur. These results are in line with those reported from Aghbashlo et al. (2013), who attributed the better retention properties of the SMP-based powders to the presence of lactose in the intrinsic chemical composition of the matrix, thus enhancing the drying behaviour by favouring the crust formation. However, in this case the authors discussed that the better wall properties could be also attributed to Maillard reaction products (e.g. protein-carbohydrate conjugates). Pourashouri et al. (2014) reported that fish oil encapsulates with GE and MD as wall material composites, over those formulated with chitosan (CS) and MD, had better retention properties (higher EE). These results were explained on the basis of lower T_g of the encapsulating material in presence of GE which favoured the crust formation leading to: (i) higher EE (85.36 ± 1.91% over 71.84 ± 1.51% for GE-MD and CS-MD microcapsules, respectively), and (ii) larger particle size (D[3,2] = 6.54 μm and D[3,2] = 5.92 for GE-MD and CS-MD microcapsules, respectively). The authors also stated that Maillard reaction products might also have favoured the formation of a tough skin. However, the previous results differ from those reported by Huang et al. (2014) who did find that the addition of trehalose to the emulsion formulation (7.0 wt% tilapia oil, 4.5 wt% gelatin, 0.5 wt% xantham gum and 13.2 wt% sucrose or sucrose:trehalose (7:3)) increased the EE by increasing the wall composites T_g. The T_g were of 83.8 °C for the sample containing only S (EE = 80.4 ± 1.13%) and 96.0 °C when trehalose was added to the formulation (EE = 87.0 ± 0.74%).

Furthermore, due to the lack of interfacial activity of low molecular weight carbohydrates, they have been used lonely as wall material constituents in presence of surface-active compounds. For instance, Drusch et al. (2009) used glucose syrup (GS, dextrose equivalent, DE38), MD (DE18), maltose (DE50) and the mixtures MD-GS or MD- maltose (DE38) as bulk agents in presence of n-octenylsuccinate-derivatised starch (n-OSA-starch) to produce fish oil microcapsules with an oil load of 40%. The drying conditions were set to 180/70 °C as the inlet and outlet temperature, respectively. The particle size (d_{90}) ranged from 49.9 to 63.1 μm and the amount of extractable oil (non-encapsulated oil) varied between 3.14% and 7.00%, meaning high EE. Other GS-based microcapsules has been produced in presence of n-OSA-starch (Serfert et al. 2009a, b), whey protein or whey protein hydrolysate (β-lactoglobulin) (Tamm et al. 2015), fish protein hydrolysates (Morales-Medina et al. 2016) and sugar beet pectin (SBP) (Polavarapu et al. 2011) as emulsifiers with different oil loads (14.33–40%) but similar encapsulation efficiencies (over 90%) (Table 10.3). Furthermore, the production of MD-based microencapsulates has also been reported. Jeyakumari et al. (2016) produced fish oil microcapsules using MD (DE16) as the bulk material from a milk-based emulsion, achieving an encapsulation efficiency of 49.34%. Moreover, Linke et al. (2020) successfully encapsulated fish oil (40% oil load) using MD (DE21) and soy protein isolate (SPI) at different protein:core ratios (0.02:1–0.22:1). The amount of emulsifier strongly influenced the EE (at the same emulsion homogenization pressure) and varied from 53.82 ± 1.07% (0.02:1 ratio) to 93.44 ± 0.05% (0.13:1 ratio). The particle size ($d_{50,3}$) in this case ranged from 86.2–115.6 μm.

By last, Binsi et al. (2017) encapsulated sardine oil (20% oil load) using NaCAS as wall polymer and gum arabic (GA) as wall co-polymer by spray-drying at 160 °C (inlet temperature). The EE after drying the fish oil emulsion ascended to 69%. Karim et al. (2016) used two different hydroxypropyl methylcellulose (HPMC) (15 or 5 cps) and their mixtures to produce menhaden fish oil microcapsules. The drying conditions were fixed to 180/80 °C temperatures pair, but the core:wall ratio varied from 1:1 to 1:0.25, thus varying the oil load. The EE ranged from 67.33 ± 0.15% to 74.75 ± 0.39%, whilst the particle size (D[4,3]) varied from 18.32 ± 0.04 to 54.67 ± 0.09 μm. The core:wall ratio of 1:1 resulted in the largest particles size and the highest EE. Patrick et al. (2013) produced single-shell and double-shell fish oil microcapsules by using GA, sodium carboxymethyl cellulose (NaCMC) and sodium polyphosphate at an inlet air temperature of 180 °C. The microcapsules total oil content were of 10.86 ± 0.33% (EE = 75.20 ± 0.73%) and 11.92 ± 0.25% (EE = 82.81 ± 0.61%) for the single-shell and double shell, respectively.

The choice of the proper encapsulating agent is of vital importance since it will further determine the microcapsules properties (e.g. particle size, bulk density, MC or PV) being EE the most important. This parameter not only quantifies the yield of the drying process regarding the oil load of the particles (core retention), but also allows to predict oxidative stability of the powder. By determining the EE, the amount of non-encapsulated oil is indirectly quantified, being the latter the amount of unprotected oil susceptible to degradation by action of pro-oxidant agents (e.g. oxygen or light). It has been assumed, in consequence, that the higher the EE, the highest the degree of protection of the wall material and the higher the oxidative stability of the encapsulated oil (lower PV and SVOP content). However, not always is possible to predict the oxidative stability of the core based on this parameter since: (i) pro-oxidant species are present in the parent emulsion (e.g. transition metals or oxygen) as a consequence of the emulsification process, and (ii) pro-oxidant species which diffuses through the wall matrix (e.g. oxygen) also affects the extent of encapsulated oil oxidation. In this regard, Linke et al. (2020) investigated the oxidation rate of the non-encapsulated and encapsulated oil fractions and their contribution to overall oxidation in fish oil microcapsules and concluded that, although non-encapsulated fraction oxidized ~7 times faster than its encapsulated counterpart, its contribution to overall lipid oxidation was negligible due to its low amount on particles surface (~ 2–18%). Moreover, oxidation of the encapsulated oil during storage time, despite being totally embedded within the wall matrix, suggested oxygen diffusion through the capsules wall. Hence, structural particle properties such as load capacity, specific surface area of oil droplets and microparticles size play a major role on encapsulated oil oxidation by affecting the oxygen supply. These results are in line with those reported by Drusch et al. (2009) and Serfert et al. (2009a) who attributed the differences in lipid oxidation course among samples to the oxygen diffusivity through the wall matrix (influenced by the differences in the molecular weight profile of the constituents) since the EE was not significantly different. Moreover, the latter authors investigated the influence of oxygen present in the drying medium (air or nitrogen) or dissolved in the parent emulsion (consequence of the shearing forces) in the microcapsules oxidative stability during storage and concluded that the course of lipid oxidation was rather determined by the oxygen present in the emulsion than in the drying gas. Hence, lipid oxidation already occurs in the microcapsules production in both the emulsion preparation (air inclusion by action of mechanical and shearing forces) and subsequent drying at high temperature.

Spray-Drying Processing Variables

Regarding the spray-drying process, there are several operational variables to be optimized such as the inlet and outlet air temperatures, feed temperature, drying air mass flow rate, feed mass flow rate and type and speed of atomization. All these variables affect the final quality of the capsules (e.g. MC, particle size, EE or PV after drying). Nonetheless, inlet drying air temperature is regarded as the most important factor for encapsulation of heat-sensitive bioactives compounds (e.g. fish oil). This parameter is mostly selected according to the type of product to be dried. Although high drying temperatures are preferred, since temperature is directly related to drying rate and final water content, there are limitations. Wang et al. (2011) encapsulated fish oil using barley protein as the encapsulating agent and reported that the highest inlet temperature studied (180 °C) led to a less uniform particle size distribution composed of particles of irregular shape. This finding was attributed to the faster particle shrinkage, produced during the early stage of drying, caused by a too high drying temperature. On the other hand, at the lowest inlet temperature studied (120 °C) the high MC of the dried capsules caused agglomeration. Different studies have been carried out to determine the influence of the drying air temperature in fish oil microcapsules properties. For instance as previously mentioned, Aghbashlo et al. (2013) produced milk protein-based microcapsules at three different inlet temperatures (140 °C, 160 °C and 180 °C) and reported that the EE, particle size and initial PV were positively correlated with the inlet drying temperature. On the other hand, as the drying temperature increased, MC and bulk density decreased. Serfert et al. (2009a) found the same trend regarding initial PV, particle size and bulk density when fish oil microcapsules were produced using GS in presence of n-OSA-starch. However, in this study the EE decreased as the inlet drying temperature increased, being this finding attributed to the formation of vacuoles and to a higher porosity as consequence of a high inlet temperature.

Stabilization by Addition of Antioxidants

Although microencapsulation itself prevents lipid oxidation, additional stabilization techniques are required since it has been demonstrated that oxidation already occurs in the early stages of microcapsules production process and during storage time. In this regard, scientists have studied the addition of antioxidants to the formulation. Serfert et al. (2009b) investigated the effect of a combination of various antioxidants of different natures (α-tocopherol, δ-tocopherol, ascorbyl palmitate, citrem, lecithin and rosemary extract) in the different process stages of microencapsulation and during storage. The capsules were produced using GS in presence of n-OSA-starch as the encapsulating agent, with 40% oil load, and the drying process was carried out at 180/70 °C temperatures setting. The main finding of this study was that the optimal combination of antioxidants depended on the microencapsulation process step, thus on the characteristics of the heterogenous system. An efficient stabilization during storage of both the emulsion and the microcapsules was achieved by using a ternary combination of antioxidants (tocopherols, ascorbyl palmitate and lecithin) in presence of rosemary extract. Jeyakumari et al. (2018) investigated the effect of adding oregano extract on the oxidative stability of microencapsulated fish during storage at three different temperatures (4 °C, 28 °C and 60 °C). The wall material consisted of MD and bovine gelatin (BG) with or without CS and the drying conditions was set to 180/90 °C as the inlet and outlet temperatures, respectively. The microcapsules oil load was of 20% and the oregano extract was dissolved in the fish oil prior to emulsification. Regardless of the storage temperature, the microcapsules containing the fish oil-oregano blend had the lowest PV and TBARS value over storage time, being this attributed to the phenolic compounds present in oregano extract, which effectively protected fish oil against oxidation. Furthermore, the results suggested that the storage temperature strongly influenced the course of lipid oxidation. The smaller values reported of PV and TBARS value were those of the sample containing oregano extract at 4 °C of storage temperature. Binsi et al. (2017) produced sardine oil microcapsules in presence of sage polyphenols (SP). The course of lipid oxidation under accelerated conditions (60 °C during 7 days) showed that, although the addition of SP enhanced oxidative stability of the microcapsules (lower PV and TBARS value over storage time), this occurred rather by a physical mechanism of protection than an antioxidative effect (e.g. radical scavenging). The authors reported that SP acted as protein crosslinkers, which stabilized the wall matrix and favoured the crust formation during the early stages of drying. This led to higher EE compared to the sample without SP (73.21% over 68.99%). Polavarapu et al. (2011) investigated the addition of extra virgin olive oil (EVOO) as antioxidant to produce fish oil microcapsules with GS and SBP as wall materials. Two different encapsulates with different oil loads were produced (25% and 50% oil) at 180/80 °C temperatures pair. The course of lipid oxidation was assessed by quantifying the propanal and hexanal content over 3 months at 25 °C and under accelerated storage conditions (80 °C and oxygen pressure of 0.5 bar). The SVOP content showed that the addition of EVOO to the formulation did not enhance fish oil oxidative stability neither in the microcapsules production nor in storage time. This finding did not correlate with the results reported under accelerated storage conditions, which indicated that the addition of EVOO to the formulation extended the IP in the microcapsules (~7.90–11.95 h). In general, the authors attributed the low oxidative stability of the microcapsules to the presence of trace metals in SBP, which could not be averted by the addition of EVOO.

Lately, the use of protein hydrolysates/peptides, having both emulsifying and antioxidant properties is gaining an increasing interest. This is because the location of the antioxidants in the oil-in-water emulsions, and thus in the dried encapsulate, determines their antioxidant activity, which is increased when the antioxidants are located at the oil-water interface. It is thought that at the oil-water or oil-encapsulating agent interfaces autoxidation process begins since pro-oxidants species (e.g. metals) contact the oil and where emulsifiers/surfactants are absorbed. Therefore, the use of emulsifying compounds with antioxidant properties (e.g. radical scavenging) would theoretically improve the oxidative stability of microencapsulated fish oil by inhibiting lipid oxidation at contact area between pro-oxidants and lipids. Moreover, protein hydrolysis enhances both hydrolysates emulsifying properties and the wall matrix protective effect since the smaller peptides size generated act rather as copolymers or fillers of the wall. Hence, protein hydrolysates are thought to improve microencapsulated fish oil oxidative stability by physical and chemical means. In this regard, different studies have been carried out. For instance, Tamm et al. (2015)

encapsulated fish oil using GS in presence of β-lactoglobulin (β-LG) or β-lactoglobulin hydrolysates (β-LGH) by spray-drying at 180/70 °C temperature pair (theoretical oil load of ~31%). The hydrolysates were produced to two different degree of hydrolysis (DH = 3% and DH = 6%) by two different enzymes; trypsin and alcalase. The EE achieved after drying was high and similar among samples (98.7% - 99.5%). Therefore, the different trend of PV curves during storage suggested an enhanced oxidative stability of the encapsulates by the addition of β-LGH (except for β-LGH, DH = 3%, alcalase). However, in this study the antioxidant activity of the hydrolysates was not assessed (e.g. DPPH scavenging activity) and the differences in the hydrolysates protective effect were attributed to the different peptide profiles as a consequence of the hydrolysates production. The lowest PV after the storage time was reported for the sample containing β-LGH, DH = 6%, produced with T (PV = 39.6 ± 5.9 mmol/kg oil). On the other hand, Morales-Medina et al. (2016) produced fish oil microcapsules with GS and fish protein hydrolysates (FPH, sardine and horse mackerel) to a DH = 5% by using alcalase or trypsin. The drying process was carried out at 180/70 °C inlet/outlet temperatures and the oil load of the encapsulates was of 14.33%. The addition of the FPH to the formulation efficiently stabilized the emulsions prior and during drying (EE = 98.0 ± 0.1%), however, it could not be established a relationship between the antioxidant activity of the hydrolysates and the course of lipid oxidation. Neither differences in the oxidative stability among samples were observed during the storage time (20 °C, 33% of relative humidity (RH), 12 weeks) nor by using different substrates (fish species) or enzymes. In this line, Unnikrishnan et al. (2019) studied the suitability of tuna protein hydrolysate (TPH, DH ~ 30%) to enhance the oxidative stability of encapsulated sardine oil. For this purpose, TPH was rather used as encapsulating agent or as core polymer. The microcapsules were produced at a core:wall ratio of 1:5 and the spray-drying process was carried out at 180/70 °C inlet/outlet temperatures (EE = 73.89 ± 1.53% - 78.73 ± 1.94%). The oxidative stability during storage (1 or 4 weeks) at three different temperatures (4 °C, 28 °C and 60 °C) was assessed by measuring the PV and TBARS value. The results showed that the presence of TPH as a wall co-polymer (in presence of NaCAS, MD and GA) or in the core, led to an increase in EE. However, the highest oxidative stability of sardine oil during storage (regardless of the temperature) was achieved when TPH was used as core material. Besides, the total replacement of NaCAS with TPH led to the less oxidatively stable encapsulates although the EE after drying was relatively high (76.64 ± 1.17%). Moreover, Intarasirisawat et al. (2015) investigated the effect of adding natural antioxidants (tannic acid, TA or oxidized tannic acid, OTA) to the microencapsulates formulation produced with skipjack roe protein hydrolysate (SRPH, DH = 5%) as the sole wall material, which also exhibit antioxidant activity. The spray-drying process was carried out at 200/108 °C temperature pair. The results showed that SRPH was not suitable as sole wall material due to the low EE achieved after drying (13%), however it was slightly improved in presence of TA and OTA (28% and 55%, respectively). The higher EE value found in presence of OTA was attributed to its crosslinking properties. Regarding the oxidative stability, it was assessed by quantifying the PV, TBARS value and SVOP content after 4 weeks of storage at

30 °C. The results suggested that the presence of OTA and TA improved the oxidative stability of the encapsulates compared to the microcapsules with SRPH alone, however, the reduced form of TA was more efficient than the oxidized form (OTA).

In the last years, the production of fish oil microcapsules using biofunctional compounds as wall materials has gained increasing interest, not only to enhance the oxidative stability of the capsules, but also to provide additional health benefits derived from the wall matrix constituents. CS is one of the most abundant natural polysaccharides and its incorporation into the diet has been related to promote cardio-protective and hypolipidemic effects (Vishnu et al. 2017). In this regard, Vishnu et al. (2017) successfully produced functional sardine oil microcapsules with high oxidative stability by using vanillic acid grafted-CS, an antioxidant-CS conjugate, as encapsulating agent. The oil load of the capsules was of 30% and the drying process was carried out at 140/70 °C inlet/outlet temperatures (EE = 84 ± 0.84%). The PV after the storage time (4 weeks at ambient temperature) became 5.5 ± 0.51 meq peroxide/kg oil and the TBARS value trend indicated that low content of secondary oxidation products was produced. Moreover, the accelerated RANCIMAT test (100 °C and 20 mL/h air flow) showed that the IP of the microcapsules was ~11 times higher than that of the sardine oil (7.67 ± 0.05 h over 0.67 ± 0.01 h) indicating a protective effect of the wall matrix. More recently, Chang et al. (2020) used thiol-modified β-lactoglobulin fibril/CS complex in presence of MD to produce fish oil microcapsules (theoretical oil load of ~37%) at different inlet air temperatures (160–180 °C). In all cases the EE achieved was high (89.40 ± 1.80% – 93.53 ± 1.57%), not being much affected by the inlet temperature. Unfortunately, the oxidative stability of the encapsulates was not investigated in this study. On the other hand, Kurek et al. (2018) encapsulated cod liver oil in a barley β-D-glucan – waxy maize modified starch blend. β-D-glucan has been related to lower glycemic and insulin responses, lipid metabolism and blood cholesterol levels (Kurek et al. 2018). The ratio of encapsulating agents varied to optimize the wall formulation, and so did the inlet drying temperatures (154–180 °C) in the experimental design leading to a total of 13 experimental runs. The results from the optimization study showed that the optimum wall matrix consisted of 85% β-D-glucan (15% waxy maize modified starch) and that the optimum inlet drying temperature was 154 °C. Under these conditions the encapsulates had an EE of 79.9% and TBARS value of 1.16 mg MDA/kg powder after drying.

Bioaccessibility and Food Enrichment

Although the purpose of producing fish oil microcapsules is to manufacture foods enriched with ω-3 PUFAs to improve peoples' health, only few studies have been carried out regarding the fate of the encapsulated oil in the food matrix (Binsi et al. 2017) and on its gastrointestinal tract (GIT) release (Vishnu et al. 2017; Binsi et al. 2017; Unnikrishnan et al. 2019) (Table 10.3). The latter is of great importance since the oral bioavailability of TAG (e.g. EPA and DHA) is determined by GIT absorption of TAG residues (e.g. monoglycerides) from gastric and pancreatic lipases hydrolysis. Furthermore, as most of the encapsulating agents used are water soluble-biopolymers (carbohydrates and proteins), little research have been conducted in order to incorporate fish oil microcapsules into water-based foodstuff (Patrick et al. 2013; Jeyakumari et al. 2016). Binsi et al. (2017) investigated the release of encapsulated oil in both the food system (using buffered saline solution) and GIT (using simulated gastric and intestinal fluids) from microcapsules produced in presence (SOE) or absence (FOE) of sage extract. The authors reported that sage extract enhanced the matrix resistance to water, thus preventing its disintegration and subsequent oil leakage when added to a food matrix (19.74% of oil release after 4 hours of incubation in the buffered saline solution over 55.22% of oil release in the FOE sample). Regarding the simulated GIT, the total oil release was higher for the FOE sample due to its poorer structural integrity (87.12 ± 0.93% over 79.27 ± 0.75% of total oil release), whereas the retention properties (EE), the oxidative stability and the water resistance were enhanced by adding SP to the formulation. In this line, Unnikrishnan et al. (2019) reported that the incorporation of TPH to the MD/GA-based wall formulation of the encapsulates, resulted in an increased structural stabilization, thus retarding the oil release in the gastric phase of digestion. On the other hand, the encapsulates produced with MD and GA in presence of NaCAS led to a higher oil release in both the gastric and intestinal phase of digestion, amounting to 88.7–93.6% of total oil release after 5 h of incubation. Moreover, Vishnu et al. (2017) reported that vanillic acid grafted-CS as the sole encapsulating agent allowed to achieve a sustained oil release in GIT, since after gastric and intestinal digestion phases, the total oil release was of 66 ± 0.21%. Karim et al. (2016) studied the GIT oil release of microencapsulates produced using blends of HPMC of different viscosities (5 or 15 cps) at different concentrations in the infeed emulsion (1.25–5%). The oil content was fixed to 5% in the emulsions. The results showed that the gastric and subsequent intestinal digestions were highly influenced by the concentration of polymer in the formulation, being both negatively correlated, since the highest percentage of oil released (81.27 ± 1.04%) corresponded to the sample with the lowest HPMC blend content in the emulsion (1.25%). On the other hand, at the same polymer concentration of the infeed emulsion (5%) the highest oil release percentage was reported for the sample containing 100% HPMC 15 cps in both gastric digestion (22.00 ± 1.45%) and gastric digestion followed by intestinal digestion (51.83 ± 2.18%), respectively.

Regarding food fortification with ω-3 PUFAs by addition of fish oil microcapsules, Patrick et al. (2013) successfully produced a probiotic fermented milk by adding single-shell (MFO1) or double-shell (MFO2) fish oil encapsulates using GA, NaCMC and sodium polyphosphate as wall materials. The PV after drying was of 2.98 ± 0.12 meq peroxides/kg oil in MFO1 sample and 2.09 ± 0.05 meq peroxides/kg oil in MFO2 sample. A sensory analysis was conducted on day 2 and day 28 of storage at 4 °C, and all types of fermented milk (including a control fermented milk without encapsulates) received an average score of 5.65/7. Unfortunately, the oxidative stability of the fortified milk over storage time was not investigated. Jeyakumari et al. (2016) produced fortified cookies by adding fish oil encapsulates to the recipe. The oxidative stability of the cookies was assessed by TBARS value over 6 months of storage at room temperature and a sensory evaluation was conducted over the fresh cookies (after baking) by a panel of 10 experts. The results showed that the cookies containing the fish oil encapsulates, regardless of the encapsulating agent (GE or MD), had the lowest TBARS value over all the storage time compared to other cookies samples containing fish oil (e.g. cookies with emulsified or neat fish oil), being this value comparable with the control cookies (without fish oil). However, only the MD-based encapsulates received a sensory evaluation score similar to that of the control cookies (3.5 ± 0.15 and 3.5 ± 0.08 for the MD-cookies and the control cookies, respectively).

10.4.2.2 Mono-Axially Electrosprayed Microencapsulates

Electrospraying is an affordable and scalable emerging technology for the production of bioactive encapsulates, which has emerged as an alternative to conventional spray-drying (Jacobsen et al. 2018). This technique, which is included within the so called electrohydrodynamic processes, allows to produce nano-microcapsules by means of high voltage. A schematic representation of the process is shown in Fig. 10.1.

Electrohydrodynamic technology principle is based on applying a high-voltage electrostatic field to charge the surface of encapsulating agents (e.g. biopolymers) solution droplets (Jacobsen et al. 2018). Hence, when the liquid feed is injected through an electrified capillary tube, as consequence of mutual charge repulsion, a meniscus of conical shape (Taylor cone) is formed at the exit. When the electric

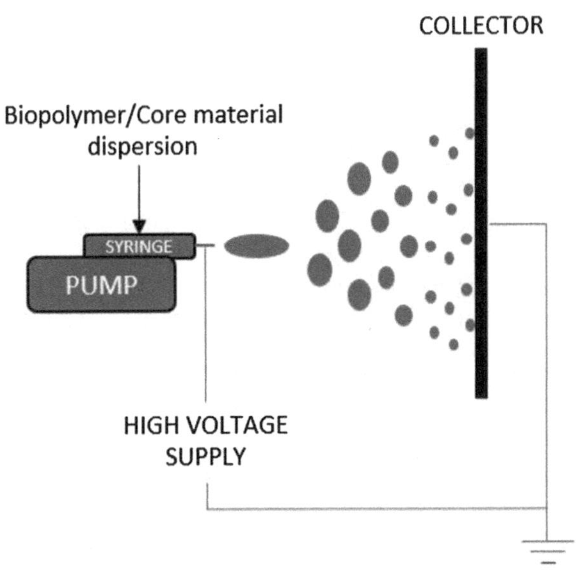

Fig. 10.1 Scheme of electrospraying process

force overcomes the solution surface tension, a charged jet is ejected from the tip of the Taylor cone and, depending on the jet stability, fibers (electrospinning) or capsules (electrospraying) are produced. At low biopolymer concentrations (low viscoelasticity of the solution) the jet destabilizes and forms a spray of charged droplets, which solidifies through solvent evaporation, and are gathered on a grounded or oppositely charged collector (García-Moreno et al. 2017). The main advantage of electrospraying technology over spray-drying, regarding heat-sensitive bioactive compounds such as ω-3 PUFAs, is that heat is not required at any point of the process. The evaporation of the solvent in which the encapsulating agent/bioactive are dispersed occurs at ambient temperature, thus thermal degradation is avoided. Furthermore, the resulting encapsulates are smaller (submicron diameter) than those produced by spray-drying (micron diameter) and present larger surface-to-volume ratio, leading to highly bioaccessible encapsulates with little detrimental organoleptic changes when incorporated into food matrices (e.g. unwanted mouthfeel).

Similar to spray-drying technology, the nano-microcapsules properties (e.g. morphology, particle size, or EE) obtained by electrospraying depend on both: (i) the electrospraying process variables (e.g. voltage, flow rate and distance from the collector) and (ii) the biopolymer solution properties (e.g. viscosity, surface tension and conductivity). The latter are of special interest since these will determine the morphology of the encapsulates (e.g. fibers, beaded fibers, capsules or fiber-interconnected capsules) and depend mainly on the biopolymer type, molecular weight, hydrophobic/hydrophilic nature and concentration, as well as on the type of solvent. Table 10.4 shows the advances made in the last decade in the encapsulation ω-3 PUFAs by electrospraying. In these studies, proteins (e.g. zein, WPC or gelatin), carbohydrates (e.g. dextran or GS) and protein-carbohydrates blends (e.g. WPI-GS or dextran blends) have been reported as suitable encapsulating biopolymers.

Type and Concentration of Wall Materials

Torres-Giner et al. (2010) produced DHA encapsulates using zein prolamine (a hydrophobic corn protein) as wall material by electrospraying means at lab scale. In this study, a biopolymer/DHA solution was prepared in aqueous ethanol (85 wt% ethanol) at a core:wall ratio of 1:2 (theoretical oil load of 33.33%). Subsequently, the solution was electrosprayed under a steady flow rate of 0.20 mL/h, 12 kV and under air atmosphere at 18 °C (RH = 40%). The collector was placed 15 cm apart from the ejecting needle. The resulting DHA-capsules presented a smoother surface than zein encapsulates without DHA, and had an average diameter of 490 ± 200 nm. However, nano-sized fibers interconnecting the capsules could be observed. Regarding the oxidative stability of the encapsulates, based on SVOP, zein efficiently protected DHA from oxidation. The content of propanal, 2,4 – heptadienal and 2,4,7 – decatrienal, which are common compounds derived from oxidation of fish oil ω-3 PUFAs, were significantly lower in the encapsulates compared to those

Table 10.4 Recent studies on ω-3 PUFAs encapsulation by electrospraying

ω-3 source	Encapsulating Agent(s)	Solvent(s)	Emulsifier(s)	Processing conditions	LC/EE[a]	Capsules size	Oxidative Stability	Food application	References
Cod liver oil	Dextran	Water	WPI Tween 20	Lab scale: 20 kV, 0.01 mL/min, 15 cm No T or RH control P-P scale: 24 needles, 42 kV, 5 mL/h, 10 cm No T or RH control	Theoretical LC: 10%; EE: 68.3–75.5%	0.1–3.3 μm	No storage analysis PV SVOP	NR	García-Moreno et al. (2017)
DHA	Zein	Ethanol/water 85/15, (v/v) %	–	12 kV, 0.20 mL/h, 15 cm 18 °C, 40%(RH)	Theoretical LC: 33.33%; EE: NR	490 ± 200 nm	18 °C/40%(RH)/7 days SVOP	NR	Torres-Giner et al. (2010)
Fish oil	Zein	Ethanol/water Isopropanol/water 70/30, (w/w) %	–	20 kV, 1 mL/h, 20 cm 21 ± 2 °C, 40%(RH)	Theoretical LC: 30%; EE: NR	400–500 nm 600–800 nm	–	NR	Moomand and Lim (2015)
ALA	WPC SPI Gelatin	Water Acetic acid, 20% (v/v)	Tween 20	15–17 kV, 0.15 mL/h, 10 cm No T or RH control	Theoretical LC: 10%; EE (based on FT-IR absorbance measurements): 23–67%	0.25–3.25 μm	80 °C/120 h FT-IR absorbance measurements	NR	Gómez-Mascaraque and López-Rubio (2016)

(continued)

Table 10.4 (continued)

ω-3 source	Encapsulating Agent(s)	Solvent(s)	Emulsifier(s)	Processing conditions	LC/EE[a]	Capsules size	Oxidative Stability	Food application	References
Cod liver oil	Dextran:Pullulan GS:Pullulan	Water	WPC Citrem	Lab scale: 17–20 kV, 0.18–0.72 mL/h, 15 cm No T or RH control EAPG: 10–15 kV, 90–108 mL/h 24 °C, 40%(RH)	Theoretical LC: 20%; EE (EAPG): 78.1–91.7%	Lab scale: 60–70% of capsules <1 μm EAPG: 70% of capsules <3 μm	20 °C/21 days PV SVOP	NR	García-Moreno et al. (2018)
MCT oil Fish oil	Dextran:Pullulan GS:Pullulan	Water	WPC Citrem	15–20 kV, 0.18–0.60 mL/h, 15 cm 24 ± 4 °C, 17–46%(RH)	Theoretical LC: 20%; EE: NR	Dextran: 1.26 ± 0.57 μm Glucose: 1.39 ± 0.52 μm	50 °C/16 days Electron spin resonance (ESR)	NR	Boerekamp et al. (2019)
Fish oil	WPI:MD	Water	Tween 20	18 kV, 1.2 mL/h, 13 cm No T or RH control	Theoretical LC: ~ 42%; EE: 60–98%	d50 = 198–974 nm	–	NR	Paximada et al. (2018)
Fish oil (DHA)	Zein	Ethanol/ water 85/15, (v/v) %	–	EAPG: 20 kV, 10 mL/min, air pressure of 10 L/min 25 °C, 40%(RH)	Theoretical LC: 33.33%; EE: 84%	1.4 ± 0.8 μm	5 °C/0%(RH)/45 days/air and darkness 23 °C/0%(RH)/45 days/ air and light 23 °C/65%(RH)/45 days/ vacuum and darkness PV	Powdered milk	Busolo et al. (2019)

Cod liver oil	Dextran:Pullulan GS:Pullulan	Water	WPC Citrem	EAPG: 15 kV, 1.8 mL/min, air pressure NR 24 °C, 40%(RH)	Theoretical LC: 20%; EE: 83.2–90.4%	Dextran: ~ 80% of capsules <2 μm Glucose: ~ 80% of capsules <3 μm	20 °C/21 days/darkness PV SVOP	Light mayonnaise	Hermund et al. (2019)
Cod liver oil	Zein	Ethanol/water 85/15, (v/v) %	–	EAPG: 10 kV, 1.4 mL/min, air pressure 10 L/min 24 °C, 40%(RH)	Theoretical LC: 20%; EE: 83%	2.4 ± 0.7 μm	20 °C/35 days/darkness PV SVOP	Low-fat mayonnaise	Miguel et al. (2019)

[a]Abbreviations: *ALA* α-linolenic acid, *DHA* docosahexaenoic, *EAPG* electrospraying assisted by pressurized gas, *EE* encapsulation efficiency, *FT-IR* Fourier transform infrared, *GS* glucose syrup, *LC* load capacity, *MD* maltodextrin, *NR* not reported, *PV* hydroperoxides content, *SPI* soy protein isolate, *SVOP* secondary volatile oxidation products, *WPC* whey protein concentrate, *WPI* whey protein isolate

reported for neat DHA. For instance, after 1 week of storage at 18 °C, the relative peak area of propanal in neat DHA amounted to 5.62%, whilst for DHA-capsules was of 0.78%. Moomand & Lim (2015) also encapsulated fish oil within a zein matrix (30% of oil load) at different protein concentrations (10 or 20 wt%) using aqueous solutions of ethanol or isopropanol (70 wt%). The electrospraying process was carried out at lab scale at a pump flow rate of 1 mL/h and 20 kV. The ambient conditions were controlled at 20 ± 2 °C and 40% RH, and the collector was placed at 20 cm from the ejecting needle. The authors reported that decreasing the zein concentration from 20 to 10 wt% led to the formation of capsules instead of fibres due to a lower viscosity of the infeed solution. In addition, the solvent used to disperse the biopolymer together with the oil phase influenced the particle size. The ethanol-based solution led to particles of diameters ranging from 400–500 nm, whilst the isopropanol-based particles varied from 600 to 800 nm.

Gómez-Mascaraque and López-Rubio (2016) carried out a comparative study between spray-drying and electrospraying using WPC, SPI and gelatin as wall materials. ALA was used as the core bioactive and the encapsulates were produced to a fixed ALA load of 10%, regardless of the drying procedure. The scanning electron microscopy (SEM) images showed that spray-dried capsules were larger and had a wider particle size distribution than those produced by electrospraying. Moreover, the fourier transform infrared (FT-IR) spectra analysis suggested that the bioactive encapsulated in spray-dried capsules was degraded during processing since the characteristic ω-3 PUFAs absorption band was not detected. The authors attributed this finding to the high inlet drying temperature used in the spray-drying process (90 °C). Conversely, electrosprayed capsules produced with WPC and SPI, which were produced at a steady flow rate of 0.15 mL/h and by applying 15–17 kV (Table 10.4), evidenced ALA presence during storage as measured by FT-IR. Thus, these results suggested an enhanced oxidative stability of ALA-loaded electrosprayed capsules produced with WPC and SPI, compared to ALA-loaded spray-dried capsules.

García-Moreno et al. (2017) optimized the biopolymer solution formulation to produce fish oil electrosprayed capsules in lab scale using dextran (molecular weight = 70 kDa, dextran70) as wall material. The polymer was dissolved in water at different concentrations (20–40 wt%), whilst the oil load of the encapsulates was fixed to 10%. It should be mentioned that the oil was introduced to the infeed solution as a fish oil-in-water emulsion stabilized with WPI. The electrospraying process was conducted at a flow rate of 0.01 mL/min, 20 kV, under inert atmosphere (nitrogen) and the collector was placed at 15 cm from the needle. In the study, the authors reported that increasing the polymer concentration resulted in the formation of fibers instead of capsules, being this fact attributed to an increase of the solution viscosity. The oxidative stability of these encapsulates is discussed below in section "Oil Emulsification Approach". More recently, the same authors (García-Moreno et al. 2018) produced fish oil encapsulates by electrospraying using dextran (molecular weight = 70 kDa, dextran70) or GS (DE38, molecular weight = 12.5 kDa) as wall materials in presence of WPC. In both cases, pullulan (molecular weight = 200 kDa) was added to the formulation as thickening agent to increase the stability of the Taylor cone at high flow rates (up to 0.012 mL/min). The oil was added to the infeed solution as neat oil to a final oil load of the encapsulates of 20%, followed by subsequent emulsification process. The electrospraying process was carried out at lab scale at a flow rate ranging from 0.003–0.012 mL/min and 17–20 kV, depending on the biopolymer solution, being the collector placed at 15 cm from the needle. In case of dextran70 solution, the authors reported that decreasing the polymers concentration (2 wt% to 1 wt% for pullulan and 20 wt% to 15 wt% for dextran) avoided capsules' fibril defects but also led to a reduction in the process flow rate. Both facts were attributed to a reduction in polymer chain entanglements and viscosity of the infeed emulsion. On the other hand, increasing the pullulan concentration from 2 wt% to 4 wt% in the GS-based emulsion (15 wt%) had no effect on the microstructures morphology since capsules were obtained in this range of pullulan content. However, at 4 wt% pullulan, although the droplets size of the emulsion increased, the electrospraying process flow rate was higher (0.010 mL/h at 4 wt% pullulan over 0.007 mL/h at 2 wt% pullulan), thus favoring microcapsules productivity. The oxidative stability of the resulting encapsulates is discussed in the next section ("Oil Emulsification Approach").

Oxygen permeability trough the wall matrix has been reported to have a major impact on encapsulates oxidative stability. Indeed, this fact is even more relevant in electrosprayed capsules since the smaller particle size leads to an increase in the surface-to-volume ratio and a reduction in the thickness of the wall matrix. In this regard, Boerekamp et al. (2019) studied the influence of the wall material in oxygen permeability and its impact on encapsulated fish oil oxidative stability by electron spin resonance (ESR). Dextran or GS were used as the wall materials in presence of pullulan and WPI, and MCT oil or fish oil were encapsulated to investigate both the oxygen permeability and oxidative stability, respectively. The electrospraying process was carried out at lab scale under controlled ambient conditions (20 ± 4 °C and 17–46% RH), at a flow rate ranging from 0.003–0.010 mL/min and an applied voltage of 15–20 kV (Table 10.4). The oxygen permeability results showed that, although both encapsulates were in a glassy state at ambient conditions, dextran-based capsules were more permeable to oxygen than GS-based encapsulates. This was attributed by the authors to a more effective packaging of GS in the wall matrix due to its lower molecular weight (12.5 kDa over 70 kDa of dextran), which resulted in less free volume, thus less oxygen permeability. Moreover, the encapsulates oxidative stability based on ESR results, showed that GS-based capsules were less oxidized over the storage time (50 °C during 16 days) that dextran-based capsules, meaning that oxygen diffusivity trough the wall matrix played a major role on oxidative stability of fish oil-loaded electrosprayed capsules.

Oil Emulsification Approach

In emulsion-based electrospraying processes (e.g. when using water-based polymers solution) both the emulsification approach and the oil introduction method in the infeed emulsion (emulsified oil or neat oil) play a major role on microcapsules properties such as particle size, EE and oxidative stability. Paximada et al. (2018) studied the influence of the emulsification process regarding the homogenization pressure (1000 or 2000 bar) and number of passes (1, 2, 4 and 8) in the fish oil-loaded electrosprayed encapsulates properties (particle size, EE). The capsules were produced using WPI as wall material in presence of MD and tween 20 at a core:wall ratio of 1:1.4 (theoretical oil load of ~42%). The electrospraying process was carried out at a constant flow rate of 0.02 mL/min, applying a voltage of 18 kV and with the collector placed at 18 cm from the needle. The authors reported that the smaller the emulsion oil droplet size (achieved by increasing the homogenization pressure and number of passes), the smaller as the electrosprayed particles size. Moreover, the EE was also affected by the emulsification conditions since the particles produced from emulsions homogenized at 1000 bar led to lower EE values (60–80%) than those homogenized at 2000 bar (EE = 70–98%). In the study, the most stable emulsion system attending to the smallest oil droplet size was the one that produced at 2000 bar after 8 passes (d_{50} = 95 nm). However, this emulsion did not produce any powder due to its high viscosity, which affected the electrospraying process. Gómez-Mascaraque & López-Rubio (2016) produced ALA-encapsulates using WPC, SPI and gelatin as wall materials. The authors reported that the emulsification approach affected both the oil droplets size distribution in the water-based (WPC and SPI) or acetic acid-based (gelatin) infeed emulsions and the EE of the resulting encapsulates (EE = 23 ± 12–67 ± 5%). Although mechanical stirring followed by ultrasonication led to smaller droplets sizes in the emulsions, the EE of the encapsulates was also lower. This fact was attributed to thermal degradation of ALA in the ultrasonication treatment, which raised the emulsion temperature up to 45 °C. The oxidative stability of the electrosprayed encapsulates was assessed by studying the degradation profile (by FT-IR) under accelerated oxidation conditions (80 °C during 2 days) and the results indicated that, although the WPC and SPI-based encapsulates presented higher stability than neat ALA, both the emulsification approach and the encapsulating agent had little impact on lipid oxidation course. On the other hand, gelatin-based encapsulates did not exert any protective effect against ALA oxidation (on the contrary accelerated its degradation), probably due to the presence of residual solvent in these capsules.

García-Moreno et al. (2018) produced carbohydrate-based (pullulan and dextran or GS) electrosprayed capsules in pilot-plant scale by using two different emulsification approaches namely, high pressure homogenization and rotor-stator homogenization. In this study the authors found that the EE of the encapsulates did not correlate with the oil droplets size in the emulsions since rotor-stator homogenization, which led to the larger oil droplets size, also led to the highest EE values of the encapsulates in both emulsions (97.1 ± 0.9% for dextran-based encapsulates and 85.7 ± 0.3% for GS-based encapsulates, respectively). However, dextran-based capsules presented smaller oil droplets size and higher EE compared to GS-based capsules regardless of the emulsification approach. In all cases, the initial PV of the encapsulated oil was higher than that of neat oil, and both the emulsification approach and the encapsulating agent had little effect. However, during storage time (20 °C during 21 days) the encapsulates produced from high-pressure homogenized emulsions presented lower oxidative stability than those produced from rotor-stator homogenized emulsions, based on PV and SVOP content trends. Moreover, although the PV trend during storage did not show significant differences regarding the encapsulating agents, differences were observed in the SVOP trends. The results showed that GS-based encapsulates were the most oxidatively stable over storage time, despite their higher extractable oil content. This finding was attributed by the authors to less free volume of the GS wall matrix due to its lower molecular weight (12.5 kDa over 70 kDa of pullulan), which reduced oxygen diffusivity. Regarding oil addition method to the infeed emulsion, García-Moreno et al. (2017) produced dextran-based electrosprayed capsules (both in lab scale and pilot-plant scale) by using two different oil-incorporation approaches: (i) emulsified fish oil (EFO) stabilized with WPI, and (ii) neat fish oil (NFO) using Tween 20 as emulsifier. The optimized biopolymer solution consisted of dextran (molecular weight = 70 kDa) at 25 wt% in water and the oil load of the particles was fixed to 10%, regardless of the oil-addition method. Although oil incorporated as NFO led to larger capsules due to the larger oil droplets size in the emulsion, the EE in these capsules was higher when compared to the encapsulates where the oil was added as EFO, even having smaller droplet sizes in the latter (75.5 ± 0.9% for NFO over 68.3 ± 0.3% for EFO). Furthermore, regarding PV after lab scale production (2 h batches), the results showed that the oil-addition approach had little impact on fish oil oxidative stability, being both encapsulates relatively oxidized after production (43.9 ± 19.8 meq O_2/kg oil for NFO and 48.1 ± 4.0 meq O_2/kg oil for EFO). However, at pilot-plant scale production, NFO-addition led to less oxidized encapsulates with PV of 21.2 ± 9.5 meq O_2/kg oil after production, when compared to EFO-based encapsulates (PV = 62.3 ± 4.5 meq O_2/kg oil).

High-Throughput Electrospraying, Addition of Antioxidants and Food Enrichment

Typically, one of the main drawbacks of electrospraying technology has been its low productivity as consequence of the low flow rates achieved during processing. To overcome this situation, Busolo et al. (2019) have recently developed a novel electrospraying technology, namely electrospraying assisted by pressurized gas (EAPG), which allows to increase the capsules production by atomizing the infeed solution into a high electrostatic field by means of compressed air. The equipment consists of an injection unit, a drying camber and a cyclone to collect the dry encapsulates as free-flowing powder. The authors had first used this technology to produce DHA-enriched fish oil encapsulates within a zein matrix (core: wall ratio of 1:2) to produce fortified milk. The electrospraying process was conducted by EAPG under controlled ambient conditions (25 °C and 40% RH) at a feed flow rate of 10 mL/min, 20 kV and an air atomizing pressure of 10 L/min, whilst the infeed solution was constantly bubbled with nitrogen. The resulting powder had a mean size of 1.4 ± 0.8 μm, an EE of 84 ± 1% and an initial PV of ~1.5 meq peroxide/kg oil. The oxidative stability of the encapsulates was studied under different storage conditions by varying the temperature (5–23 °C), RH (0–65%) and environment (dark/light, or air/vacuum) up to 45 days. In all cases, the oxidation rate of DHA neat oil was higher than that of the encapsulates, thus indicating effective protection of the zein matrix against lipid oxidation. However, the highest oxidative stability was achieved at ambient conditions (23 °C and 65% RH) at dark and under vacuum in both neat oil (18 meq peroxide/kg oil) and encapsulates (< 2 meq peroxide/kg oil), which meant that oxygen had a greater influence on lipid oxidation than temperature and humidity. In addition, regarding production of fortified milk, the encapsulates were easier to disperse in the milk preparation than neat oil and did not present any lumps or oil droplets. Moreover, the panellists who evaluated the fortified milk organoleptic properties, found little differences compared to the blank (unfortified milk) even when the milk was prepared with encapsulates stored for 45 days (at −1 °C in the dark and under vacuum).

EAPG technology has been also used to produce carbohydrate-based (Hermund et al. 2019) and protein-based (Miguel et al. 2019) encapsulates in order to produce fortified foodstuff (Table 10.4) . Hermund et al. (2019) produced fish oil encapsulates within a GS or dextran matrix in presence and absence of antioxidants (seaweed extract or δ-tocopherol and rosemary extract) to produce fortified mayonnaise. The biopolymer solution formulation consisted of WPC, pullulan and GS or dextran. The seaweed extract was added to both solutions, while commercial natural antioxidants were only added to the GS-based formulation. The oil was introduced to the infeed emulsion as an oil-in-water emulsion stabilized with citrem to a final oil load of the encapsulates of 20%. The electrospraying process was carried out at a flow rate of 1.8 mL/min and 15 kV under controlled ambient conditions (24 °C and 40% RH) leading to small capsules (< 3 μm) with high EE (83.2 ± 0.5% – 90.4 ± 0.1%). The authors reported that the addition of antioxidants, compared to other previous results, did not enhanced encapsulates oxidative stability during storage (20 °C during 21 days in the dark) based on PV. However, the addition of commercial antioxidants lowered the content of some SVOP ((E)-2-pentenal and nonanal) in GS-based capsules. Conversely, the addition of seaweed extract exerted a pro-oxidative effect on the latter encapsulates leading to a final PV of 34.6 ± 0.5 meq peroxide/kg oil after 21 days of storage. The content of some SVOP such as 1-penten-3-ol and nonanal increased as well. Nonetheless, the addition of seaweed extract to dextran-based encapsulates exhibited an antioxidative effect based on SVOP trends, except for 1-penten-ol. This fact was attributed to the pro-oxidant species present in the seaweed extract (e.g. pigments or metals), which catalyzed non-encapsulated oil oxidation more efficiently in GS encapsulates as this samples presented the lowest EE (85.7 ± 0.4% over 90.4 ± 0.1% for the dextran-based encapsulates). These authors also evaluated the oxidative stability of GS capsules-enriched mayonnaise (in presence or absence of commercial antioxidants) compared to mayonnaise fortified by adding NFO. The addition of encapsulates to an already formed mayonnaise led to an increase in the oil droplets size compared to mayonnaise fortified with NFO, indicating that the capsules could have disintegrated leading to unencapsulated oil which could flocculate/coalesce. The results also showed that GS-based encapsulates acted as a thickening agent since the apparent viscosity of capsules-enriched mayonnaise also increased. Regarding oxidative stability, all the mayonnaise samples presented a low initial PV (< 0.5 meq peroxides/kg oil), but the trends in lipid oxidation over the storage time (20 °C in the dark during 21 days) significantly differed. The results showed that the addition of encapsulates, either in presence or absence of antioxidants, did not improve the oxidative stability of the mayonnaise since both the PV and SVOP content were higher during the storage time compared to mayonnaise fortified with NFO. The authors attributed this finding to a release of oil with poorer oxidative status as consequence of the encapsulation process, caused by the encapsulates disintegration in the food matrix. Moreover, the most oxidized mayonnaise after the storage time was that fortified with encapsulates containing natural antioxidants (~ 12 meq peroxide/kg oil), which did not exert any protective effect in presence of iron released from egg yolk due to the lack of metal chelating activity. Miguel et al. (2019) encapsulated fish oil by EAPG using zein as encapsulating agent (20% oil load) with food fortification purposes (low fat mayonnaise). The EAPG process was carried out under ambient controlled conditions (24 °C and 40% RH) at an infeed solution flow rate of 1.4 mL/min, 10 kV and with an air pressure of 10 L/h (Table 10.4), resulting in encapsulates with a mean diameter of 2.4 ± 0.7 μm and an EE of 83 ± 1%. Cryo-SEM images of fortified mayonnaise showed that zein encapsulates remained intact after foodstuff production, which correlated well with its higher apparent viscosity (compared to mayonnaise fortified with NFO) due to encapsulates thickening activity. Besides, mayonnaise fortified with encapsulated fish oil presented a larger oil droplets size of approximately 2.5-fold that of NFO fortified mayonnaise (D[4,3] = 39.7 ± 0.2 μm, over D[4,3] = 16.2 ± 2.7 μm). The oxidative stability of the enriched mayonnaise was assessed by measuring the PV and SVOP content during 21 days of storage at 25 °C in the dark. The results showed that initial PV of the encapsulates-fortified mayonnaise were 4-fold that of NFO-fortified mayonnaise (2.1 ± 0.1 meq peroxide/kg oil), however this value did not change significantly over the storage time. Conversely, PV of NFO-fortified mayonnaise increased drastically after 14 days of storage. Regarding SVOP, the results suggested that the higher values found in encapsulates-fortified mayonnaise was due to: (i) volatile compounds already present in the zein protein (e.g. hexanal, 3-methyl-butanal or nonanal), and (ii) fungal spoilage of the mayonnaise (e.g. 3-methyl-1-butanol).

Bioaccessibility

Since the main purpose of designing efficient ω-3 PUFAs (e.g. fish oil) delivery systems is to increase oral EPA and DHA bioavailability, the core bioactive release in the GIT should be also studied. Moomand & Lim (2015) investigated the release profile of zein-based fish oil encapsulates (fibres or capsules) in the GIT under simulated gastric fluid (SGF) conditions (with and without pepsin) and under simulated intestinal fluid (SIF) conditions (with and without pancreatin), separately. The oil release profile under sequential exposure to SGF and SIF, in presence or absence of digestive enzymes, was also evaluated. The results showed that the release profile of encapsulated fish oil was influenced by GIT erosion behaviour of the wall material, being the release rate higher under SGF conditions (pH = 2) than under SIF conditions (pH = 6.8), in absence of digestive enzymes. Furthermore, the presence of digestive enzymes favoured oil release under both GIT conditions. These findings correlated well when sequential exposure to SGF and SIF was investigated since the release values obtained where comparable to those reported for SGF conditions alone. After sequential GIT study, the oil released in absence of enzymes ranged between 77–87%, whilst in presence of enzymes amounted to 86–95%.

10.4.2.3 Coaxially-Electrosprayed Microcapsules

Coaxial electrospraying is an emerging technology, which modifies monoaxial processing by introducing a coaxial capillary nozzle allowing to electrospray two liquids simultaneously (Jaworek 2016). There are some difference between these two technologies. Monoaxial electrospraying rely on chemical means to form a core/shell structure by dispersing the core material in the encapsulating agent solution by using emulsifiers or surface-active biopolymers, while coaxial electrospraying achieves the core/shell structure by physical separation of the solutions (Fig. 10.2). The latter leads to: (i) a centralized distribution of the encapsulated compounds (Loscertales et al. 2002), (ii) higher LC, and (iii) higher EE (Jaworek 2016). In addition, coaxial electrospraying allows producing microcapsules from two immiscible liquids without the need of preparing an emulsion. This is of great interest in terms of ω-PUFAs (e.g. fish oil) encapsulation since it has been demonstrated that the

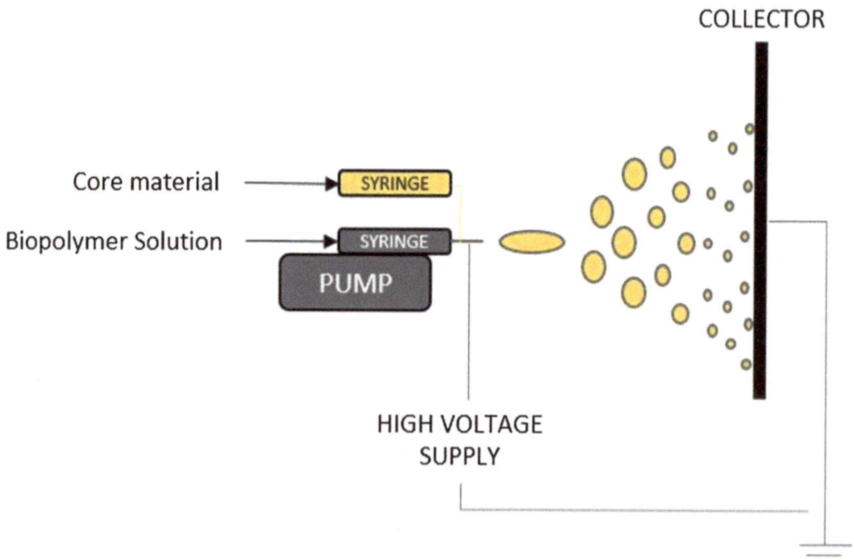

Fig. 10.2 Scheme of coaxial electrospraying process

emulsification step initiate lipid oxidation by means of mechanical stress and shear forces, which favors pro-oxidant species inclusion and/or distribution (e.g. oxygen or metal ions).

As electrohydrodynamic methods, coaxial electrospraying consist of applying high-voltage electrostatic field between the coaxial nozzle and the collector to produce core/shell-structured powder. Through the inner needle, the core material flows and through the annular gap, between the inner and the outer needle, the polymer solution flows. Both liquids are at the same potential. In this technology, the driving liquid is referred to the one with the highest conductivity, which is usually the outer liquid, although it could also be the inner solution (Loscertales et al. 2002). Process parameters (e.g. applied voltage or inner/outer flow rate ratio) and the driving liquid properties (e.g. viscosity or conductivity) will determine the process throughput (stable Taylor cone) and the nano-microcapsules properties such as particle size and wall thickness. Thus, encapsulates of poorly conductive core materials (e.g. fish oil) can be produced if the encapsulating agent solution (driving liquid) is conductive enough.

Coaxial electrospraying technology has been successfully used to encapsulate bioactive compounds for pharmaceutical purposes (e.g. drug delivery systems) (Chen et al. 2019). Nonetheless little research work is available on coaxial encapsulation of bioactives for food fortification purposes. In one important study, Yang et al. (2017) first produced fish oil-loaded fibers in presence of zein (core solution) using poly- vinylpyrrolidone as wall material (outer solution) which led to high quality encapsulates (oil load of 14.5%, EE of 96.9% and average diameter of the fibers of 560 nm) with enhanced oxidative stability and favorable oil release behavior. However, to the best of the authors' knowledge, Gómez-Mascaraque et al. (2019) reported for the first time the production of ω-3 PUFAs encapsulates by coaxial electrospraying within food-grade wall matrices and using food grade solvents. In this study, the authors successfully encapsulated ALA-zein solutions (core material) within a zein or gelatin matrix (wall material) at a final oil load of 10%. The coaxial electrospraying process was carried out at flow rates of 0.05 mL/h and 0.15 mL/h for the inner and outer solution respectively, and the voltage applied ranged from 13 kV to 18 kV. The collector was placed at 10 cm from the coaxial nozzle. Regarding morphology and size of the encapsulates, the authors reported that neither the incorporation of ALA to the zein solution nor the incorporation of an additional zein layer, affected the morphology or the size of the encapsulates. However, the addition of gelatin led to larger capsules and broader particles size distribution together with the appearance of fibril defects. Furthermore, the results showed that the EE was higher for both coaxial encapsulates than monoaxial zein-based capsules, being the highest EE greater than 80% in the zein-zein system. Nonetheless, the latter exerted a less efficient protective effect against ALA degradation under accelerated oxidation conditions (80 °C) than the gelatin-zein system despite its lower EE. This fact was attributed by the authors to the lower affinity of ALA with gelatin (outer wall layer), which could have result on less content of ALA in the capsules surface.

10.5 Conclusions and Future Perspectives

The production of fortified food-stuff containing ω-3 PUFAs in the form of fish oil is indeed possible by designing efficient delivery systems, which prevent both lipid oxidation and subsequent odor/texture deterioration of the food matrix. In this regard, the production of fish oil-in-water emulsions and fish oil microcapsules, in presence or not of antioxidants, has been widely investigated with promising outcomes. Spray-drying is currently the most used encapsulation technique in food industry, however the use of high drying temperatures together with the fact that most of the encapsulating agents are water soluble restrict its use for ω-3 PUFAs food fortification purposes. As an alternative, new technologies such as electrospraying and coaxial electrospraying, based on applying high-voltage electrostatic field to produce the encapsulates, are arising as promising encapsulation methods. Nonetheless, the main drawback of the latter technologies is the low throughput of the process, especially when food-grade biopolymers are used as encapsulating agents. Thus, further research is needed to increase the process production yield of mono- and co-axial electrospraying and to investigate the addition of the resulting electrosprayed encapsulates to complex food matrices.

Acknowledgements This work was funded by the project CTQ2017-87076-R from the Spanish Ministry of Science and Innovation. Rahmani-Manglano, N.E. acknowledges a FPI grant (PRE2018-084861) from the Spanish Ministry of Science and Innovation.

References

Aghbashlo M, Mobli H, Madadlou A, Rafiee S (2012) The correlation of wall material composition with flow characteristics and encapsulation behavior of fish oil emulsion. Food Res Int 49:379–388. https://doi.org/10.1016/j.foodres.2012.07.031

Aghbashlo M, Mobli H, Madadlou A, Rafiee S (2013) Influence of wall material and inlet drying air temperature on the microencapsulation of fish oil by spray drying. Food Bioprocess Technol 6:1561–1569. https://doi.org/10.1007/s11947-012-0796-7

Alemán M, Bou R, Guardiola F et al (2015) Antioxidative effect of lipophilized caffeic acid in fish oil enriched mayonnaise and milk. Food Chem 167:236–244. https://doi.org/10.1016/j.foodchem.2014.06.083

Berdahl DR, McKeague J (2015) Rosemary and sage extracts as antioxidants for food preservation. In: Handbook of antioxidants for food preservation. Woodhead Publishing, Cambridge, pp 177–217

Berton-Carabin CC, Ropers MH, Genot C (2014) Lipid oxidation in oil-in-water emulsions: involvement of the interfacial layer. Compr Rev Food Sci Food Saf 13:945–977. https://doi.org/10.1111/1541-4337.12097

Berton-Carabin CC, Sagis L, Schroën K (2018) Formation, structure, and functionality of interfacial layers in food emulsions. Annu Rev Food Sci Technol 9:551–587. https://doi.org/10.1146/annurev-food-030117-012405

Binsi PK, Nayak N, Sarkar PC et al (2017) Structural and oxidative stabilization of spray dried fish oil microencapsulates with gum arabic and sage polyphenols: characterization and release kinetics. Food Chem 219:158–168. https://doi.org/10.1016/j.foodchem.2016.09.126

Boerekamp DMW, Andersen ML, Jacobsen C et al (2019) Oxygen permeability and oxidative stability of fish oil-loaded electrosprayed capsules measured by Electron Spin Resonance: effect of dextran and glucose syrup as main encapsulating materials. Food Chem 287:287–294. https://doi.org/10.1016/j.foodchem.2019.02.096

Busolo MA, Torres-Giner S, Prieto C, Lagaron JM (2019) Electrospraying assisted by pressurized gas as an innovative high-throughput process for the microencapsulation and stabilization of docosahexaenoic acid-enriched fish oil in zein prolamine. Innov Food Sci Emerg Technol 51:12–19. https://doi.org/10.1016/j.ifset.2018.04.007

Calder PC (2013) Nutritional benefits of omega-3 fatty. In: Food enrichment with omega-3 fatty acids. Woodhead Publishing Limited, Cambridge, pp 3–26

Calder PC (2014) Very long chain omega-3 (n-3) fatty acids and human health. Eur J Lipid Sci Technol 116:1280–1300. https://doi.org/10.1002/ejlt.201400025

Chang C, Nickerson MT (2018) Encapsulation of omega 3-6-9 fatty acids-rich oils using protein-based emulsions with spray drying. J Food Sci Technol 55:2850–2861. https://doi.org/10.1007/s13197-018-3257-0

Chang HW, Tan TB, Tan PY et al (2020) Microencapsulation of fish oil-in-water emulsion using thiol-modified β-lactoglobulin fibrils-chitosan complex. J Food Eng 264:109680. https://doi.org/10.1016/j.jfoodeng.2019.07.027

Chen C, Liu W, Jiang P, Hong T (2019) Coaxial electrohydrodynamic atomization for the production of drug-loaded micro/nanoparticles. Micromachines 10:125. https://doi.org/10.3390/mi10020125

Decker EA (2002) Antioxidant mechanisms. In: Food lipids: chemistry, nutrition, and biotechnology. CRC Press, Boca Raton, pp 397–422

Desai KGH, Park HJ (2005) Recent developments in microencapsulation of food ingredients. Dry Technol Int J 23:1361–1394. https://doi.org/10.1081/DRT-200063478

Di Giorgio L, Salgado PR, Mauri AN (2019) Encapsulation of fish oil in soybean protein particles by emulsification and spray drying. Food Hydrocoll 87:891–901. https://doi.org/10.1016/j.foodhyd.2018.09.024

Díaz M, Dunn CM, McClements DJ, Decker EA (2003) Use of caseinophosphopeptides as natural antioxidants in oil-in-water emulsions. J Agric Food Chem 51:2365–2370. https://doi.org/10.1021/jf025984l

Drusch S, Rätzke K, Shaikh MQ et al (2009) Differences in free volume elements of the carrier matrix affect the stability of microencapsulated lipophilic food ingredients. Food Biophys 4:42–48. https://doi.org/10.1007/s11483-008-9100-9

EFSA (2009) Scientific opinion on the substantiation of health claims related to EPA, DHA, DPA and maintenance of normal blood pressure (ID 502), maintenance of normal HDL-cholesterol concentrations (ID 515), maintenance of normal (fasting) blood concentrations of tr. EFSA J 7:1–26. https://doi.org/10.2903/j.efsa.2009.1263

EFSA (2010a) Scientific opinion on the substantiation of health claims related to eicosapentaenoic acid (EPA), docosahexaenoic acid (DHA), docosapentaenoic acid (DPA) and maintenance of normal cardiac function (ID 504, 506, 516, 527, 538, 703, 1128, 1317, 1324, 1325). EFSA J 8:1–32. https://doi.org/10.2903/j.efsa.2010.1796

EFSA (2010b) Scientific opinion on the substantiation of health claims related to docosahexaenoic acid (DHA) and maintenance of normal (fasting) blood concentrations of triglycerides (ID 533, 691, 3150), protection of blood lipids from oxidative damage (ID 630), contr. EFSA J 8:1–27. https://doi.org/10.2903/j.efsa.2010.1734

EFSA (2010c) Scientific opinion on dietary reference values for fats, including saturated fatty acids, polyunsaturated fatty acids, monounsaturated fatty acids, trans fatty acids, and cholesterol. EFSA J 8:1–107. https://doi.org/10.2903/j.efsa.2010.1461

EFSA (2012) Scientific opinion on the tolerable upper intake level of eicosapentaenoic acid (EPA), docosahexaenoic acid (DHA) and docosapentaenoic acid. EFSA J 10:1–48. https://doi.org/10.2903/j.efsa.2012.2815.Available

EFSA (2014) Scientific opinion on the substantiation of a health claim related to DHA and contribution to normal brain development pursuant to article 14 of regulation (EC) no 1924/2006. EFSA J 12:1–8. https://doi.org/10.2903/j.efsa.2014.3840
Embuscado ME (2015) Spices and herbs: natural sources of antioxidants – a mini review. J Funct Foods 18:811–819. https://doi.org/10.1016/j.jff.2015.03.005
Encina C, Vergara C, Giménez B et al (2016) Conventional spray-drying and future trends for the microencapsulation of fish oil. Trends Food Sci Technol 56:46–60. https://doi.org/10.1016/j.tifs.2016.07.014
Espejo-Carpio FJ, Guadix A, Guadix EM (2013) Spray drying of goat milk protein hydrolysates with angiotensin converting enzyme inhibitory activity. Food Bioprocess Technol 7:2388–2396. https://doi.org/10.1007/s11947-013-1230-5
FAO (2010) Fats and fatty acids in human nutrition. Proceedings of the Joint FAO/WHO Expert Consultation, Rome
FDA (2019) Letter of enforcement discretion. petition for a health claim for eicosapentaenoic acid and docosahexaenoic acid and reduction of blood pressure in the general population (Docket No. FDA-2014-Q-1146). 1–79
Fernandez MT, Mira ML, Florêncio MH, Jennings KR (2002) Iron and copper chelation by flavonoids: an electrospray mass spectrometry study. J Inorg Biochem 92:105–111. https://doi.org/10.1016/S0162-0134(02)00511-1
Frankel EN (2012a) Photooxidation of unsaturated fats. In: Lipid oxidation, 2nd edn. Woodhead Publishing Limited, Sawston, pp 51–66
Frankel EN (2012b) Free radical oxidation. In: Lipid oxidation, 2nd edn. Woodhead Publishing Limited, Sawston, pp 15–24
Frankel EN (2012c) Methods to determine extent of oxidation. In: Lipid oxidation, 2nd edn. Woodhead Publishing Limited, Sawston, pp 99–127
García-Moreno PJ, Horn AF, Jacobsen C (2014) Influence of casein-phospholipid combinations as emulsifier on the physical and oxidative stability of fish oil-in-water emulsions. J Agric Food Chem 62:1142–1152. https://doi.org/10.1021/jf405073x
García-Moreno PJ, Guadix A, Guadix EM, Jacobsen C (2016) Physical and oxidative stability of fish oil-in-water emulsions stabilized with fish protein hydrolysates. Food Chem 203:124–135. https://doi.org/10.1016/j.foodchem.2016.02.073
García-Moreno PJ, Özdemir N, Stephansen K et al (2017) Development of carbohydrate-based nano-microstructures loaded with fish oil by using electrohydrodynamic processing. Food Hydrocoll 69:273–285. https://doi.org/10.1016/j.foodhyd.2017.02.013
García-Moreno PJ, Pelayo A, Yu S et al (2018) Physicochemical characterization and oxidative stability of fish oil-loaded electrosprayed capsules: combined use of whey protein and carbohydrates as wall materials. J Food Eng 231:42–53. https://doi.org/10.1016/j.jfoodeng.2018.03.005
Gharsallaoui A, Roudaut G, Chambin O et al (2007) Applications of spray-drying in microencapsulation of food ingredients: an overview. Food Res Int 40:1107–1121
Gómez-Mascaraque LG, López-Rubio A (2016) Protein-based emulsion electrosprayed micro- and submicroparticles for the encapsulation and stabilization of thermosensitive hydrophobic bioactives. J Colloid Interface Sci 465:259–270. https://doi.org/10.1016/j.jcis.2015.11.061
Gómez-Mascaraque LG, Tordera F, Fabra MJ et al (2019) Coaxial electrospraying of biopolymers as a strategy to improve protection of bioactive food ingredients. Innov Food Sci Emerg Technol 51:2–11. https://doi.org/10.1016/j.ifset.2018.03.023
Guéraud F, Atalay M, Bresgen N et al (2010) Chemistry and biochemistry of lipid peroxidation products. Free Radic Res 44:1098–1124. https://doi.org/10.3109/10715762.2010.498477
Guo L, Harnedy PA, Li B et al (2014) Food protein-derived chelating peptides: biofunctional ingredients for dietary mineral bioavailability enhancement. Trends Food Sci Technol 37:92–105. https://doi.org/10.1016/j.tifs.2014.02.007
Haahr AM, Jacobsen C (2008) Emulsifier type, metal chelation and pH affect oxidative stability of n-3-enriched emulsions. Eur J Lipid Sci Technol 110:949–961. https://doi.org/10.1002/ejlt.200800035

Hermund DB, Yeşiltaş B, Honold P et al (2015) Characterisation and antioxidant evaluation of Icelandic F. vesiculosus extracts in vitro and in fish-oil-enriched milk and mayonnaise. J Funct Foods 19:828–841. https://doi.org/10.1016/j.jff.2015.02.020

Hermund D, Jacobsen C, Chronakis IS et al (2019) Stabilization of fish oil-loaded electrosprayed capsules with seaweed and commercial natural antioxidants: effect on the oxidative stability of capsule-enriched mayonnaise. Eur J Lipid Sci Technol 121:1. https://doi.org/10.1002/ejlt.201800396

Hocman G (1988) Chemoprevention of cancer: phenolic antioxidants (BHT, BHA). Int J Biochem 20:639–651. https://doi.org/10.1016/0020-711X(88)90158-9

Horiuchi M, Umano K, Shibamoto T (1998) Analysis of volatile compounds formed from fish oil heated with cysteine and trimethylamine oxide. J Agric Food Chem 46:5232–5237. https://doi.org/10.1021/jf980482m

Horn AF, Green-Petersen D, Nielsen NS et al (2012a) Addition of fish oil to cream cheese affects lipid oxidation, sensory stability and microstructure. Agric 2:359–375. https://doi.org/10.3390/agriculture2040359

Horn AF, Nielsen NS, Jensen LS et al (2012b) The choice of homogenisation equipment affects lipid oxidation in emulsions. Food Chem 134:803–810. https://doi.org/10.1016/j.foodchem.2012.02.184

Huang H, Hao S, Li L et al (2014) Influence of emulsion composition and spray-drying conditions on microencapsulation of tilapia oil. J Food Sci Technol 51:2148–2154. https://doi.org/10.1007/s13197-012-0711-2

Intarasirisawat R, Benjakul S, Vissessanguan W et al (2015) Skipjack roe protein hydrolysate combined with tannic acid increases the stability of fish oil upon microencapsulation. Eur J Lipid Sci Technol 117:646–656. https://doi.org/10.1002/ejlt.201400247

Jacobsen C (2016) Fish oils: composition and health effects. Encycl Food Heal:686–692. https://doi.org/10.1016/B978-0-12-384947-2.00295-6

Jacobsen C, Hartvigsen K, Lund P et al (2000) Oxidation in fish oil-enriched mayonnaise. 3. Assessment of the influence of the emulsion structure on oxidation by discriminant partial least squares regression analysis. Eur Food Res Technol 211:86–98. https://doi.org/10.1007/s002179900132

Jacobsen C, Timm M, Meyer A (2001) Oxidation in fish oil enriched mayonnaise: ascorbic acid and low pH increase oxidative deterioration. J Agric Food Chem 49:3947–3956. https://doi.org/10.1021/jf001253e

Jacobsen C, García-Moreno PJ, Mendes AC et al (2018) Use of electrohydrodynamic processing for encapsulation of sensitive bioactive compounds and applications in food. Annu Rev Food Sci Technol 9:525–549. https://doi.org/10.1146/annurev-food-030117-012348

Jaworek A (2016) Electrohydrodynamic microencapsulation technology. In: Encapsulations. Nanotechnology in the agri-food industry, 2. Elsevier Oxford, pp 1–45

Jeyakumari A, Janarthanan G, Chouksey MK, Venkateshwarlu G (2016) Effect of fish oil encapsulates incorporation on the physico-chemical and sensory properties of cookies. J Food Sci Technol 53:856–863. https://doi.org/10.1007/s13197-015-1981-2

Jeyakumari A, Zynudheen AA, Parvathy U, Binsi PK (2018) Impact of chitosan and oregano extract on the physicochemical properties of microencapsulated fish oil stored at different temperature. Int J Food Prop 21:942–955. https://doi.org/10.1080/10942912.2018.1466319

Karaosmanoglu H, Kilmartin PA (2015) Tea extracts as antioxidants for food preservation. In: Handbook of antioxidants for food preservation. Woodhead Publishing, Amsterdam, pp 219–233

Karim FT, Zaidul ISM, Ghafoor K et al (2016) Microencapsulation of fish oil using hydroxypropyl methylcellulose as a carrier material by spray drying. J Food Process Preserv 40:140–153. https://doi.org/10.1111/jfpp.12591

Kolanowski W, Swiderski F, Jaworska D, Berger S (2004) Stability, sensory quality, texture properties and nutritional value of fish oil-enriched spreadable fat. J Sci Food Agric 84:2135–2141. https://doi.org/10.1002/jsfa.1770

Kurek MA, Moczkowska M, Pieczykolan E, Sobieralska M (2018) Barley β-D-glucan – modified starch complex as potential encapsulation agent for fish oil. Int J Biol Macromol 120:596–602. https://doi.org/10.1016/j.ijbiomac.2018.08.131

Laguerre M, López Giraldo LJ, Lecomte J et al (2010) Relationship between hydrophobicity and antioxidant ability of "phenolipids" in emulsion: a parabolic effect of the chain length of rosmarinate esters. J Agric Food Chem 58:2869–2876. https://doi.org/10.1021/jf904119v

Layé S, Nadjar A, Joffre C, Bazinet RP (2018) Anti-inflammatory effects of omega-3 fatty acids in the brain: physiological mechanisms and relevance to pharmacology. Pharmacol Rev 70:12–38. https://doi.org/10.1124/pr.117.014092

Lehn DN, Esquerdo VM, Dahlem Júnior MA et al (2018) Microencapsulation of different oils rich in unsaturated fatty acids using dairy industry waste. J Clean Prod 196:665–673. https://doi.org/10.1016/j.jclepro.2018.06.127

Let MB, Jacobsen C, Meyer AS (2007a) Lipid oxidation in milk, yoghurt, and salad dressing enriched with neat fish oil or pre-emulsified fish oil. J Agric Food Chem 55:7802–7809. https://doi.org/10.1021/jf070830x

Let MB, Jacobsen C, Sørensen ADM, Meyer AS (2007b) Homogenization conditions affect the oxidative stability of fish oil enriched milk emulsions: lipid oxidation. J Agric Food Chem 55:1773–1780. https://doi.org/10.1021/jf062391s

Linke A, Weiss J, Kohlus R (2020) Oxidation rate of the non-encapsulated- and encapsulated oil and their contribution to the overall oxidation of microencapsulated fish oil particles. Food Res Int 127:108705. https://doi.org/10.1016/j.foodres.2019.108705

Loscertales IG, Barrero A, Guerrero I et al (2002) Micro/nano encapsulation via electrified coaxial liquid jets. Science (80) 295:1695–1698. https://doi.org/10.1126/science.1067595

McClements DJ (2005) Food emulsions: principles, practices, and technologies. CRC Press, Boca Raton

McClements DJ, Jafari SM (2018) Improving emulsion formation, stability and performance using mixed emulsifiers: a review. Adv Colloid Interf Sci 251:55–79. https://doi.org/10.1016/j.cis.2017.12.001

Mei L, McClements DJ, Wu J, Decker EA (1998) Iron-catalyzed lipid oxidation in emulsion as affected by surfactant, pH and NaCl. Food Chem 61:307–312. https://doi.org/10.1016/S0308-8146(97)00058-7

Miguel GA, Jacobsen C, Prieto C et al (2019) Oxidative stability and physical properties of mayonnaise fortified with zein electrosprayed capsules loaded with fish oil. J Food Eng 263:348–358. https://doi.org/10.1016/j.jfoodeng.2019.07.019

Moomand K, Lim LT (2015) Properties of encapsulated fish oil in electrospun zein fibres under simulated in vitro conditions. Food Bioprocess Technol 8:431–444. https://doi.org/10.1007/s11947-014-1414-7

Morales-Medina R, Tamm F, Guadix AM et al (2016) Functional and antioxidant properties of hydrolysates of sardine (S. pilchardus) and horse mackerel (T. mediterraneus) for the microencapsulation of fish oil by spray-drying. Food Chem 194:1208–1216. https://doi.org/10.1016/j.foodchem.2015.08.122

Nguyen QV, Malau-Aduli BS, Cavalieri J et al (2019) Enhancing omega-3 long-chain polyunsaturated fatty acid content of dairy-derived foods for human consumption. Nutrients 11:1–23. https://doi.org/10.3390/nu11040743

Oikawa S, Nishino K, Oikawa S et al (1998) Oxidative DNA damage and apoptosis induced by metabolites of butylated hydroxytoluene. Biochem Pharmacol 56:361–370. https://doi.org/10.1016/S0006-2952(98)00037-9

Patrick KE, Lv Y, Muhamyankaka V et al (2013) Development of EPA-DHA microcapsules supplemented probiotic fermented milk. Akad Gıda 11:6–15

Paximada P, Howarth M, Dubey B (2018) Electrosprayed particles derived from nano-emulsions as carriers of fish oil. TechConnect briefs 2018. Adv Mater 3:1–4

Polavarapu S, Oliver CM, Ajlouni S, Augustin MA (2011) Physicochemical characterisation and oxidative stability of fish oil and fish oil-extra virgin olive oil microencapsulated by sugar beet pectin. Food Chem 127:1694–1705. https://doi.org/10.1016/j.foodchem.2011.02.044

Pourashouri P, Shabanpour B, Razavi SH et al (2014) Impact of wall materials on physicochemical properties of microencapsulated fish oil by spray drying. Food Bioprocess Technol 7:2354–2365. https://doi.org/10.1007/s11947-013-1241-2

Power O, Jakeman P, Fitzgerald RJ (2013) Antioxidative peptides: enzymatic production, in vitro and in vivo antioxidant activity and potential applications of milk-derived antioxidative peptides. Amino Acids 44:797–820. https://doi.org/10.1007/s00726-012-1393-9

Punia S, Sandhu KS, Siroha AK, Dhull SB (2019) Omega 3-metabolism, absorption, bioavailability and health benefits–a review. PharmaNutrition 10:100162. https://doi.org/10.1016/j.phanu.2019.100162

Samaranayaka AGP, Li-Chan ECY (2011) Food-derived peptidic antioxidants: a review of their production, assessment, and potential applications. J Funct Foods 3:229–254. https://doi.org/10.1016/j.jff.2011.05.006

Serfert Y, Drusch S, Schmidt-Hansberg B et al (2009a) Process engineering parameters and type of n-octenylsuccinate-derivatised starch affect oxidative stability of microencapsulated long chain polyunsaturated fatty acids. J Food Eng 95:386–392. https://doi.org/10.1016/j.jfoodeng.2009.05.021

Serfert Y, Drusch S, Schwarz K (2009b) Chemical stabilisation of oils rich in long-chain polyunsaturated fatty acids during homogenisation, microencapsulation and storage. Food Chem 113:1106–1112. https://doi.org/10.1016/j.foodchem.2008.08.079

Shimada K, Okada H, Matsuo K, Yoshioka S (1996) Involvement of chelating action and viscosity in the antioxidative effect of xanthan in an oil/water emulsion. Biosci Biotechnol Biochem 60:125–127. https://doi.org/10.1271/bbb.60.125

Simopoulos AP (2011) Evolutionary aspects of diet: the omega-6/omega-3 ratio and the brain. Mol Neurobiol 44:203–215. https://doi.org/10.1007/s12035-010-8162-0

Simopoulos AP (2016) An increase in the Omega-6/Omega-3 fatty acid ratio increases the risk for obesity. Nutrients 8:1–17. https://doi.org/10.3390/nu8030128

Song D, Khouryieh H, Abughazaleh AA et al (2011) Sensory properties and viability of probiotic microorganisms in chocolate ice cream supplemented with omega-3 fatty acids. Milchwissenschaft 66:172–175

Sørensen ADM, Haahr AM, Becker EM et al (2008) Interactions between iron, phenolic compounds, emulsifiers, and pH in omega-3-enriched oil-in-water emulsions. J Agric Food Chem 56:1740–1750. https://doi.org/10.1021/jf072946z

Tamm F, Gies K, Diekmann S et al (2015) Whey protein hydrolysates reduce autoxidation in microencapsulated long chain polyunsaturated fatty acids. Eur J Lipid Sci Technol 117:1960–1970. https://doi.org/10.1002/ejlt.201400574

Torres-Giner S, Martinez-Abad A, Ocio MJ, Lagaron JM (2010) Stabilization of a nutraceutical omega-3 fatty acid by encapsulation in ultrathin electrosprayed zein prolamine. J Food Sci 75:N69. https://doi.org/10.1111/j.1750-3841.2010.01678.x

Unnikrishnan P, Puthenveetil Kizhakkethil B, Annamalai J et al (2019) Tuna red meat hydrolysate as core and wall polymer for fish oil encapsulation: a comparative analysis. J Food Sci Technol 56:2134–2146. https://doi.org/10.1007/s13197-019-03694-w

Velez S, Nair NG, Reddy VP (2008) Transition metal ion binding studies of carnosine and histidine: biologically relevant antioxidants. Colloids Surf B Biointerfaces 66:291–294. https://doi.org/10.1016/j.colsurfb.2008.06.012

Vieira SA, Zhang G, Decker EA (2017) Biological implications of lipid oxidation products. JAOCS, J Am Oil Chem Soc 94:339–351. https://doi.org/10.1007/s11746-017-2958-2

Vishnu KV, Chatterjee NS, Ajeeshkumar KK et al (2017) Microencapsulation of sardine oil: application of vanillic acid grafted chitosan as a bio-functional wall material. Carbohydr Polym 174:540–548. https://doi.org/10.1016/j.carbpol.2017.06.076

Wang R, Tian Z, Chen L (2011) A novel process for microencapsulation of fish oil with barley protein. Food Res Int 44:2735–2741. https://doi.org/10.1016/j.foodres.2011.06.013

Yang H, Wen P, Feng K et al (2017) Encapsulation of fish oil in a coaxial electrospun nanofibrous mat and its properties. RSC Adv 7:14939–14946. https://doi.org/10.1039/c7ra00051k

Yeşiltaş B (2019) Lipid oxidation in high fat omega-3 delivery emulsions. 1–276

Yesiltas B, García-Moreno PJ, Sørensen ADM et al (2018) Effects of modified DATEMs with different alkyl chain lengths on improving oxidative and physical stability of 70% fish oil-in-water emulsions. J Agric Food Chem 66:12512–12520. https://doi.org/10.1021/acs.jafc.8b04091

Yesiltas B, García-Moreno PJ, Sørensen ADM et al (2019a) Physical and oxidative stability of high fat fish oil-in-water emulsions stabilized with sodium caseinate and phosphatidylcholine as emulsifiers. Food Chem 276:110–118. https://doi.org/10.1016/j.foodchem.2018.09.172

Yesiltas B, Sørensen ADM, García-Moreno PJ et al (2019b) Modified phosphatidylcholine with different alkyl chain length and covalently attached caffeic acid affects the physical and oxidative stability of omega-3 delivery 70% oil-in-water emulsions. Food Chem 289:490–499. https://doi.org/10.1016/j.foodchem.2019.03.087

Young AJ, Lowe GM (2001) Antioxidant and prooxidant properties of carotenoids. Arch Biochem Biophys 385:20–27. https://doi.org/10.1006/abbi.2000.2149

Chapter 11
Encapsulation of Pigmented Lipophilic Antioxidants Through Micro and Nano-emulsions

Sadia Aslam, Aqsa Akhtar, Rao Sanaullah Khan, and Nauman Khalid

11.1 Introduction

The process of oxidation is essential for both living organisms and food molecules to sustain their life activities and proper functioning of the body's cellular metabolism. Oxidation directly relates to the production of energy through the breakdown of large food molecules such as carbohydrates, proteins and fats during metabolic reactions. During the oxidation, many reactive oxygen species are produced known as free radicals. These free radicals have high-energy electron which can be able to bound other molecules, therefore, they are considered reactive species (Martín et al. 2017).

There are many exogenous factors which promote the production of free radicals such as solar radiation, ultraviolet radiation, disease condition, excess accumulation of chemicals, presence of toxic chemicals or pesticides and oxidative stress to living body or plants. One of the most important sources of the reactive oxygen species is the irreversible chemical modification in the large molecules of the cells which produce reactive compounds such as malonaldehyde and hydroperoxides capable of oxidative damage. Free radicals cause changes in the polyunsaturated lipids by producing the lipoperoxidases and cause rancidity in food molecules. These lead to cell lysis and increase the cell fluidity with the production of off-flavors, smell and toxic compounds. Free radicals cause denaturation and inactivation of the proteins, similarly, they cause mutagenesis in the genetic code of individuals by changing the basis in nucleic acids (Santos-Sánchez et al. 2019).

A small concentration of compounds classified as antioxidants helps to prevent or slow down the process of oxidation. They play a defensive role by neutralizing

---

S. Aslam · A. Akhtar · R. S. Khan · N. Khalid (✉)
School of Food and Agricultural Sciences, University of Management and Technology, Lahore, Pakistan
e-mail: nauman.khalid@umt.edu.pk

M. A. Aboudzadeh (ed.), *Emulsion-based Encapsulation of Antioxidants*, Food Bioactive Ingredients, https://doi.org/10.1007/978-3-030-62052-3_11

the free radicals which cause oxidative rancidity. They also help to reduce the proliferation of chain reactions that cause degenerative and neurological disorders. Some of the important antioxidants are vitamin A, vitamin E, docosahexaenoic acid (DHA), eicosapentaenoic acid (EPA), flavonoids, polyphenols and Q-10 enzyme (Chang et al. 2018). Antioxidants are used during food processing as food additives and increase the functional properties of the food by scavenging free radicals. Antioxidants function through seven different ways: (a) as sequestering agent (b) through chelating with metal ions (c) promoting the activity of the oxidation inhibitory enzymes (d) producing the antioxidative enzymes in medium (e) inhibiting the lipid peroxidation (f) limiting the chemical changes in sugar and proteins structure and (g) preventing the modification in the bases of nucleic acids (Carocho et al. 2018).

Pigmented lipophilic antioxidants (Fig. 11.1) are those substances that impart specific color to food substances besides acting as antioxidants. They help to scavenge the free radical by donating hydrogen ion from their hydroxyl groups and

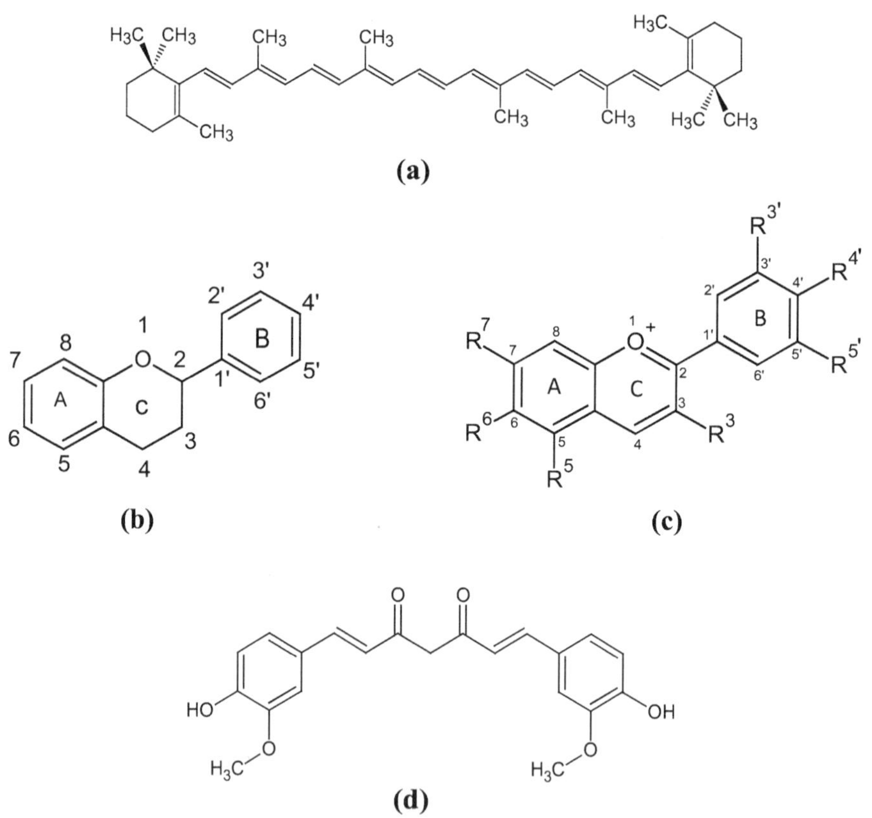

Fig. 11.1 General structures of different antioxidant pigments (**a**) carotenoid, (**b**) flavonoid, (**c**) anthocyanin and (**d**) curcumin

prevent oxidative rancidity. They play an important role in cancer prevention by chelating free radicals and increase the efficiency of the immune system of the body. They prevent human beings from cardiac and degenerative diseases. Lipophilic antioxidants belong to plants included carotenoids, chlorophylls and flavonoids which are located mainly plastids. The important plant lyophilic antioxidants include carotenoids, chlorophylls and tocopherols. These are located in the plastids of the plants and take part in different biochemical reactions in plants. Carotenoids are the molecules of photosynthetic reactions in the plant and tocopherol helps in the plant cell membrane stabilization. Humans cannot be able to synthesize these antioxidants in the body regardless of this fact, these lipophilic antioxidants play important functions in the human body. Carotenoids are a good source of vitamin A and have anti-cancer properties, plant-based foods are an excellent source of these compounds (Sircelj et al. 2018).

Encapsulation helps to increase the functional properties of the bioactive components. Encapsulants act as physical barriers and prevent these components from adverse exogenous factors such a light, temperature and heat. Pigmented lipophilic antioxidants show less stability towards these factors and change from one form to another. These substances are mainly encapsulated in order to increase their stability and functional properties and also to prevent chemical degradation (Janiszewska-Turak 2017).

11.2 Types of Pigmented Lipophilic Antioxidants

Lipophilic antioxidants are an important class of antioxidants which can increase the stability of different lipophilic foods, drugs, and other items by reducing the oxidative stress and preventing oxidative rancidity. They also help to control the formation and activity of the free radicals by scavenging the active species (Haman et al. 2017). The most important lipophilic antioxidants are pigmented (color pigment) antioxidants that impart different colors (biocolorants) in living organisms specifically in plants besides giving anti-oxidative activities. These color pigments belong to the three different sources: plants, animals and microorganisms. Plant pigments include a variety of organic chemicals such as porphyrins, carotenoids, anthocyanin which absorb different light wavelengths and exhibit different colors (Sigurdson et al. 2017). Animal origin color pigments produced in the results of nanoscale structuring which cause selective reflectance, diffraction, or scattering of certain beams of light waves. The organic color pigments produced in this way are purines, anthraquinones and melanin's (Newsome et al. 2014). Microbial pigments produced from different bacteria and fungi which include carotenoids and monascus pigments producing different color hues (Rymbai et al. 2011). Carotenoids (tetraprenoids) with conjugated double bonds impart yellow, orange and red color in different fruits, vegetables and grains. Lycopene has the efficiency to quench the destructive potential of singlet oxygen and it also helps in the prevention of oxidation of low density lipoprotein (LDL) cholesterol and reduces the risk of coronary heart diseases. Lutein, zeaxanthin, and xanthophylls can act as protective antioxidants (Corrales-Bañuelos et al. 2016). Most of the above-mentioned pigments are sensitive to heat, light, and oxygen, which results in color loss and changes in colors. These pigments also show alterations due to environmental matrices such as pH, proteins, metal ions and conjugation with organic compounds (Chaitanya Lakshmi 2014; Rodriguez-Amaya 2016).

11.2.1 Pyrrole Derivatives

Pyrroles are organic heterocyclic compounds composed of a five-membered ring that contains four carbons and one nitrogen in their structure with the general formula C_4H_4NH. They help in the formation of tetra-pyrrole rings such as chlorophylls and polypyrrole chains (Sigurdson et al. 2017). Tetrapyrrole is one of the most ancient pigments present among all living organisms including plants, blue green algae, and various species of bacteria abundantly. These pigments have a prosthetic group in their molecular structure. The most important pyrrole presents in plants are tetra-pyrroles including chlorophylls and siroheme. These compounds act as important color pigments and cofactors (Schlicke et al. 2015). Pyrrole and its associated derivatives contain active hydrogen atom (H-N) in their structure and act as natural antioxidants and help to reduce oxidative stress. Till the date, there have been hundreds of natural anti-oxidants discovered, among them all the antioxidants which have pyrrole ring moiety show an important role in the agriculture, pharmacological and food sector (Tzankova et al. 2019). As some examples, aryl and acyl substituted pyrrole derivatives showed high antioxidative activity (Kundu and Pramanik 2020).

11.2.1.1 Chlorophylls

Chlorophyll pigments are primary photosynthetic products of higher plants and impart green color in plants, vegetable and fruits. They are located in the chloroplast of the plants and help to trap the light energy from the sun which utilized during photosynthesis. 100 structures of chlorophyll molecule have been reported. There are five types of chlorophyll present normally i.e. a, b, c, d and f. Plants contain normally two types of chlorophyll a and b, which both absorb a visible range of light and help plants to prepare sugars. The chlorophyll a is more dominant in plants than chlorophyll b as the ratio of chlorophyll a to chlorophyll b is 3:1 in plants (Kang et al. 2018). The other forms of chlorophyll are c, d, and f which are found in bacteria, cyanobacteria and blue green algae. Chlorophyll compound worked as an antioxidant due to the presence of magnesium in its center which chelating the free radicals. Most of the studies reported that chlorophyll a show more antioxidative activity than the chlorophyll b (Zepka et al. 2019). Chlorophylls also act as antioxidants and prevent plants from ultraviolet rays of light. Chlorophylls are highly degradable pigments and easily degrade under high temperatures and heat. The process of degradation can mostly be affected during the storage and processing of food commodities. When plants are heated and/or exposed to acid, both the chlorophyll a and b change into the compound pheophytin which reduces the green color. During this process, the magnesium located in the center of the chlorophylls is replaced by the two hydrogens (Roshanak et al. 2016).

11.2.1.2 Structure of Chlorophyll

The chemical formula of the chlorophyll is $C_{55}H_{72}MgN_4O_5$ and its molecular weight is 893.5 g/mol (Sultan et al. 2016). Naturally, chlorophyll a and b are present in high vascular plants with the chlorophyll a to b ration of 3:1 in chloroplast of plants. It is made of the tetra-pyrrole ring (porphyrin ring) substituted with magnesium atom (Mg^{2+}) bounded in the center of the ring with highly hydrophobic esterified phytol tail ($C_{20}H_{39}$). The structural difference between chlorophyll a and chlorophyll b is that in chlorophyll a CH_3 group is attached at C7 but in chlorophyll b CHO group is attached at C7 of the chlorophyll (Kang et al. 2018). Chlorophylls are lipophilic in nature but after reacting with acid, the phytol group cleaves and changes into chlorophyll in which causes the change of state of chlorophyll from lipophilic state to hydrophilic state. Pheophytin complexes are formed when Mg^{2+} ions displace with hydrogen ions on the reaction of chlorophylls with weak acids. Pheophytin imparts olive-brown color to the fruits and vegetables. The chlorophyll pigment is treated with weak acid which displaced the magnesium ions with copper or zinc ions and forms Metallo-chlorophyll which imparts stable bright green color which is desirable for industrial purposes. These reactions are irreversible and give high stability to colors (Sigurdson et al. 2017).

11.2.2 Isoprenoid Derivatives

Isoprenoids, also known as the terpenoids, are organic compounds and the largest group of plant specialized (secondary) metabolites (Yazaki et al. 2017). They made up of two hydrocarbon chains having five carbon atoms in specific arrangements. More than hundreds of isoprenoids and their derivative have been tested till now on both molecular and biological levels for their pharmacological roles. The biological activities presented by different classes of isoprenoids (both in vivo and in vitro studies) are anti-inflammatory, anti-anti-aggregatory, anti coagulative, anti-tumor, anti-oxidative and anti-coagulative effects. Mono, di, and tri-terpenes also showed the anti-analgesic and sedative effects during in vivo studies (Kokkiripati et al. 2013; Zhao et al. 2016). Among all the activities, they are widely used as antioxidants to reduce oxidative stress by sequestering the free reactive oxygen species (Brunetti et al. 2015). Isoprenoids are also called cell's structural components as living cells of all the unicellular including bacteria, archaea, freshwater bodies and multicellular organisms have isoprenoid derivatives. They are involved in the primary and secondary metabolic activities of the living organisms too. To date, more than thousands of the isoprenoids have been identified and most of them belong to the plant kingdom (Tetali 2019). The precursors of isoprenoids are dimethylallyl pyrophosphate (DMAPP) and isopentenyl pyrophosphate (IPP). More than 50,000 isoprenoid compounds have been discovered till now (Kawamukai and biochemistry 2018).

11.2.2.1 Carotenoids

Carotenoids are lipid soluble pigmented compounds that are synthesized in plants and non-photosynthetic organisms (bacteria and fungi). Animals cannot prepare carotenoids and obtain them from food sources. The major functions of carotenoids in birds, animals, and the human body are ornamentation as they deposit different color pigments in their skin, eyes, bills, scales, and feathers. These deposited color pigments impart very striking visual appeals in different animals. Among the avian species, most of the deposited carotenoid colorations have been identified such as red beak of the zebra finch *(Taenyopigia guttata)* and red plumage of the house finch (*Carpodacus mexicanus*) (Pérez-Rodríguez et al. 2013). They act as antioxidants to keep the body safe from various types of prostate cancer through sequestering the free radical specie, involving in many intracellular signaling pathways, helping to improve the immune function, act as photoprotector to prevent the body from UV rays and act as a precursor of retinol (vitamin A). Carotenoids also can be applied in many cosmetics as a sunscreen. More than 600 carotenoids have been characterized till now. Based on structure, carotenoids are classified into two major classes. The first class is carotenes, which are hydrocarbon molecules comprising atoms of carbon; and hydrogen only including β-carotene, α-carotene, and lycopene. The other group, the xanthophylls are oxygenated carotenoids, which have a different functional group in their structure such as lutein, astaxanthin and zeaxanthin (Saini and Keum 2018).

Carotenoids are important secondary metabolites of the plants and impart orange to red color in plants. β-carotene, one of the widely available carotenoids in different fruits and vegetables, is extensively present in green leafy vegetables. The orange to red color of the β-carotene masked by the chlorophyll (green pigment) in leafy vegetables. Besides, β-carotenoid is known as the precursor of vitamin A and shows the greatest vitamin A activity which helps to improve the human vision. Besides β-carotene, zeaxanthin and astaxanthin are other important carotenoids that are present on vegetables in minor quantities. Most of the vegetables such as tomato and papaya are good sources of α-carotene and lycopene. The red color of the tomato is due to the presence of lycopene. Carotenoids could be capable of changing the color from red-orange to blue on reacting with proteins. Carotenoids play an important role in photosynthesis by being involved in the synthesis of plant food (sugar) required to perform all the biochemical activities. During photosynthetic reactions, they act as sun light-harvesting centers, help to stabilize the pigmented protein light-harvesting complexes in the thylakoid membranes of the plants (Kiokias et al. 2016; Xu and Harvey 2019). Table 11.1 represents some physio-chemical properties of carotenoids.

Lycopene is another important carotenoid that imparts bright red color in different fruits and vegetables such as tomato, watermelon, guava, apricots, red grapefruit, paprika, and pink fruit. It is lipophilic in nature and highly unsaturated acyclic compounds. The chemical formula of lycopene is $C_{40}H_{56}$ with 11 conjugated double bonds (Mozos et al. 2018). About 14 carotenoids are found in the human serum and lycopene is one of them. In nature lycopene mostly exists in trans formation. The change in the structure of lycopene occurs due to the effects of light, temperature, heat, and chemical modification into mono or poly cis form of lycopene. The cis formation of lycopene is more stable than trans formation and more soluble in human serum. Therefore, the bioavailability of the cis lycopene in the human body is more than the trans formation of lycopene (Srivastava and Srivastava 2015).

Xanthophylls are oxygenated carotenoids that are formed by the addition of oxygen in carotenes in the form of a hydroxyl group, ketone, and epoxy group. Xanthophylls are polar compounds that are mainly produced in the plants and ingested in animals after some modifications in their structure. The two major xanthophylls found in plasma are lutein and zeaxanthin, both of these carotenoids help in the prevention of age-related issues. Another most important xanthophyll is astaxanthin, which is red in color and mostly present in seafood such as shrimps. Due to the high antioxidant activity, this carotenoid is commercially produced from the algae. Fucoxanthin is also an important carotenoid present in the seaweed and consume on large scale in Asia. This carotenoid helps in gene expression and to metabolize the body fat (Amengual 2019).

Lutein is an important oxygenated pigmented carotenoid commonly found in animal products, egg yolk, fruits, vegetables, and different flowers such as marigold. They give yellow color on low concentrations and red color in high concentrations. On a commercial level, lutein isolates mainly from the marigold flowers. The

Table 11.1 Some physical properties of carotenoids

Property	Results	References
Shape	Rigid, rod-like skeleton	Mezzomo and Ferreira (2016), Craft and Furr (2019), Mercadante (2019)
Solubility in water	Insoluble	
Solubility in organic solvents	Soluble in acetone, alcohol, ether, halogenated hydrocarbons, etc.	
Nature	Hydrophobic compounds	
Volatility	Low volatile in nature	Craft and Furr (2019)
Isomerism	Formation of cis and trans isomers	
Rang of light absorbance	400–500 nm	Mercadante (2019)

presence of lutein in the macula capable of absorbing light of the blue range wavelength would indicate that they serve a protective function, preventing age-related degenerative diseases such as cataracts and macular degeneration to be developed. Lutein also acts as a strong antioxidant by quenching the reactive oxygen species or free radicals (Weigel et al. 2018).

Astaxanthins are another major antioxidant that mainly present in marine algae, crustaceans and marine animals. They exhibit high antioxidant activity among all the carotenoids. Hydroxyl (OH) group of natural astaxanthins form mono and diester with palmitic, oleic, and linoleic acids while synthetic astaxanthins are present in free form. The molecular formula and molar mass of astaxanthin are $C_{40}H_{52}O_4$ and 596.84 g/mol respectively (Khalid and Barrow 2018).

11.2.2.2 Structure of Carotenoids

Carotenoids have polyene backbone of carbon C=C (Fig. 11.1a). This feature of carotenoid helps in the development of color pigments. The structure of the carotenoid chain is rich in free electrons and double bonds, which cause carotenoids vulnerable to oxidation and structural changes during processing. Many factors cause oxidation of carotenoids such as light, temperature, moisture, metals, enzymes, peroxides, and lipids (Sigurdson et al. 2017). Carotenoids show crystalline nature and present in the form of lipid droplets in the plant. Carotenoids can make the complex with proteins, thus they can be extracted from the plants (Jeż et al. 2020). Carotenoids form by the eight isoprene molecules and are known as tetraterpenoid derivatives too. As mentioned before, there are two major groups of carotenoids, carotenes and xanthophylls. The structure of astaxanthin has two phenolic rings located at the terminals attached through polyene. Two asymmetric carbons located at the 3,3′ position of the benzenoid ring having a hydroxyl group (OH) and on the end of the molecule keto group is attached at 4,4′ position (Khalid and Barrow 2018).

11.2.2.3 Iridoids

Iridoids are the major secondary metabolites of plants and some animal species. They are made of cyclopentane ring having monoterpenoids in their structure. These organic compounds widely used in ayurvedic medicines (Patil and Laddha 2018). There are many species of plants that contain iridoids including Cannaceae, Rubiaceae, Garryaceae, Pyrolaceae, Orobanchaceae, Verbenaceae, etc. Iridoid species show antitumor, anti-inflammatory, antioxidant, anti-hepatotoxic, antiviral, anti-allergic and antimicrobial activities (Hussain et al. 2019). Iridoids are less stable towards high temperatures, sunlight, adverse acidic and alkaline conditions. Iridoids are rarely present in fruits and bitterness in fruits is associated with a sub-

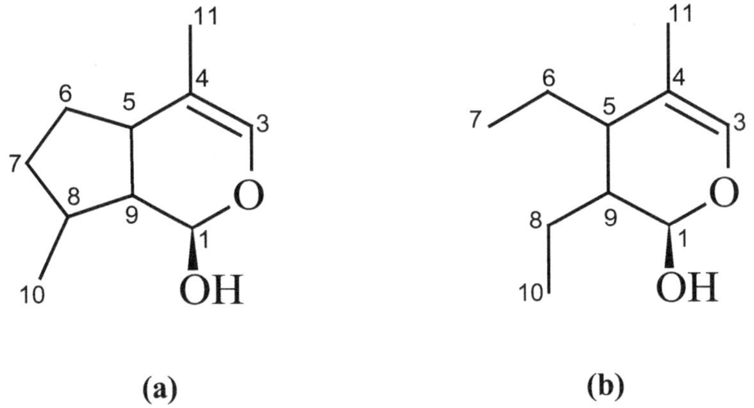

Fig. 11.2 General structures of (**a**) iridoid and (**b**) secoiridoids

class of monoterpenoid iridoid compounds named secoiridoids (Kucharska et al. 2017).

11.2.2.4 Structure of Iridoids

Iridoids (Fig. 11.2) have a basic glycosidic backbone with some variation and further divide into three groups carbocyclic iridoid, secoiridoids and plumieride. Carbocyclic iridoids comprise a *cis*-fused cyclopentane ring having β-configured substituents at C5 and C9 carbon. The changes occur in the carbocyclic skeleton of iridoid due to the addition and removal of the hydroxyl, epoxy, chloro, and olefin groups in cyclopentane ring (Hussain et al. 2019).

11.2.3 Benzopyran Derivatives

Benzopyran derivatives are a group of natural and artificial organic chemical compounds which show antitumor, antioxidant and antibiotic activities. These are bicyclic heterocyclic compounds having oxygen moiety in their structure. The naturally occurring benzopyran derivative compounds are genistein, warfarin, hesperidin and umbelliferone, coumarins, xanthenes, polyphenol, flavonoids and anthocyanin (Piyush et al. 2018).

11.2.3.1 Flavonoids

Flavonoids represent a wide group of plant secondary metabolites and impart yellow color in different plant species. They play an important role in plant physiology and are an important part of the human diet. Flavonoids are polyphenolic compounds having benzopyrone ring structures (C6-C3-C6). They occur both in the free state as aglycone or in combination with sugar as glycosidic linkage.

Flavonoids show high anti-scavenging action, therefore they can be used as antimicrobial, antioxidative, anti-inflammatory, antibacterial and vasodilatory agents. They also inhibit the activities of many enzymes such as cyclooxygenase, lipoxygenase, phospholipase, glutathione reductase and xanthine oxidase. Based on ring structure and the attached substitutes, flavonoids are further classified into different groups such as flavones, flavanols, flavanonols, isoflavones, flavanones, anthocyanidins and chalcones. Flavonoids act as antioxidants by forming a complex with a metal ion, through reducing the activity of pro-oxidant enzymes (cyclooxygenases and lipoxygenases) and scavenging the free radicals (Eghbaliferiz and Iranshahi 2016).

Normally flavonoids are present in the wall of the plant cells, seeds and tissues and extracted through chemical treatment with acid and alkali. Different types of wheat also contain high contents of flavonoids in the hull and bran of the wheat. Among them, buckwheat possesses high phenolic content including flavonoids such as quercetin, rutin, and flavone C-glycoside. Therefore, the wheat bran layer is processed further to extract flavonoid contents from the wheat and used them as antioxidants. Rutin is mostly used as an antioxidant in different human diets (Lee et al. 2016). Different types, classes and sources of important flavonoids (Panche et al. 2016) are presented in Table 11.2.

11.2.3.2 Antioxidative Mechanism of Flavonoids

Flavonoids show the great antioxidative property as compared to other antioxidants. They have the ability to immobilize reactive free radical species through a different mechanism: (1) direct scavenging of the free reactive specie, (2) activating the antioxidative enzymes such as NADPH-quinone oxidoreductase, glutathione S-transferase and UDP-glucuronosyl transferase and (3) chelating the metals such as copper and iron which promote oxidation. They also act as pro-oxidant and donate hydrogen ion to α- tocopherol present in the cell membrane. After interaction with α-tocopherol, they help to delay the oxidation of lipoproteins in the cell membrane. They can inhibit the activity of oxidase enzymes which produce superoxide radicals such as xanthine oxidase, cyclooxygenase, lipoxygenase (Procházková et al. 2011).

Table 11.2 Different types, classes and major food sources of flavonoids (Panche et al. 2016)

Flavonoid	Class	Dietary sources
Quercetin	Flavanols	Vegetables, fruits, fruit beverages, spices, soups
Rutin	Flavanols	Green tea, grape seeds, red pepper, apple, citrus, oranges, berries, peaches
Macluraxanthone	Xanthones	Hedge apple, mulberry
Genistein	Isoflavone	Fats, oils, beef, red clover, soya beans, lupin, fava beans, psoralea
Scopoletin	Coumarin	Vinegar, dandelion, coffee
Daidzein	Isoflavone	Soya beans, tofu
Taxifolin	Flavanonol	Vinegar
Naringenin	Flavanone	Grapes
Abyssinones	Flavanone	French bean seeds
Eriodictyol	Flavanone	Lemons, rosehips
Fisetin	Flavanol	Strawberries, apples, persimmons, onions, cucumbers
Theaflavin	Catechins	Tea leaves, black tea, oolong tea
Peonidin	Anthocyanidin	Cranberries, blueberries, plums, grapes, cherries, sweet potato
Diosmetin	Flavone	Vetch
Tricin	Flavone	Rice bran
Biochanin	Isoflavone	Red clover, soy, alfalfa, peanuts, chickpea
Hesperidin	Flavanone	Bitter orange, petitgrain, orange, orange juice, lemon, lime
Epicatechin	Flavan-3-ols	Milk, chocolate, commercial use
Myricetin	Flavanols	Vegetables, fruits, nuts, berries, tea, red wine
Taxifolin	Flavanonol	Citrus fruits
Kaempferol	Flavanols	Apples, grapes, tomatoes, green tea, potatoes, onions, broccoli, brussels sprouts, cucumber, lettuce, green beans, peaches, blackberries, raspberries, spinach
Luteolin	Flavones	Celery, broccoli, green pepper, parsley, thyme, dandelion, rosemary, navel oranges, oregano
Apigenin	Flavones	Milk, chocolate, reduced fat

11.2.3.3 Structure of Flavonoids

The general structure of flavonoids (Fig. 11.1b) is a skeleton of diphenylpropane, which consists of 15 carbon atoms arranged in three rings (C6-C3-C6). In other definition, it made up of two phenyl rings (A and B) which are bonded through heterocyclic ring pyran (C) located in the middle of the phenyl rings. The structural variations occur in different categories of the flavonoids due to the presence and absence of the hydroxyl group and double bonds at carbon 2 and 3 of the ring and absence of the double bond at carbon 4 (Lin et al. 2018).

11.2.3.4 Anthocyanins

Anthocyanins are one of the important bioactive pigments present in most of the plant species including fruits, vegetables and cereal grains. They are widely distributed throughout the plant and derived from the anthocyanidins and sugar molecule through a glycoside linkage. They show potential antioxidative, anti-inflammatory, anti-microbial activities. The commonly present anthocyanidins in plants are petunidin, cyanidin, pelargonidin, delphinidin, peonidin and malvidin. The commonly involved sugars in anthocyanin production are arabinose, galactose, glucose, rhamnose and xylose. The anthocyanin pigments are present in a diverse amount in different parts of plants, within various cultivars and species of the same plant (Wang et al. 2016).

The estimated daily intake of anthocyanins in human food is 12.5 mg/day as reported in the literature published in the United States (Wu et al. 2006). Anthocyanins that show major antioxidative properties can be taken through diet daily. Most of the grains show significant antioxidative properties including blue/purple wheat, blue/purple corn, and red/black varieties of the rice. Therefore, these varieties can be used as functional ingredients during the processing of different food items (Abdel-Aal et al. 2016). The anthocyanin pigments are mostly present in the outer aleurone layer or the pericarp of the wheat kernel. Therefore, the anthocyanin compounds from different varieties of wheat (especially the purple wheat variety) can be concentrated with the bran and this bran can be used as a functional ingredient in dietary fiber. The different techniques involved in the concentration of anthocyanins from bran are ethanol extraction and column chromatography for purification (Giordano et al. 2017; Li et al. 2017). A study reported that about 13 major anthocyanin compounds (Xu and Harvey 2019) have been reported in the purple wheat varieties and these anthocyanins have cyanidin 3-glucoside anthocyanidin pigment (Abdel-Aal et al. 2018).

Another study also reported that berries belong to the Rosacea family (including strawberry, raspberry and blackberry) and Ericaceae family (including blueberry and cranberry) showed a considerably large quantity of bioactive components such as anthocyanins, flavonoids and flavanols (Oancea and Calin 2016; Skrovankova et al. 2015). Anthocyanins belong to the phenolic compounds and are heterogeneous in nature having one or more aromatic rings with an attached hydroxyl group. Phenolic compounds help to reduce oxidative stress by donating a hydrogen atom to free radical which converts it into an immobilized molecule. Phenolic compounds can be present in both free forms and conjugation with other major food molecules such as carbohydrates and proteins (Koyuncu and Dilmaçünal 2010). The anthocyanins present in the berries belong to the two anthocyanidins such as cyaniding glycoside and pelargonidin cyaniding (Skrovankova et al. 2015). Anthocyanins usually impart yellow to purple color in different plants. About 600 anthocyanins have been discovered till now and being used in different industries as an antioxidant (Pervaiz et al. 2017).

11.2.3.5 Structure of Anthocyanins

Anthocyanins (Fig. 11.1c) are characterized based on sugars, hydroxyls, or methoxyl groups in the basic structure of a 2-phenylbenzopyrylium skeleton. The skeleton is composed of 15 carbon atoms located in two benzoyl rings (indicated by A and B) and a heterocyclic ring (indicated by C). Anthocyanins are the glycosylated patterns of anthocyanidin (the basic aglycone structure), and despite they are based on few numbers of aglycones, (which are cyanidin, delphinidin, malvidin, petunidin, peonidin and pelargonidin), the different substituents plus the addition of organic acids create more than 500 natural anthocyanins in plants (Calderaro et al. 2020, Sharma et al. 2018). The structure of anthocyanin has flavylium cation as its basic unit. The basic chromophore of the anthocyanin is consisted of 7-hydroxyflavylium ion and have hydroxyl substitute at the position 3 of the aromatic ring of the anthocyanin, this formation provides the thermal stability to anthocyanin structure (Yoshida et al. 2016).

11.2.4 Quinones Derivatives (Benzoquinone, Napthoquinone, Anthraquinone)

Quinones are widely present in nature and about 1200 quinones have been discovered till now (El-Najjar et al. 2011). The major plant species which contain quinones are Ranunculaceae, Asphodelaceae, Fabaceae, Ebenaceae and Rhamnaceae. The other sources are bacteria, fungi, and some classes of animals such as echinoderms. Some groups of quinones can be found in the environment through sunlight photo-oxidation of environmental contaminants such as the polycyclic aromatic hydrocarbons (Mallakin et al. 2000). Quinones are pigmented compounds and those are present in crystal form give sharp colors. The para-benzoquinone and naphthoquinone impart yellow color while ortho-benzoquinone and anthraquinone impart red color. They involve in many chemical ad biochemical reactions in plants such as photosynthesis, act as antioxidants, stabilizes the cell membrane, and maintain the metabolic processes of cells (El-Najjar et al. 2011). Anthocyanin and menadione are the two derivatives of quinone and naphthoquinone, belongs to the shell of purple sea urchin that showed significant antioxidant activity in the presence of NADPH enzyme (Deniz et al. 2015).

Naphthoquinone is the secondary metabolite and highly toxic in nature. Besides, toxic nature naphthoquinones can act as antioxidants by scavenging free radicals. A study reported that the high content of naphthoquinone is present in the Eluthrin plant which mostly used as medicinal purposes (Lestari et al. 2019).

Anthraquinones are other quinone derivatives and secondary metabolites. They are known as the anthracenediones or dioxoanthracene too. They are present in conjugated form as glycosides and free form as lichens, plants, higher filamentous fungi and insects. Anthraquinones are colored pigments and mainly used as dyes. The main dyes extracted from the plants are alizarin (shows orange to red color), purpurin (exhibits dark red color), pseudopurpurin (displays orange color), and rubiadin (exhibits yellow color). Different anthraquinone dyes that extract from the insects including carmine, carminic acid and cochineal. Anthraquinones can better act as proantioxidants and help to chelate free radical species (Fouillaud et al. 2018).

11.2.4.1 Structure of Quinones

Quinones belongs to the aromatic compounds that have an unsaturated benzene ring in which two carbonyl groups are attached. Anthraquinone has anthracene ring in its structure in which the keto group is attached at carbon 9 and 10 of the ring. The other functional groups such as OH, CH_3, CH_2OH, CHO, and COOH also attached to different positions of anthracene ring (Fouillaud et al. 2018).

11.2.5 Melanins

Melanins are dark color (black and brown) pigments that are produced during oxidative polymerization of phenolic and indolic compounds in the specialized cells melanocytes. The natural sources of melanin are animals, plants, bacterial and fungal species. The indole-type melanins are found in animal species, while catechol melanins are discovered in the plant species. The animal melanin pigments are further divided into two large groups, eumelanin (eu = good) dark brown in color and phaeomelanin (pheo = cloudy or dusky) lighter in color than eumelanin. Eumelnins are formed mainly by the oxidative polymerization of tyrosine, dihydroxyphenylalanine, dopamine and tyramine. On the other hand, phaeomelanin formed by the oxidative polymerization of cysteine and glutathione (Li et al. 2019). Allomelanins are plant melanins and normally appeared as black spots on the leaves, flowers, seeds, and fruits. Phytomelanins are black color melanin that is located in the pericarp of fruits Helianthus and Eupatorium which belongs to the Asteraceae family. Phytomelanins show antioxidant activity, antimicrobial, and UV protection activity and also are used as a coloring agent (Keles and Özdemir 2018).

Natural factors such as temperature, light affect the stability, solubility, and absorbance of the melanins. Melanins show high anti-oxidative and anti-HIV properties. Melanins are not soluble in water and commonly used organic solvents such as acetone, ethanol, chloroform. They are only soluble in an aqueous solution of alkalis and precipitated in acidic solutions having $pH < 3$ (Zou et al. 2015). Studies reported that melanins extracted from the fruiting bodies of the fungi *Exidia nigricans* showed high antioxidants, antimicrobial, and light barrier properties (Łopusiewicz 2018a), (Łopusiewicz et al. 2018). The melanin was extracted in raw

Table 11.3 Molecular formula, physical and chemical properties of melanins

Properties	Results	References
Molecular formula	$C_{18}H_{10}O_4N_2$	Sajjan et al. (2010)
Color	Brown, black	Łopusiewicz (2018a), Sajjan et al. (2010)
Solubility in water	Insoluble	
Solubility in organic solvents	Only soluble in DMSO	
Solubility in alkaline solution (1 M NaOH)	Highly soluble	
Precipitation in acidic condition	Highly precipitated	Łopusiewicz (2018a)
Reaction with bleaching agents	Decolorized on reaction with H_2O_2	
Reaction with an ammonical silver nitrate solution	Showed grey silver-colored precipitates on tube	
Reaction with polyphenols	Brown colored precipitation	
Structure	Irregular structure	Lu et al. (2014)
Charge	The negative charge on structure	Sajjan et al. (2010)

form or purified form through acid hydrolysis, continuous precipitation, and purification. The purified form of melanin showed high antioxidative properties than the raw form. On analysis with absorption spectroscopy, the extracted melanin pigments did not show any peak on UV-Visible spectroscopy at 260–280 nm which indicates that melanins have no amino acids or nucleic acids present in their moiety, so it is complicated to determine the structure of melanins (Łopusiewicz 2018a). In another study, raw and purified forms of melanin were extracted from *Armillaria mellea* rhizomorphs. The purified form of melanin obtained following the acid hydrolysis and organic solvent washing after extraction. FT-IR and Raman Spectroscopy was performed to analyze the activity of the melanin. The purified form of the melanin exhibited higher anti-oxidative and light barrier properties. Results indicated that this type of melanin can be used in different food products, food packaging material, cosmetics and pharmaceutical items to suppress oxidation and formation of free radical species (Łopusiewicz 2018b). Some important properties of melanins are highlighted in Table 11.3.

11.3 Emulsion-Based Delivery Systems for Pigmented Lipophilic Antioxidants

Many pigmented antioxidant compounds are highly lipophilic and exhibit very low solubility in water which makes their incorporation very difficult in most of the food products. Furthermore, poor water solubility means lower gastrointestinal absorption leading to less bioavailability. The use of emulsion systems for the delivery of these compounds can improve the solubility and increases the bioavailability of antioxidant pigments such as carotenoids (Sagis 2015). Encapsulation via emulsifi-

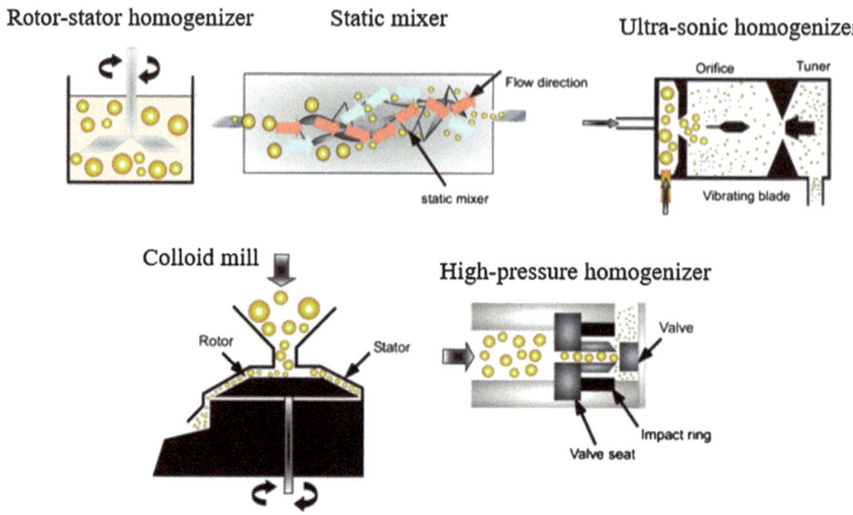

Fig. 11.3 Schematic representation of the devices used for the production of nanoemulsions through high-energy techniques

cation is a useful technique to improve the stability and bioavailability of bioactive components. Emulsion systems usually consist of two immiscible liquids dispersed in another in the form of droplets (Serdaroğlu et al. 2015). According to the size of dispersed droplets, emulsions can be distinguished as macroemulsions with droplet size 0.1–100 μm, microemulsions with a droplet size of 5–100 nm and nanoemulsions with a droplet size of 5–200 nm. Both macroemulsions and nanoemulsions are thermodynamically unstable (Devarajan and Ravichandran 2011).

11.3.1 High-Energy Methods

High-energy approaches for the formation of nanoemulsions employ mechanical devices which generate intensive forces to break the oil particles into smaller droplets (Fig. 11.3). The size of the droplet relies upon the type of the device being used, processing conditions such as temperature, time and type of the material. High-energy techniques need modern equipment such as microfluidizer (MF), high-pressure homogenizer (HPH) and ultrasonic homogenizer (USH) that all consume a large amounts of energy while operating (Gupta et al. 2010). In this manner, these are extravagant approaches, whose advantages are that they permit great control over droplet size and have a wide scope of application for a variety of bioactive components. Different research studies, in which high-energy methods have been employed for the production of pigment-encapsulated emulsion systems are presented in Table 11.4.

Table 11.4 Nanoencapsulation of lipophilic antioxidant pigments using high-energy and low-energy emulsification methods

Emulsification technique	Dispersed Phase	Surface active agents	Encapsulants	Droplet size	References
High-energy methods					
HPH	[a]SCT, MCT and LCT	β-Lactoglobulin	Curcumin	< 200 nm	Ahmed et al. (2012)
HPH	MCT	WPI	Curcumin	197–3704 nm	Li et al. (2016)
HPH	Corn oil	Tween 80	Curcumin	<200 nm	Zhou and Elias (2013)
HPH	MCT	Tween 20	Curcumin	90–122 nm	Joung et al. (2016)
HPH	Corn oil	Lactoferrin	Curcumin	149 nm	Pinheiro et al. (2013)
MF	MCT	Tween 20 and tween 80	β-Carotene	100 nm	Jo and Kwon (2014)
MF	Corn oil	Tween 20, SDS and DTAB	Curcumin	120–900 nm	Pinheiro et al. (2013)
USH	Paprika oleoresin	Tween 80, span 80	Carotenoids	50–250 nm	Pascual-Pineda et al. (2015)
USH	MCT	OSA-MS	Curcumin	140 nm	Abbas et al. (2014)
USH	MCT	WPC and tween 80	Curcumin	141 nm	Sari et al. (2015)
USH	MCT	Lecithin and tween 80	Curcumin	109–144 nm	Li et al. (2016)
Low-energy methods					
EPI	Soybean oil	Tween 20, tween 60, tween 80	Curcumin	196–462 nm	Borrin et al. (2016)
PIT	Dycapryl ether	Glycerol monostearate and cetostearyl polyoxyethylene glycol	Astaxanthin and lycopene	80 nm	Ribeiro et al. (2004)
PIT	Cupuacu butter	Remophor RH40 and span 80	β-Carotene	35 nm	Gomes et al. (2017)
SE	Grape seed oil	Tween 80	Resveratrol	140–180 nm	Davidov-Pardo and McClements (2015)
SE	Olive oil	Span 80	Crocin	10–40 nm	Mehrnia et al. (2017)

(continued)

Table 11.4 (continued)

Emulsification technique	Dispersed Phase	Surface active agents	Encapsulants	Droplet size	References
MCE	MCT	SDS, decaglycerol monolaurate (ML-750), decaglycerol monooleate (MO-7S), NaCas and ML	Astaxanthin	35–37 μm	Khalid et al. (2017b)
MCE	MCT	Tween 20	Quercetin	29 μm	Khalid et al. (2016)
MCE	Soybean oil	Sugar ester and gelatin	β-Carotene	27.6 μm	Neves et al. (2008)
ME	Palm oil	Tween 20 + WPI	Astaxanthin	100–800 nm	Ribeiro et al. (2005)

[a]Abbreviations: *LCT* long-chain triglyceride, *MCT* medium-chain triglyceride, *ML* modified lecithin, *NaCas* sodium caseinate, *OSA-MS* octenyl succinic anhydride modified starch, *SCT* short chain triglyceride, *WPC* whey protein concentrate, *WPI* whey protein isolate

11.3.1.1 High Pressure Homogenizer (HPH)

HPH is the most widely used technique for the production of nanoemulsions. In this approach, high pressure above 300 MPa is applied to the fluid, then it is forced through the close orifice of the homogenizer which generates intense turbulence and hydraulic shear disrupting the fluid into fine droplets (Fig. 11.3) (Donsì et al. 2011). The droplet size depends upon the number of cycles repeated and the level of pressure applied (Liang et al. 2012). Encapsulation of lipophilic bioactive pigments using nanoemulsions formed via HPH has been proved effective for their preservation and conveyance. Moreover, nanoemulsification via this technique can be an effective technique for improving the dispersibility, stability and bioaccessibility of natural pigment compounds.

The successful utilization of nanoemulsions for β-carotene has been demonstrated in a study. β-carotene showed excellent stability and bioavailability when encapsulated in nanoemulsion via HPH. During storage at room temperature under varying conditions of temperature, light and oxygen concentration, β-carotene showed a 50% retention rate which improved when stored under refrigeration temperature (Liang et al. 2013). In another study, β-carotene encapsulated in nanoemulsion through HPH showed higher stability when subjected to rigorous conditions of temperature and pH, which can be encountered in food systems at temperature below 37 °C (Qian et al. 2012). Another study showed that encapsulated β-carotene in sodium caseinate (NaCas), whey protein isolate (WPI) and soy protein isolate (SPI) using two-stage HPH with a particle diameter of 78, 90 and 370 nm, demonstrated higher 2,2-diphenyl-1-picrylhydrazyl (DPPH) radical scavenging activity, hydroxyl radical scavenging activity and stability under gastrointestinal conditions (Yi et al. 2015). Encapsulation of β-carotene in oil-in-water (O/W) nanoemulsions using α-lactalbumin (ALA)-catechin conjugates as emulsifiers via HPH has been proved as an effective technique to upgrade the chemical stability and antioxidant capacity of the active component (Yi et al. 2016).

Astaxanthin is the natural pigment with numerous health benefits but its utilization is limited in functional foods and beverages due to low bioavailability and low dispersibility in water. Encapsulation of higher concentration of astaxanthin in nanoemulsions prepared by HPH using NaCas and modified lecithin (ML) as emulsifying agents can significantly improve the bioaccessibility, physical and chemical stability. NaCas-incorporated emulsions exhibited higher physical and chemical stability (˃70%) during storage duration of 30 days while ML-stabilized emulsions possessed 33% bioaccessibility which is more than NaCas-stabilized emulsions that showed 6% bioaccessibility (Khalid et al. 2017a). Another study optimizes the nanoemulsions preparation conditions to stabilize the astaxanthin using HPH. Based on this study, the optimal conditions require 5 minutes' homogenization at 9000 rpm and 800 bar pressure to load astaxanthin. Astaxanthin loaded in nanoemulsions demonstrated significant stability in terms of particle size (<200 nm) and astaxanthin concentration (>20 mg/mL) during 3 months of storage under varying temperature conditions (Affandi et al. 2011). In this context, also Astaxanthin stabilized by ginseng saponin in O/W nanoemulsions with droplet diameter of 125 nm using HPH exhibited higher stability in terms of droplet diameter growth during 15 days of storage (Shu et al. 2018). Astaxanthin encapsulated in nanoemulsions prepared by HPH under 5–100 MPa exhibits significantly reduced droplet diameter of 134 nm (Sotomayor-Gerding et al. 2016). In another study, stable nanoemulsions of astaxanthin were formulated using HPH with a mean droplet diameter of 160–190 nm (Kim et al. 2012). A dual step homogenization procedure at 800 bars was successfully applied to encapsulate the 10% w/w solution of astaxanthin in nanoemulsions (Affandi et al. 2011).

It has been demonstrated that curcumin and resveratrol can also be encapsulated in nanoemulsions using HPH to improve their water dispersibility and preserve antioxidant activity. Trans-resveratrol encapsulated in peanut-oil based nanoemulsions can improve its stability by reducing chemical degradation. Curcumin trapped in solid-liquid nanoemulsions system can possess higher solubility in aqueous system and reduced particle size and crystallinity (Donsì et al. 2010). An investigation on the use of HPH to encapsulate curcumin in solid lipid nanoparticles (SLN) demonstrated that curcumin can be dispersed in water phase with mean particle size of 152.8 nm and 98% encapsulation efficiency. Curcumin loaded lipid matrix significantly improved the chemical stability and bioavailability of curcumin (Sun et al. 2013).

Saffron extract containing crocin, picrocrocin, and saffranal encapsulated in double emulsion system of water-in-oil-in-water (W/O/W) via two-stage homogenization process and stabilized with whey protein concentrate (WPC)/pectin exhibited encapsulation efficiency of 62.55%, 56.51% and 51.57% respectively. Incorporation of saffron extract into a double emulsion system has a great impact on structure and morphological characteristics of encapsulated powder with smooth-surfaced particles (Esfanjani et al. 2015). Encapsulation of lycopene in O/W emulsions via HPH at 650 bar was carried out to investigate the kinetics of lycopene degradation as a function of thermal treatment and oxygen content. Thermal treatment led to the increased concentration of 9-*cis* isomer while reduced the concentration of all-*trans* and 13-*cis* isomers of lycopene. Oxygen-free conditions decreased the lycopene degradation substantially (Ax et al. 2003). Quercetin loaded nanoemulsions can be produced with fair droplet size (152 nm) and momentous encapsulation efficiency of 93.5% using HPH technique. Quercetin loaded nanoemulsions also exhibit good physico-chemical stability and free radical scavenging activity than bare quercetin (Ni et al. 2017). Various concentrations of Tween-20 were used to stabilize the curcumin encapsulating nanoemulsion prepared by HPH. Curcumin-nanoemulsion was applied to milk system which demonstrated physical stability for 1 month at room temperature. The droplet diameter of curcumin nanoemulsion was 90–122 nm. Curcumin nanoemulsion fortified milk showed higher DPPH radical scavenging activity, 2,2′-azino-bis (3-ethylbenzothiazoline-6-sulfonic acid (ABTS) radical scavenging activity and thiobarbituric acid reactive substances (TBARS) (Joung et al. 2016). In another work, O/W nanoemulsions coated with thiol modified chitosan were prepared to encapsulate the curcumin. The resultant nanoemulsion demonstrated a fine mean droplet diameter of 110 nm, and a higher delivery rate. Pharmacokinetic profile of curcumin showed substantial anti-inflammatory potential and promoted the death of colon cancer cells (Vecchione et al. 2016). HPH is an efficient technique for encapsulation of lipophilic natural pigments; however, stability and release profile of active components may rely upon the concentration of emulsifying agent and speed of homogenizer.

11.3.1.2 Microfluidizer (MF)

A MF device comprises of high-pressure displacement pump and an interaction chamber. During the procedure, aqueous and oil phase strike with each other from two opposite microchannels at a common site of entrenchment generating high pressure causing stupendous shear. To acquire a desired size of the droplets, a raw emulsion is passed through the interaction chamber multiple times (Henry et al. 2010). The droplet size of the dispersed phase can also be controlled via adjusting the pressure of homogenization, the concentration of surfactant and the proportion of viscosities of the dispersed and continuous phase (Maali and Mosavian 2013). With these characteristics, MF technique has acquired more attention to developing nanoemulsions containing lipophilic encapsulants.

It is of high interest to encapsulate lipophilic pigment compounds in nanoemulsions using MF for application in food and beverages. Pigment emulsions consist of different types of oils or fats dispersed in water phase in the presence of a surfactant. Lipophilic pigment compounds such as β-carotene and epigallocatechin-3-gallate (EGCG) have been successfully encapsulated in nanoemulsions with MF (Jo and Kwon 2014; Zhou and Elias 2013). Encapsulation of β-carotene in nanoemulsions using MF technique has proved that droplet size depends on type and concentration of emulsifier, pressure of MF and number of cycles of homogenization. Physical and chemical stability of β-carotene is dependent on droplet size of emulsions (Jo and Kwon 2014). In another study, β-carotene was encapsulated in nanoemulsions fabricated by double-channel MF using two types of emulsifiers quillaja saponins and WPI. Resulting nanoemulsions exhibited 0.14–0.16 μm droplet diameter and demonstrated chemical and physical stability during storage period of 14 days at neutral pH under various temperatures (Luo et al. 2017). The impact of antioxidants on the chemical stability of β-carotene entrapped in nanoemulsions was evaluated in a study. β-carotene was incorporated into O/W emulsion stabilized by Twee-20 and β-lactoglobulin through MF. The degradation rate of encapsulated β-carotene decreased upon the addition of Ethylenediaminetetraacetic acid (EDTA), ascorbic acid and vitamin E (Qian et al. 2012). Research work was performed to encapsulate β-carotene into nanoemulsions stabilized by sodium caseinate by MF technique. The resulting nanodispersions were stable against particle aggregation and exhibited 17 nm mean particle size. Increased microfluidization did not show any impact on particle size but improved the polydispersity of the nanodispersions (Chu et al. 2007). Encapsulation of β-carotene in nanoemulsion using MF enhances the water dispersibility and chemical stability of encapsulated β-carotene (Chen et al. 2017).

Curcumin encapsulated nanoemulsions (O/W) emulsions were prepared using MF at 137.9 MPa to assess the bioavailability of curcumin under simulated gastrointestinal environment. Bioavailability was affected by types of emulsifier used to stabilize the emulsion, Tween 80 shows the highest bioavailability as compared to the sodium dodecyl sulphate (SDS) and dodecyl trimethyl ammonium bromide (DTAB) (Pinheiro et al. 2013). In a study, effect of concentration of EGCG, and pH was evaluated on antioxidant activity of EGCG in flaxseed oil, water in oil (W/O) emulsion prepared by MF. Overall, high concentration of EGCG can result into higher antioxidant activity, 500 μM of EGCG appears to exert best antioxidant activity at pH 5–7 (Zhou and Elias 2013).

11.3.1.3 Ultrasonic Homogenizer (USH)

In high intensity USH device, ultrasound waves with frequency of 10–100 kHz and 10–1000 W/cm^2 power are employed. High intensity ultrasound waves result in breakdown of larger oil droplets into smaller ones (Fig. 11.3) at nanoscale with upgraded stability using small amount of surface-active agent (Chemat and Khan 2011). There are two steps which participate in emulsification through USH. In first step, acoustic field makes interfacial waves that cause oil phase to scatter in the continuous phase. Furthermore, ultrasound incites acoustic cavitation which causes arrangement and breakdown of microbubbles separately because of pressure variances of sound waves. Thus, a huge degree of cavitation is created, which produces miniaturized scale implosions that breakdown large droplets into sub-micron size (Sivakumar et al. 2014). Studies have shown that high amplitude USH device can be successfully employed to prepare O/W nanoemulsions and this technique can be efficiently shifted from laboratory to pilot scale without any alteration in ultrasonic amplitude (Peshkovsky et al. 2013).

Emulsion based delivery systems can greatly increase the bioavailability of curcumin as compared to simple dispersion in water. Slightly higher bioavailability of curcumin was observed in conventional emulsions while nanoemulsions prepared via the sonication method possessed better physical stability (Ahmed et al. 2012). It was shown in a report that curcumin can also be encapsulated in medium-chain triglyceride (MCT) oil droplet nanoemulsions developed by using USH. WPC and Tween-20 were used as stabilizers, the prepared nanoemulsions exhibit a mean droplet diameter of 141.6 nm with 90.56% encapsulation efficiency. Slow-release of curcumin from nanoemulsions was observed in the simulated gastrointestinal environment with resistance to the pepsin enzyme (Sari et al. 2015). Organogel-based nanoemulsions stabilized by Tween-20 have been successfully formulated for entrapment and oral delivery of curcumin using USH. A nine-fold increase in oral bioavailability of encapsulated curcumin was noticed as compared to non-encapsulated curcumin (Yu and Huang 2012). Moreover, it was reported that the USH technique was employed to encapsulate the curcumin to improve the physical stability, loading efficiency and smaller particle size. The resultant nanoemulsion exhibited 47–55 nm mean particle diameter incorporating 23–28 mg/30 ml of curcumin with stability of 60 days at refrigeration temperature (Anuchapreeda et al. 2012). In another study, researchers utilized the optimized conditions for the development of curcumin-loded nanoemulsions using USH and octenyl succinic anhydride modified starch (OSA-MS) as stabilizer. The optimized process for the fabrication of these nanoemulsions showed more stable droplets with 6 mg/mL oil concentration of curcumin. Application of USH helped to achieve the mean droplet diameter of 159.85 nm (Abbas et al. 2014). In this context, other curcumin-loaded nanoemulsions fabricated by USH using MCT oil, Tween 80 and lecithin demonstrated 95.10% encapsulation efficiency and 0.548 mg/ml loading efficiency (Li et al. 2016).

Dietary lycopene can also be successfully encapsulated in O/W emulsions using USH. Incorporation of free radical scavengers such as propyl gallate (PG), gallic acid (GA), and α-tocopherol can significantly reduce the degradation of lycopene (Bou et al. 2011). The use of EDTA in nanoemulsions prepared by USH to preserve lycopene can also improve the stability of lycopene during storage (Boon et al. 2009). β-carotene was encapsulated in O/W emulsion using sonication technique to investigate the impact of droplet size on bioavailability. β-carotene bioavailability increased substantially with a reduction in droplet size (Salvia-Trujillo et al. 2013). Finally, in a study useful information regarding the encapsulation of paprika oleoresin in O/W emulsion through high power USH was provided (Pascual-Pineda et al. 2015).

11.3.2 Low-Energy Methods

Low-energy methods can also be employed to perform nano-emulsification producing nanoemulsions of uniform and fine droplet size. These techniques such as phase inversion temperature and (PIT) and emulsion phase inversion (EPI) use internal physical properties of the system i.e. temperature and composition to produce nanoemulsions (Setya et al. 2014). Though low-energy systems are generally more feasible to formulate smaller droplet size than high-energy methods, there are a couple of limitations for them about using the specific type of oils and emulsifiers like proteins and polysaccharides. In order to overcome this issue, high concentrations of synthetic emulsifiers are used to fabricate nanoemulsions which narrows down their application in food systems (McClements and Rao 2011). The most commonly used low-energy methods for the production of nanoemulsions are phase inversion methods, spontaneous emulsification (SE), membrane emulsification (ME) and microchannel emulsification (MCE). Several studies have proved that low-energy methods can be efficiently used for the encapsulation of natural lipophilic pigments (Table 11.4).

11.3.2.1 Phase Inversion Methods

In phase inversion methods, framework utilizes the chemical energy which is discharged during phase changes in the emulsification process. Phase transition is induced by changing the temperature of system at constant composition or changing the composition at constant temperature (Thakur et al. 2013).

Phase Inversion Temperature (PIT)

In this approach, phase inversion is achieved via rapidly changing the temperature of the oil-water-emulsifier blend below PIT while mixing. This technique engrosses the temperature-dependent solubility of surfactants which ultimately alter their lipophilic and hydrophilic characteristics. A blend of oil, water, and non-ionic surfactant in an appropriate proportion is mixed under ambient conditions forming O/W emulsion. When this initial emulsion is heated above PIT and immediately cooled under the continuous mixing process, temperature change alters the solubility and molecular shape of non-ionic surfactants. Heating the emulsion at PIT causes the dehydration of polyoxyethylene groups of non-ionic surfactants due to which hydrophile-lipophile balance (HLB) is established having very similar affinity for oil and the aqueous phase (Walker et al. 2015). When emulsion is heated at higher temperature surfactant is completely solubilized causing phase inversion from O/W emulsion to W/O emulsion, separated emulsion is balanced by rapid cooling. Rapid cooling can occur at HLB temperature or above PIT (Anandharamakrishnan 2014).

Studies have proven that PIT can be an efficient technique for the delivery of different carotenoids. Carotenoids containing nanoemulsions were prepared using dycapryl ether as dispersed phase, glycerol monostearate and cetostearyl polyoxyethylene glycol as surface active agents. Phase inversion was achieved by heating the emulsion at 74 °C and then rapid cooling at ambient temperature. The loading rate of the astaxanthin and lycopene was 0.3 and 3.5 g/L of emulsion respectively. The mean droplet diameter was 80 nm, while dispersity of emulsion was 40–200 nm. During a month of storage, no substantial degradation of carotenoids was observed (Ribeiro et al. 2004). β-carotene was encapsulated in lipid nanoparticles by PIT method using cupuacu butter as the dispersed phase, remophor RH40 and Span 80 as surface active agents. The average diameter of the obtained lipid nanoparticles was 35 nm. The static system provided the 92% bioaccessibility of β-carotene while dynamic system revealed 29% bioaccessibility of β-carotene (Gomes et al. 2017).

Emulsion Phase Inversion (EPI)

Emulsion phase inversion (EPI) known also as phase inversion composition (PIC) is a technique, where an emulsion is framed by including water to oil-emulsifier blend. It depends upon calamitous phase reversal, implying that the proportion between the oil and water phase changes, though some other changes in the surfactant's properties are excluded (Solans and Solé 2012). In this technique, the water is included under mixing in the oil phase. At first, W/O emulsion is shaped, when volume of the water increases, a phase reversal happens and the water turns into the continuous phase of scattered oil droplets, framing an O/W emulsion. In other words, the hydration level of surfactant chains increases which transforms the surfactant curvature from negative to zero. A balance between lipophilic-hydrophilic properties of the system is obtained, thus forming dynamically stable nanoemulsions having exceptionally smaller droplet size (Anandharamakrishnan 2014).

β-carotene can be stabilized in high internal phase emulsions (HIPE) using gelatin as stabilizer. Higher storage stability of β-carotene in HIPE was observed for 27 days. In vitro studies demonstrated the five-fold increase in bioaccessibility of β-carotene encapsulated in HIPE (Tan et al. 2017). In another study, pickering emulsion was fabricated by emulsion inversion point method using soybean oil as dispersed phase and Tween 20 and glycerol as surface active agents. 70% of the loaded curcumin was remained stable which demonstrates this approach as a promising technique for encapsulation of lipophilic pigments (Borrin et al. 2016). Curcumin encapsulated in pickering emulsion showed 100-fold higher stability as compared to control. In simulated gastric environment 80% retention of encapsulated was observed (Tikekar et al. 2013).

11.3.2.2 Spontaneous Emulsification (SE)

SE process depends upon chemical energy released by inclusion of water into blend of the oil phase and surfactant at consistent temperature without any change in phase. The parameters which can affect the spontaneity of the emulsification process are composition and concentration of surfactant, bulk and interfacial viscosity and interfacial tension. At the point when an oil phase is blended with a water-soluble substance, oil droplets suddenly shape. The system depends on the formation of a water dispersible component from the dispersed phase to the aqueous phase. This prompts the interfacial disturbance leading to the spontaneous formation of oil droplets (McClements and Rao 2011). In pharmaceutical industry systems using this technique for the preparation of drug delivery systems are known as self-emulsifying drug delivery systems (SEDDS) or self-nano emulsifying drug delivery systems (SNEDDS).

Saffron derived crocin was encapsulated in double nanoemulsions via SE method and droplet diameter of 429 nm and 695 nm achieved for systems containing WPC and gum arabic (GA), respectively. SE method made it possible to formulate the crocin loaded double emulsion with fine droplet size as compared to conventional emulsions (Mehrnia et al. 2017). SE approach was also used to encapsulate the grape derived resveratrol containing 10% grape seed oil as the dispersed phase, 10% Tween 80 as emulsifier and 80% aqueous phase. Smaller droplet diameter of 100 nm was observed with 120 ± 10 μg/ml dissolvability of resveratrol in oil phase. Chemical stability of resveratrol (88% retention) was substantially improved via encapsulation (Davidov-Pardo and McClements 2015). Antioxidant activity of the olive leaf extract was significantly preserved when encapsulated by SE method using soybean oil as dispersed phase. The droplet diameter for the multiple emulsion using WPC as a stabilizer was 675 nm and 1443 nm for WPC-pectin. Higher antioxidant activity of olive leaf extract can be attributed to higher solubility and controlled release (Mohammadi et al. 2016).

11.3.2.3 Microchannel Emulsification (MCE)

MCE (Fig. 11.4) is a novel technique having the potential to produce monodispersed droplets using microchannels grooved on silicon chips (Kawakatsu et al. 1997) or stainless-steel microchips (Kobayashi et al. 2012). This technique employs significant mechanical shear force to separate the dispersed phase into small droplets, which are scattered in a continuous liquid phase, bringing about emulsion droplets typically ranging from 0.1 to 100 μm and impressive polydispersity. The droplet synthesis depends on shearing by interfacial pressure, in which the dispersed phase is cut off suddenly, shaping round droplets. This continuous development prompts a low-energy contribution for emulsification (Sugiura et al. 2002). MCE has been employed to prepare lipophilic pigment nanoemulsions with improved properties e.g. quercetin (Khalid et al. 2016), astaxanthin (Khalid et al. 2017b), and β-carotene (Neves et al. 2008).

A study used the MCE method to encapsulate the different extract concentrations of astaxanthin in O/W emulsion using silicon microchannel plates. Successful emulsification was achieved with a mean droplet diameter of 35–37 μm and 98% encapsulation efficiency. O/W emulsion droplets showed substantial physical stability at room temperature and 25–27 μg/mL droplet retention during a storage period of 15 days (Khalid et al. 2017b). Another study used a MCE technique to encapsulate the quercetin employing silicon microchannel plates. Successful droplet fabrication was achieved with a mean droplet diameter of 29 μm and 0.4 mg/mL loading efficiency of quercetin using MCT oil as the dispersed phase. O/W emulsion showed stability at refrigeration and room temperature in terms of droplet coalescence during a storage period of 30 days. The formulated O/W emulsions encapsulating quercetin exhibited 80% and 70% encapsulation efficiency at refrigeration and room temperature, respectively (Khalid et al. 2016). In another study,

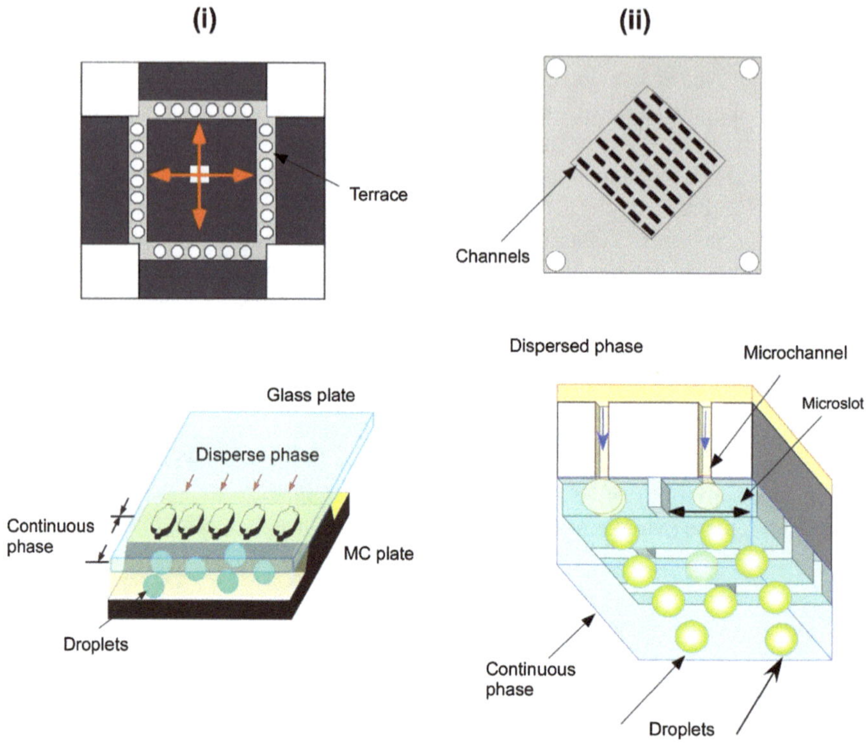

Fig. 11.4 Schematic presentation of typical microchannel emulsification (MCE), (i) grooved type MCE (ii) straight through MCE

MCE technique was employed to produce monodispersed emulsion droplets loaded with β-carotene. Soybean oil was used as a dispersed phase while sugar ester and gelatin were employed as the continuous phase. β-carotene loaded monodispersed O/W emulsion with a mean droplet diameter of 27.6 μm was achieved. β-carotene-loaded droplets showed physical stability throughout the storage period resulting in a droplet diameter of 28.2 μm after 4 months of storage in absence of light at 25 °C. β-carotene was successfully encapsulated in grooved microchannels with a droplet diameter of 9.1 μm (Neves et al. 2008).

11.3.2.4 Membrane Emulsification (ME)

ME is also categorized as a low-energy emulsification method as it needs 100 times low-energy as compared to high-energy methods (Pathak 2012). ME produces the emulsions through direct emulsification i.e. dispersed phase is pressed and passed through membrane to reduce the particle size while the continuous phase is passed through the separate channel or premix emulsification in which an already formed coarse emulsion is passed through the membrane to acquire the desired droplet size (Candéa 2013). Several different materials can be used for the production of O/W and W/O emulsions such as tubular micro-porous glass membranes and Shirasu porous glass (SPG) (Joscelyne and Trägårdh 2000), silicon nitrides (Zhu and Barrow 2005) and polytetrafluoroethylene (Yamazaki et al. 2003). Process parameters that can affect the final droplet size are type and pore size of the membrane, flow velocity, transmembrane pressure, and type of emulsifier. It is noteworthy that by increasing the transmembrane pressure, the flux of dispersed phase across the membrane, droplet size, and polydispersibility will increase too (Nazir et al. 2010).

As an example, astaxanthin-loaded O/W emulsion through premix membrane emulsification (PME) was prepared in a study. Palm oil was used as a disperses phase with two surface active agents to stabilize the O/W emulsion. Total astaxanthin concentration was decreased by 30% after 21 days of storage without any antioxidant. Mean droplet diameter was dependent on the concentration of dispersed phase which varied from 100–800 nm with varying the concentration of dispersed phase from 10% to 50% (Ribeiro et al. 2005).

11.4 Conclusion and Future Perspectives

The stability of functional food systems is at the heart of research and development in food process engineering especially for the development of bioactive functional food products. It can be effectively claimed that stability is directly linked to the efficacy of bioactive compounds in most of the food delivery systems that are intended to incorporate the novel food functions. Antioxidants that belong to the organic pigment category of compounds e.g. carotenoids, chlorophylls, and flavonoids are a better option to control oxidative stability in food systems while simultaneously imparting beneficial health impacts on human health. However, solubility of these pigments is generally very low in aqueous systems. Nanotechnology has played a decisive positive role in enhancing the stability and solubility of these pigments with the application of various emulsion making techniques. High-energy emulsification techniques (MF, HPH and USH) are finding its way in developing thermodynamically stable nanoemulsions systems in foods. Several emulsion-based systems such as single, double and nanoemulsions have been developed for encapsulation of antioxidants. It is now a well-known fact that encapsulated antioxidant pigment compounds possess greater stability (in vitro and in vivo systems) than non-encapsulated compounds. Emulsion-based encapsulation systems for the delivery of antioxidants have been proven promising techniques for their protection, physical and chemical stability, controlled release and high encapsulation efficiency. However, the industrial scale application is still limited owing to limited technical know-how and cost-effectiveness of the methods that are developed so far. Also, the food production systems are generally considered a cheaper proposition hence limiting the investment in improving the production technologies. However, the future development of food production systems that are focusing on effective delivery systems of bioactive compounds would have to invest in applying modern techniques of nanoemulsions development to effectively incorporate the natural pigments i.e., antioxidants into innovative food solutions.

References

Abbas S, Bashari M, Akhtar W, Li WW, Zhang X (2014) Process optimization of ultrasound-assisted curcumin nanoemulsions stabilized by OSA-modified starch. Ultrason Sonochem 21(4):1265–1274. https://doi.org/10.1016/j.ultsonch.2013.12.017

Abdel-Aal ESM, Hucl P, Shipp J, Rabalski I (2016) Compositional differences in anthocyanins from blue- and purple-grained spring wheat grown in four environments in Central Saskatchewan. Cereal Chem 93(1):32–38. https://doi.org/10.1094/CCHEM-03-15-0058-R

Abdel-Aal E-SM, Hucl P, Rabalski I (2018) Compositional and antioxidant properties of anthocyanin-rich products prepared from purple wheat. Food Chem 254:13–19. https://doi.org/10.1016/j.foodchem.2018.01.170

Affandi M, Julianto T, Majeed A (2011) Development and stability evaluation of astaxanthin nanoemulsion. Asian J Pharm Clin Res 4(1):142–148. https://www.researchgate.net/publication/285986315

Ahmed K, Li Y, McClements DJ, Xiao H (2012) Nanoemulsion-and emulsion-based delivery systems for curcumin: encapsulation and release properties. Food Chem 132(2):799–807. https://doi.org/10.1016/j.foodchem.2011.11.039

Amengual J (2019) Bioactive properties of carotenoids in human health. Nutrients 11(10):2388. https://doi.org/10.3390/nu11102388

Anandharamakrishnan C (2014) Techniques for nanoencapsulation of food ingredients. Springer, New York, pp 1–89

Anuchapreeda S, Fukumori Y, Okonogi S, Ichikawa H (2012) Preparation of lipid nanoemulsions incorporating curcumin for cancer therapy. J Nanotechnol 2012:1–11. https://doi.org/10.1155/2012/270383

Ax K, Mayer-Miebach E, Link B, Schuchmann H, Schubert H (2003) Stability of lycopene in oil-in-water emulsions. Eng Life Sci 3(4):199–201. https://doi.org/10.1002/elsc.200390028

Boon CS, McClements DJ, Weiss J, Decker EA (2009) Role of iron and hydroperoxides in the degradation of lycopene in oil-in-water emulsions. J Agric Food Chem 57(7):2993–2998. https://doi.org/10.1021/jf803747j

Borrin TR, Georges EL, Moraes IC, Pinho SC (2016) Curcumin-loaded nanoemulsions produced by the emulsion inversion point (EIP) method: an evaluation of process parameters and physico-chemical stability. J Food Eng 169:1–9. https://doi.org/10.1016/j.jfoodeng.2015.08.012

Bou R, Boon C, Kweku A, Hidalgo D, Decker EA (2011) Effect of different antioxidants on lycopene degradation in oil-in-water emulsions. Eur J Lipid Sci Technol 113(6):724–729. https://doi.org/10.1002/ejlt.201000524

Brunetti C, Guidi L, Sebastiani F, Tattini MJE, Botany E (2015) Isoprenoids and phenylpropanoids are key components of the antioxidant defense system of plants facing severe excess light stress. Environ Exp Bot 119:54–62. https://doi.org/10.1016/j.envexpbot.2015.04.007

Calderaro A, Barreca D, Bellocco E, Smeriglio A, Trombetta D, Laganà G (2020) Colored phytonutrients: role and applications in the functional foods of anthocyanins. In: Phytonutrients in food. Elsevier, Duxford, pp 177–119. https://doi.org/10.1016/B978-0-12-815354-3.00011-3

Candéa TV (2013) Study of membrane emulsification process as a pre-step for the microencapsulation of lipid compounds by spray drying. Faculdade de Ciências e Tecnologia. http://hdl.handle.net/10362/10412

Carocho M, Morales P, Ferreira IC (2018) Antioxidants: reviewing the chemistry, food applications, legislation and role as preservatives. Trends Food Sci Technol 71:107–120. https://doi.org/10.1016/j.tifs.2017.11.008

Chaitanya Lakshmi G (2014) Food coloring: the natural way. Res J Chem Sci 2231(8):606X. www.isca.in, www.isca.me

Chang K-H, Cheng M-L, Chiang M-C, Chen C-MJCCA (2018) Lipophilic antioxidants in neurodegenerative diseases. Clin Chim Acta 485:79–87. https://doi.org/10.1016/j.cca.2018.06.031

Chemat F, Khan MK (2011) Applications of ultrasound in food technology: processing, preservation and extraction. Ultrason Sonochem 18(4):813–835. https://doi.org/10.1016/j.ultsonch.2010.11.023

Chen J, Li F, Li Z, McClements DJ, Xiao H (2017) Encapsulation of carotenoids in emulsion-based delivery systems: enhancement of β-carotene water-dispersibility and chemical stability. Food Hydrocoll 69:49–55. https://doi.org/10.1016/j.foodhyd.2017.01.024

Chu BS, Ichikawa S, Kanafusa S, Nakajima M (2007) Preparation of protein-stabilized β-carotene nanodispersions by emulsification–evaporation method. J Am Oil Chem Soc 84(11):1053–1062. https://doi.org/10.1007/s11746-007-1132-7

Corrales-Bañuelos AB, Cuevas-Rodríguez EO, Gutiérrez-Uribe JA, Milán-Noris EM, Reyes-Moreno C, Milán-Carrillo J, Mora-Rochín S (2016) Carotenoid composition and antioxidant activity of tortillas elaborated from pigmented maize landrace by traditional nixtamalization or lime cooking extrusion process. J Cereal Sci 69:64–70. https://doi.org/10.1016/j.jcs.2016.02.009

Craft NE, Furr HC (2019) Methods for assessment of vitamin a (Retinoids) and carotenoids. In: Laboratory assessment of vitamin status. Elsevier Science, San Diego, pp 21–47. https://doi.org/10.1016/B978-0-12-813050-6.00002-4

Davidov-Pardo G, McClements DJ (2015) Nutraceutical delivery systems: resveratrol encapsulation in grape seed oil nanoemulsions formed by spontaneous emulsification. Food Chem 167:205–212. https://doi.org/10.1016/j.foodchem.2014.06.082

Deniz NG, Ibis C, Gokmen Z, Stasevych M, Novikov V, Komarovska-Porokhnyavets O, Ozyurek M, Guclu K, Karakas D, Ulukaya E (2015) Design, synthesis, biological evaluation, and antioxidant and cytotoxic activity of heteroatom-substituted 1, 4-naphtho-and benzoquinones. Chem Pharm Bull 63(12):1029–1039. https://doi.org/10.1248/cpb.c15-00607

Devarajan V, Ravichandran V (2011) Nanoemulsions: as modified drug delivery tool. Int J Compr Pharm 2(4):1–6. www.pharmacie-globale.info

Donsì F, Wang Y, Li J, Huang Q (2010) Preparation of curcumin sub-micrometer dispersions by high-pressure homogenization. J Agric Food Chem 58(5):2848–2853. https://doi.org/10.1021/jf903968x

Donsì F, Sessa M, Mediouni H, Mgaidi A, Ferrari G (2011) Encapsulation of bioactive compounds in nanoemulsion-based delivery systems. Procedia Food Sci 1:1666–1671. https://doi.org/10.1016/j.profoo.2011.09.246

Eghbaliferiz S, Iranshahi M (2016) Prooxidant activity of polyphenols, flavonoids, anthocyanins and carotenoids: updated review of mechanisms and catalyzing metals. Phytother Res 30(9):1379–1391. https://doi.org/10.1002/ptr.5643

El-Najjar N, Gali-Muhtasib H, Ketola RA, Vuorela P, Urtti A, Vuorela H (2011) The chemical and biological activities of quinones: overview and implications in analytical detection. Phytochem Rev 10(3):353–370. https://doi.org/10.1007/s11101-011-9209-1

Esfanjani AF, Jafari SM, Assadpoor E, Mohammadi A (2015) Nano-encapsulation of saffron extract through double-layered multiple emulsions of pectin and whey protein concentrate. J Food Eng 165:149–155. https://doi.org/10.1016/j.jfoodeng.2015.06.022

Fouillaud M, Caro Y, Venkatachalam M, Grondin I, Dufossé L (2018) Anthraquinones. In: Phenolic compounds in food characterization and analysis. CRC Press, Boca Raton, pp 130–170. https://hal.univ-reunion.fr/hal-01657104

Giordano D, Locatelli M, Travaglia F, Bordiga M, Reyneri A, Coïsson JD, Blandino M (2017) Bioactive compound and antioxidant activity distribution in roller-milled and pearled fractions of conventional and pigmented wheat varieties. Food Chem 233:483–491. https://doi.org/10.1016/j.foodchem.2017.04.065

Gomes GV, Sola MR, Marostegan LF, Jange CG, Cazado CP, Pinheiro AC, Vicente AA, Pinho SC (2017) Physico-chemical stability and in vitro digestibility of beta-carotene-loaded lipid nanoparticles of cupuacu butter (Theobroma grandiflorum) produced by the phase inversion temperature (PIT) method. J Food Eng 192:93–102. https://doi.org/10.1016/j.jfoodeng.2016.08.001

Gupta PK, Pandit J, Kumar A, Swaroop P, Gupta S (2010) Pharmaceutical nanotechnology novel nanoemulsion-high energy emulsification preparation, evaluation and application. The Pharma Res 3(3):117–138. http://www.thepharmaresearch.info

Haman N, Longo E, Schiraldi A, Scampicchio M (2017) Radical scavenging activity of lipophilic antioxidants and extra-virgin olive oil by isothermal calorimetry. Thermochim Acta 658:1–6. https://doi.org/10.1016/j.tca.2017.10.012

Henry JV, Fryer PJ, Frith WJ, Norton IT (2010) The influence of phospholipids and food proteins on the size and stability of model sub-micron emulsions. Food Hydrocoll 24(1):66–71. https://doi.org/10.1016/j.foodhyd.2009.08.006

Hussain H, Green IR, Saleem M, Raza ML, Nazir M (2019) Therapeutic potential of Iridoid derivatives: patent review. Inventions 4(2):29. https://doi.org/10.3390/inventions4020029

Janiszewska-Turak E (2017) Carotenoids microencapsulation by spray drying method and supercritical micronization. Food Res Int 99:891–901. https://doi.org/10.1016/j.foodres.2017.02.001

Jeż M, Błaszczak W, Zielińska D, Wiczkowski W, Białobrzewski I (2020) Carotenoids and lipophilic antioxidant capacities of tomato purées as affected by high hydrostatic pressure processing. Int J Food Sci Technol 55(1):65–73. https://doi.org/10.1111/ijfs.14231

Jo Y-J, Kwon Y-J (2014) Characterization of β-carotene nanoemulsions prepared by microfluidization technique. Food Sci Biotechnol 23(1):107–113. https://doi.org/10.1007/s10068-014-0014-7

Joscelyne SM, Trägårdh G (2000) Membrane emulsification—a literature review. J Membr Sci 169(1):107–117. https://doi.org/10.1016/S0376-7388(99)00334-8

Joung HJ, Choi MJ, Kim JT, Park SH, Park HJ, Shin GH (2016) Development of food-grade curcumin nanoemulsion and its potential application to food beverage system: antioxidant property and in vitro digestion. J Food Sci 81(3):N745–N753. https://doi.org/10.1111/1750-3841.13224

Kang Y-R, Park J, Jung SK, Chang YH (2018) Synthesis, characterization, and functional properties of chlorophylls, pheophytins, and Zn-pheophytins. Food Chem 245:943–950. https://doi.org/10.1016/j.foodchem.2017.11.079

Kawakatsu T, Kikuchi Y, Nakajima M (1997) Regular-sized cell creation in microchannel emulsification by visual microprocessing method. J Am Oil Chem Soc 74(3):317–321. https://doi.org/10.1007/s11746-997-0143-8

Kawamukai MJB (2018) Biosynthesis and applications of prenylquinones. Biosci Biotechnol Biochem 82(6):963–977. https://doi.org/10.1080/09168451.2018.1433020

Keles Y, Özdemir Ö (2018) Extraction, purification, antioxidant properties and stability conditions of phytomelanin pigment on the sunflower seeds. Int J Second Metab 5(2):140–148. https://doi.org/10.21448/ijsm.377470

Khalid N, Barrow CJ (2018) Critical review of encapsulation methods for stabilization and delivery of astaxanthin. J Food Bioactives 1:104–115. https://doi.org/10.31665/JFB.2018.1129

Khalid N, Kobayashi I, Neves MA, Uemura K, Nakajima M, Nabetani H (2016) Microchannel emulsification study on formulation and stability characterization of monodisperse oil-in-water emulsions encapsulating quercetin. Food Chem 212:27–34. https://doi.org/10.1016/j.foodchem.2016.05.154

Khalid N, Shu G, Holland BJ, Kobayashi I, Nakajima M, Barrow CJ (2017a) Formulation and characterization of O/W nanoemulsions encapsulating high concentration of astaxanthin. Food Res Int 102:364–371. https://doi.org/10.1016/j.foodres.2017.06.019

Khalid N, Shu G, Kobayashi I, Nakajima M, Barrow CJ (2017b) Formulation and characterization of monodisperse O/W emulsions encapsulating astaxanthin extracts using microchannel emulsification: insights of formulation and stability evaluation. Colloids Surf B: Biointerfaces 157:355–365. https://doi.org/10.1016/j.colsurfb.2017.06.003

Kim DM, Hyun SS, Yun P, Lee CH, Byun SY (2012) Identification of an emulsifier and conditions for preparing stable nanoemulsions containing the antioxidant astaxanthin. Int J Cosmet Sci 34(1):64–73. https://doi.org/10.1111/j.1468-2494.2011.00682.x

Kiokias S, Proestos C, Varzakas T (2016) A review of the structure, biosynthesis, absorption of carotenoids-analysis and properties of their common natural extracts. Curr Res Nutr Food Sci J 4(Special Issue Carotenoids March 2016) 25–37. https://doi.org/10.12944/CRNFSJ.4.Special-Issue1.03

Kobayashi I, Wada Y, Hori Y, Neves MA, Uemura K, Nakajima M (2012) Microchannel emulsification using stainless-steel chips: oil droplet generation characteristics. Chem Eng Technol 35(10):1865–1871. https://doi.org/10.1002/ceat.201200044

Kokkiripati PK, Kamsala RV, Bashyam L, Manthapuram N, Bitla P, Peddada V, Raghavendra AS, Tetali SD (2013) Stem-bark of Terminalia arjuna attenuates human monocytic (THP-1) and aortic endothelial cell activation. J Ethnopharmacol 146(2):456–464. https://doi.org/10.1016/j.jep.2012.12.050

Koyuncu MA, Dilmaçünal T (2010) Determination of vitamin C and organic acid changes in strawberry by HPLC during cold storage. Notulae Botanicae Horti Agrobotanici Cluj-Napoca 38(3):95–98. https://doi.org/10.15835/nbha3834819

Kucharska A, Sokół-Łętowska A, Oszmiański J, Piórecki N, Fecka I (2017) Iridoids, phenolic compounds and antioxidant activity of edible honeysuckle berries (Lonicera caerulea var. kamtschatica Sevast.). Molecules 22(3):405. https://doi.org/10.3390/molecules22030405

Kundu T, Pramanik A (2020) Expeditious and eco-friendly synthesis of new multifunctionalized pyrrole derivatives and evaluation of their antioxidant property. Bioorganic Chem 103734. https://doi.org/10.1016/j.bioorg.2020.103734

Lee L-S, Choi E-J, Kim C-H, Sung J-M, Kim Y-B, Seo D-H, Choi H-W, Choi Y-S, Kum J-S, J-DJJocs P (2016) Contribution of flavonoids to the antioxidant properties of common and tartary buckwheat. J Cereal Sci 68:181–186. https://doi.org/10.1016/j.jcs.2015.07.005

Lestari D, Kartika R, Marliana E (2019) Antioxidant and anticancer activity of Eleutherine bulbosa (Mill.) Urb on leukemia cells L1210. Journal of Physics: Conference Series (Vol. 1277, No. 1, p. 012022). IOP Publishing. https://doi.org/10.1088/1742-6596/1277/1/012022

Li J, Hwang I-C, Chen X, Park HJ (2016) Effects of chitosan coating on curcumin loaded nano-emulsion: study on stability and in vitro digestibility. Food Hydrocoll 60:138–147. https://doi.org/10.1016/j.foodhyd.2016.03.016

Li Q, Somavat P, Singh V, Chatham L, de Mejia EG (2017) A comparative study of anthocyanin distribution in purple and blue corn coproducts from three conventional fractionation processes. Food Chem 231:332–339. https://doi.org/10.1016/j.foodchem.2017.03.146

Li X, Wang J, Liu H, Chen L, Jiang A (2019) Comparation of structure and properties of sea cucumber melanin before and after degradation. J Biotech Res. [ISSN: 1944-3285 10:283–292

Liang R, Xu S, Shoemaker CF, Li Y, Zhong F, Huang Q (2012) Physical and antimicrobial properties of peppermint oil nanoemulsions. J Agric Food Chem 60(30):7548–7555. https://doi.org/10.1021/jf301129k

Liang R, Shoemaker CF, Yang X, Zhong F, Huang Q (2013) Stability and bioaccessibility of β-carotene in nanoemulsions stabilized by modified starches. J Agric Food Chem 61(6):1249–1257. https://doi.org/10.1021/jf303967f

Lin M, Zhang J, Chen X (2018) Bioactive flavonoids in Moringa oleifera and their health-promoting properties. J Funct Foods 47:469–479. https://doi.org/10.1016/j.jff.2018.06.011

Łopusiewicz Ł (2018a) Isolation, characterisation and biological activity of melanin from Exidia nigricans. World Scientific News 91:111–129

Łopusiewicz Ł (2018b) The isolation, purification and analysis of the melanin pigment extracted from Armillaria mellea rhizomorphs. World Scientific News 100:135–153

Łopusiewicz Ł, Jędra F, Bartkowiak A (2018) The application of melanin modified gelatin coatings for packaging and the oxidative stability of pork lard. World Scientific News 101:108–119

Lu Y, Ye M, Song S, Li L, Shaikh F, Li J (2014) Isolation, purification, and anti-aging activity of melanin from Lachnum singerianum. Appl Biochem Biotechnol 174(2):762–771. https://doi.org/10.1007/s12010-014-1110-0

Luo X, Zhou Y, Bai L, Liu F, Deng Y, McClements DJ (2017) Fabrication of β-carotene nanoemulsion-based delivery systems using dual-channel microfluidization: physical and chemical stability. J Colloid Interface Sci 490:328–335. https://doi.org/10.1016/j.jcis.2016.11.057

Maali A, Mosavian MH (2013) Preparation and application of nanoemulsions in the last decade (2000–2010). J Dispers Sci Technol 34(1):92–105. https://doi.org/10.1080/01932691.2011.648498

Mallakin A, Dixon DG, Greenberg BM (2000) Pathway of anthracene modification under simulated solar radiation. Chemosphere 40(12):1435–1441. https://doi.org/10.1016/S0045-6535(99)00331-8

Martín J, Kuskoski EM, Navas MJ, Asuero AG (2017) Antioxidant capacity of anthocyanin pigments. In: Flavonoids-from biosynthesis to human health, 3, 205–255. BoD-Books on Demand. https://doi.org/10.5772/67718

McClements DJ, Rao J (2011) Food-grade nanoemulsions: formulation, fabrication, properties, performance, biological fate, and potential toxicity. Crit Rev Food Sci Nutr 51(4):285–330. https://doi.org/10.1080/10408398.2011.559558

Mehrnia M-A, Jafari S-M, Makhmal-Zadeh BS, Maghsoudlou Y (2017) Rheological and release properties of double nano-emulsions containing crocin prepared with Angum gum, Arabic gum and whey protein. Food Hydrocoll 66:259–267. https://doi.org/10.1016/j.foodhyd.2016.11.033

Mercadante AZ (2019) Carotenoid esters in foods: physical, chemical and biological properties, vol 13. Royal Society of Chemistry

Mezzomo N, Ferreira SR (2016) Carotenoids functionality, sources, and processing by supercritical technology: a review. J Chem 2016:1–16. https://doi.org/10.1155/2016/3164312

Mohammadi A, Jafari SM, Esfanjani AF, Akhavan S (2016) Application of nano-encapsulated olive leaf extract in controlling the oxidative stability of soybean oil. Food Chem 190:513–519. https://doi.org/10.1016/j.foodchem.2015.05.115

Mozos I, Stoian D, Caraba A, Malainer C, Horbańczuk JO, Atanasov AG (2018) Lycopene and vascular health. Front Pharmacol 9:521. https://doi.org/10.3389/fphar.2018.00521

Nazir A, Schroën K, Boom R (2010) Premix emulsification: a review. J Membr Sci 362(1–2):1–11. https://doi.org/10.1016/j.memsci.2010.06.044

Neves MA, Ribeiro HS, Kobayashi I, Nakajima M (2008) Encapsulation of lipophilic bioactive molecules by microchannel emulsification. Food Biophys 3(2):126–131. https://doi.org/10.1007/s11483-008-9056-9

Newsome AG, Culver CA, Van Breemen RB (2014) Nature's palette: the search for natural blue colorants. J Agric Food Chem 62(28):6498–6511. https://doi.org/10.1021/jf501419q

Ni S, Hu C, Sun R, Zhao G, Xia Q (2017) Nanoemulsions-based delivery systems for encapsulation of quercetin: preparation, characterization, and cytotoxicity studies. J Food Process Eng 40(2):e12374. https://doi.org/10.1111/jfpe.12374

Oancea S, Calin F (2016) Changes in total phenolics and anthocyanins during blackberry, raspberry and cherry jam processing and storage. Romanian Biotechnol Letters 21(1):11232–11237

Panche A, Diwan A, Chandra SR (2016) Flavonoids: an overview. J Nutr Sci 5:e47. https://doi.org/10.1017/jns.2016.41

Pascual-Pineda LA, Flores-Andrade E, Jiménez-Fernández M, Beristain CI (2015) Kinetic and thermodynamic stability of paprika nanoemulsions. Int J Food Sci Technol 50(5):1174–1181. https://doi.org/10.1111/ijfs.12750

Pathak M (2012) Numerical simulation of droplet dynamics in membrane emulsification systems. In: Numerical simulation—from theory to industry:415–438. BoD – Books on Demand, Germany. https://doi.org/10.5772/50180

Patil SP, Laddha KS (2018) Dyes derived from Iridoids and Secoiridoids: a brief review. The Natural Products J 8(4):268–274. https://doi.org/10.2174/2210315508666180522081456

Pérez-Rodríguez L, Martínez-Padilla J, Mougeot F (2013) Carotenoid-based ornaments as signals of health status in birds: evidences from two galliform species, the red-legged partridge (Alectoris rufa) and the red grouse (Lagopus lagopus scoticus). In: Carotenoids: food sources, production and health benefits. Nova Science Publishers, New York

Pervaiz T, Songtao J, Faghihi F, Haider MS, Fang J (2017) Naturally occurring anthocyanin, structure, functions and biosynthetic pathway in fruit plants. J Plant Biochem Physiol 5(2):1–9. https://doi.org/10.4172/2329-9029.1000187

Peshkovsky AS, Peshkovsky SL, Bystryak S (2013) Scalable high-power ultrasonic technology for the production of translucent nanoemulsions. Chem Eng Process Process Intensif 69:77–82. https://doi.org/10.1016/j.cep.2013.02.010

Pinheiro AC, Lad M, Silva HD, Coimbra MA, Boland M, Vicente AA (2013) Unravelling the behaviour of curcumin nanoemulsions during in vitro digestion: effect of the surface charge. Soft Matter 9(11):3147–3154. https://doi.org/10.1039/C3SM27527B

Piyush K, Kuldeep S, Azizur RM, Misbahul HS, Pranay W (2018) A review of benzopyran derivatives in pharmacotherapy of breast cancer. Asian J Pharm Clin Res 11(7):43–46. https://doi.org/10.22159/ajpcr.2018.v11i7.26017

Procházková D, Boušová I, Wilhelmová NJF (2011) Antioxidant and prooxidant properties of flavonoids. Fitoterapia 82(4):513–523. https://doi.org/10.1016/j.fitote.2011.01.018

Qian C, Decker EA, Xiao H, McClements DJ (2012) Inhibition of β-carotene degradation in oil-in-water nanoemulsions: influence of oil-soluble and water-soluble antioxidants. Food Chem 135(3):1036–1043. https://doi.org/10.1016/j.foodchem.2012.05.085

Ribeiro H, Briviba K, Rechkemmer G, Schubert H (2004) Zelluläre Aufnahme von flüssigen Lycopin-Formulierungen. German Nutr Soc 6:43

Ribeiro HS, Rico LG, Badolato GG, Schubert H (2005) Production of O/W emulsions containing astaxanthin by repeated premix membrane emulsification. J Food Sci 70(2):E117–E123. https://doi.org/10.1111/j.1365-2621.2005.tb07083.x

Rodriguez-Amaya DB (2016) Natural food pigments and colorants. Curr Opin Food Sci 7:20–26. https://doi.org/10.1016/j.cofs.2015.08.004

Roshanak S, Rahimmalek M, Goli SAH (2016) Evaluation of seven different drying treatments in respect to total flavonoid, phenolic, vitamin C content, chlorophyll, antioxidant activity and color of green tea (Camellia sinensis or C. assamica) leaves. J Food Sci Technol 53(1):721–729. https://doi.org/10.1007/s13197-015-2030-x

Rymbai H, Sharma R, Srivastav M (2011) Bio-colorants and its implications in health and food industry–a review. Int J Pharmacol Res 3(4):2228–2244. Corpus ID: 55493011

Sagis LM (2015) Microencapsulation and microspheres for food applications. Elsevier Science. https://doi.org/10.1016/C2013-0-18624-2

Saini RK, Keum Y-SJFC (2018) Carotenoid extraction methods: a review of recent developments. Food Chem 240:90–103. https://doi.org/10.1016/j.foodchem.2017.07.099

Sajjan S, Kulkarni G, Yaligara V, Kyoung L, Karegoudar T (2010) Purification and physiochemical characterization of melanin pigment from Klebsiella sp. GSK. J Microbiol Biotechnol 20(11):1513–1520. https://doi.org/10.4014/jmb.1002.02006

Salvia-Trujillo L, Qian C, Martín-Belloso O, McClements D (2013) Influence of particle size on lipid digestion and β-carotene bioaccessibility in emulsions and nanoemulsions. Food Chem 141(2):1472–1480. https://doi.org/10.1016/j.foodchem.2013.03.050

Santos-Sánchez NF, Salas-Coronado R, Villanueva-Cañongo C, Hernández-Carlos B (2019) Antioxidant compounds and their antioxidant mechanism. In: Antioxidants. Springer Nature. https://doi.org/10.5772/intechopen.85270

Sari T, Mann B, Kumar R, Singh R, Sharma R, Bhardwaj M, Athira S (2015) Preparation and characterization of nanoemulsion encapsulating curcumin. Food Hydrocoll 43:540–546. https://doi.org/10.1016/j.foodhyd.2014.07.011

Schlicke H, Richter A, Rothbart M, Brzezowski P, Hedtke B, Grimm B (2015) Function of tetrapyrroles, regulation of tetrapyrrole metabolism and methods for analyses of tetrapyrroles. Procedia Chem 14:171–175. https://doi.org/10.1016/j.proche.2015.03.025

Serdaroğlu M, Öztürk B, Kara A (2015) An overview of food emulsions: description, classification and recent potential applications. Turkish J Agric Food Sci Technol 3(6):430–438. https://doi.org/10.24925/turjaf.v3i6.430-438.336

Setya S, Talegaonkar S, Razdan B (2014) Nanoemulsions: formulation methods and stability aspects. World J Pharm Pharm Sci 3(2):2214–2228. Corpus ID: 98462885

Sharma P, Ghosh C, Rathore JS (2018) Anthocyanin: structure, synthesis and its beneficial role in health. Res Rev 5(1):45–55. https://doi.org/10.37591/rrjofst.v5i1.424

Shu G, Khalid N, Chen Z, Neves MA, Barrow CJ, Nakajima M (2018) Formulation and characterization of astaxanthin-enriched nanoemulsions stabilized using ginseng saponins as natural emulsifiers. Food Chem 255:67–74. https://doi.org/10.1016/j.foodchem.2018.02.062

Sigurdson GT, Tang P, Giusti MM (2017) Natural colorants: food colorants from natural sources. Annu Rev Food Sci Technol 8:261–280. https://doi.org/10.1146/annurev-food-030216-025923

Sircelj H, PETKOVSEK MM, Veberic R, Hudina M, Slatnar AJTJoA, Forestry (2018) Lipophilic antioxidants in edible weeds from agricultural areas. Turkish J Agric Forestry 42(1):1–10. https://doi.org/10.3906/tar-1707-25

Sivakumar M, Tang SY, Tan KW (2014) Cavitation technology–a greener processing technique for the generation of pharmaceutical nanoemulsions. Ultrason Sonochem 21(6):2069–2083. https://doi.org/10.1016/j.ultsonch.2014.03.025

Skrovankova S, Sumczynski D, Mlcek J, Jurikova T, Sochor J (2015) Bioactive compounds and antioxidant activity in different types of berries. Int J Mol Sci 16(10):24673–24706. https://doi.org/10.3390/ijms161024673

Solans C, Solé I (2012) Nano-emulsions: formation by low-energy methods. Curr Opin Colloid Interface Sci 17(5):246–254. https://doi.org/10.1016/j.cocis.2012.07.003

Sotomayor-Gerding D, Oomah BD, Acevedo F, Morales E, Bustamante M, Shene C, Rubilar M (2016) High carotenoid bioaccessibility through linseed oil nanoemulsions with enhanced physical and oxidative stability. Food Chem 199:463–470. https://doi.org/10.1016/j.foodchem.2015.12.004

Srivastava S, Srivastava AK (2015) Lycopene; chemistry, biosynthesis, metabolism and degradation under various abiotic parameters. J Food Sci Technol 52(1):41–53. https://doi.org/10.1007/s13197-012-0918-2

Sugiura S, Nakajima M, Seki M (2002) Prediction of droplet diameter for microchannel emulsification. Langmuir 18(10):3854–3859. https://doi.org/10.1021/la0255830

Sultan AA, Al-Mosawi MHA, Mousa HK (2016) Mechanical properties of polypropylene plastelized by chlorophyll. Kufa J Eng 7(3):55–65

Sun J, Bi C, Chan HM, Sun S, Zhang Q, Zheng Y (2013) Curcumin-loaded solid lipid nanoparticles have prolonged in vitro antitumour activity, cellular uptake and improved in vivo bioavailability. Colloids Surf B: Biointerfaces 111:367–375. https://doi.org/10.1016/j.colsurfb.2013.06.032

Tan H, Zhao L, Tian S, Wen H, Gou X, Ngai T (2017) Gelatin particle-stabilized high-internal phase emulsions for use in oral delivery systems: protection effect and in vitro digestion study. J Agric Food Chem 65(4):900–907. https://doi.org/10.1021/acs.jafc.6b04705

Tetali SD (2019) Terpenes and isoprenoids: a wealth of compounds for global use. Planta 249(1):1–8. https://doi.org/10.1007/s00425-018-3056-x

Thakur N, Walia MK, Kumar SLH (2013) Nanoemulsion in enhancement of bioavailability of poorly soluble drugs: a review. Pharmacophore 4(1):15–25. http://www.pharmacophorejournal.com

Tikekar RV, Pan Y, Nitin N (2013) Fate of curcumin encapsulated in silica nanoparticle stabilized Pickering emulsion during storage and simulated digestion. Food Res Int 51(1):370–377. https://doi.org/10.1016/j.foodres.2012.12.027

Tzankova D, Vladimirova S, Peikova L, Georgieva M (2019) Synthesis and preliminary antioxidant activity evaluation of new pyrrole based aryl hydrazones. Bulg Chem Commun 51:179–185

Vecchione R, Quagliariello V, Calabria D, Calcagno V, De Luca E, Iaffaioli RV, Netti PA (2016) Curcumin bioavailability from oil in water nano-emulsions: in vitro and in vivo study on the dimensional, compositional and interactional dependence. J Control Release 233:88–100. https://doi.org/10.1016/j.jconrel.2016.05.004

Walker R, Decker EA, McClements DJ (2015) Development of food-grade nanoemulsions and emulsions for delivery of omega-3 fatty acids: opportunities and obstacles in the food industry. Food Funct 6(1):41–54. https://doi.org/10.1039/C4FO00723A

Wang Y, Zhu J, Meng X, Liu S, Mu J, Ning C (2016) Comparison of polyphenol, anthocyanin and antioxidant capacity in four varieties of Lonicera caerulea berry extracts. Food Chem 197:522–529. https://doi.org/10.1016/j.foodchem.2015.11.006

Weigel F, Weiss J, Decker EA, McClements DJJF (2018) Lutein-enriched emulsion-based delivery systems: influence of emulsifiers and antioxidants on physical and chemical stability. Food Chem 242:395–403. https://doi.org/10.1016/j.foodchem.2017.09.060

Wu X, Beecher GR, Holden JM, Haytowitz DB, Gebhardt SE, Prior RL (2006) Concentrations of anthocyanins in common foods in the United States and estimation of normal consumption. J Agric Food Chem 54(11):4069–4075. https://doi.org/10.1021/jf0603001

Xu Y, Harvey PJ (2019) Carotenoid production by Dunaliella salina under red light. Antioxidants 8(5):123. https://doi.org/10.3390/antiox8050123

Yamazaki N, Naganuma K, Nagai M, Ma GH, Omi S (2003) Preparation of W/O (water-in-oil) emulsions using a PTFE (polytetrafluoroethylene) membrane—a new emulsification device. J Dispers Sci Technol 24(2):249–257. https://doi.org/10.1081/DIS-120019976

Yazaki K, Arimura G-i, Ohnishi T (2017) 'Hidden'terpenoids in plants: their biosynthesis, localization and ecological roles. Plant Cell Physiol 58(10):1615–1621. https://doi.org/10.1093/pcp/pcx123

Yi J, Lam TI, Yokoyama W, Cheng LW, Zhong F (2015) Beta-carotene encapsulated in food protein nanoparticles reduces peroxyl radical oxidation in Caco-2 cells. Food Hydrocoll 43:31–40. https://doi.org/10.1016/j.foodhyd.2014.04.028

Yi J, Fan Y, Zhang Y, Zhao L (2016) Characterization of catechin-α-lactalbumin conjugates and the improvement in β-carotene retention in an oil-in-water nanoemulsion. Food Chem 205:73–80. https://doi.org/10.1016/j.foodchem.2016.03.005

Yoshida K, Cheynier V, Quideau S (eds) (2016) Recent advances in polyphenol research. Wiley, Somerset

Yu H, Huang Q (2012) Improving the oral bioavailability of curcumin using novel organogel-based nanoemulsions. J Agric Food Chem 60(21):5373–5379. https://doi.org/10.1021/jf300609p

Zepka LQ, Jacob-Lopes E, Roca M (2019) Catabolism and bioactive properties of chlorophylls. Curr Opin Food Sci 26:94–100. https://doi.org/10.1016/j.cofs.2019.04.004

Zhao D-D, Jiang L-L, Li H-Y, Yan P-F, Zhang Y-L (2016) Chemical components and pharmacological activities of terpene natural products from the genus Paeonia. Molecules 21(10):1362. https://doi.org/10.3390/molecules21101362

Zhou L, Elias RJ (2013) Antioxidant and pro-oxidant activity of (−)-epigallocatechin-3-gallate in food emulsions: influence of pH and phenolic concentration. Food Chem 138(2–3):1503–1509. https://doi.org/10.1016/j.foodchem.2012.09.132

Zhu J, Barrow D (2005) Analysis of droplet size during crossflow membrane emulsification using stationary and vibrating micromachined silicon nitride membranes. J Membr Sci 261(1–2):136–144. https://doi.org/10.1016/j.memsci.2005.02.038

Zou Y, Hu W, Ma K, Tian M (2015) Physicochemical properties and antioxidant activities of melanin and fractions from Auricularia auricula fruiting bodies. Food Sci Biotechnol 24(1):15–21. https://doi.org/10.1007/s10068-015-0003-5

Chapter 12
Characterization Techniques for Emulsion-Based Antioxidant Carriers with Biomedical Applications

Gloria María Pontes-Quero, Eva Espinosa-Cano, Daniel Fernández-Villa, Miguel Huerta-Madroñal, María Rosa Aguilar, and Blanca Vázquez-Lasa

12.1 Introduction

Depending on the application, emulsions can be developed to have different physical or chemical properties. Their benefit as vehicles of bioactive components such as antioxidants is becoming popular in order to improve, for instance, the stability or bioavailability of the encapsulated compound. An emulsion is a mixture of two immiscible liquids where one of them (i.e. internal or discontinuous phase) is dispersed as droplets into the other liquid (i.e. continuous phase) (Gurpreet and Singh 2018). In the food industry or in biological applications, it is normally interesting to have an aqueous continuous phase while in the cosmetic industry an oil continuous phase is preferred. In order to achieve an efficient encapsulation of antioxidant

Gloria María Pontes-Quero, Eva Espinosa-Cano, Daniel Fernández-Villa, Miguel Huerta-Madroñal, contributed equally to the work.

G. M. Pontes-Quero (✉)
Group of Biomaterials, Department of Polymeric Nanomaterials and Biomaterials, Institute of Polymer Science and Technology, ICTP-CSIC, Madrid, Spain

Alodia Farmacéutica SL, Madrid, Spain

Networking Biomedical Research Centre in Bioengineering, Biomaterials and Nanomedicine, CIBER-BBN, Madrid, Spain

E. Espinosa-Cano · M. R. Aguilar · B. Vázquez-Lasa
Group of Biomaterials, Department of Polymeric Nanomaterials and Biomaterials, Institute of Polymer Science and Technology, ICTP-CSIC, Madrid, Spain

Networking Biomedical Research Centre in Bioengineering, Biomaterials and Nanomedicine, CIBER-BBN, Madrid, Spain

D. Fernández-Villa · M. Huerta-Madroñal
Group of Biomaterials, Department of Polymeric Nanomaterials and Biomaterials, Institute of Polymer Science and Technology, ICTP-CSIC, Madrid, Spain

M. A. Aboudzadeh (ed.), *Emulsion-based Encapsulation of Antioxidants*, Food Bioactive Ingredients, https://doi.org/10.1007/978-3-030-62052-3_12

molecules, it is important to assess their polarity and stability in either water or oil internal phases. Hydrophobic antioxidants are currently encapsulated into oil/water (O/W) type emulsions whereas hydrophilic ones in water/oil (W/O) or water/oil/water (W/O/W) emulsions depending on the optimal external phase for the application.

Antioxidants are widely used as dietary supplements and have also industrial applications such as food and cosmetic preservatives, or the prevention of metal corrosion. The medical use of antioxidants dates back to the ancient Egyptians although the term *antioxidant* appears in the medical literature in the early nineteenth century; and were first thought to prevent oxidation of unsaturated fats. However, it was in the 1990s when oxidative stress was discovered to play a key role in cell death and, as a consequence, antioxidants became important in medicine (Lobo et al. 2010). The application of antioxidant delivery systems based on emulsions is primarily found in the biomedical and food industries. The antioxidant carrier can be the emulsion itself, having the antioxidant as the dispersed phase, or solid particles fabricated by an emulsification process. In any case, to ensure the successful performance of these antioxidant carriers, knowledge of analytical methods and instruments to characterize them is needed. In this chapter, different characterization techniques have been discussed. The chapter has been organized attending to the physicochemical property of the antioxidant carrier under study. Although there are numerous emulsion-based systems that could be described, this chapter makes emphasis on emulsion-based carriers used in the biomedical field (Fig. 12.1).

Regarding the fact that emulsions are thermodynamically unstable in nature, it is necessary to study their long-term stability as one of the most important properties governing their shelf-life. Here, emphasis has also been placed on droplet size, morphology and structure as they are crucial parameters that can dramatically affect factors such as stability of antioxidant emulsion systems. The study of the antioxidant activity of the system is also essential to ensure the antioxidant performance of the system. Moreover, the rheological properties of the antioxidant emulsion depend on both the dispersed and continuous phases, and can affect the emulsion stability, making necessary its study and understanding. Finally, both sterility and cytotoxicity issues are of pivotal relevance when the carriers have been designed to be used on humans, in order to ensure the safety of the antioxidant system.

12.2 Size and Zeta Potential

One of the key parameters of any antioxidant carrier is the assessment of the particle size. Most emulsion droplets have a mean diameter of >1 μm but mini- and nanoemulsions can be also formed with droplet sizes in the 100–500 nm range. The most commonly used technique for size analysis of antioxidant emulsion droplets is light scattering (static or dynamic light scattering). This method measures droplet sizes by detecting the percentage and angle of back-scattered light when a

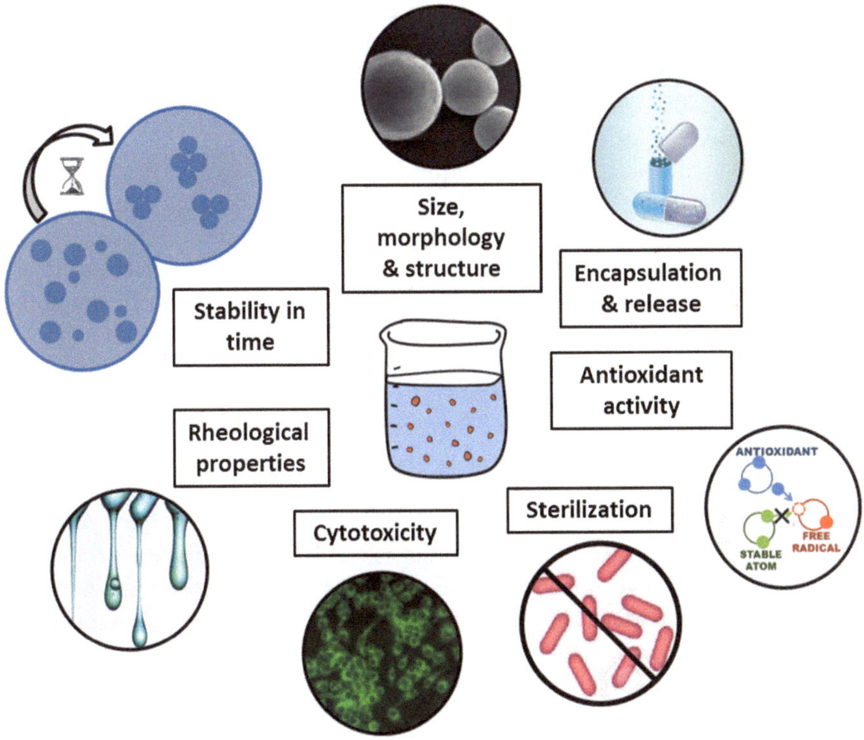

Fig. 12.1 Scheme of the main properties of emulsion-based antioxidant carriers

monochromatic beam of near infrared light is directed through the sample (Hu et al. 2017). In particular, static light scattering (SLS) measures the intensity of the scattered light as a function of the scattering angle or concentration. On the contrary, dynamic light scattering (DLS) measures the time-dependent fluctuations in the scattered light intensity, which allows the determination of the translational diffusion coefficients (i.e. Brownian motion). Hence, by assuming that smaller particles show faster Brownian motion than larger ones and create a larger rate of intensity fluctuations, it is possible to determine particle/molecular size (Stetefeld et al. 2016). SLS allows determining the particle size within the range of 100 nm to 1000 μm, whereas DLS is used to detect particles of 1 nm to 5 μm in size (McClements 2007). As a clear example of the usefulness of light scattering methods to characterize emulsions and optimize formulations, Acevedo-Fani et al. measured the size distribution as a function of the surfactant concentration by DLS. They formulated single layer and multilayer emulsions based on lactoferrin and alginate containing the antioxidant resveratrol. They used DLS to optimize the size distribution of emulsion droplets to avoid aggregation phenomena by modifying lactoferrin concentration (Acevedo-Fani et al. 2017). The main disadvantage of traditional light scattering methods is that they do not provide accurate measurements for concentrated emulsions (Hu et al. 2017), and some authors draw on diluting the emulsions before measuring (Acevedo-Fani et al. 2017; Tian et al. 2019). In order to avoid artifacts because of multiple scattering, 3D cross-correlation light scattering can be used. This technique performs two light scattering experiments at the same time, with the same scattering vector and sample volume in order to obtain common information to both (Block and Scheffold 2010).

Other less common techniques for droplets size analysis include electrical pulse counting, nanoparticle tracking analysis (NTA), single particle optical sizing (SPOS) and ultrasonic spectrophotometry. They present some advantages when compared to light scattering techniques. On the one hand, electrical pulse counting, NTA and SPOS allow measuring size of individual droplets, one at a time, leading to a more precise size distribution analysis but requiring significant dilution of samples. On the other hand, ultrasonic spectrophotometry does not require sample dilution, providing an interesting alternative when highly concentrated samples are under investigation (Hu et al. 2017). Microscopic techniques such as scanning electron microscopy (SEM) and transmission electron microscopy (TEM) are also used to measure size of antioxidant carriers. These techniques will be discussed in detail in Sect. 12.3.3.

The surface charge (zeta potential) of the carrier is also an important parameter to be studied. The microelectrophoretic technique is the most commonly used method to measure surface charge of emulsion droplets encapsulating antioxidants. In this technique, a certain voltage is applied across two oppositely charged electrodes at either end of a cell containing the emulsion. Then, charged droplets start moving towards oppositely charged electrode at a certain velocity (v). The charge sign is determined by the direction of the motion and the charge magnitude by the velocity. Light scattering measurements allow determining both, direction and velocity of particles displacement by observing the Doppler shift, which is defined as the change in the frequency of light scattered by moving particles; the faster the motion, the higher the shift in frequency in the scattered light (Fig. 12.2) (Uskoković 2012).

At low Reynolds numbers, electrophoretic mobility, U_e, is defined as the ratio between measured particle velocity (v) and the applied electric field (E) (Eq. 12.1):

$$\mathrm{Ue} = \frac{\mathrm{v}}{\mathrm{E}} \tag{12.1}$$

From that, the zeta potential is computed using Smoluchowski's mobility equation (Eq. 12.2):

$$\mathrm{Ue} = \frac{\varepsilon_{\mathrm{r}}\varepsilon_0\xi}{\eta} \tag{12.2}$$

where ε_r is the dielectric constant of the dispersion medium, ε_0 is the relative permittivity of free space and η is the dynamic viscosity of the dispersion medium (Ja'afar et al. 2015). The main disadvantage of this technique is the same as for other light

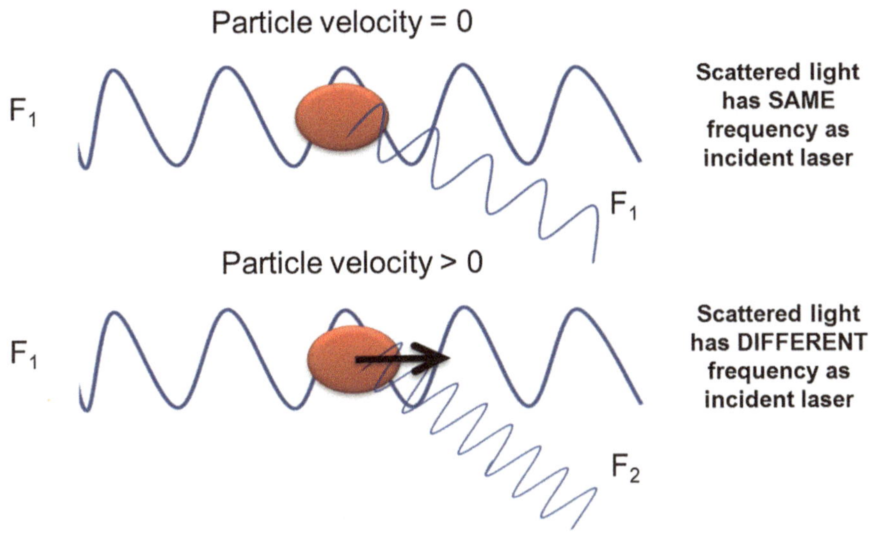

Fig. 12.2 Scheme showing that the frequency of scattered light (F_1) will be the same as the incident laser (F_1) for stationary particles, but different (F_2) if the particles are moving

scattering techniques; diluted samples are required to avoid multiple scattering effects (Hu et al. 2017).

12.3 Morphology and Structure

12.3.1 Optical Microscopy

The study of the morphology and the structure of the particles is one of the key parameters when characterizing emulsion-based antioxidant carriers. Optical microscopy is the traditional form of microscopy which employs lens and visible light to closely observe a sample. Optical microscopes, also known as light microscopes, use one or a series of lenses that are placed between the eye of the observer and the specimen to magnify the image of the probe. It is the most used technique to analyze morphology and structure of emulsions because of its simplicity, reduced cost and availability in most research facilities. However, it presents several limitations, most of them related to its relatively limited resolution when compared to other microscopy techniques and poor image contrast (Table 12.1) (Di Gianfrancesco 2017). For instance, using optical microscopy, it could be sometimes difficult to distinguish the nature of emulsion systems (whether they are made of proteins, polysaccharides…). Moreover, due to the similar refractive indexes of different components, contrast among them may be often too low. To solve these issues,

Table 12.1 Main characteristics of optical microscopy, SEM and TEM

	Optical microscope	SEM	TEM
Source	Light beam	Electron beam	Electron beam
Resolution limit	~2 μm	~2 nm	~0.2 nm
Thickness of specimen	Thin	Small enough to fit in the chamber of the scope and coated in metal atoms	Ultra-thin and coated in metals atoms
Image	2D, including intracellular visualization	3D and surface image	2D, including intracellular visualization
Sample	Alive or dead	Dead	Dead
Other requirements	Do not need vacuum and reduced cost	Need vacuum, expensive	Need vacuum, expensive

stains or dyes, as well as optical microscopes with specialized features, can be used to improve image contrast. Dyes or chemical stains are often added to emulsions or dissolved in one of the phases before imaging, allowing to highlight a certain component of an emulsion and to facilitate its observation. Nevertheless, interactions between a compositional material and the coloring agent may occur affecting the emulsion so, chemical stains are not always suitable. Moreover, phase contrast or differential interference contrast microscopy can solve this problem converting small differences in the refractive index into meaningful differences in light intensity (Murphy 2001).

12.3.2 *Confocal Laser Scanning Microscopy*

This microscopy technique was developed to improve the optical resolution and contrast of traditional wide-field fluorescence microscope. Employing a point-illumination operation mode at the confocal plane, confocal laser scanning microscopy can eliminate out-of-focus light overcoming the limitation of conventional fluorescence microscopy (Pawley 2006). In addition, it allows obtaining three-dimensional images and surface profiles of samples using its scanning feature. However, this microscopy requires longer exposure time for enough signal intensity, being therefore not desirable, for instance, for samples that are sensitive to light (Hu et al. 2017). Samples that do not naturally fluoresce require the incorporation of a fluorescent agent in order to clearly differentiate continuous and disperse phases in this type of microscopy (Fig. 12.3).

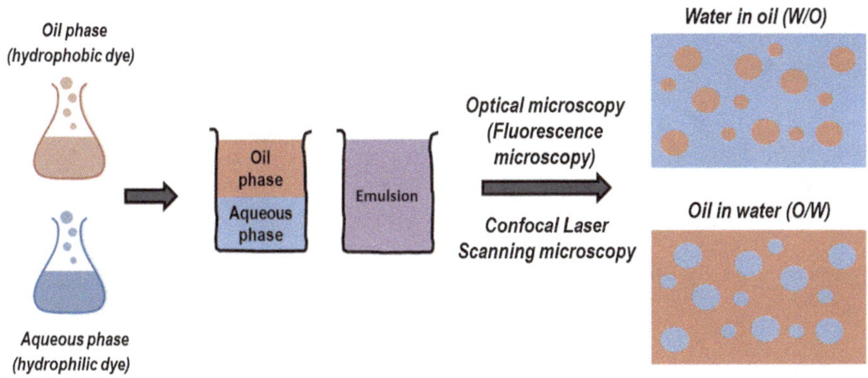

Fig. 12.3 Emulsion microstructure characterization by confocal or fluorescent microscopy using a fluorescent dye

12.3.3 Electron Microscopy

Electron microscopy is one of the most powerful imaging techniques used to investigate the morphology and microstructure of particles formed by emulsification. There are two types of electron microscopy: SEM and TEM. Both of them use a beam of electrons to obtain high-resolution images of the specimen even if the working principle is different (see Fig. 12.4). SEM is used to scan the sample with an electron beam that interacts with the atoms of the specimen, giving rise to information about its surface topography and composition. On the other hand, TEM relies on the transmission of electrons through the sample, which allows microstructure fine details and crystalline state study. Because of this, samples studied by TEM should be thin enough not to hinder electron transmission (Meroni and Raikos 2018).

In the next subsection, a more detail information about the working principle and pros/cons of SEM and TEM are given.

12.3.3.1 Scanning Electron Microscopy (SEM)

SEM is one of the most used techniques for the analysis of antioxidant delivery or carrier systems as it can produce high-resolution images (limited to ~0.5 nm) of the surface of a sample. SEM images are formed by back-scattered electrons and secondary electrons generated during the interaction of primary electrons with the sample. Back-scattered electrons are incident high-energy electrons that are scattered after interaction with the sample, while secondary electrons are produced because of excitation of specimen atoms during sample irradiation. Back-scattered electrons provide both, compositional and topographic information of the sample. On the other hand, secondary electrons are used for topographic analysis, being able to resolve surface structures down to the order of 10 nm or better (Zhou et al. 2007). Since the basis of SEM relies on surface processes rather than electron

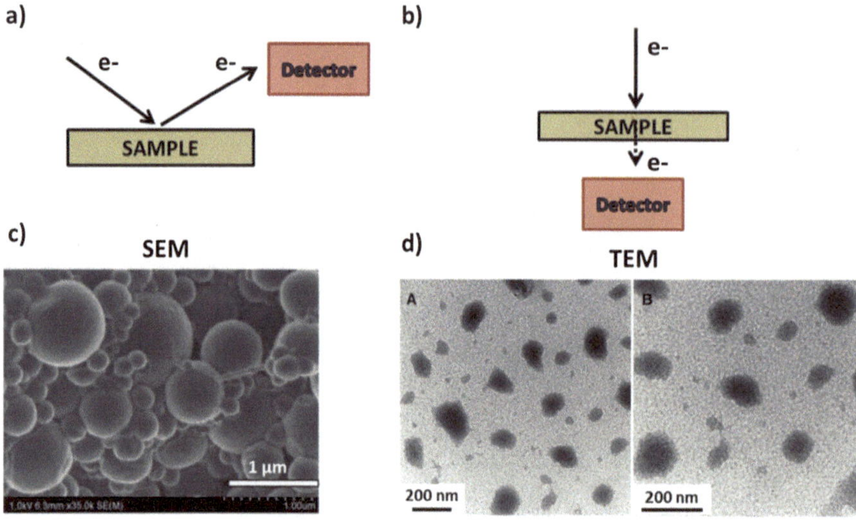

Fig. 12.4 Schematics of (**a**) SEM and (**b**) TEM electron-based image formation and (**c**) SEM micrograph of curcumin-loaded PLGA (poly(D,L-lactide-co-glycolide) nanoparticles prepared by single emulsion solvent evaporation (**d**) TEM micrograph of an O/W nanoemulsion containing curcumin. (Reprinted from the references (Umerska et al. 2018) and (Joung et al. 2016) with permission)

transmission, it offers the possibility to obtain three-dimensional images, useful feature for topographic sample analysis. Another advantage is the large field of analysis that can be examined at a glance by SEM. Nevertheless, it is an expensive technique that requires high vacuum and sample conductivity (see Table 12.1).

Xiuhua et al. used SEM to observe and compare raw silymarin and silymarin nanoparticles. While the first one displayed irregular form and different particle size, silymarin nanoparticles were nearly spherical and smaller in size (Xiuhua et al. 2016). Hee et al. studied the influence of different preparation conditions on the shape and morphology of virgin coconut oil microcapsules. They concluded that morphology was independent of the amount of antioxidant loaded since microcapsules with the highest and lowest encapsulation efficiencies were almost identical. In addition, they showed that virgin coconut oil was dispersed in the wall matrix as small droplets and had a homogeneous core distribution (Hee et al. 2017).

12.3.3.2 Transmission Electron Microscopy (TEM)

TEM is the other electron microscopy normally used to characterize delivery or carrier systems. In this case, the electron beam is achieved by focusing electrons with metal apertures and electromagnetic lens, which allow only electrons within a small range of energy to pass through. Then, the electron beam is applied to the specimen and transmitted electrons are collected on a screen to form the image (Tang and

Yang 2017). The main advantage of electrons is their small wavelength that guarantees a good resolution (0.2 nm) and, consequently, TEM is a preferable option for detailed characterization of nanoscale samples (see Table 12.1). However, TEM has several disadvantages. The specimen may be damaged by the electron beam and as small sample thickness is required to allow electron transmission, which in some cases could be expensive and time-consuming. Additionally, the atmosphere in a TEM column should maintain a high vacuum to enhance the electron transmission (Hu et al. 2017). There are several examples in literature in which TEM is used to examine and characterize the size and shape of the dispersed phase of antioxidant emulsions. Xiuhua et al., Zhang et al. and Hatanaka et al. used TEM to corroborate nanoparticle size profiles determined by DLS, as well as their spherical shape (Hatanaka et al. 2010; Xiuhua et al. 2016; Zhang et al. 2017b).

Cryo-TEM, which is a type of TEM where samples are frozen to cryogenic temperatures, can also be used to study antioxidant carriers. For example, Cheng et al. studied the structural details in the interaction of molecules and emulsions using this technique (Cheng et al. 2014). In this case, they used cryo-TEM to confirm the adsorption of antioxidant peptides at the interface of soybean O/W emulsions.

12.3.4 Atomic Force Microscopy (AFM)

AFM is a microscopy technique that allows high-resolution images of flat sample surfaces. It is based on the raster scanning of a sample previously immobilized with a force-sensing cantilever ended in a sharp tip. AFM uses the force that acts between the tip and sample as the imaging signal (Silva et al. 2012). There are two operating modes of AFM: static and dynamic. In static AFM, the force is translated into a deflection of the cantilever, while in the dynamic mode, the cantilever is deliberately vibrating. Since experimental realization of static AFM is difficult, the dynamic operating mode allows the topographic study of samples without direct contact. Dynamic AFM can be based on amplitude (AM-AFM or tapping mode) or frequency modulation (FM-AFM) in which variations on the amplitude or frequency of the oscillation tip respectively, are used as the feedback signal to image sample topography (Etzler and Drelich 2012; Giessibl 2003). The most common AFM imaging mode is the tapping mode that gives two types of images, a height or topographic image and a phase image, which identifies surface features that cannot be identified with the topographic image (Etzler and Drelich 2012).

AFM offers some advantages compared to SEM and TEM. Sample preparation is not needed, for example, in terms of sample metallization or sample thickness requirements; and the same probe can be manipulated and used several times. In addition, the dynamic mode based on amplitude modulation of AFM allows imaging under physiological conditions, since micrographs of liquid samples in air can be obtained. However, sample damage can appear when operating in the static mode with soft or sticky samples, as the tip is in direct contact with the surface of the probe (Hu et al. 2017).

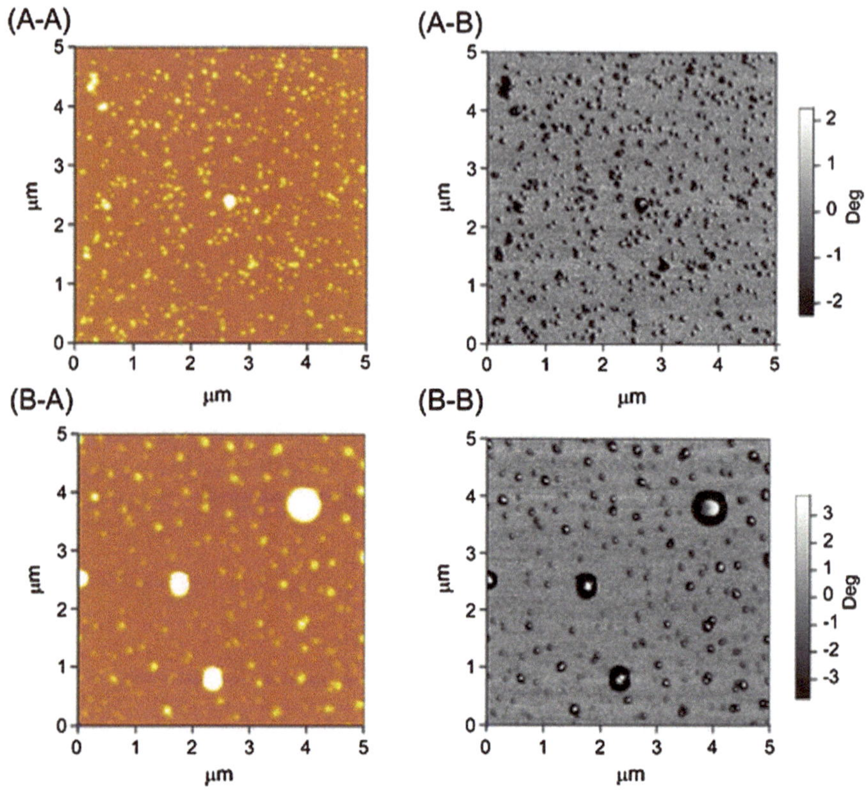

Fig. 12.5 AFM images of fresh soybean O/W emulsions (10% w/w oil) prepared with 11.1 mg/mL Tween 20 + 20 mg/mL PPH using height retrace mode (A-A) and phase retrace mode (A-B) or with Tween 20 only using height retrace mode (B-A) and phase retrace mode (B-B). (Reprinted with permission from (Cheng et al. 2014). Copyright 2014 American Chemical Society)

For instance, Cheng et al. showed the influence on size and distribution of oil droplets when using Tween 20 or combining it with potato protein hydrolysate (PPH), demonstrating that their cooperativity enabled better oil droplets distribution in aqueous phase (Fig. 12.5) (Cheng et al. 2014). Nikolic et al. used AFM to determine morphological properties, perform microstructural studies and confirm mean droplet size of curcumin-loaded nanoemulsions (Nikolic et al. 2018). Also, spherical shape and sizes of 200 nm were observed by Del Prado et al. using AFM on nanoparticles encapsulating curcumin (Del Prado et al. 2019).

12.3.5 Fourier Transform Infrared (FTIR) Spectroscopy

In order to study the chemical structure and composition of antioxidant carriers, the FTIR technique could be used. FTIR is based on an infrared radiation (IR) passing through a sample, where most of it is absorbed, while some of it is transmitted. The infrared region goes from 12800 to 10 cm^{-1} and is divided into near-IR (14000–400 cm^{-1}), mid-IR (4000–400 cm^{-1}) and far-IR (400–10 cm^{-1}). Molecules absorb these specific frequencies of light since they correspond to the frequency of vibration of the bonds in the molecules. Each sample presents characteristic absorption peaks that correspond to the frequencies of vibration between the bonds of the atoms of the material. As a result of IR absorption, changes in the dipole moment of molecule's bonds occurs, leading to vibrational transitions which give rise to characteristic absorption peaks. The intensity of IR absorption bonds give information about molecular components and structure (Pallua et al. 2011).

The size of the absorption peaks depends on the quantity of a specific bond in a sample, and since materials are composed by a unique combination of atoms, there are not two exact absorption spectrums. For this reason, this technique can be used not only to identify the chemical composition of a compound, but also to detect the amount of different species in a mixture, as well as the interaction between components through particular functional groups. In the same manner, changes in chemical structures, bond formation or cleavage, can be also monitored by FTIR, as they will give rise to different absorption peaks. Therefore, the major advantages of FTIR are the possibility to identify and distinguish compounds in a mixture and evaluate their structure, the small time required for the analysis and its high sensitivity and reproducibility (Jin et al. 2016).

There are several modes in IR spectroscopy. Traditionally, the transmission mode, in which the signal intensity is often expressed as absorbance, was used to obtain surface information, although it requires sample preparation. Solid materials normally have to be diluted with the IR-inactive KBr and pressed to form the KBr pellets, while liquid samples need to be filled into a liquid cell with a suitable path length. Nowadays, IR-measurements are mainly performed in the ATR (Attenuated Total Reflection) mode since this technique is simpler and faster to use. All types of samples are placed undiluted on the ATR crystal, avoiding spectra variations due to sample preparation. Another method is specular reflectance IR, a non-contact and non-destructive mode that works on the principle that every sample has a refractive index that varies with the frequency of light to which it is exposed (Amma et al. 2015; Beasley et al. 2014). Consequently, by examining the change in the refractive index at different frequency bands, users can make assumptions regarding the absorbency of the sample.

Because of this, FTIR has been widely used in the last years, for instance, to characterize antioxidant particles formed by emulsification processes. Su et al. showed that the primary interactions between β-lactoglobulin nanoparticles and (-)-Epigallocatechin-3-gallate occurred via hydrogen bonding and hydrophobic effects by FTIR (Su et al. 2020). Also, Shaddel et al. took advantage of FTIR to confirm the chemical cross-linking reaction in the encapsulation of anthocyanins with gelatin and gum Arabic (Shaddel et al. 2018). Finally, conformational changes of phosvitin-resveratrol complexes in microemulsions were observed by Duan et al. employing this infrared spectroscopy (Duan et al. 2016).

12.3.6 X-Ray Diffraction (XRD)

Other technique used to investigate the structural characteristics of antioxidant carriers is X-Ray Diffraction (XRD). XRD is a non-destructive analytical technique used to obtain information about the crystallographic structure, chemical composition and physical properties of materials. XRD is based on the observation of the scattered intensity of an X-ray beam hitting a sample as a function of incident and scattered angles, polarization and wavelength (Jin et al. 2016). X-rays are a form of light with wavelengths in the range of 0.01–10 nm. When X-rays scatter from a substance with structure at the nanoscale, interference can take place, giving rise to a pattern of higher and lower intensities. Thus, the result is a diffraction pattern that does not superficially resemble the underlying structure but provides information about the internal structure on length scales from 0.1 to 100 nm.

XRD is mostly used for the identification of crystalline compounds by their diffraction pattern, but in the context of antioxidant carriers formed by emulsification processes, it is a powerful technique that can be used to report the encapsulation of a molecule comparing the diffraction pattern of loaded and unloaded systems, perform stability studies or analyze crystallographic structures of different formulations. Several examples can be found in the literature. Behbahani et al. confirmed curcumin encapsulation in solid lipid nanoparticles produced by micro-emulsion and ultrasonication (Behbahani et al. 2017); Xiuhua et al. evaluated the occurrence of structural changes of silymarin nanoparticles upon modification of several parameters of the nanoemulsion experiment (Xiuhua et al. 2016); and Shaddel et al. analyzed the stability of microencapsulated anthocyanins performing crystallinity studies by XRD (Shaddel et al. 2018). In this last study, gelatin-gum arabic microcapsules showed a semi-crystalline structure in contrast to the core material, which was amorphous as it did not show any peak between 2Theta of 10° and 80°. In fact, microcapsules exhibited a reduced peak in the same range as gelatin and gum arabic confirming its semi-crystalline structure and higher stability than non-encapsulated anthocyanins.

12.4 Emulsion Stability

As previously mentioned, emulsions are conformed by two immiscible liquids (continuous and dispersed phases). The polarity difference of internal and external phase liquids makes emulsions thermodynamically unstable, meaning that if the two phases are allowed to stand for long enough time, they will eventually separate (Weiss 2002). Depending on the application, instability of emulsions may lead to different undesirable effects. For instance, unstable emulsions used as drug delivery systems may lead to severe adverse side effects (Ujhelyi et al. 2018) whereas in the food industry instability can decrease the product quality and shorten shelf life (Akbari 2018). Therefore, it is crucial to understand the mechanism that causes

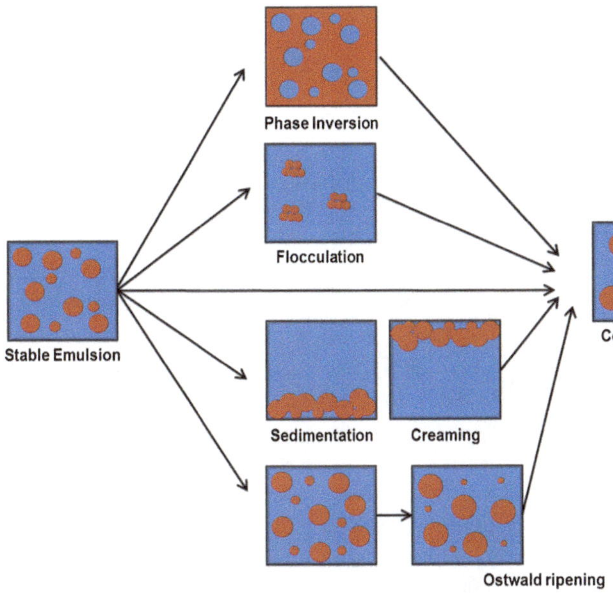

Fig. 12.6 Schematic representation of instability mechanisms in the emulsion system

emulsion instability and carefully evaluate the stability of such systems. The main mechanisms that cause emulsion instability are gravitational separation (creaming/ sedimentation), flocculation and coalescence (aggregation phenomena), Ostwald ripening (change of an inhomogeneous structure over time), and phase inversion and they are summarized in Fig. 12.6 (Hu et al. 2017).

Emulsions stability is influenced by droplet concentration, size, morphology, surface charge, phase-phase interactions, and rheological behavior (Hu et al. 2017). In addition, emulsion stability can be affected by a number of external factors like temperature, pressure or pH (Weiss 2002). This section focuses on the most common used methods to assess emulsion stability.

12.4.1 Visual Observation

Visual observation of the emulsion is the simplest, quickest and cheapest method to assess gravitational separation (i.e. sedimentation or creaming). Sedimentation occurs if the internal phase presents higher density than the continuous phase causing droplets to accumulate at the bottom, whereas creaming occurs when droplets move upwards because they present lower density than the continuous phase. For example, Carpenter et al. clearly observed creaming effects after 21 days storage of curcumin-loaded emulsions (Carpenter et al. 2019). However, visual observation does not allow studying other instability phenomena as well as droplets smaller than

100 μm. Thus, the study of emulsion stability usually requires more expensive analytical instruments.

12.4.2 Microscopy

Droplets and instability phenomena that cannot be observed at naked eye (i.e. diameter below 100 um) are observed by microscopic techniques. For example, flocculation can be easily identified when droplets start to get close to each other without merging into unique larger ones. By contrast, Ostwald ripening and coalescence are observed as droplets start fusing together leading to larger droplets or droplets with heterogeneous size distributions. Normally, to test the antioxidant emulsion stability, microscopy is combined with other techniques. The study of the evolution of particle size and size distribution in time using microscopy in combination with particle-sizing techniques like DLS will give more precise information about instability mechanisms. For instance, optical microscope images at 0 and 21 days were taken to confirm stability of rosemary extract-loaded emulsions regarding Ostwald ripening and coalescence that was already observed by DLS size analysis (Erdmann et al. 2015).

12.4.3 Particle Size, Polydispersity and Concentration Analysis in Time

Gravitational separation (i.e. sedimentation and creaming) can be studied by determining time evolution of the size and concentration of droplets at different regions of the sample under specific external conditions (temperature, time, pH) (Hu et al. 2017). Flocculation or coalescence is characterized by an increase in mean size and polydispersity of sample size distribution with time. Several examples of stability evaluation by measuring size at different time points are found in the antioxidant emulsion literature. For example, Chen et al. described curcumin-loaded O/W emulsion with improved stability in time when compared to the free drug. Significant changes in size occurred at longer time points when curcumin was loaded on emulsion droplets as compared to free curcumin (Chen et al. 2016).

12.4.4 Surface Charge or Zeta Potential Analysis

Surface charge is another key factor affecting stability of emulsions. Emulsion droplets with surface charge or zeta potential of equal sign experience electrostatic repulsive forces (Hu et al. 2017). A higher surface charge leads to stronger electrostatic repulsion and, hence, reduces the probability of aggregation. Flocculation is more likely to happen in the presence of weak attractive forces whereas strong attractive forces can lead to fusion of droplets or coalescence. A zeta potential greater than |30 mV| is considered sufficient for stabilization (Gurpreet and Singh 2018).

pH and ionic strength of the emulsion are key variables affecting ionization and surface charge of droplets, especially if ionizable groups are exposed on their surface (Emerenciano et al. 2019). Consequently, the influence of these variables in stability is commonly reported (Uskoković 2012). In some cases, electrostatic stabilization is not sufficient for droplet stabilization and they need to be stabilized via steric or electrosteric mechanisms, usually by means of surfactants. For instance, Peng et al. demonstrated that pH had a significant effect on mean particle size of capsaicin-loaded O/W emulsions (Peng et al. 2018). Moreover, they demonstrated that the effect on size was dependent on the ionic nature of the employed surfactant. Using non-ionic surfactants, they observed an increase in size at lower pH while a decrease in size was observed when using lecithin, a surfactant containing ionizable anionic phospholipid groups at acidic pH. As previously said, antioxidant emulsions are primarily used in the food industry or biomedical applications and, therefore, surface charge and stability are studied at pH values close to physiological pH (Chen et al. 2016; Emerenciano et al. 2019).

Normally, authors determine antioxidant emulsion stability by combining visual observation, size analysis techniques and/or zeta potential determination at different time points and at specific conditions depending on the application. For instance, Acevedo-Fani et al. monitored the mean droplet size and zeta-potential of a resveratrol-containing emulsion at room temperature and applied light to simulate common food storage conditions (Acevedo-Fani et al. 2017). Similarly, when antioxidant emulsions are used for medical applications size and zeta-potential changes are evaluated in time at physiological conditions (Rinaldi et al. 2017).

12.4.5 Emulsion Stability Index – Volumetric Method and BS (Back-Scattering) Method

Emulsion stability index (ESI) can reflect the ability to resist instabilities of O/W or W/O/W emulsions (Choi et al. 2014; Tian et al. 2019). It gives an estimation of emulsion stability after a determined period of time (Tian et al. 2019). It can be either determined by the firstly described volumetric method (Eq. 12.3) which is based on visual observation or by, the formerly described, back-scattering (BS) method (Eq. 12.4) which is based on the absorbance and back-scattering of light, as it passes through the colloidal suspension. Good correlation has been reported between both methods (Choi et al. 2014):

$$\mathrm{ESI}(\%) = \left(1 - \frac{V_w}{V_e}\right) \times 100 \tag{12.3}$$

where V_e is the volume of the o/w emulsion and V_w is the volume of the separated bottom layer after the desired storage period.

$$\mathrm{ESI}(\mathrm{min}) = \frac{\mathrm{A}_0}{\Delta A}\mathrm{t} \tag{12.4}$$

where A_0 is the absorbance of the emulsion right after homogenization, ΔA is the change in absorbance after time t (i.e. $(A_t - A_0)$) and t is the time interval.

The main advantage of determining ESI by the BS method is that it is an optical non-destructive method and, therefore, no sample dilution is needed; and it provides useful information about aggregation phenomena (flocculation and coalescence) during destabilization methods (Choi et al. 2014). Tian et al. recently used BS method to determine ESI of a tea polyphenol-containing O/W emulsion on a 60 minutes time interval (Tian et al. 2019). They demonstrated that emulsion stability was highly dependent on droplet concentration.

12.4.6 Thermal Stability

Temperature has also an influence on antioxidant emulsion stability and antioxidant encapsulation efficiency, degradation or loss of function. That is why stability upon temperature change is also studied by the already described visual, microscopic or hydrodynamic properties monitoring methods. As an example, Sunee et al. visually demonstrated stability of xanthone-loaded microemulsion at room temperature, 45 °C and after 2 months of temperature cycles of 4 °C (48 hours) to 45 °C (48 hours). However, they observed precipitation when storing the emulsion at 4 °C attributed to the lower solubility of xanthone at low temperatures (Sunee et al. 2017).

A common technique also used to study antioxidant emulsion thermal stability is Differential Scanning Calorimetry (DSC) (Silva et al. 2012). DSC is a thermoanalytical technique that allows measuring the thermodynamic parameters associated with thermally induced phase transitions of a sample when compared to a reference material, which does not undergo a phase transition within the temperature range under investigation. Initially, an equal amount of heat is linearly applied to the sample cell and the reference cell, maintaining at zero their differential temperature. Whenever the sample undergoes a temperature induced phase transition, a portion of the heat applied is absorbed or released, leading to a temperature differential between sample and reference material. This difference is detected by the instrumental control system that supplies higher or lower amount of heat to the sample cell to maintain the temperature equal to that of the reference cell (McElhaney 1986). Several examples of the use of DSC to study the thermal stability of emulsion-based antioxidant carriers can be found in the literature. Shaddel et al. proved high thermostability and long residual action of anthocyanins after emulsion-based encapsulation (Shaddel et al. 2018), while a decrease in thermostability of resveratrol-phosvitin microparticles was assessed by Duan et al. comparing DSC profiles of different formulations (Duan et al. 2016).

12.4.7 Accelerated Stability Tests

Sometimes, stability of emulsions is tested in accelerated conditions in order to shorten the long term storage studies. The most commonly used methods in the field of antioxidant emulsions are centrifugation and dilution stability tests. Centrifugation stability tests, also called phase separation methods to assess stability, consist of centrifugation of the emulsion at a specific frequency of rotation for a specific time (e.g. 10,000 rpm for 10 minutes) and, then, the investigation of the phase (Rashid et al. 2018). Dilution stability tests are usually needed to ensure that the stability of the emulsion is not compromised by a phase inversion when dilution of the system occurs, for instance, when in contact with blood in case of drug delivery systems (Emerenciano et al. 2019).

12.5 Rheological Properties

Rheological analysis, which is the science of material deformation and flow of antioxidant carrier emulsions, is essential for the design of these systems, as it plays a major role in product stability and performance (Derkach 2009). This characterization is performed with different types of rheometers such as shear rheometers, which control the applied shear stress or strain; or extensional rheometers, that control extensional stress or strain. They provide information about flow behavior, viscosity, yield stress or elastic and loss moduli. The composition and characteristics of both dispersed and continuous phases of an emulsion, play a key role in the rheological behavior of the emulsion (Fig. 12.7) (Kim and Mason 2017). Viscosity and other flow parameters of emulsions can relate to other properties such as the dispersed phase volume fraction, the nature of the continuous phase, the droplet size and morphology, the presence of emulsifiers/surfactant or the colloidal interactions (Tatar et al. 2017).

The volume fraction of the antioxidant droplets in the emulsion is one of the most critical aspects determining the emulsion viscosity. As the antioxidant volume fraction increases, the viscosity increases too, since the packaging of more molecules makes flow more difficult (Tatar et al. 2017). This effect has been observed in multiple antioxidant-containing emulsions (Gomes et al. 2016; Gouda et al. 2017; Lonni et al. 2016; Sellimi et al. 2015). Particle size should not have any effect on the viscosity if there are not attractive or repulsive forces between the particles

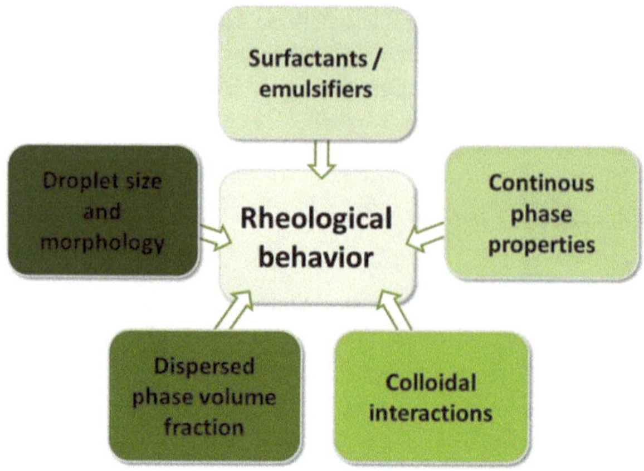

Fig. 12.7 Possible factors influencing the rheological behavior of antioxidant emulsions

(Tatar et al. 2017). Nonetheless, when particle-particle interactions occur, viscosity may be affected. For instance, various authors explain a reduction or increase in the viscosity of an antioxidant emulsion because of the increased or reduction of droplet sizes, respectively (Chang et al. 2017; Goncalez et al. 2015).

Loading of an antioxidant into the dispersed phase of an emulsion may also affect the rheological behavior of the system. For example, Goncalez et al. observed that the incorporation of the lipophilic antioxidant molecule kojic dipalmitate in a multiple emulsion altered the flow behavior from shear-thinning to low-viscosity Newtonian, due to the droplet size reduction occurring with the antioxidant loading, which thinned the flow (Goncalez et al. 2015). However, other studies have shown that the antioxidant did not exert any change in the flow behavior (Cefali et al. 2015). The amount and nature of surfactants that can be added to stabilize the emulsion system can also alter the emulsion rheology as shown by Peng et al. (2018), who observed that an increase in the water content of the aqueous phase produced a decreased of the emulsion viscosity.

12.6 Antioxidant Emulsion Encapsulation and Release

After performing the physicochemical characterization of the emulsion system in terms of size, morphology, structure, rheology and surface charge, it is of crucial importance to confirm the successful antioxidant encapsulation and release of the antioxidant. To evaluate the efficiency of an encapsulating emulsion system, parameters like encapsulation efficiency, release profile and storage stability need to be determined. According to Sunee et al., the antioxidant should be loaded before complete formulation is reached, in order to ensure the highest encapsulation efficacy (Sunee et al. 2017). The amount of antioxidant incorporated within the system is quantified relative to the original input mass or volume (Eqs. 12.5 and 12.6) (Hu et al. 2017).

$$\text{Encapsulation Efficiency}(\%) = \frac{\text{antioxidant incorporated}}{\text{initial amount of antioxidant}} \times 100 \quad (12.5)$$

$$\text{Loading capacity}(\%) = \frac{\text{mass of antioxidant in the emulsion}}{\text{total mass of emulsion}} \times 100 \quad (12.6)$$

UV/visible spectrophotometry and chromatography are the most common techniques used to determine the antioxidant content in the system and the release kinetics of this compound (i.e. amount of antioxidant within the system at different time points).

12.6.1 UV/Visible Spectrophotometry

Antioxidant compounds are normally polyphenols that absorb ultraviolet/visible (UV/VIS) light and can be easily detected by spectroscopic techniques in the range of 200–800 nm. Therefore, UV/VIS spectrophotometry is a non-destructive and rapid technique that can be used for qualitative analysis of compounds that can absorb energy from electromagnetic waves of 200–800 nm, and excite electrons on their surroundings from ground state to excited state. Thereafter, these electrons release the energy to return to their ground state. Depending on their chemical structure, antioxidants would absorb and emit at different specific light wavelengths and, thus, could be differentiated according to their characteristic absorption spectrum. When using this technique, it is important to take caution when choosing the solvent because the absorption spectrum may change depending on it. The UV-VIS absorption is directly related to concentration by the Beer-Lambert law (Eq. 12.7) (Mantele and Deniz 2017).

$$A = \varepsilon \times l \times c \quad (12.7)$$

where ε is the molar absorbance coefficient of the antioxidant (L/mol cm), l is the path length (cm), and c is the concentration of the antioxidant compound (molarity). Therefore, an increase in absorbance implies an increase in the concentration (Fig. 12.8).

Tang et al. evaluated the release kinetics of resveratrol (λ_{abs} = 305 nm) from an O/W emulsion system at different pH values (Tang et al. 2019). They observed non-significant differences on the amount of released drug at different pH values but a significant improvement was detected when comparing with resveratrol on its free form. In another work, the successful encapsulation of lutein on O/W emulsion systems was assessed by UV spectrophotometry (λ_{abs} = 450 nm), obtaining an encapsulation efficacy ~99.75% and a loading capacity ~48.78%.

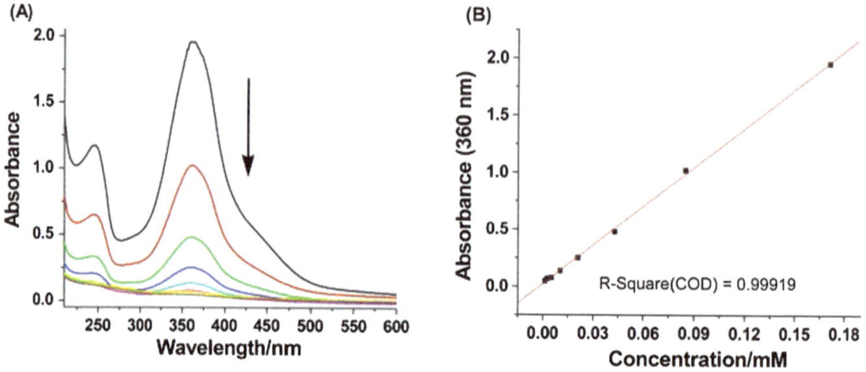

Fig. 12.8 (**a**) UV-VIS absorbance of CDAS (a derivative of β-cyclodextrin) in H_2O at 298 K under different concentrations, and (**b**) plot of the UV-vis absorbance against the concentration. The absorbance of CDAS was measured at $\lambda_{max} = 360$ nm. (Reproduced with permission from the reference (Zhu et al. 2012) (Attribution license: https://creativecommons.org/licenses/by/4.0/))

12.6.2 Chromatography

Chromatography is the most commonly used technique to determine the concentration of an antioxidant in an emulsion and its release over time. Basically, chromatographic techniques separate compounds in a mixture (mobile phase) by differential retention times on a stationary phase. Compounds can be separated attending to different forces: partition, absorption, ion exchange, size exclusion, or affinity. The mobile phase could be a liquid (liquid chromatography, LC), a gas (gas chromatography, GC) or a supercritical fluid (Coskun 2016).

Among all the available chromatographic instruments, high performance LC (HPLC) is the most commonly used for the quantitative and qualitative analysis of encapsulated compounds. The mobile phase is a liquid, the stationary phase, a solid, and the compounds are separated according to their polarity reaching the detector at different elution times (Hu et al. 2017). Detectors could be (a) universal, which measure any global change in the emerging liquid mobile phase (e.g. infrared, IR, detector), or (b) selective that measure a specific property of the eluting compound (e.g. UV-VIS or fluorescence detector) (Swartz 2010). In UV-VIS detectors, which are the most commonly used, detection of compounds as they come out of the stationary phase is usually done at a specific wavelength (Fig. 12.9). However, several antioxidant compounds can absorb at the same wavelength and separation should be performed in order to quantify different compounds. Therefore, the selection of the stationary/mobile phase is essential for the accuracy of this method. Flow rate is also a crucial variable to achieve proper peak separation (Coskun 2016). At lower flow rates, better peak separation and higher accuracy is achieved. HPLC quantification of antioxidant release or encapsulation efficacy is widely used in literature (Meroni and Raikos 2018). For instance, Sunee et al. introduced xanthone into a microemulsion and successfully evaluated its in vitro release by HPLC

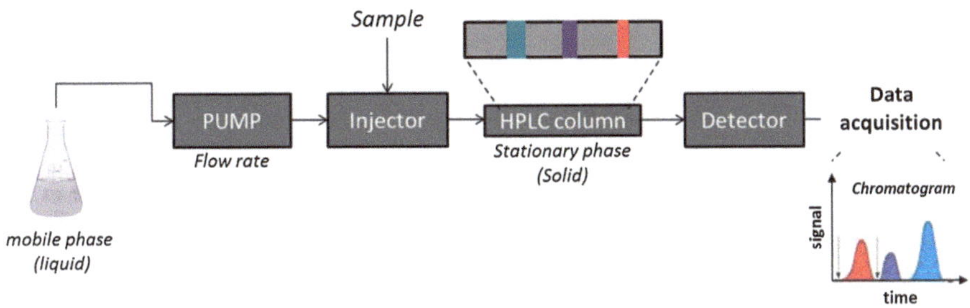

Fig. 12.9 Scheme of the HPLC technique

measurements with a mobile phase of 100% acetonitrile, flow rate of 1 mL/min at room temperature. The retention time was monitored at λ = 320 nm (Sunee et al. 2017).

12.6.3 DSC

DSC can also be used to test the encapsulation efficiency. For instance, Rutz et al. used this technique to confirm the high encapsulation efficiency of palm oil and β-carotene into chitosan-based microparticles. Moreover, with thermographs, they suggested a core material protection because of the absence of endothermic events of the encapsulated compounds and predominance of the thermal profiles of the wall materials (Rutz et al. 2016). Evaluation of structural changes at the crystal level of an emulsion is another possible application of DSC. Behbahani et al. demonstrated curcumin solubilization in stearic acid and tripalmitin nanoparticles as the endothermic peak showed by crystalline curcumin disappeared in the nanoparticulated system (Behbahani et al. 2017). Similarly, Xiuhua et al. used DSC to confirm the decrease of crystallinity in silymarin nanoparticles compared to raw silymarin (Xiuhua et al. 2016).

12.7 Antioxidant Properties

Oxidative stress can be defined as an imbalance between the production of reactive oxygen and nitrogen species (ROS/RNS) and their elimination by the antioxidant defenses of a biological system. This imbalance can lead to irreversible molecular and cellular damage by the attack of free radicals to the cells (Tan et al. 2018). In humans, oxidative stress is linked to many different diseases such as cancer, obesity, cardiovascular, and degenerative diseases (de Araújo et al. 2016; Matschke et al. 2019). In order to counteract this imbalance, the organism has several endogenous and exogenous antioxidant molecules of enzymatic (superoxide dismutase, catalase

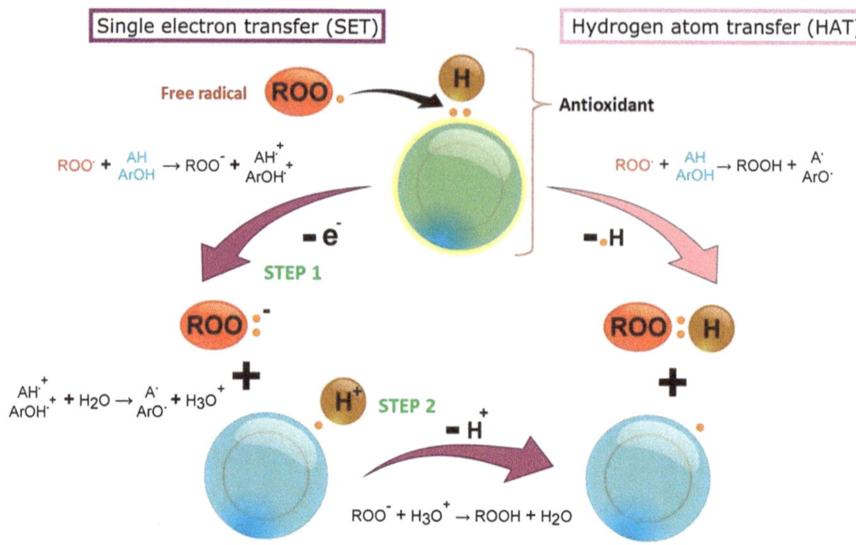

Fig. 12.10 Schematic representation of single electron or hydrogen transfer methods

or glutathione peroxide) or non-enzymatic (vitamin A, vitamin C, vitamin E or β-carotene) nature (Liguori et al. 2018).

Various methods can be used to study and compare the antioxidant activity of emulsion-based antioxidant carriers, having each one a specific target. These methods can be classified into two main groups: methods based on a single electron transfer (SET) reaction, where the oxidant is reduced (ABTS and FRAP assays); and hydrogen atom transfer (HAT) methods, where the antioxidant molecule and the substrate compete for free radicals (DPPH, ORAC) (see Fig. 12.10) (Alam et al. 2013; Shivakumar and Yogendra Kumar 2018). Furthermore, in order to get a closer and more realistic insight into the performance of emulsion systems, cellular assays can also be used. For instance, the cellular production of ROS/RNS can be measured using different fluorescent dye-based assays and indirect methods can be used to study damage to DNA, lipids and proteins (Zhang et al. 2017a). In this context, the principal assays used for measuring of the antioxidant capacity of emulsions are described below.

12.7.1 DPPH

The DPPH test was first introduced by Blois (1958) and developed by Brand-Williams et al. (1995) and is one of the simplest and quickest methods for the analysis of the radical scavenging activity of a system. The assay measures the ability of an antioxidant system (RH) to reduce the stable free radical

1,1-diphenyl-2-picrylhydrazyl (DPPH*) through a hydrogen transfer, as represented in the following reaction:

$$\text{DPPH}* + \text{RH} \rightarrow \text{DPPH}-\text{H} + \text{R}*$$

DPPH*, normally dissolved in methanol, ethanol or isopropyl alcohol, is deep purple in color and turns into pale yellow when radical scavenging occurs. This color change is measured by spectrophotometry at 515–517 nm (λ_{max} of DPPH*) where a lower absorbance reading of the sample shows stronger radical scavenging activity (RSA). Finally, the RSA is calculated as stated in Eq. 12.8,

$$\text{RSA}(\%) = \frac{\left(\text{A}_{\text{DPPH}} - \text{A}_{\text{SAMPLE}}\right)}{\text{A}_{\text{DPPH}}} \cdot 100 \tag{12.8}$$

where A_{DPPH} and A_{SAMPLE} correspond to the absorbance of DPPH with and without the antioxidant sample, respectively.

12.7.2 ABTS

In the ABTS (2,2′-azinobis-(3-ethylbenzothiazoline-6-sulfonate)) or trolox equivalent antioxidant capacity (TEAC) decolorization assay, the relative scavenging ability of an antioxidant system is measured and compared with trolox (an analogue of vitamin E) (Re et al. 1999). In the assay, the radical cation ABTS* is formed by reacting the ABTS salt with a strong oxidizing compound (e.g. sodium persulfate, potassium permanganate, potassium persulfate). This radical, which is blue-green in color (with absorption maxima at 415, 645, 734 and 815 nm) is converted back to its colorless form when is reduced by a hydrogen-donating antioxidant. The ABTS method is a simple and rapid test, applicable for both hydrophilic and lipophilic antioxidant systems and has good repeatability. As a consequence, it is widely reported for emulsion-based antioxidant systems.

12.7.3 FRAP

FRAP (Ferric Reducing Antioxidant Power) is another method to determine the antioxidant capacity of emulsion systems; and was first used to evaluate the antioxidant properties of plasma (Benzie and Strain 1996). The method is based on the reduction at low pH of Fe^{+3} tripyridyltriazine to Fe^{+2} tripyridyltriazine, a blue complex that gives an absorbance peak at 593 nm, which is proportional to the total antioxidant activity of the system. FRAP has very similar principles to the ABTS assay except that the former is developed under acidic conditions and the latter at neutral pH.

12.7.4 ORAC

The Oxygen Radical Absorbance Capacity (ORAC) method is also a spectrophotometric test to quantify the ability of various compounds to quench free radicals. The assay measures the oxidative degradation of a fluorescence probe such as fluorescein in the presence of free radicals generated by azo-initiators like AAPH (2,2′-azobis(2-amidino-propane)dihydrochloride) (Re et al. 1999). The higher the degree of inhibition of fluorescence loss, the higher is the antioxidant activity. The degree of antioxidant-mediated protection is quantified using trolox as a standard.

In Table 12.2, a comparison between the characteristics of the different methods is presented. Usually, characterizing the antioxidant activity of emulsions is not limited to only one method, but is a combination of them, since each one acts by different mechanisms. The DPPH is the most used assay to have a first approach of the antioxidant potential of emulsion-based systems, as it is the simplest and fastest, followed by ABTS and ORAC, and finally FRAP. The value obtained in each assay is different. For example, in the research of Gallego et al. the order, from higher to lower, of antioxidant activity of O/W emulsions containing ethanolic extracts of *Caesalpinia decapetala* was ORAC > TEAC > DPPH > FRAP, values compared to Trolox, since each test measures the antioxidant activity in a different way (Gallego et al. 2017).

Many recent studies have assessed the antioxidant capacity of emulsion-based carriers. For example, Hee et al. concluded by DPPH and ABTS that the antioxidant capacity of virgin coconut oil was not affected by its microencapsulation with supercritical carbon dioxide spray-drying. (Hee et al. 2017). Similar results were obtained in the research of Tirado et al. where the antioxidant activity of astaxanthin was not affected when it was encapsulated in ethyl cellulose by supercritical extraction (Tirado et al. 2019). On the other hand, Giménez-Rota et al. obtained different antioxidant properties of β-carotene when it was encapsulated into *poly-lactic-co-glycolic acid* (PLGA) or poly-lactic acid (PLA) microcarriers, where higher values of antioxidant activity by DPPH in the second type of encapsulation was achieved (Gimenez-Rota et al. 2019).

Table 12.2 Comparison of different spectrophotometric antioxidant assays for the characterization of emulsion-based antioxidant carriers

Assay	Reaction mechanism	Type of assay	Chromophores	λ (nm)	pH
DPPH	HAT	Absorbance	DPPH* radical	515	7–7.4
ABTS/TEAC	SET	Absorbance	ABTS* radical	415, 645, 734, 815	7.4 (in PBS)
FRAP	SET	Absorbance	Ferrous tripyridyltriazine	593	3.6
ORAC	HAT	Fluorescence	Fluorescein	484–520	7.4

12.7.5 EPR or ESR

Another less used analytical technique to test the antioxidant activity of emulsion systems is Electron Paramagnetic Resonance (EPR) or Electron Spin Resonance (ESR), which is especially suitable since free radicals are species having unpaired electrons. It is a form of magnetic resonance spectroscopy such as nuclear magnetic resonance but, while in the former atomic nuclei interact with electromagnetic resonance under an external magnetic field, in the case of EPR unpaired electrons are the ones that interact with the radiation. Briefly, electrons have a *spin* which gives them a magnetic moment. Under an external magnetic field, electrons can orient parallel or antiparallel to the direction of the field, creating two energy levels for unpaired electrons. This technique detects and measures the transition between these energy levels. The antioxidant status of an emulsion system can be monitored by this technique as the elimination of stable free radicals such as DPPH or 4-Hydroxy-2,2,6,6-tetramethylpiperidine 1-oxyl (TEMPOL), which have a well-defined EPR spectrum. For example, Aboudzadeh et al. proved that the radical scavenging activity of O/W microemulsions encapsulating α-tocopherol increased in time, demonstrating at the same time a slow release of α-tocopherol from the droplets (Aboudzadeh et al. 2018). In a similiar O/W microemulsion of carotenoids, Chaari et al. demonstrated that the antiradical properties of carotenoids where not affected by their micro- and nano-encapsulation (Chaari et al. 2018).

12.7.6 Cellular Assays

Cellular antioxidant assays are used for biomedical applications where emulsion systems are designed to be in contact with cells and tissues and are normally complemented with other physicochemical techniques. Cellular cultures are increasingly being used since the previously described methods do not reflect the cellular environment. Cells are continuously exposed to different oxidizing agents and stimuli, making possible to study the antioxidant capacity of a system using different oxidative markers (Marrocco et al. 2017; Zhang et al. 2017a). Antioxidant carriers can be either added to the cell culture simultaneously with the stressor or incubated with the antioxidant system to be incorporated into the cells. The direct measurement of cellular ROS/RNS production via binding to a fluorescent dye is one of the most used cellular antioxidant assays. One of the most common dyes is 2′, 7′-dichlorodihydrofluorescin diacetate (DCFH-DA), a fluorescent dye which is intracellularly deacetylated by esterases to non-fluorescent 2′, 7′-Dichlorodihydrofluorescin (DCFH) and oxidized by free radicals to fluorescent 2′, 7′-dichlorodihydrofluorescein (DCF). This dye was used, for example, in the research of Gu et al. to test the effect of encapsulating β-carotene with catechin–egg white protein conjugates on the antioxidant activity of the system (Gu et al. 2018). Indirect methods can also be used such as evaluating DNA/RNA damage, lipid oxidation, or protein oxidation/

nitration caused by oxidative stress (Choudhry et al. 2016; Duan et al. 2016). Measurement of specific analytes or endogenous antioxidant molecules such as glutathione can also test the antioxidant properties of emulsions (Murphy and Lampe 2018).

12.8 Sterilization

Sterilization requirements are of pivotal importance when manufacturing products that are designed to be in contact with the human body. There is not an always-functioning sterilization method so, it is necessary to select the most suitable one in a case-by-case manner, taking into consideration both cargo and emulsion compositions.

At the manufacturing stage, there are two well-differentiated operating modes. The first one consists of working on aseptic conditions and it is usually employed with injectable emulsions. This is especially useful when some characteristics of the final product are altered when sterilizing by conventional methods. To solve this, emulsion-based antioxidant carriers are prepared in clean rooms (Class-100 environments), the equipment is previously autoclaved, and the initial substrates are sterilized by different techniques such as filtration or heat sterilization (Toh and Chiu 2013). However, this is a highly regulated process that must meet the good manufacturing practices (GMP) imposed by the U.S. Food and Drug Administration (FDA) or the European Medicines Agency (EMA) (Food and Drug Administration. Guidance for Industry: Sterile Drug Products Produced by Aseptic Processing - Current Good Manufacturing Practice 2004), and therefore, it is usually one of the last options to be considered (Hippalgaonkar et al. 2010). On the contrary, instead of working in aseptic conditions, there are diverse techniques that can be used when the production stage has been completed, named *terminal sterilization techniques*, which are going to be reviewed hereafter.

12.8.1 *Filtration*

Filtration is one of the simplest methods for sterilizing aqueous products. When the emulsion is filtered, it goes through a membrane with pores of 0.22 μm in diameter that blocks particles over that size, including bacteria, yeasts and even spores ("World Health Organization. Methods of sterilization," 2019). However, this diameter is not small enough to avoid virus penetration into the sterile filtrate. At the laboratory scale, filtration is a very useful method because of its quickness and the availability of different commercial sterile filters. Depending on the hydrophilicity of the emulsion system, different filters can be employed such as polyvinylidene fluoride (hydrophilic), polycarbonate (hydrophilic) or cellulose acetate (hydrophobic). Nevertheless, at the industrial scale, filtration has several limitations including the necessity of aseptic filters and the high pressures to work under, which results in huge costs (Toh and Chiu 2013). Regarding emulsion-based antioxidant carriers, the narrow pores of the membranes limit the applicability of this technique to sterilize emulsions with droplets over 200 nm. However, for nanoemulsions that can be filtered, it is a great option when the carrier or the encapsulated antioxidant is thermolabile (Lidgate et al. 1992).

12.8.2 Terminal Heat Sterilization

Terminal heat sterilization is one of the most commonly employed techniques for sterilizing samples both at the preclinical and manufacturing stages due to its effectivity and convenience. In fact, the World Health Organization (WHO) recommends its use whenever possible because of its reliability ("World Health Organization. Methods of sterilization," 2019). There are two different heat-based sterilizations depending on how the heat transference is carried out.

12.8.2.1 Saturated Steam Sterilization (Also Known as Autoclaving)

In this method, the samples are exposed to saturated steam under pressure in order to denature irreversibly microbial proteins. The process requires the control of temperature, pressure and time, being recommended processes of 15 minutes at 121–124 °C and 200 kPa. Depending on the products, alternative conditions can be proposed. For instance, when sterilizing products containing thermolabile molecules, temperatures can be below 121 °C, although the combination of time, pressure and temperature must be previously validated ("World Health Organization. Methods of sterilization," 2019).

12.8.2.2 Dry Heat Sterilization

Contrary to steam sterilization, dry heat sterilization is based on the oxidation of cell constituents. Thus, it has been used more for sterilizing non-aqueous, thermoresistant samples rather than antioxidant ones. Compared to the previous method, applied temperatures are higher and times range are between 30 and 180 minutes (Toh and Chiu 2013; "World Health Organization. Methods of sterilization," 2019). However, taking into account the sterilization mechanisms underlying these two techniques, both of them might have detrimental effects on the preparations because material hydrolysis directly correlates with temperature and agglomeration, and breakdown and deformation of polymers have been described when applied temperatures were higher than the glass transition temperatures of the polymers (Dubey 2014; Toh and Chiu 2013).

12.8.3 High-Pressure Processes

High-pressure processes were developed in order to overcome the problems involved in thermolability method. Although this parameter has not been as largely studied as heat sterilization, high pressure techniques can strongly influence different kinds of biomolecules such as proteins or polysaccharides via modifying both their electrostatic and hydrophobic interactions, thus constituting an excellent tool for denaturing microbial proteins (Gharibzahedi et al. 2019). For this reason, high-pressure methods have been used for both, pasteurization and sterilization processes. Pasteurization conditions are milder when compared to sterilization ones, being not intense enough to completely eliminate bacterial spores (e.g. 600 MPa, 5 minutes, 20 °C), and, therefore, it is more used in the food industry. On the other hand, high-pressure sterilization is required for biomedical purposes and these more severe conditions involve increasing the temperature (e.g. 800 MPa, 5 minutes, 80 °C) (van de Ven et al. 2007). However, the high cost of these techniques makes them to be only used in laboratories or pilot plants in small volumes.

Regarding emulsion-based antioxidant carriers, it has been demonstrated that high-pressure processes do not have a detrimental effect on the encapsulated compounds, for example, not being degraded or losing their stability (Young et al. 2018). Moreover, it has been reported that these processes have achieved an increase on the bioactivity of encapsulated antioxidants due to the denaturation of proteins that acted as emulsifiers by increasing the thickness of the interfacial layer between phases and the packing density of proteins (Wan Mohamad et al. 2018).

12.8.4 Irradiation Techniques

Irradiation techniques include gamma (ϒ) and UV-irradiation. The former is sometimes applied on certain drugs and surgical equipment while UV-irradiation is restricted to surfaces sterilization because of its lower penetrance. However, economic and technical limitations are common for both, including the necessity of well-trained staff, specially designed installations and expensive equipment ("World Health Organization. Methods of sterilization," 2019).

They are not the most suitable methods for sterilizing emulsion-based antioxidant carriers because the degradation of bacterial DNA and membranes occurs via a free radical formation mechanism. While a complete sterilization of the sample would be achieved, the antioxidant cargo might lose its biological activity because of the oxidant nature of the newly-formed free radicals (Toh and Chiu 2013). In fact, it has been reported that adding antioxidants into the emulsions can reduce the effect of these free radicals, being ϒ-irradiation commonly used for encapsulating cargos with other applications. Another way to reduce the generation of these free radicals consists of freeze-drying the emulsions, avoiding the formation of hydroxyl radicals by eliminating the water phase. However, depending on the carrier, some coordinated water might remain within the vehicle, and therefore oxidize the cargo (Mohammed et al. 2006; Toh and Chiu 2013).

12.8.5 *Ethylene Oxide Sterilization*

Ethylene oxide (C_2H_4O) is a gaseous compound used as a sterilizing product due to its high solubility in water and electrophile behavior. Specifically, one of the carbons of the molecule is attacked by the nitrogen atom of the amino groups of both bacterial nucleic acids and proteins, leading to the alkylation of these molecules and, consequently, to microorganisms death (including virus) (Lonni et al. 2016). However, as other sterilizing techniques, its action mechanism might not be the most adequate for emulsion-based antioxidant carrier sterilization. According to Forman et al., "antioxidants are nucleophilic reductants that directly react with oxidants", and, therefore, ethylene oxide would reduce their antioxidant activity (Forman et al. 2014). If considered for other applications, ethylene oxide may be a good option to take into account when there are thermolabile carriers involved. In addition, it has been demonstrated not to increment lipid vesicles size and not to aggregate polymeric microspheres when the initial crystallinity of the polymers is high enough (Ah et al. 2001; Choi et al. 2001).

12.8.6 *Comparison of Different Sterilization Methods*

All the aforementioned techniques have been schematized in Table 12.3. To conclude this section, some works comparing the three more appropriate sterilizing techniques for emulsion-based antioxidant carriers are going to be reviewed.

Table 12.3 Main advantages and limitations of different methods for sterilizing emulsion-based antioxidant carriers

Method	Advantages	Disadvantages	Cost	Utility
Filtration	Admits thermolabile products	Droplets under 0.22 μm and economic limitations	High	Med.
Saturated steam	Cost and utility	Thermolability, degradation, agglomeration, etc.	Low	High
Dry heat	Cost and utility	Free radical formation, thermolability, degradation…	High	Low
High-pressure process	Admits thermolabile products and various operating conditions	Still in development, costs and maintenance	High	High
ϒ-irradiation	High penetration	Free radical formation, degradation, agglomeration...	High	Low
UV-irradiation	Low cost and availability	Low penetration and free radical formation	Low	Low
Ethylene oxide	Admits thermolabile products	Free radical formation	Med.	Low

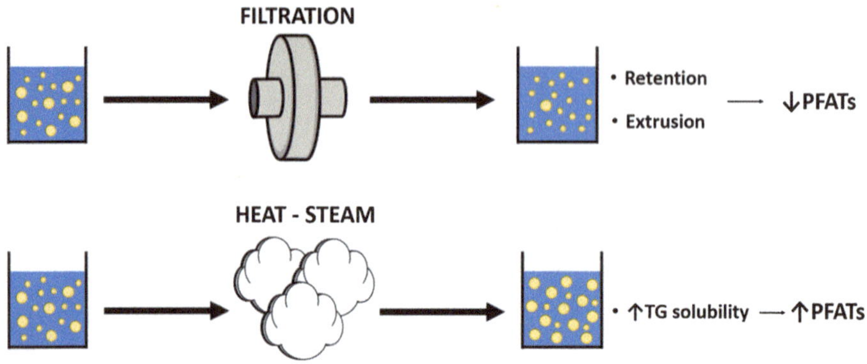

Fig. 12.11 Effect of the chosen sterilization method on particle size, evaluated as PFATs values. PFATs stands for percentage of fat droplets greater than a determined particle size, and TG, for triglycerides

To conclude this section, some works comparing the three more appropriate sterilizing techniques for emulsion-based antioxidant carriers are reviewing. Recently, Cappellani et al. carried out a study where they compared the properties of a dye-loaded injectable nanoemulsion sterilized by filtration or steam heat sterilization (Rosi Cappellani et al. 2018). Attending to the in vitro physicochemical characterization, the most interesting feature of this study is the complementary use of two different techniques for analyzing the size dispersion of the droplets: DLS and SPOS (see Sect. 12.2). In this study, the mean hydrodynamic diameter of filtered emulsions did not significantly vary while it slightly increased when the emulsions were autoclaved. However, these changes were relatively small when compared with the changes on the percentage of the volume of oil droplets with a diameter larger than 1.79 ($PFAT_{1.79}$) and 5 μm ($PFAT_5$). Therefore, while filtration strongly reduced both PFATs, by retention of larger droplets or by rupture during extrusion, heat sterilization had the opposite effect, which may be due to the increase in triglycerides (TG) water solubility, leading to a higher rate of Ostwald ripening (see Sect. 12.4) (Fig. 12.11). These PFATs values are of particular importance when the emulsions are intended for parenteral administration. According to United States Pharmacopoeia, $PFAT_5$ cannot be over 0.05% in volume because it could lead to pulmonary capillaries obstruction inducing fat embolism syndrome.

In addition, regarding antioxidant carriers, it is important to check also the cargo of the vehicles after the sterilization process. Specifically, β-carotene has been shown to undergo oxidation or isomerization when exposed to heat. As shown by Borba et al., their β-carotene-loaded nanoemulsion was highly stable after the thermal treatment but storing conditions (among other factors) led to its degradation (Borba et al. 2019).

12.9 Cytotoxicity

It is also essential to test if these emulsion systems can cause any damage due to their composition or their biological fate inside the body. Nowadays, there is not a specific guideline for testing toxicity of emulsions, but analogous analyses of biomaterials toxicity can be performed.

12.9.1 *Emulsions Droplet Fate*

Depending on the application, form of administration and physicochemical characteristics of the tested emulsion, the fate of the emulsion droplets may have different relevance. For example, there has been concern about the use of nanoemulsions because its reduced dimensions may alter the absorption, distribution, metabolism, and excretion processes, thus promoting toxicity (McClements and Rao 2011). However, proper in vitro and in vivo preclinical studies can support the safety of the systems. For instance, when intended for oral administration, emulsions are usually digested along the gastrointestinal tract and absorbed in the intestine. These emulsions are generally made of biodegradable materials that do not present further complications. However, they can also be made of indigestible oils, such as hydrocarbons or mineral oils, or even having their droplets coated with indigestible shells of dietary fibers, which would avoid its digestion. Therefore, the droplets could be directly absorbed and accumulated in some tissues (McClements and Rao 2011).

Regarding its composition, some of the compounds used during the preparation of the emulsion might be toxic if they have not been previously removed (e.g. organic solvents), or if they are in high concentrations (e.g. surfactants as emulsifiers) (Kralova and Sjöblom 2009). In addition, an increase in the bioavailability of a determined compound does not necessarily correlate with a beneficial effect, especially those that may cause adverse effects at high doses (McClements and Rao 2011). All these factors are highly dependent on the composition and characteristics of the tested emulsion, and for this reason it is so important to fully know the concrete physicochemical characteristics of the system before performing these assays.

12.9.2 *Cytotoxicity Testing*

In order to test biocompatibility, cytotoxic tests are carried out for emulsions as for any other kind of delivery systems or biomaterials. These tests analyze the in vitro cellular response of different cell types in order to identify vulnerable or altered cells or toxic concentrations, and include: morphology assessment, cell viability, mutagenicity, and oxidative damage tests.

Examining cell morphology is one of the simplest ways to identify cytotoxicity. Every cell type has a characteristic shape and appearance and any modification could mean a toxic insult. Apoptotic and necrotic cells have characteristic signals which indicate cell death such as cytoplasmic vacuolation, granularity and even detachment from the substrate. These features can be observed with an optical microscope but more details can be obtained with further biochemical, immunohistochemical and microscopic evaluation.

Regarding cell viability and cell proliferation, there are many tests that allow us to check the metabolic state of the cell cultures treated with emulsions. The principal ones are commented below.

12.9.2.1 Dyes Exclusion

Membrane integrity is a differential feature between living and dead cells. While the first ones have well-regulated systems to avoid some dyes entrance, dead cells have pores or disruptions that allow these dyes to penetrate inside. One example is trypan blue, which is routinely used in cell culture labs to test the viability of cell suspensions. Trypan blue is not able to cross intact membranes and for this reason, it only stains the background and the dead cells, thus showing living cells as refracting dots (Strober 2001).

Other methods are more eye-catching such as the simultaneous staining with fluorescein diacetate and propidium iodide (PI), where viable cells fluorescence in bright-green and dead cells, in bright-red. In this assay, fluorescein passes through living membranes and accumulates inside after being hydrolyzed by intracellular esterases, while PI only enters dead cells, thus causing this color divergence (Jones and Senft 1985).

12.9.2.2 Spectrophotometric/Colorimetric Assays

Other assays take advantage of color or absorbance emission at characteristic wavelengths in order to check cell viability status. The lactate dehydrogenase (LDH) assay is a commonly used assay in drug testing that allows to measure necrotic effects (Montenegro et al. 2011; Tzankova et al. 2016). In this sense, necrotic cells release LDH to the culture medium. When the reaction cocktail is added to the supernatants, a reduction reaction takes place with varying intensity, depending on the LDH concentration (Yoon et al. 2018). However, when testing emulsion-based antioxidant carriers, it is important to check if it can bias the colorimetric measurement. Other tests analyze the mitochondrial activity of the culture such as MTT or Alamar Blue (Dhakar et al. 2019; Shenoy et al. 2017; Tzankova et al. 2016). In the first one, a yellow tetrazolium salt (3-(4,5-dimethylthiazol-2-yl)-2,5-diphenyltetrazolium bromide) is reduced by living cells to blue formazan crystals which must be solubilize in another solvent for quantification. In contrast, the Alamar Blue assay has a non-fluorescent dye which is reduced by living cells to a

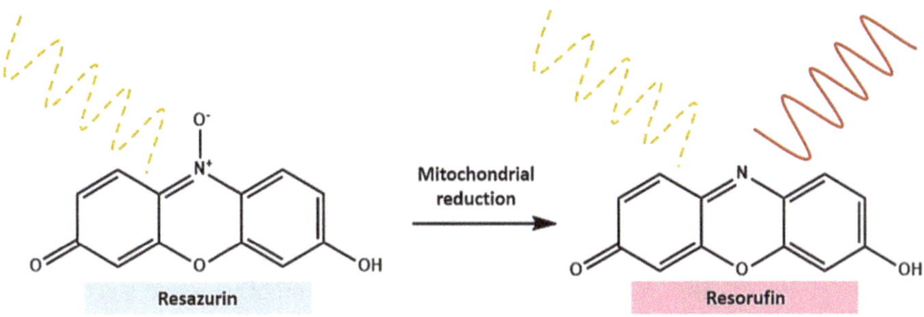

Fig. 12.12 The non-fluorescent Alamar Blue substrate (resazurin) is reduced by viable cells to a pink-colored, bright red-fluorescent product (resorufin)

pink fluorescent molecule (Fig. 12.12). Both MTT and Alamar Blue assays give an idea of cytotoxic or proliferative effects of a determined compound but some differences can be found. For example, MTT is more cumbersome and can produce false positives when antioxidants are tested (Bruggisser et al. 2002). On the other hand, Alamar Blue is simpler as it can be performed over time, because of its non-cytotoxicity (Bopp and Lettieri 2008). It is important to note that some drugs with high antioxidant properties can indeed cause false-negative results by interfering with the reducing property of viable cells. In these cases, washing with phosphate buffered saline before adding the incubating solution is recommended (Shenoy et al. 2017).

12.9.2.3 Flow Cytometry Analyses

Flow cytometry is a perfect technique for assessing cell viability status in a cell-by-cell manner. A cell suspension is passed through the cytometer and cells are aligned and separated so that the cytometer laser can individually analyze them, giving information about size and complexity of each cell. DNA content in each cell can be measured if the suspension is previously permeabilized and PI is added. In this sense, the emitted fluorescence of PI would be directly proportional to DNA content and, therefore, graphics of the cell cycle of the culture could be obtained. Apoptotic cells appeared below G_1 phase (the lapse of the cycle between two mitotic divisions), where there is less DNA due to its degradation (Darzynkiewicz et al. 2010). Different methods can be applied using other markers such as annexin V or clusters of differentiation tagged with fluorescent molecules for more complex analyses.

Tests related to mutagenicity and carcinogenicity issues are not as commonly used as the previous ones. In fact, they are more used in in vivo models which are beyond the scope of this chapter. These tests study the possible genotoxic and carcinogenic effects that the samples can have on the living systems, including gene mutations, DNA aberrations and chromosomal alterations. Among these assays, Comet and Ames tests are the most frequently used due to their fastness and simplicity (Omidi et al. 2017). Finally, the oxidative damage of the emulsions should also be tested, although for antioxidant carriers, it would give idea of the potency of the developed product (see Sect. 12.7).

12.10 Conclusions

The performance of an emulsion-based antioxidant carrier will be intimately related to its properties. Therefore, an appropriate characterization of the system is required to ensure its suitability for the desired final application. In this regard, this chapter highlights the main characteristics affecting antioxidant carrier quality and effectiveness; and summarizes the most commonly used techniques for their characterization. The examples referenced in this chapter allow the reader to understand the difference between the analytical methods used to characterize these systems. Choosing the most appropriate techniques for each situation depends on three main factors: the actual characteristics of the analyzed system, the cargo and its future application. It is important to assess the size, morphology and structure of the vehicles because they influence the rheological properties and stability of the antioxidant emulsion. Regarding the cargo, it has to be demonstrated that its loading within the emulsion-based carriers does not affect its structure, functions and antioxidant properties. Finally, when intended to be in contact with humans, cytotoxic and sterility issues need to be addressed. Thus, by knowing the principles underlying these characterization methods and techniques, the reader will be able to choose among the different options attending to their advantages and limitations, and to obtain as much information as possible from ongoing experiments.

References

Aboudzadeh MA, Mehravar E, Fernandez M, Lezama L, Tomovska R (2018) Low-energy encapsulation of alpha-tocopherol using fully food grade oil-in-water microemulsions. ACS Omega 3(9):10999–11008. https://doi.org/10.1021/acsomega.8b01272

Acevedo-Fani A, Silva HD, Soliva-Fortuny R, Martín-Belloso O, Vicente AA (2017) Formation, stability and antioxidant activity of food-grade multilayer emulsions containing resveratrol. Food Hydrocoll 71:207–215. https://doi.org/10.1016/j.foodhyd.2017.05.007

Ah YC, Choi Y, Kim SY, Kim SH, Lee KS, Byun Y (2001) Effects of ethylene oxide gas sterilization on physical properties of poly(L-lactide)-poly(ethylene glycol)-poly(L-lactide) microspheres. J Biomater Sci Polym Ed 12(7):783–799. https://doi.org/10.1163/156856201750411666

Akbari S (2018) Emulsion types, stability mechanisms and rheology: a review. Int J Innov Res Sci Stud 1(1):14–21

Alam MN, Bristi NJ, Rafiquzzaman M (2013) Review on in vivo and in vitro methods evaluation of antioxidant activity. Saudi Pharm J 21(2):143–152. https://doi.org/10.1016/j.jsps.2012.05.002

Amma S-i, Luo J, Pantano CG, Kim SH (2015) Specular reflectance (SR) and attenuated total reflectance (ATR) infrared (IR) spectroscopy of transparent flat glass surfaces: a case study for soda lime float glass. J Non-Cryst Solids 428:189–196. https://doi.org/10.1016/j.jnoncrysol.2015.08.015

Beasley MM, Bartelink EJ, Taylor L, Miller RM (2014) Comparison of transmission FTIR, ATR, and DRIFT spectra: implications for assessment of bone bioapatite diagenesis. J Archaeol Sci 46:16–22. https://doi.org/10.1016/j.jas.2014.03.008

Behbahani ES, Ghaedi M, Abbaspour M, Rostamizadeh K (2017) Optimization and characterization of ultrasound assisted preparation of curcumin-loaded solid lipid nanoparticles: application of central composite design, thermal analysis and X-ray diffraction techniques. Ultrason Sonochem 38:271–280. https://doi.org/10.1016/j.ultsonch.2017.03.013

Benzie I, Strain J (1996) The ferric reducing ability of plasma (FRAP) as a measure of "antioxidant power" the FRAP assay. Anal Biochem 239:70–76

Block ID, Scheffold F (2010) Modulated 3D cross-correlation light scattering: improving turbid sample characterization. Rev Sci Instrum 81(12):123107. https://doi.org/10.1063/1.3518961

Blois MS (1958) Antioxidant determinations by the use of a stable free radical. Nature 181(4617):1199–1200. https://doi.org/10.1038/1811199a0

Bopp SK, Lettieri T (2008) Comparison of four different colorimetric and fluorometric cytotoxicity assays in a zebrafish liver cell line. BMC Pharmacol 8:*8*. https://doi.org/10.1186/1471-2210-8-8

Borba CM, Tavares MN, Macedo LP, Araujo GS, Furlong EB, Dora CL, Burkert JFM (2019) Physical and chemical stability of beta-carotene nanoemulsions during storage and thermal process. Food Res Int 121:229–237. https://doi.org/10.1016/j.foodres.2019.03.045

Brand-Williams W, Cuvelier ME, Berset C (1995) Use of a free radical method to evaluate antioxidant activity. LWT Food Sci Technol 28(1):25–30. https://doi.org/10.1016/S0023-6438(95)80008-5

Bruggisser R, von Daeniken K, Jundt G, Schaffner W, Tullberg-Reinert H (2002) Interference of plant extracts, phytoestrogens and antioxidants with the MTT tetrazolium assay. Planta Med 68(5):445–448. https://doi.org/10.1055/s-2002-32073

Carpenter J, George S, Saharan VK (2019) Curcumin encapsulation in multilayer oil-in-water emulsion: synthesis using ultrasonication and studies on stability and antioxidant and release activities. Langmuir 35(33):10866–10876. https://doi.org/10.1021/acs.langmuir.9b01523

Cefali LC, Souza-Moreira TM, Corrêa MA, Salgado HRN, Isaac VLB (2015) Development and evaluation of an emulsion containing lycopene for combating acceleration of skin aging. Braz J Pharm Sci 51(3):579–590. https://doi.org/10.1590/s1984-82502015000300010

Chaari M, Theochari I, Papadimitriou V, Xenakis A, Ammar E (2018) Encapsulation of carotenoids extracted from halophilic Archaea in oil-in-water (O/W) micro-and nano-emulsions. Colloids Surf B Biointerfaces 161:219–227. https://doi.org/10.1016/j.colsurfb.2017.10.042

Chang C, Li X, Li J, Niu F, Zhang M, Zhou B et al (2017) Effect of enzymatic hydrolysis on characteristics and synergistic efficiency of pectin on emulsifying properties of egg white protein. Food Hydrocoll 65:87–95. https://doi.org/10.1016/j.foodhyd.2016.11.004

Chen LC, Chen YC, Su CY, Wong WP, Sheu MT, Ho HO (2016) Development and characterization of lecithin-based self-assembling mixed polymeric Micellar (saMPMs) drug delivery systems for curcumin. Sci Rep 6:37122. https://doi.org/10.1038/srep37122

Cheng Y, Chen J, Xiong YL (2014) Interfacial adsorption of peptides in oil-in-water emulsions costabilized by Tween 20 and antioxidative potato peptides. J Agric Food Chem 62(47):11575–11581. https://doi.org/10.1021/jf5038135

Choi Y, Kim SY, Moon MH, Kim SH, Lee KS, Byun Y (2001) Poly(ethylene glycol)-poly(L-lactide) diblock copolymer prevents aggregation of poly(L-lactide) microspheres during ethylene oxide gas sterilization. Biomaterials 22(9):995–1004. https://doi.org/10.1016/s0142-9612(00)00265-9

Choi SJ, Won JW, Park KM, Chang P-S (2014) A new method for determining the emulsion stability index by backscattering light detection. J Food Process Eng 37(3):229–236. https://doi.org/10.1111/jfpe.12078

Choudhry QN, Kim MJ, Kim TG, Pan JH, Kim JH, Park SJ et al (2016) Saponin-based nanoemulsification improves the antioxidant properties of vitamin A and E in AML-12 cells. Int J Mol Sci 17(9). https://doi.org/10.3390/ijms17091406

Coskun O (2016) Separation techniques: chromatography. North Clin Istanb 3(2):156–160. https://doi.org/10.14744/nci.2016.32757
Darzynkiewicz Z, Halicka HD, Zhao H (2010) <nihms-247430.pdf>. Adv Exp Med Biol 676:137–147
Araújo, Rosngela F. F. de, Martins, Danyelly Bruneska G., & Borba, Maria Amélia C. S. M. (2016). Oxidative stress and disease. doi:https://doi.org/10.5772/65366
Del Prado M, Magaña J, Mejía-Contreras BA, Borbolla-Jiménez FV, Giraldo D, Pina-Barba M et al (2019) In vitro cell evaluation of curcumin-loaded PCL/F68 nanoparticles for potential application in neuronal diseases. J Drug Delivery Sci Technol 52:905–914. https://doi.org/10.1016/j.jddst.2019.05.042
Derkach SR (2009) Rheology of emulsions. Adv Colloid Interf Sci 151(1–2):1–23. https://doi.org/10.1016/j.cis.2009.07.001
Dhakar NK, Caldera F, Bessone F, Cecone C, Pedrazzo AR, Cavalli R et al (2019) Evaluation of solubility enhancement, antioxidant activity, and cytotoxicity studies of kynurenic acid loaded cyclodextrin nanosponge. Carbohydr Polym 224:115168. https://doi.org/10.1016/j.carbpol.2019.115168
Di Gianfrancesco A (2017) Technologies for chemical analyses, microstructural and inspection investigations:197–245. https://doi.org/10.1016/b978-0-08-100552-1.00008-7
Duan X, Li M, Ma H, Xu X, Jin Z, Liu X (2016) Physicochemical properties and antioxidant potential of phosvitin-resveratrol complexes in emulsion system. Food Chem 206:102–109. https://doi.org/10.1016/j.foodchem.2016.03.055
Dubey R (2014) Controlled-release injectable microemulsions: recent advances and potential opportunities. Expert Opin Drug Deliv 11(2):159–173. https://doi.org/10.1517/17425247.2014.870151
Emerenciano DP, Baracho BBD, de Medeiros ML, Rocha HAO, Xavier FH Jr, da Veiga VF Jr, Maciel MAM (2019) Physicochemical characterizations and antioxidant property of copaiba oil loaded into SNEDDS systems. J Braz Chem Soc 30:234–246
Erdmann ME, Zeeb B, Salminen H, Gibis M, Lautenschlaeger R, Weiss J (2015) Influence of droplet size on the antioxidant activity of rosemary extract loaded oil-in-water emulsions in mixed systems. Food Funct 6(3):793–804. https://doi.org/10.1039/c4fo00878b
Etzler, Frank M., & Drelich, Jaroslaw. (2012). Atomic force microscopy for characterization of surfaces, particles, and their interactions. 307–331. https://doi.org/10.1016/b978-1-4377-7883-0.00006-7
Food and Drug Administration. Guidance for industry: sterile drug products produced by Aseptic Processing - Current Good Manufacturing Practice. (2004). Available: https://www.fda.gov/media/71026/download
Forman HJ, Davies KJ, Ursini F (2014) How do nutritional antioxidants really work: nucleophilic tone and para-hormesis versus free radical scavenging in vivo. Free Radic Biol Med 66:24–35. https://doi.org/10.1016/j.freeradbiomed.2013.05.045
Gallego MG, Skowyra M, Gordon MH, Azman NA, Almajano MP (2017) Effect of leaves of Caesalpinia decapetala on oxidative stability of oil-in-water emulsions. Antioxidants (Basel) 6(1). https://doi.org/10.3390/antiox6010019
Gharibzahedi SMT, Hernández-Ortega C, Welti-Chanes J, Putnik P, Barba FJ, Mallikarjunan K et al (2019) High pressure processing of food-grade emulsion systems: antimicrobial activity, and effect on the physicochemical properties. Food Hydrocoll 87:307–320. https://doi.org/10.1016/j.foodhyd.2018.08.012
Giessibl FJ (2003) Advances in atomic force microscopy. Rev Mod Phys 75(3):949–983. https://doi.org/10.1103/RevModPhys.75.949
Gimenez-Rota C, Palazzo I, Scognamiglio MR, Mainar A, Reverchon E, Della Porta G (2019) β-Carotene, α-tocoferol and rosmarinic acid encapsulated within PLA/PLGA microcarriers by supercritical emulsion extraction: encapsulation efficiency, drugs shelf-life and antioxidant activity. J Supercrit Fluids 146:199–207. https://doi.org/10.1016/j.supflu.2019.01.019

Gomes A, Costa ALR, de Assis Perrechil F, da Cunha RL (2016) Role of the phases composition on the incorporation of gallic acid in O/W and W/O emulsions. J Food Eng 168:205–214. https://doi.org/10.1016/j.jfoodeng.2015.07.041

Goncalez ML, Marcussi DG, Calixto GM, Correa MA, Chorilli M (2015) Structural characterization and in vitro antioxidant activity of kojic dipalmitate loaded w/o/w multiple emulsions intended for skin disorders. Biomed Res Int 2015:304591. https://doi.org/10.1155/2015/304591

Gouda M, Zhang S, Liu Y, Sheng L, Ma M (2017) Effects of four natural antioxidant phenyl terpenes on emulsifying and rheological properties of egg yolk. LWT Food Sci Technol 83:59–67. https://doi.org/10.1016/j.lwt.2017.04.075

Gu L, Pan C, Su Y, Zhang R, Xiao H, McClements DJ, Yang Y (2018) In vitro bioavailability, cellular antioxidant activity, and cytotoxicity of beta-carotene-loaded emulsions stabilized by catechin-egg white protein conjugates. J Agric Food Chem 66(7):1649–1657. https://doi.org/10.1021/acs.jafc.7b05909

Gurpreet K, Singh SK (2018) Review of nanoemulsion formulation and characterization techniques. Indian J Pharm Sci 80(5):781–789

Hatanaka J, Chikamori H, Sato H, Uchida S, Debari K, Onoue S, Yamada S (2010) Physicochemical and pharmacological characterization of alpha-tocopherol-loaded nano-emulsion system. Int J Pharm 396(1–2):188–193. https://doi.org/10.1016/j.ijpharm.2010.06.017

Hee YY, Tan CP, Rahman RA, Noranizan M, Smith RL, Chong GH (2017) Production of virgin coconut oil microcapsules from oil-in-water emulsion with supercritical carbon dioxide spray drying. J Supercrit Fluids 130:118–124. https://doi.org/10.1016/j.supflu.2017.07.037

Hippalgaonkar K, Majumdar S, Kansara V (2010) Injectable lipid emulsions-advancements, opportunities and challenges. AAPS PharmSciTech 11(4):1526–1540. https://doi.org/10.1208/s12249-010-9526-5

Hu Y-T, Ting Y, Hu J-Y, Hsieh S-C (2017) Techniques and methods to study functional characteristics of emulsion systems. J Food Drug Anal 25(1):16–26. https://doi.org/10.1016/j.jfda.2016.10.021

Ja'afar F, Leow CH, Garbin V, Sennoga CA, Tang MX, Seddon JM (2015) Surface charge measurement of SonoVue, definity and optison: a comparison of laser Doppler electrophoresis and micro-electrophoresis. Ultrasound Med Biol 41(11):2990–3000. https://doi.org/10.1016/j.ultrasmedbio.2015.07.001

Jin W, Xu W, Liang H, Li Y, Liu S, Li B (2016) Nanoemulsions for food: properties, production, characterization, and applications. In: Grumezescu AM (ed) Emulsions. Academic, London, pp 1–36

Jones KH, Senft JA (1985) An improved method to determine cell viability by simultaneous staining with fluorescein diacetate-propidium iodide. J Histochem Cytochem 33(1):77–79. https://doi.org/10.1177/33.1.2578146

Joung HJ, Choi MJ, Kim JT, Park SH, Park HJ, Shin GH (2016) Development of food-grade curcumin nanoemulsion and its potential application to food beverage system: antioxidant property and in vitro digestion. J Food Sci 81(3):N745–N753. https://doi.org/10.1111/1750-3841.13224

Kim HS, Mason TG (2017) Advances and challenges in the rheology of concentrated emulsions and nanoemulsions. Adv Colloid Interf Sci 247:397–412. https://doi.org/10.1016/j.cis.2017.07.002

Kralova I, Sjöblom J (2009) Surfactants used in food industry: a review. J Dispers Sci Technol 30(9):1363–1383. https://doi.org/10.1080/01932690902735561

Lidgate DM, Trattner T, Shultz RM, Maskiewicz R (1992) Sterile filtration of a parenteral emulsion. Pharm Res 9(7):860–863. https://doi.org/10.1023/a:1015836512890

Liguori I, Russo G, Curcio F, Bulli G, Aran L, Della-Morte D et al (2018) Oxidative stress, aging, and diseases. Clin Interv Aging 13:757–772. https://doi.org/10.2147/CIA.S158513

Lobo V, Patil A, Phatak A, Chandra N (2010) Free radicals, antioxidants and functional foods: impact on human health. Pharmacogn Rev 4(8):118–126. https://doi.org/10.4103/0973-7847.70902

Lonni AA, Munhoz VM, Lopes GC, Longhini R, Borghi-Pangoni FB, Dos Santos RS et al (2016) Development and characterization of multiple emulsions for controlled release of Trichilia catigua (Catuaba) extract. Pharm Dev Technol 21(8):933–942. https://doi.org/10.3109/1083745 0.2015.1081611

Mantele W, Deniz E (2017) UV-VIS absorption spectroscopy: Lambert-Beer reloaded. Spectrochim Acta A Mol Biomol Spectrosc 173:965–968. https://doi.org/10.1016/j.saa.2016.09.037

Marrocco I, Altieri F, Peluso I (2017) Measurement and clinical significance of biomarkers of oxidative stress in humans. Oxidative Med Cell Longev 2017:6501046. https://doi.org/10.1155/2017/6501046

Matschke V, Theiss C, Matschke J (2019) Oxidative stress: the lowest common denominator of multiple diseases. Neural Regen Res 14(2):238–241. https://doi.org/10.4103/1673-5374.244780

McClements DJ (2007) Critical review of techniques and methodologies for characterization of emulsion stability. Crit Rev Food Sci Nutr 47(7):611–649. https://doi.org/10.1080/10408390701289292

McClements DJ, Rao J (2011) Food-grade nanoemulsions: formulation, fabrication, properties, performance, biological fate, and potential toxicity. Crit Rev Food Sci Nutr 51(4):285–330. https://doi.org/10.1080/10408398.2011.559558

McElhaney RN (1986) Differential scanning calorimetric studies of lipid-protein interactions in model membrane systems. Biochim Biophys Acta Rev Biomembr 864(3):361–421. https://doi.org/10.1016/0304-4157(86)90004-3

Meroni E, Raikos V (2018) Physicochemical stability, antioxidant properties and bioaccessibility of beta-carotene in orange oil-in-water beverage emulsions: influence of carrier oil types. Food Funct 9(1):320–330. https://doi.org/10.1039/c7fo01170a

Mohammed AR, Bramwell VW, Coombes AG, Perrie Y (2006) Lyophilisation and sterilisation of liposomal vaccines to produce stable and sterile products. Methods 40(1):30–38. https://doi.org/10.1016/j.ymeth.2006.05.025

Montenegro L, Campisi A, Sarpietro MG, Carbone C, Acquaviva R, Raciti G, Puglisi G (2011) In vitro evaluation of idebenone-loaded solid lipid nanoparticles for drug delivery to the brain. Drug Dev Ind Pharm 37(6):737–746. https://doi.org/10.3109/03639045.2010.539231

Murphy DB (2001) Fundamentals of light microscopy and electronic imaging. Wiley-Liss, Inc, New York

Murphy NP, Lampe KJ (2018) Fabricating PLGA microparticles with high loads of the small molecule antioxidant N-acetylcysteine that rescue oligodendrocyte progenitor cells from oxidative stress. Biotechnol Bioeng 115(1):246–256. https://doi.org/10.1002/bit.26443

Nikolic I, Jasmin Lunter D, Randjelovic D, Zugic A, Tadic V, Markovic B et al (2018) Curcumin-loaded low-energy nanoemulsions as a prototype of multifunctional vehicles for different administration routes: physicochemical and in vitro peculiarities important for dermal application. Int J Pharm 550(1–2):333–346. https://doi.org/10.1016/j.ijpharm.2018.08.060

Omidi, Meisam, Fatehinya, Atena, Farahani, Masomeh, Akbari, Zahra, Shahmoradi, Saleheh, Yazdian, Fatemeh, et al. (2017). Characterization of biomaterials. 97–115. doi:https://doi.org/10.1016/b978-0-08-100961-1.00007-4

Pallua JD, Pezzei C, Huck-Pezzei V, Schonbichler SA, Bittner LK, Bonn GK et al (2011) Advances of infrared spectroscopic imaging and mapping technologies of plant material. Curr Bioact Compd 7(2):106–117. https://doi.org/10.2174/157340711796011179

Pawley JB (2006) Fundamental limits in confocal microscopy. In: Pawley JB (ed) Handbook of biological confocal microscopy. Springer US, Boston, pp 20–42

Peng C, Svirskis D, Lee SJ, Oey I, Kwak HS, Chen G et al (2018) Design of microemulsion system suitable for the oral delivery of poorly aqueous soluble beta-carotene. Pharm Dev Technol 23(7):682–688. https://doi.org/10.1080/10837450.2017.1287729

Rashid M, Wani T, Mishra N, Sofi H, Sheikh F (2018) Development and characterization of drug-loaded self-solidnano-emulsified drug delivery system for treatment of diabetes. Mater Sci Res India 15:01–11. https://doi.org/10.13005/msri/150101

Re R, Pellegrini N, Proteggente A, Pannala A, Yang M, Rice-Evans C (1999) Antioxidant activity applying an improved ABTS radical cation decolorization assay. Free Radic Biol Med 26(9–10):1231–1237. https://doi.org/10.1016/s0891-5849(98)00315-3

Rinaldi F, Hanieh PN, Longhi C, Carradori S, Secci D, Zengin G et al (2017) Neem oil nanoemulsions: characterisation and antioxidant activity. J Enzyme Inhib Med Chem 32(1):1265–1273. https://doi.org/10.1080/14756366.2017.1378190

Rosi Cappellani M, Perinelli DR, Pescosolido L, Schoubben A, Cespi M, Cossi R, Blasi P (2018) Injectable nanoemulsions prepared by high pressure homogenization: processing, sterilization, and size evolution. Appl Nanosci 8(6):1483–1491. https://doi.org/10.1007/s13204-018-0829-2

Rutz JK, Borges CD, Zambiazi RC, da Rosa CG, da Silva MM (2016) Elaboration of microparticles of carotenoids from natural and synthetic sources for applications in food. Food Chem 202:324–333. https://doi.org/10.1016/j.foodchem.2016.01.140

Sellimi S, Younes I, Ayed HB, Maalej H, Montero V, Rinaudo M et al (2015) Structural, physicochemical and antioxidant properties of sodium alginate isolated from a Tunisian brown seaweed. Int J Biol Macromol 72:1358–1367. https://doi.org/10.1016/j.ijbiomac.2014.10.016

Shaddel R, Hesari J, Azadmard-Damirchi S, Hamishehkar H, Fathi-Achachlouei B, Huang Q (2018) Double emulsion followed by complex coacervation as a promising method for protection of black raspberry anthocyanins. Food Hydrocoll 77:803–816. https://doi.org/10.1016/j.foodhyd.2017.11.024

Shenoy N, Stenson M, Lawson J, Abeykoon J, Patnaik M, Wu X, Witzig T (2017) Drugs with anti-oxidant properties can interfere with cell viability measurements by assays that rely on the reducing property of viable cells. Lab Investig 97:494–497. https://doi.org/10.1038/labinvest.2017.18

Shivakumar A, Yogendra Kumar MS (2018) Critical review on the analytical mechanistic steps in the evaluation of antioxidant activity. Crit Rev Anal Chem 48(3):214–236. https://doi.org/10.1080/10408347.2017.1400423

Silva HD, Cerqueira MÂ, Vicente AA (2012) Nanoemulsions for food applications: development and characterization. Food Bioprocess Technol 5(3):854–867. https://doi.org/10.1007/s11947-011-0683-7

Stetefeld J, McKenna SA, Patel TR (2016) Dynamic light scattering: a practical guide and applications in biomedical sciences. Biophys Rev 8(4):409–427. https://doi.org/10.1007/s12551-016-0218-6

Strober W (2001) Trypan blue exclusion test of cell viability. Curr Protoc Immunol, Appendix 3, Appendix 3B. https://doi.org/10.1002/0471142735.ima03bs21

Su J, Guo Q, Chen Y, Mao L, Gao Y, Yuan F (2020) Utilization of β-lactoglobulin- (−)-Epigallocatechin- 3-gallate(EGCG) composite colloidal nanoparticles as stabilizers for lutein pickering emulsion. Food Hydrocoll 98:105293. https://doi.org/10.1016/j.foodhyd.2019.105293

Sunee C, Panee S-a, Ruttiros K (2017) Preparation, characterization and antioxidant activity of xanthone-loaded making (Hodgsonia heteroclita) microemulsions. Int J Pharm Pharm Sci 9(3). https://doi.org/10.22159/ijpps.2017v9i3.16584

Swartz M (2010) Hplc detectors: a brief review. J Liq Chromatogr Relat Technol 33(9–12):1130–1150. https://doi.org/10.1080/10826076.2010.484356

Tan BL, Norhaizan ME, Liew WP, Sulaiman Rahman H (2018) Antioxidant and oxidative stress: a mutual interplay in age-related diseases. Front Pharmacol 9:1162. https://doi.org/10.3389/fphar.2018.01162

Tang, C. Y., & Yang, Z. (2017). Transmission Electron Microscopy (TEM). 145–159. doi:https://doi.org/10.1016/b978-0-444-63776-5.00008-5

Tang H, Xiang S, Li X, Zhou J, Kuang C (2019) Preparation and in vitro performance evaluation of resveratrol for oral self-microemulsion. PLoS One 14(4):e0214544. https://doi.org/10.1371/journal.pone.0214544

Tatar, B. C., Sumnu, G., & Sahin, S. (2017). Rheology of emulsions. 437–457. doi:https://doi.org/10.1016/b978-0-08-100431-9.00017-6

Tian B, Wang Y, Wang T, Mao L, Lu Y, Wang H, Feng Z (2019) Structure and functional properties of antioxidant nanoemulsions prepared with tea polyphenols and soybean protein isolate. J Oleo Sci 68:689–697. https://doi.org/10.5650/jos.ess19067

Tirado DF, Palazzo I, Scognamiglio M, Calvo L, Della Porta G, Reverchon E (2019) Astaxanthin encapsulation in ethyl cellulose carriers by continuous supercritical emulsions extraction: a study on particle size, encapsulation efficiency, release profile and antioxidant activity. J Supercrit Fluids 150:128–136. https://doi.org/10.1016/j.supflu.2019.04.017

Toh M-R, Chiu GNC (2013) Liposomes as sterile preparations and limitations of sterilisation techniques in liposomal manufacturing. Asian J Pharm Sci 8(2):88–95. https://doi.org/10.1016/j.ajps.2013.07.011

Tzankova V, Gorinova C, Kondeva-Burdina M, Simeonova R, Philipov S, Konstantinov S et al (2016) In vitro and in vivo toxicity evaluation of cationic PDMAEMA-PCL-PDMAEMA micelles as a carrier of curcumin. Food Chem Toxicol 97:1–10. https://doi.org/10.1016/j.fct.2016.08.026

Ujhelyi Z, Vecsernyés M, Fehér P, Kósa D, Arany P, Nemes D et al (2018) Physico-chemical characterization of self-emulsifying drug delivery systems. Drug Discov Today Technol 27:81–86. https://doi.org/10.1016/j.ddtec.2018.06.005

Umerska A, Gaucher C, Oyarzun-Ampuero F, Fries-Raeth I, Colin F, Villamizar-Sarmiento MG et al (2018) Polymeric nanoparticles for increasing oral bioavailability of curcumin. Antioxidants (Basel) 7(4). https://doi.org/10.3390/antiox7040046

Uskoković V (2012) Dynamic light scattering based microelectrophoresis: main prospects and limitations. J Dispers Sci Technol 33(12):1762–1786. https://doi.org/10.1080/01932691.2011.625523

van de Ven C, Courvoisier C, Matser A (2007) High pressure versus heat treatments for pasteurisation and sterilisation of model emulsions. Innovative Food Sci Emerg Technol 8(2):230–236. https://doi.org/10.1016/j.ifset.2006.12.001

Wan Mohamad WAF, McNaughton D, Augustin MA, Buckow R (2018) Characterisation of beta-carotene partitioning in protein emulsions: effects of pre-treatments, solid fat content and emulsifier type. Food Chem 257:361–367. https://doi.org/10.1016/j.foodchem.2018.03.027

Weiss J (2002) Emulsion stability determination. Curr Protocol Food Anal Chem 3(1):D3.4.1–D3.4.17. https://doi.org/10.1002/0471142913.fad0304s03

World Health Organization (2019) Methods of sterilization. Ninth. Retrieved from http://apps.who.int/phint/pdf/b/7.5.9.5.8-Methods-of-sterilization.pdf

Xiuhua Z, Yiping D, Ying Z, Yuangang Z, Bolin L, Mingfang W et al (2016) Silymarin nanoparticles through emulsion solvent evaporation method for oral delivery with high antioxidant activities, bioavailability, and absorption in the liver. RSC Adv 6(95):93137–93146. https://doi.org/10.1039/C6RA12896C

Yoon HJ, Zhang X, Kang MG, Kim GJ, Shin SY, Baek SH et al (2018) Cytotoxicity evaluation of turmeric extract incorporated oil-in-water nanoemulsion. Int J Mol Sci 19(1). https://doi.org/10.3390/ijms19010280

Young S, Basiana E, Nitin N (2018) Effects of interfacial composition on the stability of emulsion and encapsulated bioactives after thermal and high pressure processing. J Food Eng 231:22–29. https://doi.org/10.1016/j.jfoodeng.2018.02.022

Zhang H, Yin M, Huang L, Wang J, Gong L, Liu J, Sun B (2017a) Evaluation of the cellular and animal models for the study of antioxidant activity: a review. J Food Sci 82(2):278–288. https://doi.org/10.1111/1750-3841.13605

Zhang J, Zhang X, Wang X, Huang Y, Yang B, Pan X, Wu C (2017b) The influence of maltodextrin on the physicochemical properties and stabilization of beta-carotene emulsions. AAPS PharmSciTech 18(3):821–828. https://doi.org/10.1208/s12249-016-0572-5

Zhou W, Apkarian R, Wang ZL, Joy D (2007) Fundamentals of scanning electron microscopy (SEM). In: Zhou W, Wang ZL (eds) Scanning microscopy for nanotechnology: techniques and applications. New York, Springer New York, pp 1–40

Zhu L, Yan H, Zhao Y (2012) Cyclodextrin-based rotaxanes on gold nanoparticles. Int J Mol Sci 13(8):10132–10142. https://doi.org/10.3390/ijms130810132

Abbreviation List

A

AAPH	2,2'-azobis(2-amidino-propane)dihydrochloride
AAS	Atomic absorption spectroscopy
ABTS	2,2'-azino-bis(3-ethylbenzothiazoline-6-sulfonic acid
ADME	Absorption, distribution, metabolism, and excretion
AFM	Atomic force microscopy
AI	Adequate intake
ALA	α-Linolenic acid
AM-AFM	Amplitude modulation atomic force microscopy
AMVN	2,2′-azobis(2,4-dimethylvaleronitrile)
AO	Antioxidant
AOM	Active oxygen method
API	Active pharmaceutical ingredient
ARA	Arachidonic acid
ATR	Attenuated total reflection
AV	Anisidine value
ASX	Astaxanthin

B

BAC	Biologically active compounds
BG	Bovine gelatin
BHA	Butylated hydroxyanisole
BHT	Butylated hydroxytoluene
BITC	Benzyl isothiocyanate
BS	Backscattering
BSA	Bovine serum albumin

M. A. Aboudzadeh (ed.), *Emulsion-based Encapsulation of Antioxidants*, Food Bioactive Ingredients, https://doi.org/10.1007/978-3-030-62052-3

C

CA	Caffeic acid
CA (or CGA)	Chlorogenic acid
C-AC	Carbon alkyl chains
CA/LF	Chlorogenic acid/lactoferrin
CA/LF/PD	Chlorogenic acid/lactoferrin/polydextrose
CCD	Central composite design
CCTG	Caprylic/capric triglyceride
CD	Conjugated diene
CHD	Coronary heart disease
CITREM	Citric acid esters of mono- and diglycerides of fatty acids
CM	Colloid mill
CMC	Carboxymethyl cellulose
CPP	Critical packing parameter
CS	Chitosan
CTAB	Cetrimonium bromide
CT/EWP	Catechin/egg white protein
CV	Cyclic voltammetry
CVD	Cardiovascular diseases

D

DATEM	Diacetyl tartaric acid ester of mono- and diglycerides
DBOS-MS	Debranched octenyl succinic anhydride modified starch
DCF	2’, 7’-dichlorodihydrofluorescein
DCFH	2’, 7’-dichlorodihydrofluorescin
DCFH-DA	2’, 7’-dichlorodihydrofluorescin diacetate
DD	Degree of deacetylation
DDI	Dipole-dipole interactions
DE	Dextrose equivalent
DG	Dodecyl gallate
DH	Degree of hydrolysis
DHA	Docosahexaenoic acid
DLS	Dynamic light scattering
DM	Degree of methylation
DMAPP	Dimethylallyl pyrophosphate
DME	Direct membrane emulsification
DML	Decaglycerol monolaurate
DMO	Decaglyceryl monooleate
DMSO	Dimethyl sulfoxide
DNA	Deoxyribonucleic acid
DPPH	2,2-diphenyl-1-picrylhydrazyl
DPPP	Diphenyl-1-pyrenylphosphine
DPPP-O	Diphenyl-1-pyrenylphosphine oxide
DRI	Daily reference intake

DS	Degree of substitution
DSC	Differential scanning calorimetry
DTAB	Dodecyltrimethylammonium bromide

E

EA	Encapsulating agent
EAPG	Electrospraying assisted by pressurized gas
EDTA	Ethylenediaminetetraacetic acid
EE	Encapsulation efficiency
EFSA	European food safety authority
EFO	Emulsified fish oil
EGCG	Epigallocatechin-3-gallate
ELM	Emulsion liquid membranes
EMA	European medicines
EO	Extractable oil
EOF	Electroosmotic flow
EPA	Eicosapentaenoic acid
EPI	Emulsion phase inversion
EPR	Electron paramagnetic resonance
ESI	Emulsion stability index
ES	Encapsulation stability
ESR	Electron spin resonance
EVOO	Extra virgin olive oil
EWP	Egg white protein

F

FA	Fatty acid
FAO	Food and agriculture organization of the united nations
FDA	Food and drug administration
FE	French oak hydro-glycolic fruit extract
FFA	Free fatty acid
FITC-BSA	Fluorescein isothiocyanate-conjugated bovine serum albumin
FM-AFM	Frequency modulation atomic force microscopy
FO	Fish oil
FPH	Fish protein hydrolysates
FRAP	Ferric reducing antioxidant power
FRS	Free-radical scavengers
FT-IR	Fourier transform infrared

G

GA	Gallic acid
GA	Gum Arabic
GC	Gas chromatography
GC–MS	Gas chromatography–mass spectrometry
GE	Fish gelatin

GIT	Gastro-intestinal tract
GMP	Good manufacturing practices
GS	Gravitation separation
GS	Glucose syrup
GS	Gemini surfactant
GSO	Grape seed oil

H

H^+	Hydrogen atom
HAT	Hydrogen atom transfer
HB	Hydrogen bonding
HDBM	Hydroxy dimethoxy benzyl malonate
HEPES	4-(2-hydroxyethyl) piperazine-1-ethanesulfonic acid
HHE	4-hydroxy-2-hexenal
HIPE	High internal phase emulsions
His	Histidine
HLB	Hydrophilic-lipophilic balance
HMP	High-methylester-pectin
HPH	High-pressure homogenizer
HPKO	Hydrogenated palm kernel oil
HPLC	High performance liquid chromatography
HPO	Hydrogenated palm oil
HPMC	Hydroxypropyl methylcellulose
HSM	High-speed mixer
HT	Hydroxytyrosol

I

IFT	Interfacial tension
IMF	Intermolecular force
IP	Induction period
IPP	Isopentenyl pyrophosphate
IR	Infrared radiation
ISIS	Isostearyl isostearate

L

$L^\bullet$	Lipid free radical (alkyl radical)
LA	Linoleic acid
LAE	Lauric arginate
LC	Load capacity
LC	Low molecular weight chitosan
LC	Liquid chromatography
LC/EGCG	Low molecular weight chitosan/EGCG conjugates
LCT	Long-chain triglyceride
LC ω-3 PUFAs	Long chain polyunsaturated fatty acids
LDF	London dispersive forces
LDH	Lactate dehydrogenase

LDL	Low density lipoproteins
LEO	Lemongrass essential oil
LF	Lactoferrin
LH	Unsaturated lipid
LMP	Low methyl pectin
LO	Lavender essential oil
LO•	Alkoxyl radicals
LOO•	Peroxyl radicals
LOOH	Hydroperoxides
LPSC	Luminol-chemiluminescence based peroxyl radical scavenging capacity
LUT	Lutein
LYC	Lycopene
Lys	Lysine

M

MC	Moisture content
MCT	Medium-chain triglyceride
MD	Maltodextrin
MDA	Malonaldehyde
ME	Microemulsion
ME	Membrane emulsification
MEEKC	Microemulsion electrokinetic chromatography
Met	Methionine
MF	Microfluidizer
ML	Modified lecithin
m.o	Months old
MPC	Milk protein concentrate
MTT	3-(4,5-Dimethyl-2-thiazolyl)-2,5-diphenyl-2H-tetrazolium bromide
MW	Molecular weight

N

NaCas	Sodium caseinate
NaCMC	Sodium carboxymethyl cellulose
N/A	Not given
NCh	Nanochitin
ND	Neurodegenerative disorder
NFC	Nanofibrillated cellulose
NFO	Neat fish oil
NO	Nitric oxide radical
n-OSA-starch	n-octenylsuccinate-derivatised starch
NPs	Nanoparticles
NR	Not reported
NSET	Nanometal surface energy transfer
NTA	Nanoparticle tracking analysis

O

OG	Octyl gallate
OH	Hydroxyl radical
OH^-	Hydroxyl ions
O_2^-	Superoxide anion
ORAC	Oxygen radical absorbance capacity
OSA	Octenyl succinic anhydride
OSA-MS	Octenyl succinic anhydride modified starch
OSQS	Octenyl succinate quinoa starch
OTA	Oxidized tannic acid
O/W	Oil-in-water
O/W/O	Oil-in-water-in-oil

P

PDI	Polydispersity index
PE	Pickering emulsion
PEG	Polyethylene glycol
POEs	Palm oil esters
PFO	Passion fruit oil
PG	Propyl gallate
PGPR	Polyglycerol polyricinoleate
Phe	Phenylalanine
PHSO	Partially hydrogenated sunflower oil
PV	Hydroperoxides content
PGA	Propylene glycol alginate
PHC	L-α-phosphatidylcholine
PI	Propidium iodide
PIC	Phase inversion composition
PIT	Phase inversion temperature
PLA	poly(lactic acid)
PLGA	poly(D,L-lactide-co-glycolide)
PME	Premix membrane emulsification
PPI	Pea protein isolate
PPH	Potato protein hydrolysate
PTSA	p-toluenesulfonic acid
PUFA	Polyunsaturated fatty acid
PWP	Polymerised WPI
PUFA	Polyunsaturated fatty acid

Q

QS	Quillaja saponin

R

RBD	Refined, bleached and deodorised

RBO	Rice bran oil
RE	Red raspberry hydro-glycolic fruit extract
RH	Relative humidity
RI	Refractive index
RO	Red raspberry seed oil
RO_2	Peroxyl radical
ROS	Reactive oxygen species
RPO	Red palm oil
RNS	Reactive nitrogen species
RSA	Radical scavenging activity
RSM	Response surface methodology

S

SBP	Sugar beet pectin
SCT	Short chain triglyceride;
SDS	Sodium dodecyl sulfate
SE	Spontaneous emulsification
SEDDS	Self-emulsifying drug delivery systems
SEM	Scanning electron microscopy
SER	Surfactant-to-emulsion ratio
SET	Single electron transfer
SETPT	Single electron transfer-proton transfer
SFP	Sunflower protein
SGF	Simulated gastric fluid
SGPI	Scallop gonad protein isolates
SIF	Simulated intestinal fluid
SLN	Solid lipid nanoparticles
SLS	Static light scattering
SMEDDS	Self-microemulsifying drug delivery systems
SNEDDS	Self-nanoemulsifying drug delivery systems
SMLS	Static multiple light scattering
SMO	Spearmint oil
SMP	Sucrose monopalmitate
SMP	Skim milk powder
SOR	Surfactant to oil ratio
SOW	Surfactant-oil-water
SP	Sage polyphenols
SPI	Soybean protein isolate
SPG	Shirasu porous glass
SPLET	Sequential proton loss-electron transfer
SPOS	Single particle optical sizing
SRPH	Skipjack roe protein hydrolysate
SSIF	Simulated small intestinal fluids
SSL	Sodium stearoyl lactate
SSPS	Soybean soluble polysaccharides

STPP	Sodium tripolyphosphate
SVOP	Secondary volatile oxidation products

T

TA	Tannic acid
TAG	Triacylglycerols
TBARS	Thiobarbituric acid reactive substance
TBHQ	Tertiary butyl hydroquinone
TEAC	Trolox equivalent antioxidant capacity
TEM	Transmission electron microscopy
TEMPOL	4-Hydroxy-2,2,6,6-tetramethylpiperidine 1-oxyl
TG	Triglycerides
T_g	Glass transition temperature
TNF-α	Tumor necrosis factor alpha
TOTOX	Total oxidation
TPH	Tuna protein hydrolysate
TPP	Tripolyphosphate
TPTZ	2,4,6-tripyridyl-s-triazine
Trp	Tryptophan
TS	Transmitting
Tyr	Tyrosine

U

UDP	Uridine 5'-diphospho-glucuronosyltransferase
UFP	*Ulva fasciata* polysaccharide
USH	Ultrasonic homogenizer
UV	Ultraviolet
UV-VIS	Ultraviolet-visible

V

VA	Vitamin A
vdW	van der Waals
VE	Vitamin E
UV/VIS	Ultraviolet/visible

W

WBO	Wheat bran oil
WGNP	Wheat gluten nanoparticle
WHO	World health organization
W/O	Water-in-oil
W/O/W	Water-in-oil-in-water
WP	Whey permeate
WPC	Whey protein concentrate
WPH	Whey protein hydrolysate
WPI	Whey protein isolate

X

XRD	X-Ray diffraction
XG	Xanthan gum

Y

y.o.	Years old

Z

ZS	Zwitterionic surfactant

α

α-La	alpha-lactalbumin
α-TOC	α-tocopherol

β

βC	beta-carotene
β-LG	beta-lactoglobulin
β-LGH	β-lactoglobulin hydrolysates

ω

ω-3 PUFAs	Omega-3 polyunsaturated fatty acids

Nomenclature and Symbols Index

A	Absorbance at 517 nm of sample and DPPH
A_0	Absorbance at 517 nm of DPPH without sample
A_b	Absorbance at 517 nm of sample without DPPH
C_0	Expected concentration of the encapsulated compound
$C_{recovered}$	Concentration of the compound in the recovered aqueous phase
C_t	Concentration of marker found in the external aqueous after storage time (t)
F_B	Buoyancy force due to the density difference between the dispersed and continuous phases
F_D	Drag force, due to the continuous phase flow, parallel to the membrane surface
F_G	Gravitational force
F_L	Dynamic lift force due to asymmetric shear stress profile at membrane surface
F_{SP}	Static pressure difference between the dispersed and continuous phases
$F\gamma$	Capillary force
Ry	Recovery yield of the encapsulated compound
T	Temperature
T_{wb}	Wet bulb temperature
γ	Surface tension
γ_{MO}	Wetting of the oil phase on the membrane
γ_{MW}	Wetting on the water phase on the membrane
γ_{WO}	Interfacial tension between water and oil
ΔA	Change of the interfacial area
ΔG_f	Free energy of emulsion formation
ΔP_{disrup}	Disruption pressure required to overcome interfacial tension between oil and water in order to reduce the emulsion size and go through the membrane pores
ΔP_{flow}	Flow pressure required to overcome membrane resistance to the flow
ΔP_{tm}	Transmembrane pressure
ΔS	Change of entropy

M. A. Aboudzadeh (ed.), *Emulsion-based Encapsulation of Antioxidants*, Food Bioactive Ingredients, https://doi.org/10.1007/978-3-030-62052-3

Kolibri 360 GmbH
Ferdinand-Jühlke-Straße 7
99095 Erfurt, Deutschland
produktsicherheit@kolibri360.de